Frontiers of Multifunctional Nanosystems

NATO Science Series

A Series presenting the results of scientific meetings supported under the NATO Science Programme.

The Series is published by IOS Press, Amsterdam, and Kluwer Academic Publishers in conjunction with the NATO Scientific Affairs Division

Sub-Series

I. Life and Behavioural Sciences	IOS Press
II. Mathematics, Physics and Chemistry	Kluwer Academic Publishers
III. Computer and Systems Science	IOS Press
IV. Earth and Environmental Sciences	Kluwer Academic Publishers
V. Science and Technology Policy	IOS Press

The NATO Science Series continues the series of books published formerly as the NATO ASI Series.

The NATO Science Programme offers support for collaboration in civil science between scientists of countries of the Euro-Atlantic Partnership Council. The types of scientific meeting generally supported are "Advanced Study Institutes" and "Advanced Research Workshops", although other types of meeting are supported from time to time. The NATO Science Series collects together the results of these meetings. The meetings are co-organized bij scientists from NATO countries and scientists from NATO's Partner countries – countries of the CIS and Central and Eastern Europe.

Advanced Study Institutes are high-level tutorial courses offering in-depth study of latest advances in a field.
Advanced Research Workshops are expert meetings aimed at critical assessment of a field, and identification of directions for future action.

As a consequence of the restructuring of the NATO Science Programme in 1999, the NATO Science Series has been re-organised and there are currently Five Sub-series as noted above. Please consult the following web sites for information on previous volumes published in the Series, as well as details of earlier Sub-series.

http://www.nato.int/science
http://www.wkap.nl
http://www.iospress.nl
http://www.wtv-books.de/nato-pco.htm

Series II: Mathematics, Physics and Chemistry – Vol. 57

Frontiers of Multifunctional Nanosystems

edited by

Eugenia Buzaneva
Kyiv National Taras Shevchenko University,
Ukraine

and

Peter Scharff
TU Ilmenau,
Institut für Physik/Fachgebiet Chemie,
Germany

Kluwer Academic Publishers

Dordrecht / Boston / London

Published in cooperation with NATO Scientific Affairs Division

Proceedings of the NATO Advanced Research Workshop on
Frontiers in Molecular-Scale Science and Technology of Fullerene, Nanotube, Nanosilicon, Biopolymer (DNA, Protein) Multifunctional Nanosystems
Kyiv, Ukraine
9–12 September 2001

A C.I.P. Catalogue record for this book is available from the Library of Congress.

ISBN 1-4020-0560-1

Published by Kluwer Academic Publishers,
P.O. Box 17, 3300 AA Dordrecht, The Netherlands.

Sold and distributed in North, Central and South America
by Kluwer Academic Publishers,
101 Philip Drive, Norwell, MA 02061, U.S.A.

In all other countries, sold and distributed
by Kluwer Academic Publishers,
P.O. Box 322, 3300 AH Dordrecht, The Netherlands.

Printed on acid-free paper

Printed in the Netherlands.

TABLE OF CONTENTS

Part III. Fundamental Properties of Nanosystems

PREFACE

During the last years of the 20th and in the first year of the 21st century the molecular-scale science and nanotechnology showed to be a rapidly developing field, resulting in the construction of nanosystems from carbon molecules (fullerenes), supramolecules (nanotubes, polymers, biopolymers – DNA, proteins and their complexes) as well as metal and semiconductor nanocrystals (nano-Si, Ge, CdS, CdSe, ZnS, SiO_x, oxides of transition metals, etc.).
Now the novel multifunctional nanosystems based on such materials and their nanotechnology are starting in their development.

Therefore the development, characterisation, and an exploitation of novel multifunctional nanosystems from materials, based on the assembly of molecular and nanocrystal components, is an exceptionally active and rapidly expanding field. For this reason, the topic of molecular and nanocrystal-based multifunctional nanosystems was chosen as the subject of a Workshop, sponsored by the NATO Scientific Affairs Division, NATO International Science Exchange Programme, which was organized by us at September 2001 in Kiev as a NATO Advanced Research Workshop.

This Workshop follows the Kiev ARW in May 2000 on the problem of nano-optoelectronics. The interest of scientists, whether they are physicists, chemists or biochemists, in nano-optoelectronics systems and in other functional nanosystems is very widespread, which added a really international and multicultural flavor to Kiev ARW 2001.

The coupling of molecular and nano-crystalline nanosystems, and both components with nanoelectronics, nano-optoelectronics, magnetoelectronics and sensorics, has produced a boiling pot of ideas, problems, discoveries and applications. This statement is abundantly illustrated in the present book.

The first year of the new millennium provides an obvious transition point at which many aspects of molecular-scale science and technology of functional nanosystems can be assessed with respect to the research progress, made in the preceeding decades, and to the challenges that lie ahead in the upcoming decades. This book was planned to mark this with the objective of presenting a collection of papers from experts, which provide broad perspectives on the state-of-the art in the various disciplines of molecular-scale science and technology, and on the directions for future research.

The contents of this book providės a snapshot, at a particular point in time, on different subjects such as:
Modelling and Simulation of Nanoparticle and Molecular Nanosystems (Part I)
Nanotechnology of Nanoparticle and Molecular Nanosystems (Part II)
Fundamental Properties of Nanosystems (Part III)
Single and Assembled Molecules Experiments (Part IV)
Multifunctional Nanosystems (Part V).

The theoretical modelling and numerical simulation of nanosystems, including methods of atom-based quantum mechanic, integrated molecular orbital molecular mechanics, molecular dynamics, other methods, suitable for modelling particles synthesis, assembling and interaction processes, and modelling of carbon nanosystems properties have been described. The tight binding and the wave packet dynamical STM simulation methods are reviewed with applications to carbon nanotubes images. Two models (discrete DM and continuous CM) of the stress-strain state of long multiwall carbon nanotubes, subjected to internal pressure, are described. The solutions of DM equations and the main CM equation (the Eulers differential equation) define the distribution of the stresses between the layers, the forces in the layers, the deformation of the layers, and the strain across the wall of a tube (Part I).

The methods to obtain nanosystems from fullerenes, nanotubes and polymers/biopolymers – DNA proteins, metals (Au, Ag, Pt, etc.), and semiconductor nanocrystals (nano-Si, Ge, CdS, CdSe, ZnS, SiO_x, oxides of transition metals, etc.) including synthesis, assembly, reactivity, stability, process of the formation of recognition templates for biomaterials with engineered bioreactivity, are reported. Some recent "hot topics" in research are covered, such as the DNA nanotechnology, or metal nanowires self-assembly and layer-by-layer assembly of nanotubes, recognition of templates for biomaterials, as well as non-thermal plasma synthesis of nanocarbons. The nanosystems discussed encompass a diverse set of compositions and structures, including clusters, supramolecular assemblies, assemblies incorporating semiconductor nanocrystals and biomolecule-based components, and metal nanowires self-assembly. This field of research is clearly still in its infancy, and we are only beginning to understand in detail some of the parameters that govern the self-assembly of nanosystems. The ultimate future goal is to control these processes in a similar way, as chemists are able to control and predict the formation of molecular nanosystems (Part II).

The specific concepts, properties, phenomena, and processes related to nanosystems from molecular/cluster networks with quantum dots, quantum wires and other quantum structures, are presented in Part III. The nanosystems from clusters, supramolecular assemblies, and assemblies incorporating fullerenes are discussed. A full range of potentially interesting nanosystems properties, including electronic, magnetic, optical, mechanical, structural and chemical characteristics are considered. Some very actual scientific results are covered, such as the magnetic properties of undoped fullerene polymers (data included in a paper published in NATURE, October 2001), as well as studies on protein electron carrier complexes and on DNA modifications by novel antitumor platinum drugs. Thus, studies of the physical, chemical and, which is the most novel, biological properties of molecular nanosystems, having at least one dimension, ranging from molecular-scale to submicron, are going on, exhibiting improved and novel properties, which are a direct result of their molecular/nano-size dimensions and self-forming bonds between them in nanonetworks.

For the characterization of biomolecules new principles of scanning probe microscopy and spectroscopy are realized. Also here, some recent "hot topics" in research are covered, such as images of biomacromolecules – virus, and Th. terrestris spores surface (Part IV).

Realisation and application of novel multifunctional nanosystems in biomedicine, electronics, optics, magnetoelectronics, spintronics, pharmaceutical biocomplexes and biosensor nanosystems are shown (Part V). Key attributes are the ability in nanosystems to access the additional dimension of multiple length scales, and available structural complexity via organic chemistry synthetic methodologies, and the innovative assembly of such diverse components.

The enormous variation among fullerene, nanotube, nanosilicon, metal and semiconductor nanocrystals, polymer/biopolymer components, processes, and targeted structures available to scientists and engineers engaged in innovative multifunctional nanosystem research, leads to the clear conclusion that developments in the field demand interdisciplinary efforts. These include established cross-disciplinary efforts, such as those involving both organic and inorganic chemists as well as cooperative efforts with biochemists and biotechnologists, and the less established multidisciplinary collaborations between chemists and physicists, as well as engineers. The panel on the Workshop concluded that encouraging interaction and joint efforts among these groups is essential for future development of the field. As will be evident from the articles below, there are in fact ample opportunities in a wide range of disciplines for important contributions to the development of multifunctional nanosystems and nanotechnology.

It should be noted that the success of our multidisciplinary meeting and of this Book was guaranteed in advance by the enthusiastic work of: Dario Braga (Italy), Angel Rubio (Spain), Sergiy Romanov (Germany), Michael Waring (UK), Andrzej Huczko and Wladyslaw Torbich (Poland), Mark Geza (Hungary). We express our gratitude to all those, who have contributed to the workshop, especially to Key Speakers: Viktor Brabec (Czech Republic), Mathias Brust (UK), Yury Gogotsi (USA), Juergen Horbach (Germany), Nina Kovtyukhova (USA), Buddy Ratner (USA), Sergiy Romanov (Germany), Yuri Shirshov (Ukraine), Michael Waring (UK), Zeno Gaburro (Italy), Kars-Michiel H. Lenssen (The Netherlands).

We thank all the authors who responded eagerly to this demanding task and promptly delivered high quality manuscripts. This book would not have been possible without the efficient and cheerful help of our scientific secretary Alexandr Gozchinskiy and editorial assistant, Yuri Prylutskyy. We are most grateful to them for all their efforts.

We gratefully acknowledge the generous support of the NATO Scientific Affairs Division and our Universities, which made the workshop and this resulting book possible, and to Kluwer for efficient publication.

Eugenia Buzaneva
Kiev, Ukraine, November 2001

Peter Scharff
Ilmenau, Germany, November 2001

Photograph of the participants of the NATO ARW in Kiev (September 2001)

MULTISCALE COMPUTER SIMULATIONS IN PHYSICS, CHEMISTRY, AND BIOLOGY: THE EXAMPLE OF SILICA

JÜRGEN HORBACH, CLAUS MISCHLER, KURT BINDER_
Institut für Physik, Johannes Gutenberg–Universitðt Mainz,
Staudinger Weg 7, D–55099 Mainz, Germany; 2
WALTER KOB
Laboratoire des Verres, Université de Montpellier II,
Place E. Bataillon Case 069, 34095 Montpellier, France

Abstract.

We show to what extent molecular dynamics simulations (MD) can explore struc-tural and dynamic properties of atomic systems whereby the system under consideration is amorphous silica (SiO2). Two studies are presented: (i) a large scale simulation of the dynam-ics of a SiO2 melt and (ii) the investigation of free silica surfaces where a mixture of a classical MD and a Car–Parrinello molecular dynamics is used.

1. Introduction

Computer simulations are nowadays a very useful tool to get insight in the microscopic behavior of molecular systems in very different fields such as condensed matter physics, physical and theoretical chemistry, biology, etc. [1]. The big advantage of computer simulations is that they provide a large freedom regarding the system that can be studied. E.g., in the case of the very complex problem of protein folding very different systems are considered in
simulations ranging from simple lattice models to models that include chemi-cal details. In a simulation one can also investigate systems in thermodynamic states which are difficult to realize experimentally, such as negative pressure and high temperatures. Moreover, quantities can be determined which are not accessible in an experiment or which require assumptions on the measuring process.

The most important simulation methods are the Monte Carlo (MC) method and molecular dynamics (MD) computer simulations [2]. In this paper we are considering only the latter method. In a classical MD one solves Newton's equations of motion for an atomic or molecular system. This results in the full microscopic information in form of the particle trajectories (which is of course only true if quantum mechanical effects can be neglected). Thus, especially if one is interested in the dynamics a MD is the method of choice. Despite its advantages the MDmethod has also several drawbacks. Firstly one is very restricted in the time scales and the system sizes that can be simulated. Thus, for the example of the dynamics of a silica melt we discuss in Sec. 2 to what extent the dynamics of an atomic system can be explored in a MD. Another drawback of a classical MD is that one has to model the interactions

E. Buzaneva and P. Scharff (eds.), Frontiers of Multifunctional Nanosystems, 1–15.

between the atoms by a potential function. One can circumvent this problem by using an *ab initio* method, the so–called Car–Parrinello molecular dynamics (CPMD). But this method is restricted to rather short time scales and small system sized as compared to conventional MD. In Sec. 3 we present a way to combine a classical MD with a CPMD, and apply it to the investigation of a free silica surface. Eventually, we summarize a few conclusions (Sec. 4).

2. Case 1: The Dynamics of Amorphous Silica

Understanding the glass transition of supercooled liquids and the glassy state of materials has been a longstanding problem of condensed matter physics and materials science [3, 4, 5, 6, 7]. In a rather narrow temperature range, the shear viscosity, $\eta(T)$, may increase by 15 orders of magnitude (from 10^{-2} to 10^{13} Poise) before the liquid falls out of equilibrium and the structural changes cease. But the instantaneous structure of the liquid, as revealed by diffuse elastic scattering of X–rays or neutrons, hardly differs from that of the corresponding amorphous solid. In order to compare different glassformers one often defines a glass transition temperature T_g as the temperature at which the viscosity η has a value of 10^{13} Poise (at this temperature typical relaxation times are of the order of 10^2 sec). If one plots the logarithm of $\eta(T)$ versus the scaled inverse temperature T_g/T (the so–called Angell–Plot [8]) one finds that the shape of the different curves is not universal but depends on the material. For the two extreme cases of the shape, Angell has introduced the terms "strong" and "fragile" glassformers [8]. Strong glassformers are liquids for which $\eta(T)$ shows an Arrhenius law ($\eta \propto \exp[E_A/(k_B T)]$) in the whole accessible temperature range, i.e. in the Angell plot the corresponding curve is a straight line. In contrast to that, for fragile glassformers one observes a strong bending of the $\eta(T)$ curve in an Angell plot.

In the following we present the results of an investigation of the dynamics of a viscous silica melt. This is a very interesting issue because silica is the prototype of a strong glassformer. Apart from this it is also very important from a technological point of view because it is the basis material of window glass. Moreover, amorphous silica occurs also in many natural rocks, and thus is of great scientific interest in geology.

As a model to describe the interactions between the atoms in silica we use a pair potential which was proposed by van Beest *et al.* [9]. It contains a pseudo–Coulomb part and a short–ranged part,

$$\phi(r) = \frac{q_\alpha q_\beta e^2}{r} + A_{\alpha\beta} \exp\left(-B_{\alpha\beta} r\right) - \frac{C_{\alpha\beta}}{r^6} \quad \alpha, \beta \in [\mathrm{Si}, \mathrm{O}], \qquad (1)$$

where r is the distance between an atom of type α and an atom of type β. The effective charges are $q_O = -1.2$ and $q_{Si} = 2.4$, and the parameters $A_{\alpha\beta}$,

$B_{\alpha\beta}$, and $C_{\alpha\beta}$ can be found in the original publication. They were fixed by using a mixture of *ab initio* calculations and classical lattice dynamics simulations. The long–ranged Coulomb forces (and the potential) were evaluated by means of the Ewald summation technique. As an integrator for the simulation we used a velocity Verlet algorithm [2] with a time step of 1.6 fs. More details on the potential and the calculation of the forces can be found in Ref. [10].
Our simulations were done at a constant mass density, $\rho_m = 2.37$ g/cm^3, which is close to the experimental one. We equilibrated the system at each temperature by doing a NVT simulation (the temperature was kept constant by means of a stochastic collision algorithm) and started then, from the fully equilibrated melt, production runs at constant energy. Systems of 8016 particles were simulated in a cubic box of size (48.37 Å)^3. The size was chosen because for smaller sizes there is a pronounced evidence of finite size effects in the dynamics [11, 12]. In order to improve the statistics we averaged at each temperature over two independent runs. We could equilibrate the system for temperatures $T \geq 2750$ K, carrying out runs up to 20 ns length. Due to the relatively large system size and the long time scales it was necessary to use a parallelized code on a Cray T3E. The overall computational effort (equilibration plus production) for the lowest temperature, $T = 2750$ K, was about 13 CPU years of single processor time (or 2.5 months on 64 processors of a Cray T3E).
In order to test how reliably the structure of silica is reproduced by our simple model we have also cooled the melts down to $T = 300$ K with the slowest rate acceptable for a simulation (still as high as 10^{12} K/s). At this temperature neutron scattering experiments have been done in which the static structure factor $S_n(q)$ of silica in its glass state was determined. Here $S_n(q)$ represents the intensity of the radiation observed with a momentum transfer $\hbar q$ and can be calculated in the simulation by

$$S_n(q) = \frac{1}{N_{\rm Si} b_{\rm Si}^2 + N_{\rm O} b_{\rm O}^2} \sum_{k,l=1}^{N} b_k b_l \langle \exp[i\vec{q} \cdot (\vec{r}_k - \vec{r}_l)] \rangle \ , \qquad (2)$$

where $N = N_{\rm Si} + N_{\rm O}$ denotes the number of atoms and b_k, $k \in \{\rm Si, O\}$, are the neutron scattering lengths which we have taken from the literature [13]. Note that no adjustable parameter is needed to calculate $S_n(q)$ by Eq. (2). Fig. 1 shows $S_n(q)$ from our simulation in comparison with a neutron scattering result of Price and Carpenter [14]. We recognize that qualitatively, as well as quantitatively, the two curves are in very good agreement, in that not only the location, but also the height of the various peaks is reproduced very well.
The static structure factor is a good quantity to study the structure at larger length scales than the distance between nearest neighbors which is in our case that of Si–O bonds, i.e. around 1.6 Å [10]. Indeed, the first peak in $S_n(q)$ at $q = 1.7$ Å^{-1} (the so–called first sharp diffraction peak) reflects the

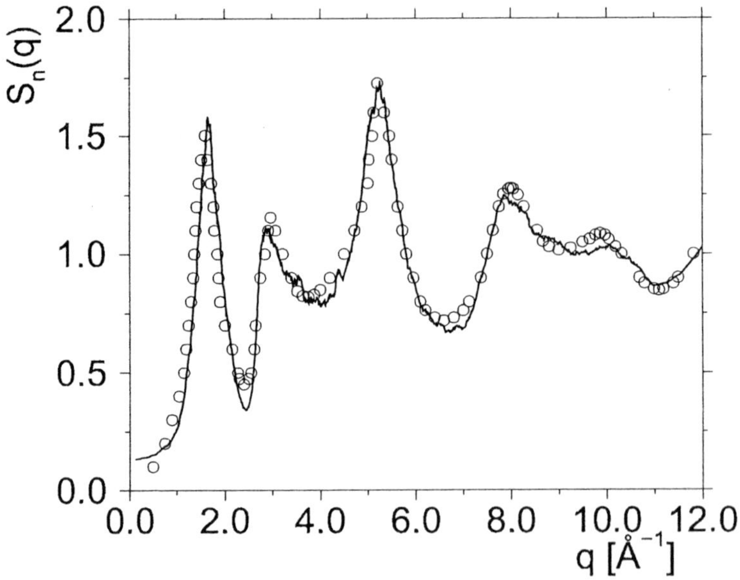

Figure 1. Static structure factor of amorphous silica (SiO_2) at room temperature ($T = 300$ K) plotted vs. wavenumber q (q is related to the scattering angle θ as $q = (4\pi/\lambda)\sin\theta$, λ being the wavelength of the radiation used for the scattering experiment). The full curve is the molecular dynamics simulation, using the measured neutron scattering length of Si and O atoms [13], and the circles are from a neutron scattering experiment [14].

intermediate range order in silica, arising from connected tetrahedra in a disordered tetrahedral network. Especially, the good agreement of this feature with experiment (see Fig. 1) demonstrates that it is not necessary to consider three–body terms in the potential to give an accurate description of the network structure in silica since the BKS potential (Eq. 1), which is used in this study, is a pure pair potential.
A simple quantity to study the dynamics of a fluid system on a microscopic level is the mean–squared displacement (MSD) of a tagged particle (of type α), which is given by

$$\langle r^2(t) \rangle = \frac{1}{N_\alpha} \sum_{l=1}^{N_\alpha} \langle |\vec{r}_l(t) - \vec{r}_l(0)|^2 \rangle \ . \tag{3}$$

The time dependence of this quantity in the case of oxygen is shown in Fig. 2 for all temperatures investigated. For high temperatures, curves to the left, two temperature regimes can be distinguished. At short times the motion of the particles is ballistic and the MSD is proportional to t^2. For long times the motion is diffusive and hence we have $\langle r^2(t) \rangle \propto t$. These two regimes are also seen at low temperatures, curves to the right. But at these temperatures these two regimes are separated by a third one in which the MSD forms a

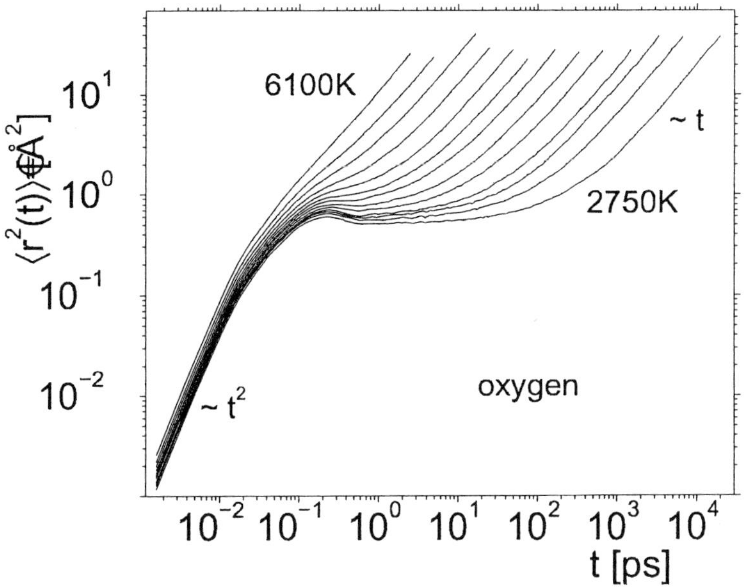

Figure 2. Time dependence of the mean squared displacement of oxygen for different temperatures.

plateau which becomes more pronounced the lower the temperature is. The microscopic reason for this behavior is the so–called cage effect, i.e., the fact that in this time regime the tagged particle is temporarily trapped by the particles surrounding it. The particle can only leave the cage if a neighbor has left its cage, in turn the latter particle can only leave its cage if a neighbor of it has left its cage, etc. Thus, the mechanism, which is connected with the escape of a particle from its cage, is a highly cooperative process. It is also noticeable that at low temperatures the MSD exhibits a peak around 0.2 ps. We have shown elsewhere [12] that this feature is related to the so–called boson peak, a vibrational feature at frequencies around 1 THz whose precise origin is currently still a matter of debate (see [12] and references therein). Note that the MSD for silicon shows qualitatively the same behavior as the one for oxygen in Fig. 2 (the MSD for silicon can be found in Ref. [10]).
Using the Einstein relation, $\lim_{t\to\infty} < r^2(t) > /(6t) = D$, one can calculate the diffusion constants D from the MSD. These are plotted in Fig. 3 as a function of the inverse temperature in the upper left part. As expected for a strong glassformer, we find that at low temperatures the diffusion constants show an Arrhenius dependence, $D \propto \exp[-E_{\mathrm{A}}/(k_B T)]$. The activation energies E_{A} are 4.66 eV and 5.18 eV for oxygen and silicon, respectively. These numbers compare very well with the ones determined in experiments, namely 6 eV for silicon [15] and 4.7 eV for oxygen [16]. Such an excellent agreement between simulation and experiment is not at all trivial as was shown recently

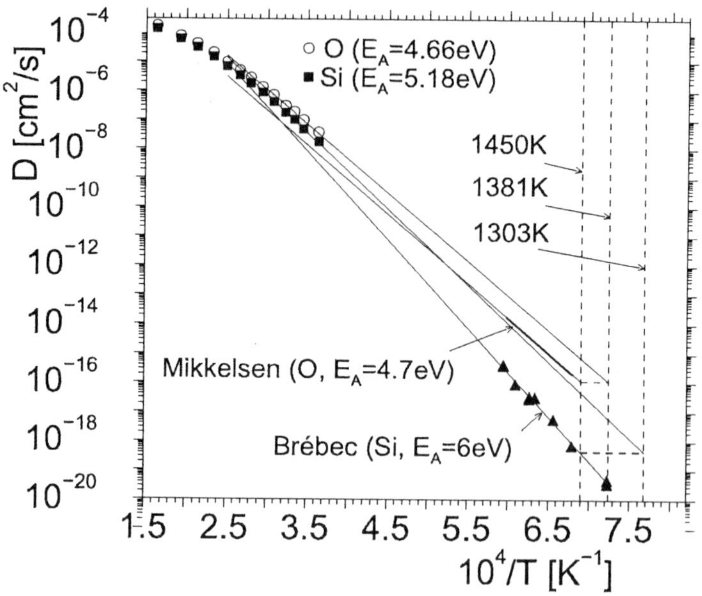

Figure 3. Arrhenius plot of the diffusion constants as determined from our simulation (open circles, oxygen; solid squares, silicon) and the experimental values of oxygen (bold line) and silicon (filled triangles). Thin solid lines: extrapolation of our data to low temperatures and extrapolation of the experimental data to high temperatures. See text for the discussion of the other lines.

by Hemmati and Angell [17] who found that various models of silica can give very different results for D. In the lower right part of Fig. 3 also the experimental data from Refs. [15] and [16] are shown. Although there is big gap of about seven orders of magnitude between the diffusion constants from the simulation and the experiment, one can, however, extrapolate the Arrhenius laws from the simulation to the temperature regime which is accessible to the experiment. We see that such an extrapolation overstimates the experimental diffusion constants by about one decade in the case of oxygen and by about two decades in the case of silicon. Furthermore, one can estimate from such an extrapolation the glass transition temperature T_g at which our model would have a viscosity of 10^{13} Poise (one could directly determine this temperature in a simulation if one would have access to computers which are 10^{10} times faster than the ones available). For this we read off the experimental values of the diffusion constants at the experimental $T_g = 1450$ K (horizontal dashed lines) and determine the temperature at which the Arrhenius extrapolations of our data cross these values. From this construction we obtain a $T_{g,sim}$ of 1381 K and 1303 K for oxygen and silicon, respectively (vertical dashed lines) which underestimates the experimental value for T_g only within 10%.

At high temperatures, say for $T > 3500$ K, we see in Fig. 3 that there are significant deviations from an Arrhenius behavior. An explanation to rationalize these deviations is offered by the so–called mode coupling theory (MCT) [4] which predicts a power law for the temperature dependence of the diffusion constant,

$$D \propto (T - T_c)^{\gamma}, \tag{4}$$

with a critical temperature T_c and a critical exponent γ. This power law is the result of the *ideal version* of the theory in which activated processes (hopping processes) in the dynamics are fully excluded. In the scenario of the (idealized) MCT relaxation times diverge if one approaches the critical temperature T_c from above. For $T < T_c$ the system gets "stuck" and e.g., the mean squared displacement would only increase up to a plateau value (after the short time regime) and never reach a diffusive regime. The physical reason for this behavior is the so–called cage effect which we have explained before by means of Fig. 2. However, the theory implies that the diffusion constants become zero as T_c is approached which disagrees with experiment. The reason for this disagreement might be that hopping processes are excluded from the theory and the solution might be the so–called extended MCT where such hopping processes are taken into account. In the latter version of MCT the glass transition at T_c is replaced by a smooth crossover from a power law behavior as given by Eq. 4 at high temperatures to a simple Arrhenius behavior at low temperatures. Indeed, we can interpret the behavior of the diffusion constants from our simulation within this physical picture. We have shown elsewhere [10] that the high temperature regime for the diffusion constants can be well described by Eq. (4) with $T_c = 3330$ K and $\gamma \approx 2.1$, and also many other predictions of the theory are valid [18].
Another important transport quantity, which can be determined experimentally also for much higher temperatures than the diffusion constants, is the shear viscosity η. η can be calculated in the simulation by means of a so–called Green–Kubo relation, i.e. essentially by the time integral of the autocorrelation function of the off–diagonal elements of the pressure tensor (for a definition see, e.g., Ref. [19]). It is quite demanding to compute η with high accuracy in a simulation since this quantity is a collective one. Thus, in order to gain a reasonable statistics the simulation runs were ten times longer than the usual ones and the smallest temperature for which we have determined η was 3000 K. Fig. 4 shows an Arrhenius plot of η in which we compare the simulation to experiments by Urbain *et al.* [20]. In the latter experiment the viscosity was measured up to temperatures as high as 2800 K (therefore the gap between simulation and experiment is now much smaller than that for the diffusion constants). At low temperatures our simulation data can be described by an Arrhenius law whereas for high temperatures strong deviations to such a behavior are found (similar to the case of the diffusion constants). We find an activation energy of 5.19 eV which is in very good agreement

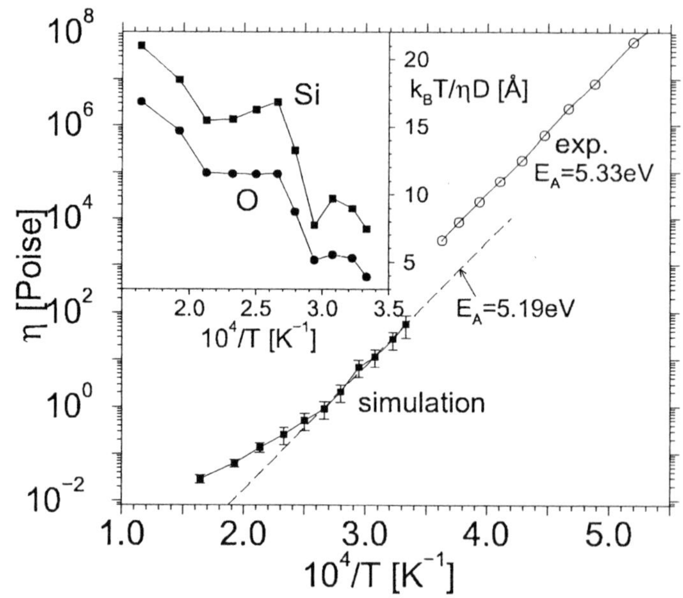

Figure 4. Main figure: Arrhenius plot of the shear viscosity from the simulation (solid squares). The dashed line is a fit with an Arrhenius law to our low–temperature data. The open circles are experimental data from Urbain *et al.* [20]. Inset: temperature dependence of the left hand side of Eq. (5) to check the validity of the Stokes–Einstein relation.

with the experimental value of 5.33 eV. The extrapolation of the Arrhenius law from the simulation to lower temperatures shows that the experimental viscosity as measured by Urbain *et al.* (open circles), is understimated by about a factor of 10 which is not much because the experimental value has an uncertainty in the prefactor of the Arrhenius law of about a factor of 2 [21]. Since we have determined the diffusion constants and the viscosity at the same time, we can check the validity of the Stokes–Einstein relation given by

$$\frac{k_B T}{\eta D} = \lambda = \text{const.}\,, \qquad (5)$$

where the constant λ has, in the phenomenological Eyring theory, the meaning of a length of an elementary diffusion step [22]. Eq. (5) is often used to convert the viscosity into diffusivities and vice versa. From the inset of Fig. 4 we see that Eq. (5) is not valid for our system in that λ depends strongly on temperature changing from values around 20 Å at high temperatures to values around 5 Å at the lowest temperature. Thus we can conclude that for our system the Stokes–Einstein relation is not well suited to extract the viscosity from the diffusion constants.

3. Case 2: The Structure of Free Silica Surfaces

We have seen in the preceding paragraph that the BKS potential is able to reproduce many structural and dynamical properties of real silica very well such as the static structure factor, the glass transition temperature, and the activation energies of the viscosity and the diffusion constants. The latter quantities are for the bulk since the system was simulated with periodic boundary conditions. It is much less obvious that the BKS potential gives also a reliable description of silica surfaces. The reason is that it was optimized to reproduce bulk properties such as the experimental elastic constants of α quartz. A similar fitting of the potential parameters to surface properties of real silica is difficult because of the lack of experimental data in this case. Although it is straightforward to investigate the properties of free surfaces of SiO_2 by means of a MD using the BKS potential it is not clear how one could test whether the results have anything in common with real silica.

One possibility to circumvent this problem is to use Car–Parrinello molecular dynamics (CPMD) [23] in which, different from a classical MD, the electronic degrees of freedom are taken into account via a density functional theory. Therefore, in contrast to a classical MD an effective potential between the ions is calculated self–consistently on the fly, i.e. the instantaneous geometry of the ions is always taken into account and one does not have to model the potential energy between the ions by a given function such as Eq. (1). However, it is not yet possible to replace the classical MD method by the CPMD, since due to the huge computational burden only relatively short time scales, a few ps, as well as small systems, a few hundred particles, can be simulated. In contrast to that we have demonstrated in the preceding paragraph that classical MD allows to simulate thousands of particles over several ns.

Since the time scale which is accessible in a CPMD is very restricted the idea of our approach is to combine a classical MD and a CPMD. To this end, we first equilibrate the system with a classical MD in which we again model the interactions between the atoms by the BKS potential, Eq. (1). Then, we use these configurations as the starting point of a CPMD. The goal of this investigation is twofold: Firstly, we want to see whether the classical configurations are stable in CPMD. And if this is the case we can secondly compare the structural differences as obtained by the two methods. So we get an idea of how accurate the BKS potential is able to describe silica surfaces, and we have hints of how this potential energy model could be improved.

In order to investigate a free silica surface one could consider a film geometry, i.e. periodic boundary conditions (PBC) in two directions and an infinite free space above and below the remaining third direction. Unfortunately, this is not a very good solution since the Ewald summation technique for the long ranged Coulomb interactions becomes unefficient in this case [2]. Therefore,

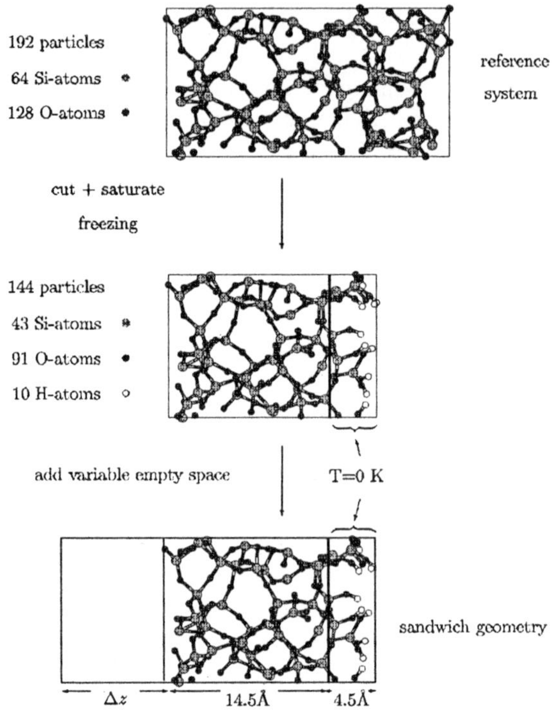

Figure 5. Procedure to create the used sandwich geometry.

we have adopted the following strategy which is illustrated in Fig. 5: i) We start with a system at $T = 3400$ K with PBC in three dimensions (box dimensions: $L_x = L_y = 11.51$ Å and $L_z' = 23$ Å). ii) We cut the system perpendicular to the z–direction into two pieces. This is only done for oxygen–silicon bonds such that we get only free oxygen atoms at this interface. iii) These free oxygen atoms are now saturated by hydrogen atoms. The place of these hydrogen atoms is chosen such that each of the new oxygen–hydrogen bonds is in the same direction as the oxygen–silicon bonds which were cut and have a

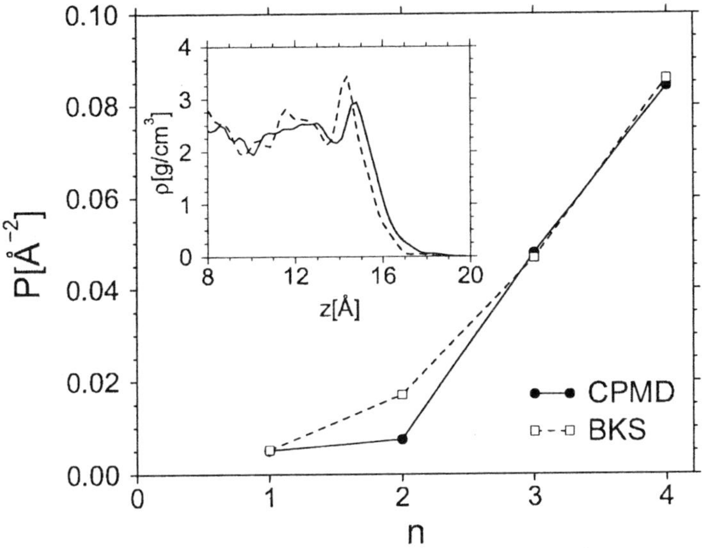

Figure 6. Probability to find a ring of size *n*. Inset: *z*–dependence of the mass density.

length of approximately 1 Å. The interaction between the hydrogen and the oxygen atoms as well as the silicon atoms are described only by a Coulombic term. The value of the effective charge of the hydrogen atoms is set to 0.6 which ensures that the system is still (charge) neutral. iv) We make atoms which have a distance from the interface that is less than 4.5 Å completely immobile whereas atoms that have a larger distance can propagate subject to the force field. v) We add in *z*–direction an empty space $\Delta z = 6.0$ Å and thus generate a free surface at around 14.5 Å. With this sandwich geometry we can use periodic boundary conditions in all three directions. We have made sure that the value of Δz is sufficiently large that the results do not depend on it anymore. Eventually, we have a system of 91 oxygen, 43 silicon, and 10 hydrogen atoms in a simulation box with $L_x = L_y = 11.51$ Å and $L_z \approx 25$ Å. We have fully equilibrated 100 independent configurations for about 1 ns. Using a subset of these configurations as starting points we subsequently started CPMD simulations [24]. We used conventional pseudopotentials for silicon and oxygen and the BLYP exchange functions [25, 26]. The electronic wave–functions were expanded in a plane wave basis set with an energy cutoff of 60 Ry and the equations of motion were integrated with a time step of 0.085 fs for 0.2 ps. In the analysis of the CPMD run only those configurations were taken into account that were produced later than 5 fs after the start of the CPMD run in order to allow the system to equilibrate at least locally [27]. From the 100 BKS configurations we have picked up those for which one of

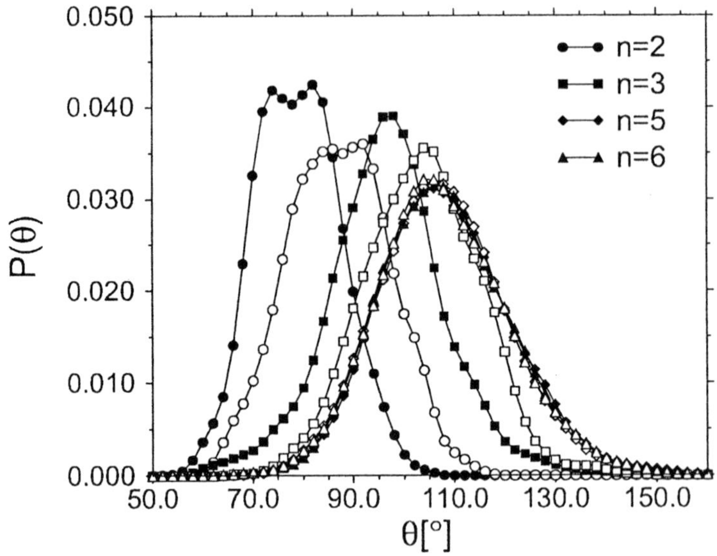

Figure 7. Distribution of O–Si–O angles for different ring sizes (BKS: filled symbols, CPMD: open symbols).

the three following cases holds: i) The surface exhibits no defects, i.e. all Si and O atoms are four and two–fold coordinated, respectively. ii) The surface has an undercoordinated oxygen atom and an undercoordinated silicon atom. iii) The surface has an overcoordinated oxygen atom and an undercoordinated silicon atom. We used two BKS configurations for each of the cases i)–iii) and started the CPMD runs. The computational effort for the latter was immense: To simulate 1 ps one needs 11000 CPU hours of single processor time (or about one month on 16 processors of a Cray T3E).

A quantity which is appropriate to characterize the network structure is the ring distribution. A ring is defined as a closed loop of n consecutive Si–O segments. The largest differences between the results of the classical and that of the CPMD simulation is found for the short rings, i.e. $n < 5$. Fig. 6 shows the probability to find a ring of size n for the case of the BKS simulation and the CPMD. We see that with the BKS model the frequency with which a ring of size two occurs is overestimated by about a factor of two as compared to the CPMD result. In agreement with this observation we find that the overshoot in the z–dependence of the mass density profile near the surface is less pronounced for the CPMD than for the classical MD with the BKS potential (see inset of Fig. 6), since two–membered rings are relatively dense.

Another interesting result is the dependence of the distribution of angles O–Si–O on the ring size n (Fig. 7). For large n, $n > 4$, i.e. for rings which are normally found in the bulk, the results of the two different methods are in

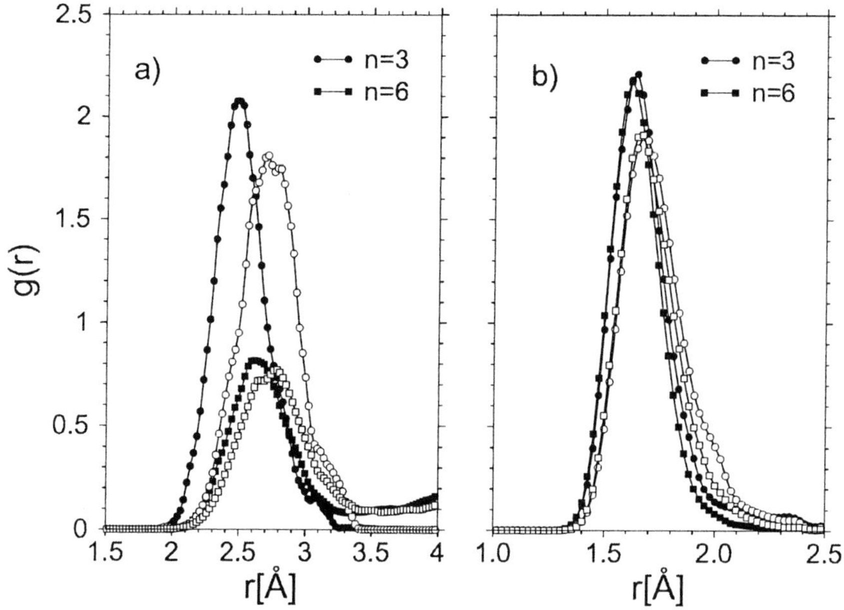

Figure 8. Radial distribution function for different ring sizes, (a) O–O and (b) Si–O (BKS: filled symbols, CPMD: open symbols).

good agreement [27]. For smaller n, however, the mean O–Si–O angle from CPMD is shifted to larger values in comparison to that of the classical MD. This shift becomes more pronounced with decreasing n. Furthermore also the shape of the distributions starts to become different if n is small.

This effect can be understood better if one analyzes the partial radial distribution functions $g(r)$ which are shown in Fig. 8. We see that for the Si–O correlation the curves from the CPMD are shifted to larger distances by about 0.04 Å and that this shift is independent of n. Also the $g(r)$ for the O–O correlation are shifted to larger r, but now we observe different shifts for different n. In particular we note that the O–O distance is nearly independent of n for the CPMD whereas it increases with n in the case of the BKS result. These effects result in the difference of the distribution of the O–Si–O angles if n is small.

In conclusion we find that for the structure on larger length scales the BKS simulation and the CPMD yield to similar results whereas the details of the structural elements on short scales (for instance, the short rings) are different in both methods. This shows that it is probably necessary to use *ab initio* methods like CPMD to reproduce quantitatively the properties of silica surfaces on short length scales.

4. Concluding remarks

In this report, we have shown that large scale MD simulations are able to give a lot of insight into the microscopic dynamics of atomic systems. We have demonstrated this for the example of the dynamics of a silica melt near the glass transition. Our model for silica accounts for the interaction between the atoms via simple pair potentials. Nevertheless, this model is able to give quite a realistic description of the structural and dynamic properties of real silica. Furthermore, we have shown that a more detailed understanding of the dynamics of silica can be gained than it is possible in experiments. So for temperatures $T \geq 2750$ K no measurements can be done, but the dynamics shows very interesting features at these temperatures in that the mechanism for the transport changes from the one as described by the ideal version of mode coupling theory (high temperatures) to an activated dynamics (low temperatures). But there is certainly still a considerable gap between the range of time scales accessed in experiments and in the simulations (see Fig. 3). Ideas for new simulation methods are necessary to bridge this time–scale gap.

Although effective potentials as used in classical MD simulations often seem to decribe atomic materials very realistically (as we have demonstrated for silica), there may be subtle features in the local structure which can be only quantitatively described by *ab initio* methods such as CPMD. This has been illustrated in this report by the investigation of free silica surfaces. The drawback of CPMD is that it is restricted to small time–scales and small system sizes. But if one uses a combination of classical MD (in our case with the BKS potential) and CPMD one obtains more reliable answers than if one would use each of both methods without consideration of the other one to investigate the structure of atomic materials.

Acknowledgements

We gratefully acknowledge the financial support by the SCHOTT Glaswerke Fond, DFG under SFB 262, and the BMBF under grant No. 03N6015. We thank the NIC Jülich for a generous grant of computing time.

References

1. K. Binder, G. Ciccotti (eds.), Monte Carlo and Molecular Dynamics simulations of Condensed Matter Systems, Societá Italiana di Fisica, Bologna, 1996.
2. There are several excellent textbooks on MC and MD methods, e.g. M. P. Allen and D. J. Tildesley, *Computer Simulations of Liquids* (Clarendon, Oxford, 1987); D. Frenkel and B. Smit, *Understanding Molecular Simulation* (Academic Press, San Diego, 1996);

K. Binder and D. P. Landau, *A Guide to Monte Carlo Simulations in Statistical Physics* (Cambridge University Press, Cambridge, 2000).

3. J. Jäckle, Rep. Progr. Phys. **49**, 171 (1986).
4. W. Götze and L. Sjögren, Rep. Progr. Phys. **55**, 241 (1992); W. Götze, J. Phys.: Condens. Matter **11**, A1 (1999).
5. J. Zarzycki (Ed.), *Material Science and Technology, Vol. 9: Structure and Properties of Glasses* (VCH, Weinheim, 1991).
6. K. L. Ngai (Ed.), *Proceedings of the Third International Discussion Meeting on Relaxations in Complex Systems*, J. Non–Cryst. Sol. **235–237** (1998).
7. M. Tokuyama, J. Oppenheim (Eds.), *AIP Conference Proceedings, Vol. 469, Slow Dynamics in Complex Systems* (American Institute of Physics, Woodbury, 1999).
8. C. A. Angell, in *Relaxation in Complex Systems*, ed. by K. L. Ngai and G. B. Wright (US Dept. Commerce, Springfield, 1985).
9. B. W. H. van Beest, G. J. Kramer, and R. A. van Santen, Phys. Rev. Lett. **64**, 1955 (1990).
10. J. Horbach and W. Kob, Phys. Rev. B **60**, 3169 (1999).
11. J. Horbach, W. Kob, K. Binder, and C. A. Angell, Phys. Rev. E **54**, R5897 (1996).
12. J. Horbach, W. Kob, and K. Binder, Eur. Phys. J. B **19**, 531 (2001).
13. S. Susman, K. J. Volin, D. G. Montague, and D. L. Price, Phys. Rev. B **43**, 11076 (1991).
14. D. L. Price and J. M. Carpenter, J. Non–Cryst. Sol. **92**, 153 (1987).
15. G. Brébec, R. Seguin, C. Sella, J. Bevenot, and J. C. Martin, Acta Metall. **28**, 327 (1980).
16. J. C. Mikkelsen, Appl. Phys. Lett. **45**, 1187 (1984).
17. M. Hemmati and C. A. Angell, in *Physics meets Geology*, ed. by H. Aoki and R. Hemley (Cambridge University Press, Cambridge, England, 1998).
18. J. Horbach and W. Kob, Phys. Rev. B **64**, 041503 (2001).
19. J. P. Boon and S. Yip, *Molecular Hydrodynamics* (Dover, New York, 1980).
20. G. Urbain, Y. Bottinga, and P. Richet, Geochim. Cosmochim. Acta **46**, 1061 (1982).
21. O. V. Mazurin, M. V. Streltsina, and T. P. Shvailko–Shvailkovskaya, *Handbook of Glass Data, Part A: Silica Glass and Binary Silicate Glasses* (Elsevier, Amsterdam, 1983).
22. S. Glasstone, K. J. Laidler, and H. Eyring, *The Theory of Rate Processes* (McGraw–Hill, New York, 1941).
23. R. Car and M. Parrinello, Phys. Rev. Lett. **55**, 2471 (1985).
24. CPMD Version 3.3a, J. Hutter, P. Ballone, M. Bernasconu, P. Focher, E. Fois, S. Goedecker, M. Parrinello, and M. Tuckermann, MPI für Festkörperforschung and IBM Research 1990–2000.
25. N. Trouiller and J. L. Martins, Phys. Rev. B **43**, 1993 (1991).
26. C. Lee, W. Yang, and R. G. Parr, Phys. Rev. B **37**, 785 (1988).
27. M. Benoit, S. Ispas, P. Jund, and R. Jullien, Eur. Phys. J. B **13**, 631 (2000).

APPLICATION OF THE IMOMM (INTEGRATED MOLECULAR ORBITAL MOLECULAR MECHANICS) METHOD FOR BIOPOLYMERS

I. KOMÁROMI AND L. MUSZBEK
University of Debrecen, Department of Clinical Biochemistry and Molecular Pathology & Research Group for Thrombosis and Haemostasis of the Hungarian Academy of Sciences
H-4032 Nagyerdei krt.98, Debrecen, Hungary

Abstract.

An IMOMM (Integrated Molecular Orbital Molecular Mechanics) -as special case of Morokuma's ONIOM (Our N-layered Integrated Molecular Orbital Molecular Mechanics) method- implementation will be presented emphasizing the importance of proper treatment of the QM/MM borderline region. Based on calculations carried out on small peptides the effect of different kinds of coupling schemes was analysed. Introducing electronic embedding between layers requires attention, especially when coupling similar to the "traditional" QM/MM (Quantum Mechanics/ Molecular Mechanics) is applied for electrostatics, in order to count all the important electrostatics energy terms and avoid (or decrease) double counting. A selected example (the study of the transglutamination reaction of blood coagulation FactorXIII) will be given in order to demonstrate the performance of the method.

Key words: Molecular orbital/ Molecular mechanics/ Hybrid QM/MM methods/ONIOM/ IMOMM/ FactorXIII

1. Introduction

During the last decades computational chemistry/biology has become one of the respected research tools, which can be fruitfully applied to explain different kinds of measured quantities and to construct working (i.e. predictive) models. It is especially true for high level (correlated) quantum chemistry computation for small systems

E. Buzaneva and P. Scharff (eds.), Frontiers of Multifunctional Nanosystems, 17–28.

promising results could be obtained using molecular mechanics/dynamics simulations [1-3] for large biological systems allowing us to have a new insight into the nature of the dynamics of biopolymers. However, until now only a relatively few examples could be found in the literature for applying such methods to simulate biochemical processes where chemical bonding pattern or electronic structure changes during the simulating period. (Typical examples are e.g. the enzyme reactions and the electronic excitations.) The reason is that the empirical force fields applied in these simulations are usually based on frozen electronic structure and to introduce parameters to take into account the electronic degrees of freedom is not straightforward.

As a consequence of this, we need quantum chemical calculations only on the part of the total system where the electronic structure of the system significantly changes and empirical force field quality computation is sufficient on the rest of the system [4-9]. Although the basic idea is simple, practical implementations of reliable hybrid methods are not always straightforward. In this paper we will review the existing hybrid quantum chemical molecular mechanical methods [10-19] emphasizing the ONIOM type QM/MM method, the IMOMM [20-22].

2. Hybrid QM/MM (quantum mechanical molecular mechanical) methods

2.1 *GENERAL CONSIDERATIONS*

The landmark paper in the field of QM/MM theory and its application to interpret enzyme reaction was published 25 years ago by Warshel and Levitt [4]. They modeled the reaction of lysozyme. Since then several kinds of method have been developed.

The common feature of these methods is that the total system can be composed from the inner (or model) part and the outer part. When the system is decomposed, usually chemical bonds are cut. In order to perform quantum chemical calculations on the inner subsystem, one needs either to saturate these dangling bonds or form a so-called boundary region with centers participating in both regions. (All the methods will be discussed here are based on link atom concept.) The inner region with saturated dangling bonds and will be called *model system*. The difficulties coming from introducing link atoms are analyzed in details and solutions are proposed in the paper of Bakowies and Thiel [19].

The first difficulty is to construct an energy expression for the whole system. From this point of view these hybrid methods can be divided into two main groups [10]. The first one uses the so-called additive scheme to construct the total energy expression for the entire (real) system :

$$E_{QM/MM,\mathrm{Real}} = E_{QM,\mathrm{model}} + E_{MM,outer} + E_{QM/MM,Inner\leftrightarrow outer} + E_{corr(link)} \quad (1)$$

At this case the total energy can be decomposed into four parts: the quantum mechanical energy of the model system, the molecular mechanics energy of the outer region, an interaction between the inner and outer regions and corrections because new (link) atoms were introduced. Most of the existing implementations are based on this scheme.

The other possibility is to use the subtractive scheme for the total energy expression of the system:

$$E_{QM/MM,real} = E_{MM,real} + E_{QM,model} - E_{MM,model} \tag{2}$$

This expression can be regarded as a correction (or extrapolation) scheme, which adds a correction calculated only for the inner region to the energy of the total system. In this case three different calculations (molecular mechanics for the whole system and both quantum mechanics and molecular mechanics for the model system) are done. A practical implementation based on this scheme is e.g. the IMOMM and (since it can be generalized) the ONIOM methods. The ONIOM and IMOMM methods will be discussed in detail later on.

The hybrid methods can also be classified on the basis of how the interactions between inner and outer regions are formulated[19]. When the interaction is calculated at molecular mechanical level only, it is called mechanical embedding.

In the case when the inter-region (interlayer) energies are calculated –at least partially- at quantum mechanics level the electronic embedding expression is used. This calculation is usually carried out via a modified one electron Hamiltonian when the partial charges of the outer region act as perturbation.

The polarization embedding - in which some polarization effect on the molecular mechanical atoms are taken into account - will not be discussed here.

2.2 *THE GENERAL ONIOM SCHEME*

The ONIOM extrapolation scheme has been developed by Morokuma and his coworkers[21,23-25].For two layer calculations it corresponds to Equation 2. The ONIOM energy usually called extrapolated energy and from this point of view the $E_{QM/MM,real}$ energy can be regarded as an extrapolated $E_{QM,real}$ energy

The extrapolation scheme outlined above can be easily generalized. Starting from the smallest system and performing the highest level calculation only on it, the energy of the next system can be extrapolated using the extrapolation formulae, which is given above. Continuing this extrapolation scheme, one can obtain the highest level energy approximation for the System(2), System(3),..., finally for the whole system (System(N)) carrying out 2N-1 calculations:

$$E_{Method\,(1),\,System\,(N)} = E_{Method\,(N),\,System\,(N)} + \sum_{i=1}^{N-1}\left(E_{Method\,(i),\,System\,(i)} - E_{Method\,(i+1),\,System\,(i)}\right) \tag{3}$$

Of course, the energy value from this successive extrapolation scheme can be very far from the energy would be obtained at highest level for the whole system. The aim of the correction is given in equations 2-3 is to obtain an (approximately) parallel energy

Of course, the energy value from this successive extrapolation scheme can be very far from the energy would be obtained at highest level for the whole system. The aim of the correction is given in equations 2-3 is to obtain an (approximately) parallel energy (hyper)surface to those, which would have been obtained by the highest level calculation on the whole system.

Due to the definitions (to the subtractive scheme) given above, the gradients and Hessians and many more useful quantities can be easily calculated.

2.3 *THE IMOMM METHOD*

The Integrated Molecular Orbital Molecular Mechanics method in its original form has been developed by Maseras and Morokuma[20]. We will outline the IMOMM method focusing on the problems arising at the borderline regions between model and real systems calculated at different levels of theories.

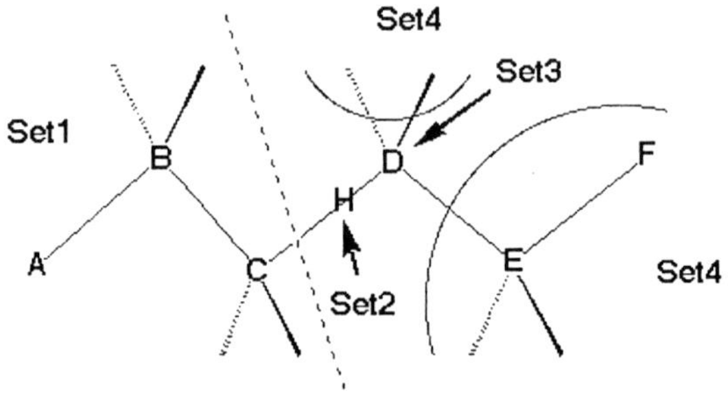

*Figure 1.*Schematic representation of real and model system at the borderline region

On the Figure 1 a schematic representation is drawn for the borderline region where the layers connected by chemical bond.

Assuming a given system partitioning which is mainly determined by the chemical intuition and the cost of computation, we will pay most of our attention to the problems of interaction between the model-only and real-only subsystems. We will examine in detail how this can be modelled in the IMOMM scheme.

The atoms on the Figure 1 can be divided into four sets [4,12]: The Set1 atoms form the model-only while the Set4 atoms the real only subsystems, respectively. The Set2 atoms are present only when the calculation is performed on the model system (they fill the dangling bonds) and are replaced by Set3 atoms at calculations on the real system. On the Figure 1 the H belong to the Set2 and D to the Set3 atoms, while the A, B and C to the Set1 and E and F to the Set4, respectively.

In the paper the following notations will be used: M=model system *with* substituted dangling bonds (Set1+Set2); M'=model system *without* substituted dangling bonds (=Set1); M''=model system (M') *with* D atoms (=Set1+Set3); R=real system (the whole one, *including* M') (= Set1+Set3+Set4); R'=real system *without* M' (= Set3+Set4); R''=real system *without* M' and *without* D atoms (=Set4).

depending on how it is implemented, the "traditional" (additive) QM/MM can predict different energy than the IMOMM does.

Using electronic embedding the "traditional" QM/MM schemes [e.g. 13] and the integrated scheme [20-22] usually yield different results.

In the ONIOM scheme, one should calculate the model system at high level involving the effect of surrounding charges via the one electron Hamiltonian (4).

$$H_{H,M} = H^{0}_{H,M} + \sum_{R''} \tau_{R''} q_{R''} \left(\sum_{M} \frac{Z_M}{r_{R''M}}\right) - \sum_{R''} \tau_{R''} q_{R''} \left(\sum_{\mu} \frac{1}{r_{R''\mu}}\right) \tag{4}$$

In this equation, $H^{0}_{H,M}$ stands for the unperturbed Hamiltonian of the model system. Scale factors for qR'' charges, R'' are introduced. These scale factors , or can be useful e.g. when at the borderline region a highly charged atom can be found and the wavefunction would be "overpolarized". In the present installation [22,26] three different scale factors included in the Equation 4., depending on how "far" is the atom with the corresponding partial charge (one = , two = , or three or more = bonds) from the "Set3" (D on the Fig. 1.) atom. The partial charge on the atom D is excluded from this summation because it is replaced by atom H in the model system.

The corresponding low-level calculation for the same system is a classical molecular mechanical calculation, where the Coulomb term between the surrounding charges and the model system atoms are also calculated according to Equation 5. For an obvious reason, if scale factors were used at Equation 4., the same factors have to be used in Equation 5.

$$E_{L,M} = E^{0}_{L,M} + \sum_{R''} \tau_{R''} q_{R''} \left(\sum_{M} \frac{q_M}{r_{R''M}}\right) \tag{5}$$

In the Equation 5 the $E^{0}_{H,M}$ stands for the unperturbed low level energy of the model (M) system. The low level real system calculation is the same as for the mechanical embedding, a pure molecular mechanics calculation for the real system. However, at the borderline region the same interactions are calculated three different way: via the modified one electron Hamiltonian (4) for the high level model system, using classical Coulomb term in the low level model system calculation (5) and angle bending or torsional terms *plus* scaled (or unscaled) electrostatic terms (e.g. for the C-D-E-F atoms) in the low level real system calculation. These three different kinds of calculation not necessarily lead to the right correction in the combined energy (2). The other disadvantage is that the Coulomb terms should be evaluated (at least formally) three times. The advantage of this method is that all the necessary energy terms are evaluated without double-counting any of them.

In the "additive" QM/MM scheme within the framework of IMOMM the electrostatic interactions between the inner (M') and real-only (R'') parts is calculated *only* at high level. The high level energy for the model system is the same as for

ONIOM-type electronic embedding while the low level model system energy is the same as for mechanical embedding. The low level real system energy differs substantially from either of the previous calculations. Since the electrostatic interaction between layers calculated via the modified one electron Hamiltonian, the corresponding term must not be calculated at low level:

$$E_{L,R(\text{electrostatics})} = \frac{1}{2} \sum_{i=R''} (\sum_{j=R''} (\frac{\tau_{ij} * q_i * q_j}{r_{i,j}})) + \frac{1}{2} \sum_{i=M''} (\sum_{j=M''} (\frac{\tau_{ij} * q_i * q_j}{r_{i,j}})) \tag{6}$$

The combined energy is coming from the Equation 1. In this case electrostatics energies between certain atoms three or two bond far from each other (e.g. C-D-E-F and C-D-E on Fig.1) calculated fully at high level. However, the corresponding torsional and bending terms are expected to take into account partially these interactions, yielding double counting in certain extent. On the other hand, the complete negligation of these charges will neglige important electrostatic effects on the model system. A possible way to solve this problem is scaling (by the factor ,) these charges to get the "optimal" solution. Using electronic embedding, the ONIOM and "additive" QM/MM methods, when the real and model part is not connected by chemical bond, can yield exactly the same results. The difference between them is in the treatment of the borderline region:

- The ONIOM scheme adds the difference of the electrostatic energies obtained at high and low levels for the model system to the real one.
- In the additive QM/MM scheme the electrostatic energy between two atoms belonging different partitions are not calculated at low level.

3. Test Calculations

Selected examples will be given in order to demonstrate how the IMOMM method works[22,29]. The results for small compounds will be compared the corresponding quantum chemical calculation for the entire system. Only the most characteristic feature of these calculations will be mentioned, more details will be published elsewhere. All the calculations have been carried out by means of the Gaussian'99 package [26]

In the first example the rotational energy barrier and the rotational transition state geometry will be determined for a model peptide bond using molecular mechanics (AMBER), ab initio (HF-6-31G) quantum chemical and hybrid (IMOMM) methods. It is immediately apparent from the figure, that the CNCC dihedral angle is quite similar for all of these methods. For the amide bond length, however, the pure AMBER predicts substantially larger value than the others. The reason is, that during the rotation the partially double bonded amide bond loses its double bond character and in the transition state it can be regarded as a single bond. Since the IMOMM calculation practically predicts the same value as the pure quantum chemical one it demonstrates clearly the advantage of using mixed method, when during a process the electronic structure is changing.

state it can be regarded as a single bond. Since the IMOMM calculation practically predicts the same value as the pure quantum chemical one it demonstrates clearly the advantage of using mixed method, when during a process the electronic structure is changing.

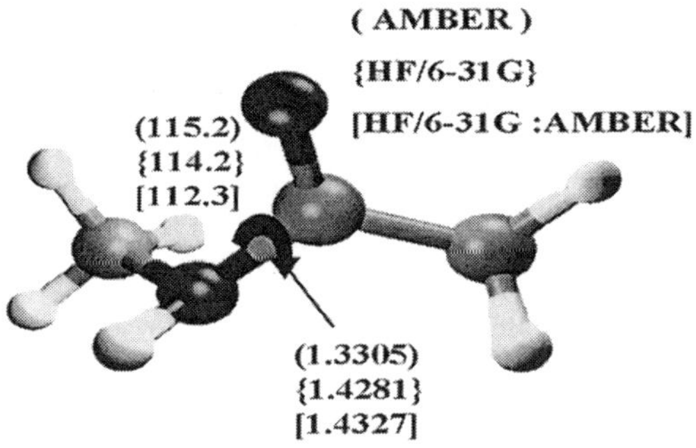

Figure 2.: Amide bond rotational transition state for N-methyl-acetamide. The bond length and torsional angles for the amide bond are marked.

It should be noted that for the rotational energy barrier all these methods predicts approximately the same values (18-20 kcal/mol)

The second example is focused on the effect of charge distribution on the borderline region and the effect of the ONIOM-type and traditional electronic embedding schemes. Series of model calculations have been performed on the diglycine molecule using different charge scaling factors for both the traditional (i.e. the interlayer electrostatic energy calculated only at QM level) and ONIOM-type electronic embedding. Selected computed parameters can be found in the Table 1. The charge sets and scale factors are depicted on the Figure 3.

Which immediately apparent from the Table is, that the effect of scale factors are not uniformly bad or good for all the values. It is hard to predict in advance the effect of the applied charge scale factor.

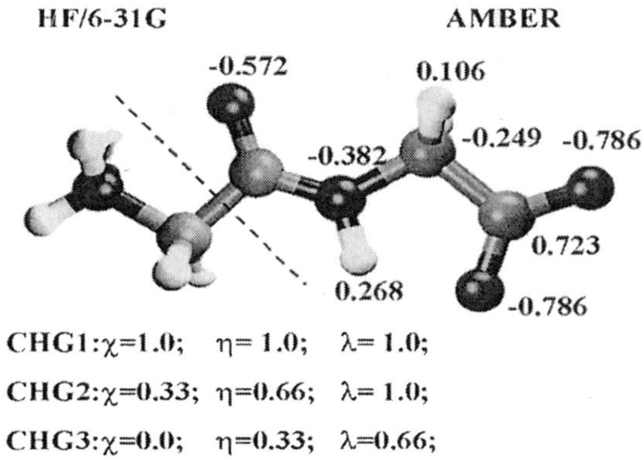

Figure 3.: Partial charges and scale factors applied in QM/MM calculations using electronic embedding scheme for diglycine molecule

at no charge scaling the ONIOM and "traditional" QM/MM calculation predict quite similar geometry parameters for diglycine. Contrarily, when charge scaling was applied, the result of ONIOM-type electronic embedding seems to be more acceptable. These results proved that at the borderline region the charge distribution has strong influence on the geometrical parameters.

Table 1.: *Geometry of the diglycine molecule obtained from two-layer geometry optimization at the HF/6-31G:MM/AMBER level of theory. (The effect of "surrounding" charges is taken into effect through electronic embeddings (EE), using three different charge sets.) For comparison the HF/6-31G:MM/AMBER with mechanical embedding (ME.), "pure" HF/6-31G and MM/AMBER geometries are also given. QM/MM(EE) means "traditional" QM/MM-like calculations while ONIOM(EE) ONIOM-like calculation. Charge sets CHG1, CHG2 and CHG3 were used for the electrostatics between layers within the electronic embedding framework.*

Paramete	HF/6-31G	MM/AMBER	HF/6-31G AMBER (ME)	HF/6-31G:MM/AMBER(EE)					
				QM/MM			ONIOM		
				CHG1	CHG2	CHG3	CHG1	CHG2	CHG3
N1-C1	1.511	1.472	1.515	1.541	1.532	1.529	1.542	1.526	1.517
C1-C2	1.531	1.510	1.498	1.502	1.514	1.523	1.507	1.499	1.498
C2-O1	1.248	1.228	1.228	1.207	1.225	1.232	1.215	1.226	1.228

The charges on the molecular mechanics can be seen on Figure 3

The third model calculations have been performed on the guanidinium formiate complex. This type of interactions is relatively frequent between amino acid side chains like e.g. arginine and aspartate or glutamate. The geometry could obtained from a restricted geometry optimization (using C2v symmetry) can be seen on Figure 4.

At this case the "traditional" QM/MM like and the IMOMM scheme with electronic embedding yields exactly the same results.

It can be expected (and it is found) that the electronic structure and geometry of the guanidinium cation strongly depend on the presence of the anion. This is due to its delocalised electronic structure, even if the cations are usually less polarizable than the

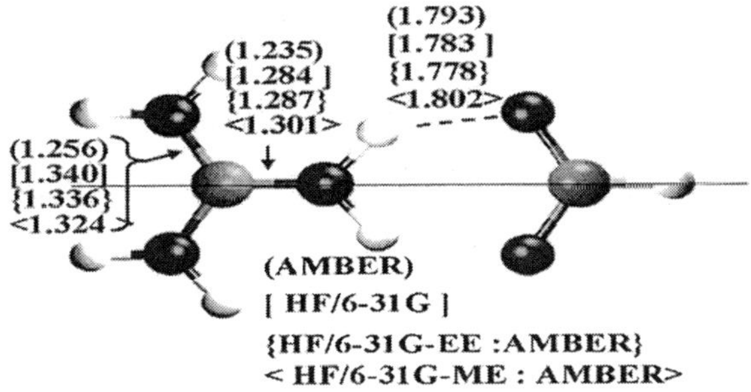

Figure 4.: Characteristic geometry parameters of the guanidinium-formiate complex obtained from different computation. (EE=electronic-, ME= mechanical embedding)

anions. It can be immediately seen that regardless from the applied method the intermolecular distances are quite similar (are in the 1.778 and 1.802 A range).

It can be observed, however, quite different geometrical parameters predicted for the guanidinium cation. It can be seen from the Figure 4. how important can be the electronic embedding scheme in IMOMM (QM/MM) calculations. This is due to the fact that the wavefunction of the guanidinium ion is strongly polarized by the quite large partial charges located on the formiate anion, even if the cations are usually less polarizable than the anions.

The next example (calculations carried out on the part of the blood coagulation Factor XIII) will demonstrate the applicability of the IMOMM method as a research tool in theoretical study of enzyme reaction.

Although this is far more complex task than the previous ones, some important conclusions can be drawn. Since proteins have large number geometrical variable meaningful calculations on them usually requires molecular dynamics simulations and sampling. Due to the short dynamics has been performed on the Factor XIII and its Michaelis complexes these results should be regarded as preliminary ones. However even these results can demonstrate how the IMOMM calculations can be applied for theoretical calculation on large systems, e.g. on biopolymers.

Many experimental studies have been carried out on the factor XIII using natural and artificial substrates [27 and references cited therein]. A reaction mechanism – similar to those of proteinases (but in reverse direction since the FXIII in contrast to the proteinases makes peptide bond) – also has been proposed[28]. Our goal was to find a computational evidence for this effect and also to predict geometries and relative energies for the transition state structures and reaction intermediates.

The calculations have been carried out using HF-6-31G:AMBER IMOMM method with ONIOM type electronic embedding. The full details of the computation and the obtained results will be given in another publication[29]

It has been also postulated that the Trp279 residue plays an important role as a component of the oxyanion hole that stabilize reaction intermediates (Figure 5).

The most important results we have reached so far (based on molecular dynamics simulations and IMOMM calculations) can be summarized:

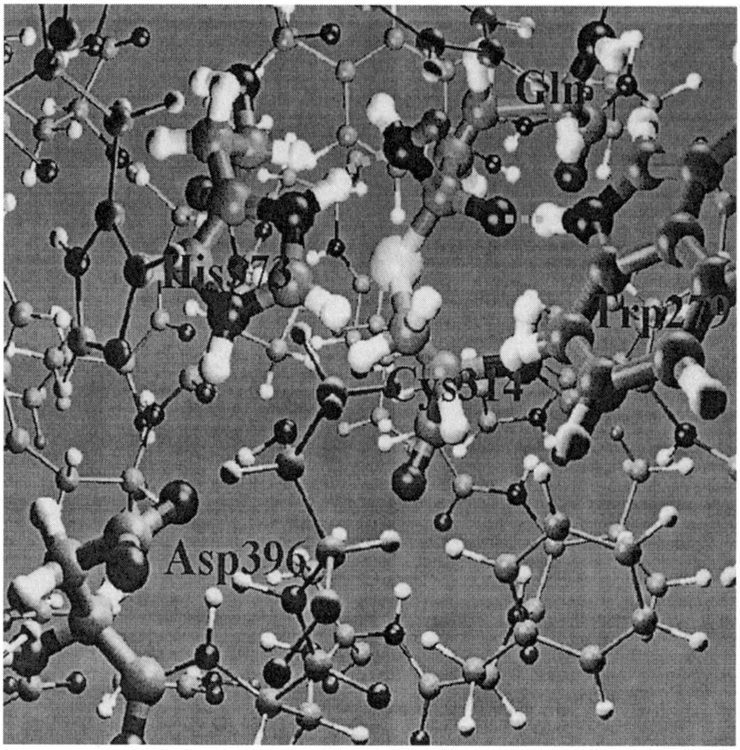

Figure 5.: The catalytic triad and the neighboring residues of the FXIII molecule in the predicted conformation at the tetrahedral intermediate state

The active site geometry for both of the possible tetrahedral intermediate configuration and the geometries of the corresponding Michaelis complexes were minimized using QM/MM method. It could be showed that only one of them can have the proper conformation in which the HNE2 atom of the protonated histidine can migrate to the NH2 group of the glutamine side chain (Figure 5.) and the next step of the reaction can take place. From our calculation it could be pointed out, that this tetrahedral intermediate forms more easily from the corresponding Michaelis complex than the other one (by ~16 kcal/mol)The crucial role of the Trp279 residue also has been shown in stabilizing polar intermediates during the reaction. This is done either by means of a strong hydrogen bond or by a contribution to the electrostatic field at the position of charged atoms. The contribution of the Trp279 amino acid is also crucial in destabilizing the Cys(0) His(0)-Asp(-) form of the catalytic triad and stabilizing the Cys(-)-His(+)-Asp(-) form. For this destabilization energy 3.2 kcal/mol could be obtained from Trp279Gly hypothetical mutation. However, not only the Trp279 residue but also the surrounding protein environment stabilizes the polar reaction intermediates as it showed by IMOMM calculations taking into the surrounding charges via the modified one electron Hamiltonian.

4. Conclusions

In this paper an overview on hybrid QM/MM methods, especially on the IMOMM method has been given. Special attention has been paid and the equations were given for the electronic embedding schemes can be applied together with the subtractive ONIOM method. Using selected examples the performance of the integrated scheme is demonstrated.

References:

1. Brooks, B.R. Bruccoleri, R.E. Olafson, B.D. States, D.J. Swaminathan, S. and Karplus M., (1983), "CHARMM - A program for macromolecular energy, minimization, and dynamics calculations" J. Comp. Chem. **4**(2), 187-217
2. Cornell, W.D. Ciepak, P. Bayly, C.I. Gould, I.R. Merz, K. Ferguson, D.M. Spellmeyer, D.C. Fox, T. Caldwell, J.W. and Kollman, P., (1995), "A 2nd generation force-_eld for the simulation of proteins, nucleic-acids, and organic-molecules" J. Am. Chem. Soc. **117**, 5179
3. van Gunsteren W.F. and Mark A.E. (1992), "On the interpretation of biochemical data by molecular dynamics computer simulation"Eur. J. Biochem. **204** 947-961
4. [Warshel A. and Levitt M. (1976), "Theoretical Studies of Enzymic Reactions: Dielectric, Electrostatic and Steric Stabilisation of the Carbonium Ion in the Reaction of Lysozyme" J. Mol. Biol., **103**, 227
5. [Waszkowycz B., Hillier, I.H. Gensmantel N., and Payling, D.W. (1991), "Combined quantum mechanical-molecular mechanical study of catalysis by the enzyme phospholipase-A2 - an investigation of the potential-energy surface for amide hydrolysis" J. Chem. Soc., Perkin Trans. **2**, 2025-2032
6. Lee, Y.S. Hodoscek, M. Brooks, B.R. Kador, P.F. (1998), "Catalytic mechanism of aldose reductase studied by the combined potentials of quantum mechanics and molecular mechanics" Biophys. Chem. 70(3), 203-216
7. Lyne, P.D. Hodoscek M. and Karplus, M. (1999), "A hybrid QM-MM potential employing Hartree-Fock or density functional methods in the quantum region" J. Phys. Chem. A **103**(18), 3462-3471
8. Eurenius, K.P. Chatfield, D.C. Brooks, B.R. and Hodoscek, M. "Enzyme mechanisms with hybrid quantum and molecular mechanical potentials .1. " Theoretical considerations, Int. J. Quant. Chem. 60(6), 1189-1200 (1996).
9. Turner, A.J. Moliner, V. and Williams, I.H., (1999). "Transition-state structural re_nement with GRACE and CHARMM: Flexible QM/MM modelling for lactate dehydrogenase" Phys. Chem. Chem. Phys. **1**(6), 1323-1331
10. Sherwood P. (2000), "Hybrid quantum mechanics/molecularmechanics approaches" in *Modern Methods and Algorithms of Quantum Chemistry*, J. Grotendorst (Ed.), pp. 257-277 John von Neumann Institute for Computing, Vol. 1
11. Gao J., (1996), "Methods and applications of combined quantum mechanical and molecular mechanical potentials", p.119-185 in Reviews in Computational Chemistry, (D. Boyd and ??? eds.) VCH Publishers, Inc., New York
12. Mordasini T.Z. and Thiel W. (1998), "Combined quantum mechanical and molecular mechanical approaches" Chimia, **52**(6), 288-291
13. Field M.J., Bash P.A., and Karplus M. (1990), "A Combined Quantum Mechanical and molecular mechanical potential for molecular dynamics simulation" J. Comp. Chem. **11**(6), 700-733
14. Singh U.C. and Kollman P.A. (1986), "A combined ab initio quantum mechanical and molecular mechanical method for carrying out simulations on complex molecular systems: Applications to the CH3Cl+Cl_ exchange reaction and gas-phase protonation of polyethers" J. Comp. Chem. **7**(6), 718-730

15. Thery, V. Rinaldi, D. Rivail, J.-L. Maigret, B. and Ferenczy, G.G. (1994), "Quantum mechanical computations on very large molecular systems: The local selfconsistent field method" J. Comp. Chem. **15**(3), 269-282

16. Gao, J. Amara, P. Alhambra, C. and Field, M.J. (1998), "A Generalised Hybrid Orbital (GHO) Method for the treatment of boundary atoms in combined QM/MM calculations" J. Phys. Chem. A **102**, 4714-4721

17. Thompson, M.A. (1996), "QM/MMpol: A consistent model for solute/solvent polarization. Application to the aqueous solvation and spectroscopy of formaldehyde, acetaldehyde, and acetone" J. Phys. Chem. **100**(34), 14492-14507

18. Philipp D.M. and Friesner, R.A. (1999), "Mixed ab initio QM/MM modeling using frozen orbitals and tests with alanine dipeptide and tetrapeptide" J. Comp. Chem **20(14)**, 1468-1494

19. Bakowies D. and Thiel W. (1996), "Hybrid models for combined quantum mechanical and molecular mechanical approaches" J. Phys. Chem. **100**(25), 10580-10594

20. F. Maseras and K. Morokuma, IMOMM: A new integrated ab-initio + molecular mechanics geometry optimisation scheme of equilibrium structures and transition states, J. Comp. Chem. 16(9), 1170-1179 (1995).

21. Dapprich, S.. Komáromi, I Byun, K. S., Morokuma, K. Frisch M. J. (1999) "A new ONIOM implementation in Gaussian98. Part I. The calculation of energies, gradients, vibrational frequencies and electric field derivatives" (Theochem) **462**, 1-22

22. Komáromi, I. Vreven, T. Dapprich, S., Byun, K. S. Morokuma, K. Frisch M. J., "A new ONIOM implementation in Gaussian98. Part II. IMOMM calculations" in preparation

23. Matsubara, T. Maseras, F. Koga N. and Morokuma, K. (1996). "Application of the New "Integrated MO + MM" (IMOMM) Method to the Organometallic Reaction: Pt(PR3)2 + H2 (R=H, Me, t-Bu and Ph)" J. Phys. Chem., **100**, 2573-2580

24. Matsubara, T. Sieber S. and Morokuma K., (1996), "A Test of the New "Integrated MO + MM" (IMOMM) Method for the Conformational Energy of Ethane and n-Butane" Int. J. Quantum Chem., **60**, 1101-1109

25. Svensson, M. Humbel, S. Froese, R. D. J. Matsubara, T. Sieber, S. and Morokuma, K. (1996), "ONIOM: A Multilayered Integrated MO + MM Method for Geometry Optimizations and Single Point Energy Predictions. A Test for Diels-Alder Reactions and $Pt(P(t\text{-}Bu)_3)_2$ + H_2 Oxidative Addition" J. Phys. Chem., **100**, 19357-19363

26. Gaussian 99, Development Version (Revisions 0.1 and B1) (1998) Frisch, M. J. Trucks, G. W. Schlegel, H. B. Scuseria, G. E.. Robb, M A. Cheeseman, J. R. Zakrzewski, V. G. Montgomery, J. A. Stratmann, R. E. Burant, J. C. Dapprich, S. Millam, J. M. Daniels, A. D. Kudin, K. N. Strain, M. C. Farkas, O. Tomasi, J. Barone, V. Cossi, M. Cammi, R. Mennucci, B. Pomelli, C. Adamo, C. Clifford, S. Ochterski, J. Petersson, G. A. Ayala, P. Y. Cui, Q. Morokuma, K. Malick, D. K. Rabuck, A. D. Raghavachari, K. Foresman, J. B. Cioslowski, J. Ortiz, J. V. Stefanov, B. B. Liu, G. Liashenko, A. Piskorz, P. Komaromi, I. Gomperts, R. Martin R. L. Fox, D. J. Keith, T. Al-Laham, M. A. Peng, C. Y. Nanayakkara, A. Gonzalez, C. Challacombe, M. Gill, P. M. W. Johnson, B. Chen, W. Wong, M. W. Andres, J. L. Head-Gordon, M. Replogle, E. S. and Pople, J. A. Gaussian, Inc., Pittsburgh PA,

27. Muszbek, L. Yee, V.C. Hevessy Zs. (2000), "Blood Coagualtion Factor XIII: Structure and Function" Thrombosis Research **94**, 271-305

28. Pedersen, L.C. Yee, V.C. Bishop, P.D. Le Tong, I. Teller, D.C. Stenkamp, R.E. (1994) "Transglutaminase factor XIII uses proteinase-like catalytic triad to crosslink macromolecules" Protein Sci. **3** 331-338

29. Komáromi I., Muszbek, L., in preparation

MOLECULAR ORBITAL SIMULATION OF SEMICONDUCTOR AND METAL CLUSTERS

V.S. Gurin
Physico-Chemical Research Institute, Belarusian State University, 220080, Minsk, Belarus

Abstract:

Within discussion of a bridge between inorganic chemistry concepts and molecular design of large structures some results of quantum chemical calculations based on the molecular orbital (MO) approach are considered. Four different homologous series of Cd_xS_y-cores for construction of an example of binary clusters are proposed and the calculation results are illustrated for Cd_xS_y ($x \leq 17, y \leq 32$) including also additionally joined hydrogen atoms and organic moieties. A dependence of geometry and electronic structure on cluster sizes and the nature of terminating groups is studied. The calculations of homonuclear metallic Ag_n clusters on the example of Ag_8 family are shown for a number of geometries. Properties of the clusters depend strongly on the optimized geometry, and each cluster can be predicted to behave like distinct molecules with own physical and chemical features. Thus, MO calculations are shown to open frontiers of direct theoretical studies of complex structures in cluster-based systems of the different chemical nature.

1. Introduction

In the inorganic branch of molecular-scale science and nanotechnology there are a lot of objects those are comparable with organic chemistry in the diversity of their features. Non-carbon compounds constructed from chemical elements in the context of conventional chemistry and textbook valence rules give usually much less feasibilities for diversity than organic chemistry with carbon. However, under the growth of molecular-scale building units from atoms to clusters and aggregates the number of bonds and new chemical forms is enlarged enormously. E.g., the number of structure

E. Buzaneva and P. Scharff (eds.), Frontiers of Multifunctional Nanosystems, 29–42.

isomers of homonuclear and binary transition metal clusters in the range of X_n, n>10, is comparable with that for corresponding hydrocarbons taking into account possibility of multivalence of X instead of constant four-valence for carbon. Moreover, a very rich chemistry of ligated and multinuclear clusters [1] can provide a variety of inorganic compounds and hybrid organic-inorganic ones comparable with the natural organic chemistry. In such compounds metal and metal-nonmetal cores allow us to consider each central atom as a 'base' for construction of own variety of compounds, and own homologous series. The more building units the more pathways for creation of new compounds occur. A scale of modern nanotechnology is approaching to the above range of cluster sizes, and the latter can successfully play a role in quantum device creation (e.g., [2]). Within the framework of old discussion on a bridge between inorganic chemistry concepts and molecular design of large structures in the present work we consider some results of quantum chemical calculations based on the molecular orbital (MO) approach. This approach describes on equal footings both simple molecules, complexes and clusters of arbitrary size, and its usage is illustrated for heteronuclear clusters with binary cores M_nX_m, where (M = Cd, X = S), and for homonuclear metallic Ag_n clusters. Within the homologous series we analyse cluster features and factors determined their experimentally observable properties. These series of the clusters are examples of constructions those result in diversity of chemical species within inorganic chemistry concepts.

2. Models and calculation technique

A cluster geometry in the general case can be arbitrary, and no *a priopi* information are about feasibility of certain geometries, and many different compositions appears with various stoichiometry [3]. However, one can assume often, at least as a first trial, the bulk-like geometry for model construction, and in the case of Cd and Zn chalcogenides geometry this assumption is really true: the small clusters those are experimentally observable are similar to the bulk fragments of sphalerite [4,5] or (less often) to wurtzite [6]. Cadmium chalcogenide clusters (i.e. species with Cd_xS_y cores) exist in the broad range from simple diatomics in gaseous phase and mononuclear thiolates, $Cd(SR)_4$, through large clusters $Cd_xS_yR_z$ with x,y,z>50 in which Cd_xS_y core is a part of the bulk-like lattice [7]. Illustrating the calculation results here for the family of the cadmium chalcogenide clusters we have divided it onto following homologous series distinguished by the type of Cd_xS_y-core construction (Figures 1-6):

(i) a central Cd atom which bonds with four S atoms with the tetrahedral environment (examples: CdS_4, $Cd_{13}S_4$; $Cd_{13}S_{16}$, $Cd_{17}S_{16}$, $Cd_{17}S_{28}$, etc.);

(ii) a central S atom which analogously coordinates four Cd atoms also with the tetrahedral environment (examples: Cd_4S, Cd_4S_{13}, $Cd_{16}S_{13}$, Cd_8S_{13} and Cd_8S_{17}, the two latter appear in the less ideal model to avoid large excess of Cd atoms);

(iii) a cage with the composition of Cd_4S_6 (called as adamantaneous one) with the symmetry centre in the centre of this cage rather than at an atom in contrast with items (i) and (ii). This construction generates clusters Cd_4S_{10}, $Cd_{10}S_{16}$, $Cd_{10}S_{20}$, etc.;

(iv) the analogous adamantaneous cage with interchange of Cd and S atoms, Cd_6S_4, also with the symmetry centre in the centre of this cage: $Cd_{10}S_4$, $Cd_{10}S_{16}$, isomeric with that in item (iii), $Cd_{16}S_{10}$, etc.

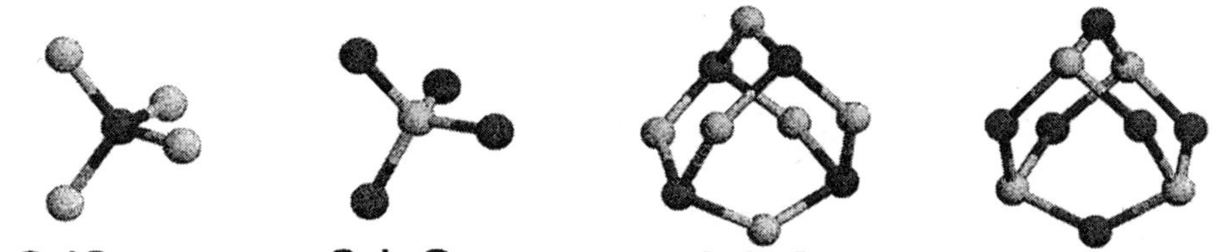

CdS_4 (i) Cd_4S (ii) Cd_4S_6 (iii) Cd_6S_4 (iv)

Figure 1. Structures of the building units for the four families of Cd_xS_y clusters

Starting from these building units a step-by-step construction of the clusters leads to a number of model cores Cd_xS_y. Many of them exist really as ligand bound species, Cd_xS_y–R, meanwhile some members are new ones, but they can be predicted from the calculations. An inevitable outcome of this construction inherent to the cluster cores containing two (or more) types of atoms (Cd and S in our case) is the occurrence of dangling bonds due to low-saturated surface atoms. These atoms bound with the ligands, and this feature is one of affinity of such clusters with the multinuclear cluster compounds. The dangling bonds were simulated in the calculations with hydrogen atoms, methyl- and simplest alcohol remnants. The constructed series of clusters were used as starting points to generate geometries in the calculations performed by the method based on the MO LCAO (molecular orbitals as linear combination of atomic orbitals) approach for ground states with closed electronic shells within the restricted Hartree-Fock (HF) self-consistent field (SCF) scheme. Basis sets used were effective core potentials (ECP) for internal core orbitals (28 electron core for Cd and 10 electron core for S) by Hay and Wadt (LANL2DZ in the Gaussian notation) [8] were taken for the both elements. Then, the geometry optimization was attained with the T_d symmetry or with arbitrary distortion. The calculations were realized with RS/6000 workstation and Pentium II-IV computers using GAMESS(US) package [9].

A family of metal clusters is considered here on the example of silver Ag_n that is rather popular metal in the field of cluster science, however, a lot of unresolved questions remains due to complicacy of the electronic structure of silver atom and variety of possible geometries even in the range $n<10$. Features of the metal-metal bond in homonuclear metal clusters provide much more choices in atom bonding than in the case of heteronuclear clusters with covalent and ionic bonds. Thus, *a priori* silver clusters can possess various geometries, and each one can be treated as proper chemical species since electronic structure and, consequently, chemical and physical properties of the clusters are to depend strongly on size (number of atoms) and geometry. Experimental information on geometry of few-atomic silver clusters (at $n>4$) is scarce [10], and there is great feasibility to coexist many isomers in any real system under non-zero temperatures.

We have considered on the example of the family with $n=8$ a diversity of cluster structures and their properties. The nuclearity of silver clusters with eight atoms was picked out because Ag_8 has been shown to be magic cluster in some systems [11,12], and 8-atom nuclearity is rather attractive from geometrical viewpoint [13]. Meanwhile, 8 atoms provide already many geometrical isomers supporting the idea to treat clusters as variety of different chemical species. There were calculated the structures of different symmetry of some probable geometries related to the following point groups: O_h (cube), T_d (2 types of tetrahedrons), D_{4h} (rectangular parallepiped), C_{2v} (different 3D polyhedrons), C_{2h}(a rhomb with two couples of atoms over and under its plane), $D_{\infty h}$ (linear), and their structures are depicted in Figure 7. The four slightly different cases were taken for the C_{2v} type constructed by taking two-atomic rods of different lengths and placing them in such position that the center of each rod aligned along a vertical axis (z), whilst retaining the perpendicular position of one with respect to the others and to the z-axis - C_{2v}(1-3); C_{2v}(4) is the two rhombs in parallel planes rotated at 90°. Any distortion from the initial symmetry was allowed and took place usually for the high-symmetrical models. For the whole set of the clusters under consideration we present here the data for ground states of Ag_8. The calculations were carried out by restricted and unrestricted (for open shells in the case of odd *n*) SCF HF methods using HONDO7 [14] and GAMESS(US) package [9]. As well as in the case of the above Cd_xS_y clusters the ECP were used with 19 valence electrons and the 28-electron core for silver atoms [8]. Such more extended basis is of preference for atoms of d-elements as compared with possible 1- and 11- electron shells [15].

3. Clusters with Cd_xS_y-cores

The first homologous series in the classification of the prevoius Section is constructed by the subsequent addition of atomic layers of Cd and S to the elementary unit CdS_4:

$CdS_4 + 12Cd \rightarrow Cd_{13}S_4$;

$Cd_{13}S_4 + 12S \rightarrow Cd_{13}S_{16}$;

$Cd_{13}S_{16} + 4Cd \rightarrow Cd_{17}S_{16}$;

$Cd_{17}S_{16} + 12S \rightarrow Cd_{17}S_{28}$;

$Cd_{17}S_{28} + 4S \rightarrow Cd_{17}S_{32}$, etc.

The whole charge of these clusters obtained from the assumption of Cd^{2+} and S^{2-}ions in the construction appears to be large, unrealistic, but this fact is inevitable under this intermediate construction since the tetrahedral coordination results in more number of sulfur atoms as compared with cadmium. These charges can be neutralized by hydrogen or another groups like hydrocarbon or hydrocarbon-hydroxyl remnants, $-C_nH_{2n}$-OH, but that construction rises enormously the number of possible clusters under consideration.

3.1. *SIMPLEST CdS_4-UNITS*

The simplest construction in this series, CdS_4^{6-}(Figure 2) belongs more to classical complexes with one central atom and tetrahedral environment of the ligands, S^{2-}, leading to the big total charge with anomalous charge at the central atom; the bond Cd-S appears to be very polar, and the cluster instable against interaction with any reagent and decomposition since the value of bond length is much more than the equilibrium Cd-S in the bulk material (3.199 Å and 2.52 Å, respectively). However, the effect of hydrogen addition to this cluster is very significant, hydrogen neutralizes the high charge at S, and the Cd-S distance decreases down to 2.629 Å. It is noticeably that any arbitrary distortion under the geometry optimization procedure starting from the perfect T_d for both CdS_4^{6-} and $CdS_4H_4^{2-}$ does not change the parameters of final optimized

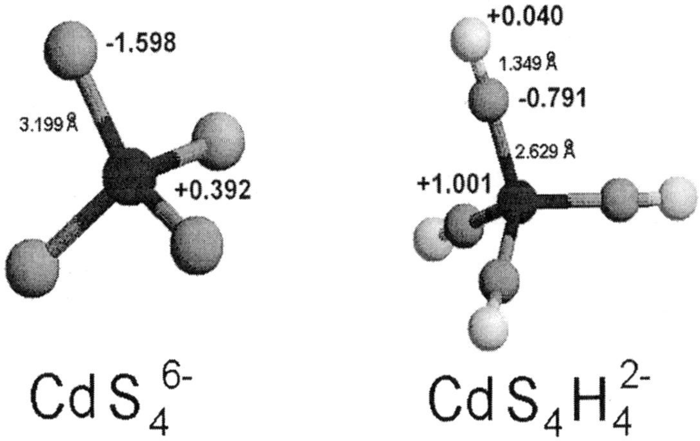

Figure 2. Geometry and effective charges calculated for the clusters with CdS_4-core

structure. Analogously, an initially distorted geometry relaxes to the above symmetrical ones (with difference in the bond length $\leq 0.03\%$). This couple of the clusters, CdS_4^{6-} and $CdS_4H_4^{2-}$, has the gap between HOMO-LUMO (highest occupied MO - lowest unoccupied MO) very large, 8.66 eV and 9.50 eV for CdS_4^{6-} and $CdS_4H_4^{2-}$, respectively. p-AO of sulfur atoms (p_z in the first and $p_{x,y}$ in the second case) have the most contribution, and the first electronic transition (HOMO-LUMO) is allowed via dipole mechanism, while in the case of $CdS_4H_4^{2-}$it is not allowed, and the nearest allowed transition appears at 9.86 eV. These gaps are far from the experimental data for some CdS_4 - R_4 examples (R=thiophenolate or ethanedithiolate), but the effect of termination is very strong and no any data for the calculated CdS_4^{6-} and $CdS_4H_4^{2-}$ are available. Interatomic distances in $CdS_4H_4^{2-}$, in contrast, are close to that observed in these thiolate Cd compounds (2.5 – 2.6 Å). Thus, the calculations show that the hydrogen-terminated CdS_4 clusters are the working building unit for the family of the larger ones, and on their basis we constuct the next members increasing the diversity of Cd_xS_y–species.

3.2. *$Cd_{13}S_4$ AND $Cd_{13}S_{16}$-CORES*

As the result of addition of 12 Cd atoms to CdS_4^{6-} the structure $Cd_{13}S_4^{18+}$ is produced with very high excess charge and reveals already the different values of Cd-S distances for inner 'shell' and external one: 2.546 Å and 2.748 Å, respectively (under same starting values during the geometry optimization). Scarcely it would be stable without termination with other atoms, and there are no evidences on its experimental existence to date. Being the intermediate species in the cluster growth simulated by the present construction $Cd_{13}S_4$ is to join easily sulfur atoms with formation of the next member, $Cd_{13}S_{16}^{6-}$.

The construction of clusters with composition $Cd_{13}S_{16}$ can be done by the two ways resulting in 'open' and 'compact' structures (Figure 3). They both retain the full T_d

TABLE 1. *Calculated energetic characteristics of the clusters*

Composition	Energy of the first allowed electronic transition and types of MO, eV		Binding energy per one Cd-S bonding unit, BE/n, eV*)	
	Compact Structure	Open Structure	Compact structure	Open structure
CdS_4^{6-}	8.66 ($T_2 \rightarrow A_1$)		2.47	
$CdS_4H_4^{2-}$	9.86 ($E \rightarrow A_1$)			
$Cd_{13}S_4^{18+}$	9.41 ($T_2 \rightarrow A_1$)		< 0	
$Cd_{13}S_{16}^{6-}$	6.20 ($T_2 \rightarrow A_1$)	5.03 ($T_2 \rightarrow A_1$)	1.72	1.33
$Cd_{13}S_{16}H_{12}^{6+}$	8.90 ($T_2 \rightarrow A_1$)	5.60 ($T_2 \rightarrow A_1$)		
$Cd_{17}S_{28}^{22-}$	2.74 ($T_2 \rightarrow T_1$)		4.13	
$Cd_{17}S_{28}H_{24}^{2+}$	6.69 ($T_2 \rightarrow T_1$)			
$Cd_{17}S_{28}(CH_3)_{24}^{2+}$	3.78 ($A_2 \rightarrow A_1$)			
$Cd_{17}S_{28}(CH_2OH)_{24}^{2+}$	4.19 ($A_2 \rightarrow A_1$)			
$Cd_{17}S_{32}^{30-}$	6.35 ($T_1 \rightarrow T_2$)		0.76	
$Cd_{17}S_{32}H_{28}^{2-}$	3.22 ($T_1 \rightarrow T_2$)			
$Cd_4S_{13}^{18-}$	4.60**)		< 0	
$Cd_4S_{13}H_{12}^{6-}$	8.88**)			
$Cd_8S_{13}^{10-}$	5.33***)		1.30	
$Cd_8S_{17}^{18-}$	7.30***)		< 0	
$Cd_4S_6^{4-}$	7.18 ($T_1 \rightarrow A_1$)		3.92	
$Cd_6S_4^{4-}$	8.13 ($T_2 \rightarrow A_1$)		3.64	

*) *BE/n was defined as difference in total energies of a cluster and corresponding separated ions divided on the number of Cd-S bonds. To avoid ambiguousity in the definition of BE with the terminating groups the data are given here only for the clusters without termination.*

**) *Geometry optimization without symmetry constraints resulted in the lower energy (given).*

***) *Clusters have no full T_d symmetry and were optimized within C_1 group.*

symmetry and *a priori* have equal footings to exist, they have same number of 2-coordinated sulfur atoms and 3-coordinated Cd atoms; internal Cd and S atoms are 4-coordinated in the both cases. The clusters of this composition are almost stoichiometric (Cd/S atomic ratio ~ 1). One can expect the less external effects upon their properties as well the less effect from the type of termination than for the previous members of the family.

The compact structure $Cd_{13}S_{16}$ have the more binding energy (Table 1) than the open one and the both types are quite stable. Meanwhile, the inner Cd-S bonds in the compact $Cd_{13}S_{16}$ clusters are noticeably longer than external ones, whilst in the open clusters the situation is reversed: the internal Cd-S distances are shorter on 0.17 Å than Cd-S bonds in the next layer. Also, the bond lengths in the external and pre-external layer in the compact structures are almost same and rather close to experimental distances mêasured in CdS-clusters terminating with thiolate ligands [16]. Energies of the first electronic transition can be estimated by values of HOMO-LUMO difference, and for $Cd_{13}S_{16}$ composition they are 5.853 eV (6.204 eV - dipole allowed) and 5.032 eV. These values are essentially less than those for the above homologs with less number of atoms. This fact reflects the familiar general trend to increase energy gap with decreasing cluster size due to quantum confinement effect.

A termination simulated by hydrogen influences the properties of $Cd_{13}S_{16}$ clusters significantly only for compact version: HOMO-LUMO gap increases up to 8.901 eV. The H-terminated compact clusters are more stable again. Evidently, the addition of

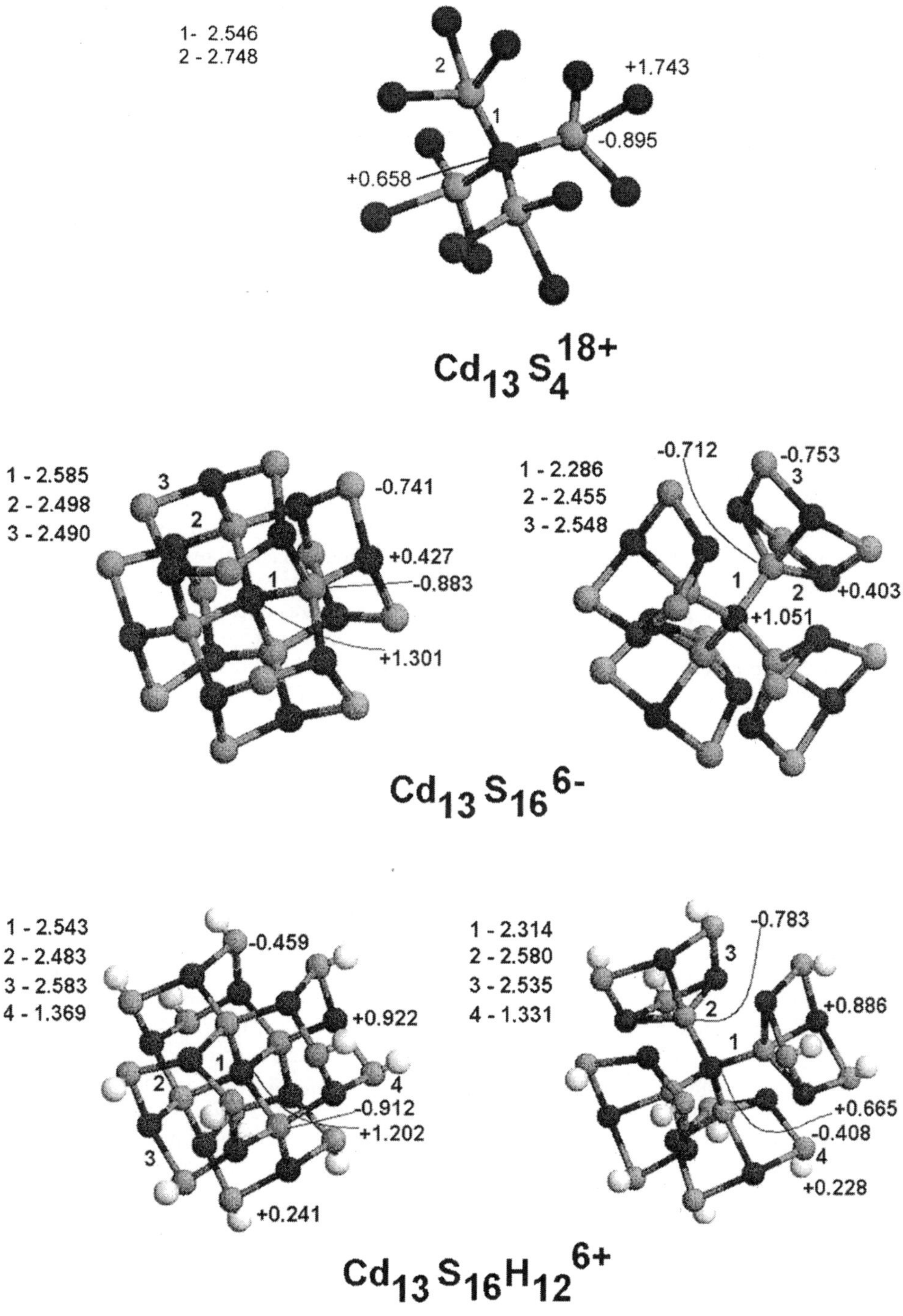

Figure 3. Geometry and effective charges calculated for the clusters with $Cd_{13}S_{16}$-core

hydrogen atoms neutralizes effective atomic charges of external part, but inner charges vary a little in spite of the overall charge of the clusters grows on +12e. The above features in the difference of interatomic distances remain valid after the hydrogen addition, and the very small distance for internal Cd-S bonds occurs for the open structures (2.314 Å against 2.380 Å and 2.535 Å for the second and external shells, respectively). From the other hand, the compact $Cd_{13}S_{16}H_{12}^{6+}$ clusters are characterized by somewhat longer Cd-S distances for the internal bonds as compared with the second layer on ~0.05 Å; and the next (external) layer has the longer bonds (2.583 Å) again. These observations indicate that these clusters with the $Cd_{13}S_{16}$-core have each own property determined by composition, bond length and type of the construction. They could be detected experimentally if to stop the growth process at the corresponding steps, however, really, up to date the next homologs are detectable easily (see below).

3.3. $Cd_{17}S_{28}$- AND $Cd_{17}S_{32}$-CORES

The addition of one more layers of Cd atoms and S atoms to the above $Cd_{13}S_{16}$ core results in clusters with $Cd_{17}S_{28}$ composition (Figure 4). Equilibrium interatomic distances are not same for different layers in the bare clusters $Cd_{17}S_{28}^{22-}$, the interior one has the minimum value, and the values are rather close to the abovementioned bulk Cd-S distance. That indicates more similarity of this structure with the bulk material (sphalerite-type lattice) than the cores of less size, and this structure appears to be more stable (estimated by the BE/n, Table 1) than the corresponding $Cd_{13}S_{16}$-ones that shows the true trend of growing binding energy with size for cores of the same series. Effective charges at Cd atoms decrease from interior to exterior, but no such regularity for S atoms; the external S atoms possess large negative charges, they are two-coordinated and expected to be very reactive. The large full charge of the bare core is neutralized by termination with hydrogen or more complicated groups (Figures 4,5). $Cd_{17}S_{28}$-cores bound with H have the slightly changed Cd-S distances, and only bonds of external Cd-S layer prolong strongly. The termination by CH_3- increases the Cd-S bond lengths noticeably, that again evidences on the dramatic effect of ligand type upon these series of clusters; meanwhile CH_2OH- groups make Cd-S distances somewhat less again, in particular, in the external layer (Figure 4). Big atomic charges are concentrated in the interior, and they become less with CH_3- and CH_2OH-termination. Thus, the proper termination by organic groups strongly diminishes reactivity of the cluster surface that explains their protective action in this context.

The HOMO-LUMO transitions for these clusters with $Cd_{17}S_{28}$-cores also evidences on dramatic effect of the type of termination (Table 1): H-termination increases this value, but CH_3-termination decreases it again: 3.78eV (λ=324 nm), and CH_2OH- gives 4.19eV (λ=293 nm). Symmetry types of MO between which these transitions occur are different for these changes of termination indicating on the ligand effect upon the electronic structure of inorganic cores rather than the simple steric stabilization. It should be emphasized that the latter is in quite reasonable correspondence with the experimental ones for the dimeric cluster which includes such $Cd_{17}S_{28}$-core, $[Cd_{17}S_{30}R_{26}]_2$, R = CH_2CH_2OH [17]. These results indicate that the cores with ligands behave like multinuclear complexes in which a ligand-metal interaction is very usual.

The $Cd_{17}S_{32}$-cores are constructed from $Cd_{17}S_{28}$-ones by addition of four S atoms to the corner 3-coordinated Cd atoms (Figure 4). These new S atoms are rather distinct from other, evidently due to the lower coordination number. They are to possess the higher reactivity, which can be lowered by hydrogen termination, i.e. in $Cd_{17}S_{32}H_{28}^{2-}$.

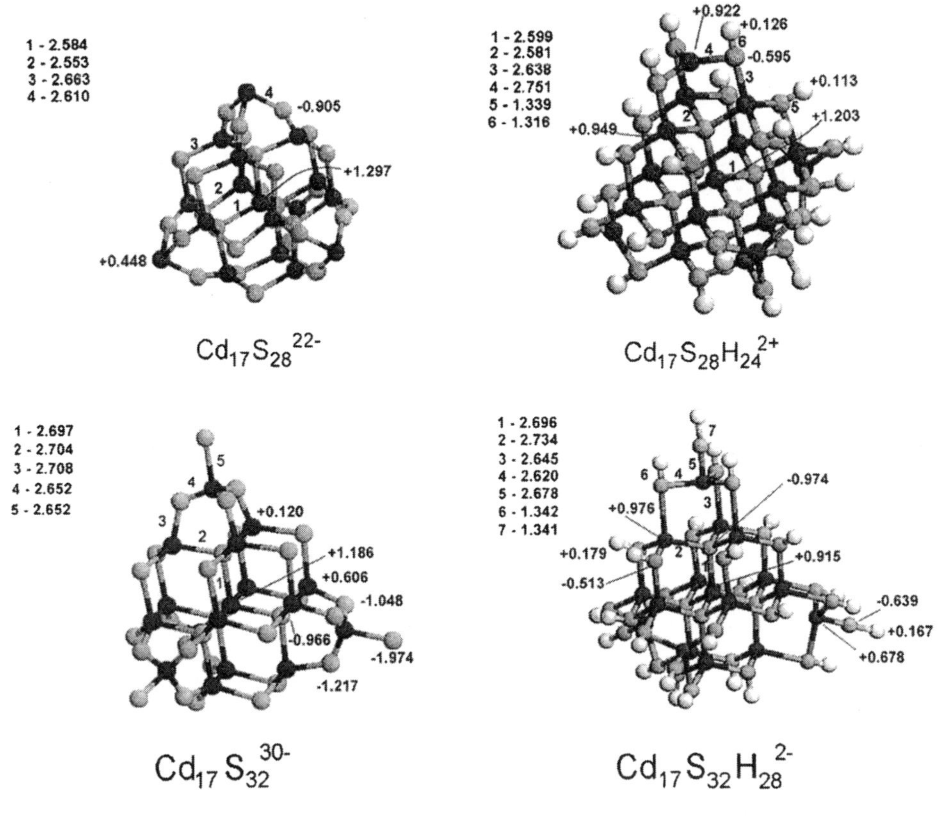

Figure 4. Geometry and effective charges calculated for the clusters with $Cd_{17}S_{28}$ and $Cd_{17}S_{32}$-cores

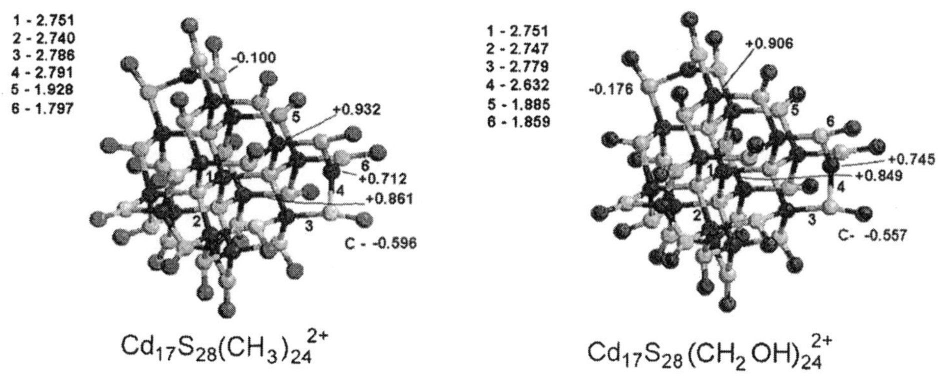

Figure 5. Geometry and effective charges calculated for CH_3- and CH_2OH-terminated $Cd_{17}S_{28}$-cores

Under formation of the above polymeric $[Cd_{17}S_{30}R_{26}]_n$ these four S atoms play a role of intercluster bonding units. Thus, this example indicate how enormously can be enlarged a variety of the clusters in this family with admission of intercluster interaction, bonding, and polymerization.

A calculation of the neutral $Cd_{17}S_{30}H_{26}$ was not available under the theory level used (SCF RHF) since it loses a symmetry that greatly increased computing requirements. However, from the above data on clusters with $Cd_{17}S_{28}$- and $Cd_{17}S_{32}$-cores one may believe that the value of HOMO-LUMO transition is to be middle between those for $Cd_{17}S_{28}H_{24}^{2+}$ and $Cd_{17}S_{32}H_{28}^{2-}$. Thus, the experimental value for the first absorption maximum, 4.23 eV, fits this interval. MO taking part in these transitions are formed dominantly with p(S)-orbitals and s(Cd) and p(Cd)-orbitals (LUMO). In the cluster $Cd_{17}S_{28}H_{24}^{2+}$ the termimating Cd atoms contribute into LUMO, while HOMO is formed from maximum contribution of interior. In $Cd_{17}S_{32}H_{28}^{2-}$, in contrast, the terminating S atoms contribute into HOMO, but the main part of LUMO is from interior. In the other words, these electronic transitions in clusters with $Cd_{17}S_{28}$- and $Cd_{17}S_{32}$-cores are accompanied by charge transfer effect in different direction, from the core interior to the surface, and from the surface inward, respectively. The type of termination, thus, is very significant for behaviour of electronic excitation illustrating the sensitivity of physical properties of Cd_xS_y-based clusters with respect to their environment in line with their geometry and chemical composition.

3.4. *ON CLUSTERS WITH Cd_xS_y-CORES FROM OTHER SERIES*

The other families of the Cd_xS_y clusters can be constructed accordingly to the classification given in Section 2, and a number of structures were obtained step-by-step by subsequent addition of Cd and S atomic shells retaining the tetrahedral coordination in Cd-S_x and S-Cd_x. The clusters with Cd_4S_{13}-core begin the family which allows addition of many terminating groups because the big excess of S atoms over Cd ones (Figure 6) [18] that results in more realistic parameters of bond lengths and effective charges than the clusters without termination. Calculations with more complicated terminating groups forwards these clusters with rather low number of Cd atoms to comparison with experiment [19].

An addition of the layer of Cd atoms to Cd_4S_{13}-cores would produce the unrealistic structures with many excessive Cd atoms, so some other, more complicated and lowering the whole symmetry clusters can be built, Cd_8S_{13}–cores and, again with addition of the four corner S atoms, Cd_8S_{17}-cores. These structures also were detected under the growth process of CdS nanoparticles [19].

The clusters built on the basis of cages with composition of Cd_4S_6 and Cd_6S_4 are very different from the above considered family with a central Cd atom only for these lowest members: there are difference in the number of atoms having incomplete 2-, 3-coordination (Figure 6), however, under growth one can obtain cores like $Cd_{10}S_{16}$ those are very distributed in real cadmium thiolate clusters [7]. Some calculation data for selected members of these families of Cd_xS_y are presented in Table 1.

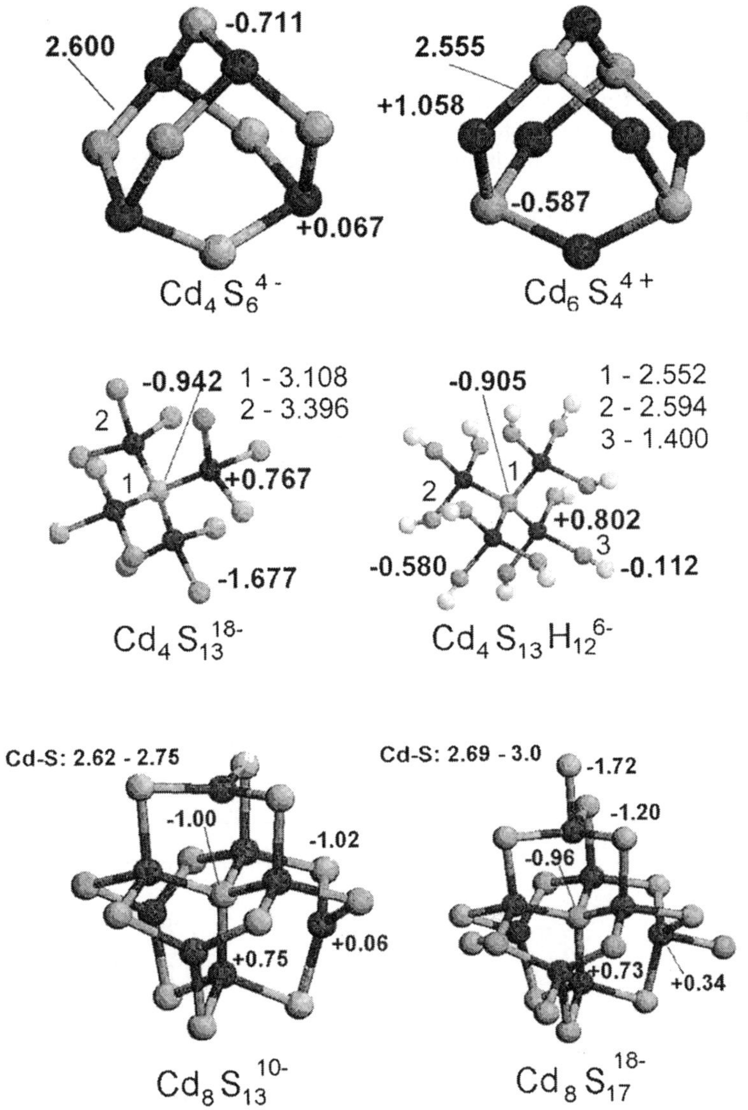

Figure 6. Geometry and effective charges calculated for some Cd_xS_y-clusters of different families

4. Silver clusters

A selected series of Ag_8 clusters (Figure 7) was calculated, however, the whole set of possible isomers of Ag_8 is very numerous. Some models were considered earlier also at *ab initio* MO LCAO RHF level with account one valence electron from each silver atom [15].

Main calculation data are summarized in Table 2 and Figure 7 as binding energies per one atom and interatomic distances, respectively. According to these results, there is very significant effect of geometry upon properties, and each isomer may be treated as chemical species with own reactivity and physical features. Values of relative stability of these clusters vary not very strongly (in the range of 1 eV): O_h and C_{2v} clusters are of maximum stability, however, the difference in the stability is not large, and one can assume a coexistence of many isomers under ambient temperature. Thus, from the purely energetic positions there is probability of different isomers of Ag_8 of rather variable structure. The geometric factor can acquire a greater importance in some special situations (e.g. to fit zeolite cavities [20]) and then it could help us to select the Ag_8 isomers.

TABLE 2. *Characteristics of the Ag_8 clusters calculated*

Geometry of the clusters	Binding energy per atom, BE/n, eV	HOMO-LUMO transition, eV
O_h	2.30	6.49
T_d (1)	1.95	4.97
T_d (2)	2.24	6.85
D_{4h}	1.81	2.62
$D_{\infty h}$(linear)	1.93	4.37
C_{2v} (1)	2.26	6.27
C_{2v} (2)	2.27	6.54
C_{2v} (3)	2.28	5.83
C_{2v}(4)	2.04	5.78
C_{2h}	1.69	2.28

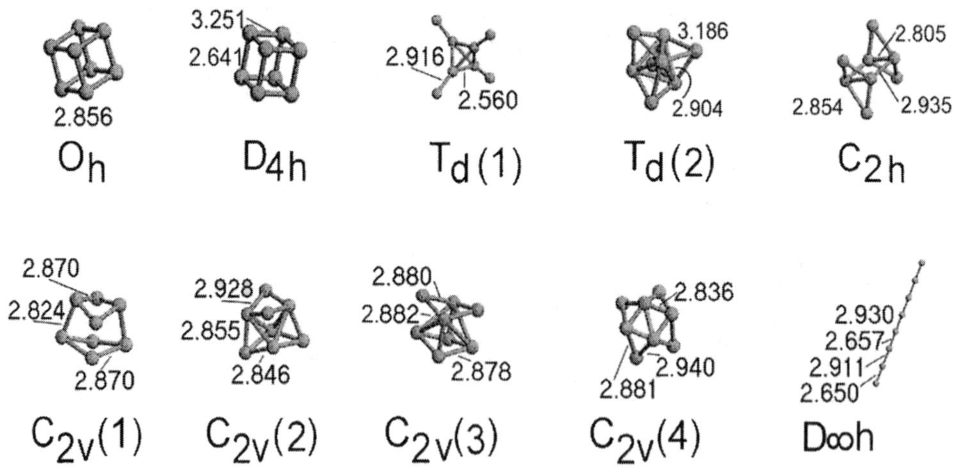

Figure 7. Geometry for some Ag_8 clusters under calculation

It should be noticed that the calculation data for neutral clusters do not argue any distortion of the starting geometries, however, if to calculate the charged clusters (+1 e and −1 e), some high-symmetrical structures, O_h, T_d, and D_{4h} can be distorted, in general down to C_1 symmetry, and an open electronic shells appears in them. That provides also increased number of allowed transitions and more complicated spectra due to symmetry breaking. Meanwhile, this distortion is not too high to change size of clusters significantly.

The energetic spectrum of all clusters in a series with various geometry is also rather different and illustrated here by the values of the first electronic transition (Table 2). A position of HOMO varies from 4.0 through 6.1 eV, and it has the maximum contribution from s-orbitals for all these clusters. Thus, properties of the clusters depend strongly on the optimized geometry, and each cluster is to behave like distinct molecules rather than Ag_8 composition would determine their features.

5. Conclusions

On the basis of the results of *ab initio* MO LCAO calculations of several series of cadmium sulfide clusters, Cd_xS_y-R($x \leq 17$, $y \leq 32$), we simulate a large set of clusters and monitor their construction within the framework of Cd-S coordination rules. There is the pronounced effects of geometry, coordination and the nature of terminating groups (-H and -CH_3, -CH_2OH) upon properties of the clusters: redistribution of electronic density, variation of equilibrium interatomic distances and orbital energies. The calculated properties of some clusters are correlated with the experimental data for nearest analogs, and some structures can be predicted as probable candidates in the series of growing clusters. A tendency of the energy gap variation due to quantum confinement in semiconductor nanoclusters is reproduced for the families under study with large non-monotonous oscillations.

The calculations at the similar level for Ag_8 clusters show a series of geometrical isomers with different electronic properties and support the idea to consider each cluster of certain geometry as different species opening a great variety of homonuclear cluster-based compounds with variable *n*, geometry of cores and terminating groups.

6. Acknowledgments

The author would like to thank Drs. A. Eychmüller, N. Bogdanchikova and V. Petranovskii for fruitful discussions. The work was performed partially under support of Volksvagen Foundation (Hannover) and Ministry of Education of Belarus. The author also acknowledges the funding for a part of this research by CONACYT, Mexico.

References

1. Krebs, B. and Henkel, G. (1991) "Transition-metal thiolates: from molecular fragmnents of sulfidic solids to models for active centers in biomolecules", Angew. Chem. Int. Ed. Engl. **30**, 769-788.
2. Schmid, G. and Chi, L.F. (1998) "Metal clusters and colloids", *Adv. Mater.* **10**, 515-526.

3. Castleman, A.W.jr. and Bowen, K.H.jr. (1996) "Clusters: Structure, energetics, and dynamics of intermediate states of matter", *J. Phys. Chem.* **100**, 12911-12944.
4. Lover, Th., Bowmaker, G.A., Seakins, J.M. and Cooney, R.P. (1997) "Vibrational spectroscopic study of thiophenolate-capped nanoclusters of CdS and of cadmium thiophenolate complexes", *Chem. Mater.* **9**, 967-975.
5. Lee, G.S.H., Fisher, K.J., Vassallo, A.M., Hanna, J.V., Dance, I.G. (1993) "Solid-State ^{113}Cd NMR of three structural isomers of $[S_4Cd_{10}(SPh)_{16}]^{4-}$ ", *Inorg. Chem.* **32**, 66-72.
6. Hosokawa, H., Fujiwara, H., Murakoshi, K., Wada, Y., Yanagida, Sh., Satoh, M. (1996) "In-situ EXAFS observation of the surface structure of colloidal CdS nanocrystallites in N,N-dimethylformamide", *J. Phys. Chem.* **100**, 6649-6656.
7. Dance, I.G. (1994) "Metal chalcogenide cluster chemistry", *Progr. Inorg. Chem.* **41**, 637-803
8. Hay, P.J. and Wadt, W.R. (1985) "Ab initio effective core potentials for molecular calculations. I. Potentials for the transition metal atoms Sc to Hg." *J. Chem. Phys.* **82**, 270-283.
9. Schmidt, M.W., Baldridge, K.K. and Boatz, J.A. et al. (1993) "GAMESS", *J. Comput. Chem.* **14**, 1347-1363.
10. Henglein, A. (1995) "Electronics of colloidal nanometer particles", *Ber. Bunsenges. Phys. Chem.* **99**, 903-913.
11. Ershov, B.G.; Janata, E. and Henglein, A. (1993) "Growth of silver particles in aqueous solution: long-lived "magic" clusters and ionic strength effect", *J. Phys. Chem.* **97**, 339-343.
12. Federmann, F., Hoffmann, K., Quaas, N. and Toennies, J.P. (1999) "Spectroscopy of extremely cold silver clusters in helium droplets", *Eur. Phys. J.* **D9**, 11-14.
13. Dubov, P.L., Korolkov, D.V. and Petranovskii, V.P. (1995) *Clusters and matrix-isolated cluster superstructures (in Russian)*, Publisher of St.Peterburg State University: St. Peterburg.
14. Dupuis, M., Watts, J.D., Villar, H.O., Hurst, G.J.B. (1987) *"HONDO: Ab Initio HFMO Calculations (Version 7.0), QCPE Program No. 544"*. IBM Corp. Sci. Eng. Comp. Kingston, New York.
15. Bonacic-Koutecky, V., Cespiva, L., Fantucci, P. and Koutecky, J. (1993) "Effective core potential-configuration interaction study of electronic structure and geometry of small neutral and cationic Ag_n clusters: predictions and interpretation of measured properties", *J. Chem. Phys.*, **98**, 7981-7994.
16. Dance, I.G. (1986) "The structural chemistry of metal thiolate complexes", *Polyhedron*, **5**, 1037-1104.
17. Vossmeyer, T., Reck, G., Katsikas, L., Haupt, E.T.K., Schulz, B., Weller, H., (1995) "A double-diamond superlattice built up of $Cd_{17}S_4(SCH_2CH_2OH)_{26}$ clusters", *Science* **267**, 1476-1479.
18. Gurin, V.S. (2000) "Large clusters with cadmium sulfide cores: simulation of thiolate ligands in surface by ab initio MOLCAO calcualation", *Surface Review and Letters* **7**, 161-169.
19. Nosaka, Y., Shigeno, H. and Ikeuchi, T. "Formation of polynuclear cadmium-thiolate complexes and CdS clusters in aqueous solution studied by means of stopped-flow and NMR spectroscopies", *J. Phys. Chem.*, **99**, 8317-8322.
20. Gurin, V.S., Bogdanchikova, N.E. and Petranovskii, V.P. (2000) "Few-atomic silver clusters in zeolites: ab initio MO LCAO calculation and optical spectroscopy", *J. Phys. Chem.* **B104**, 12105-12110.

MODELING AND INTERPRETATION OF STM IMAGES OF CARBON NANOSYSTEMS

G. I. MÁRK [1;2], L. P. BIRÓ [2], AND PH. LAMBIN[1]
[1] *Département de Physique, Facultés Universitaires N.D.P.*
61 Rue de Bruxelles, B 5000 Namur, Belgium
[2] *Research Institute for Technical Physics and Materials Science*
H-1525 Budapest P.O.Box 49, Hungary
EMail: mark@sunserv.kfki.hu

Abstract.

Scanning tunneling microscopy (STM) is the only tool making it possible to study both the topography and electronic structure of carbon nanosystems in sub-nanometer resolution. The interpretation of the STM images of carbon nanosystems is more complicated than in the case of at, single crystalline surfaces. Because of this computer simulation is a helpful tool in understanding the experimental data. In this paper the image formation in STM and the particularities of imaging supported carbon nanosystems are discussed. The tight binding and the wave packet dynamical STM simulation methods are reviewed with applications showing their complementary merits. These methods are simple enough to make feasible in the near future their application to more complex carbon nanosystems like coiled nanotubes and nanotube crossings.

1. Introduction

The discovery of the fullerene [1] opened a new era of carbon science: the science of carbon nanostructures. For the time being a large variety of these systems [2] are known: fullerenes [1,3], carbon nanotubes [3,4], peapods [5], nanotube junctions [6], Y-connections [7], aggregations of nanotubes, carbon onions, etc. Apart from being very interesting from the theoretical point of view, carbon nanostructures have a wide range of potential applications owing to their remarkable mechanical and electrical properties. There are not many tools available to characterize isolated carbon

E. Buzaneva and P. Scharff (eds.), Frontiers of Multifunctional Nanosystems, 43–58.

the possibility to study both the topography and the electronic structure of these nanosystems in sub-nanometer resolution. However, this very property of STM makes it not easy to interpret STM images because the influence of the topography and the electronic structure has to be separated. Several other factors, from which the STM tip geometry is the most important, also influence the imaging mechanism. Because of all these features image simulation is inevitable for a correct interpretation of STM images.

2. Scanning Tunneling Microscopy

In the STM instrument [8] an atomically sharp, conducting tip is brought within a distance of a few tenths of a nanometer to a conducting surface (sample) while an U_t bias of the order of $1\,V$ is applied between the tip and the sample. Because of the small separation a quantum mechanical tunnel current I_t of the order of $1\,nA$ is flowing through the tip-sample gap. In the most common, so called topographic mode of the STM the tip is mechanically scanned above the sample (in the X-Y plane) by piezo actuators while a feedback loop controls the height of the tip above the sample (Z coordinate of the tip) to maintain the tunnel current at a preselected constant $I_{setpoint}$ value. The so generated 3D motion of the tip defines the topographic STM image. In practical STM instruments the $Z(X,Y)$ topographic image is constructed from the U_X, U_Y, U_Z voltages applied to the piezos.

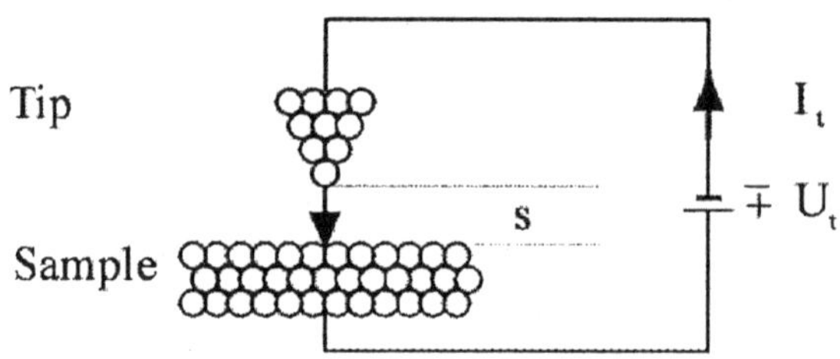

Figure 1. Concept of the STM. Small circles symbolize the atoms. s is the tip-sample tunnel gap.

The simplest explanation of STM operation is based on 1D tunneling theory [9]. When a quantum mechanical particle of energy E is incident on a potential barrier of height V_0 and width s the transmission probability is

$$T \propto e^{-2\kappa s}\,, \tag{1}$$

where

$$\kappa = \sqrt{2m/\hbar^2(V_0 - E)} \,. \tag{2}$$

This expression is valid in the asymptotic limit of $\kappa s >> 1$ (i.e. small transmission).

In an STM instrument, however, one of the electrodes is not a plane, but a sharp tip (see Fig. 1). Because of the exponential dependence of the tunneling probability with distance the majority of the tunnel current flows in a narrow channel between the tip apex and the sample plane. The typical minimum achievable half width at half maximum HW_T of the tunneling channel is $0.1 - 0.2\,nm$ with sharp tips [10]. This narrow tunneling channel is responsible for the atomic resolution of the STM.

2.1. STM ON CARBON NANOSTRUCTURES

Carbon nanostructures are 3D objects with characteristic radii of curvature in the nm range. This value is comparable on the one hand to the characteristic value of the tip-sample tunnel gap (which is of the order of $0.5\,nm$ in typical STM experiments) and on the other hand to the apex radius of the STM tip. This makes the interpretation of STM images of carbon nanostructures different from that of flat surfaces.

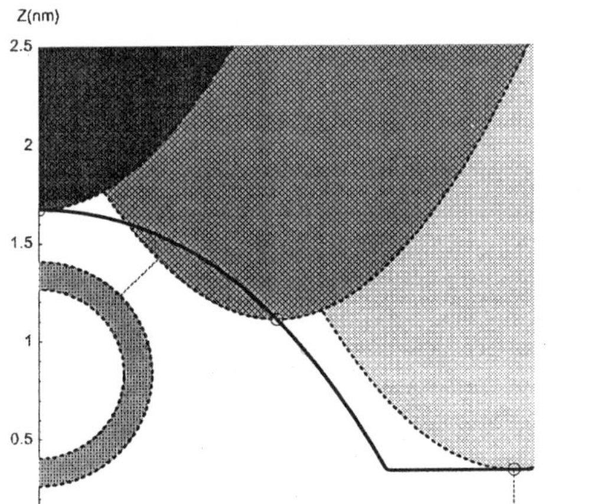

Figure 2. Geometric line cut. The thick line is the geometric line cut drawn by the apex of the tip (marked by open circle). The shaded lower half plane, middle ring and upper hyperbolas show the vertical cross sections of the sample, nanotube, and tip, respectively. The electrodes are bounded by their jellium effective surfaces (broken lines). Several, typical tip positions are shown, with different shades of grey. For each tip position shown, the nearest tip and sample points are joined by a thin dashed line.

The effect of the comparable characteristic radii of the sample and the tip is generally called "tip convolution effect". The tip convolution causes a broadening of the sample features as demonstrated on Fig. 2. for the case of a nanotube scanned by a hyperbolic tip. Because of its exponential distance dependence the tunneling current tends to follow the shortest path, which is normal to the tube, and the corresponding "off-z" component of the current increases as the tip moves aside to the tube [11]. These shortest path segments are shown on Fig. 2. by dashed lines. During the lateral scanning I_t is kept constant by the feedback loop. Assuming a uniform local density of states (LDOS) on the sample and tip surfaces the effect of the feedback loop can be taken into account approximately by keeping the length of the tunneling path at a constant value. The curve drawn by the apex point of the tip when using this simple approximation is called a *geometric line cut*. As it is demonstrated in Fig. 2., when the tip follows the curvature of the nanotube, the tunneling point moves away gradually from the tip apex to its side. However, when the tip apex approaches the support surface the tunneling is switched back from the side of the tip to the apex. After this switching the tip follows a line parallel to the support surface. As demonstrated in Sec. 5.3. this simple geometric model is justified by wave packet dynamical calculations.

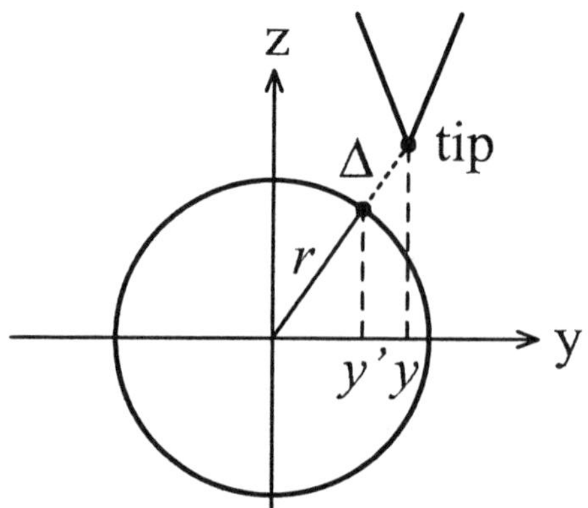

Figure 3. Relation between the position of the STM tip (y) and that of an imaged atom (y') of the nanotube when the tunneling current follows the shortest path between tip and nanotube (dashed line).

The effect of the comparable values of the characteristic radius of the sample and the tunnel gap is demonstrated in Fig. 3. It causes a stretching of the image of the atomic lattice in the direction parallel to the sample curvature. To study this effect the tip radius was assumed to be zero throughout the subsequent discussion. Using the geometric approximation discussed above we assume the tip-sample gap to be a constant Δ value. (The effect of the atomic lattice causes only a small variation on the order of $0.1\,nm$ to this constant gap.) Hence when the tip scans above a nano object with radius of curvature r its apex moves on a circle of radius $r+\Delta$. As a consequence the features on surface of the radius r are projected onto the radius $r+\Delta$ surface and are therefore stretched by a factor of $(r+\Delta)/r$. In the special case of a carbon nanotube, the radius of curvature is small only in the direction perpendicular to the tube axis (say, y direction) but it is infinite in the direction parallel to the tube axis (x direction). This causes an asymmetric stretching in the atomic resolution STM images of small diameter nanotubes: the atomic lattice is inflated in the y direction. As demonstrated in Sec. 5.3. this stretching is reproduced in tight-binding calculated atomic resolution STM images of carbon nanotubes. It is also seen in experimental STM images taken on single wall nanotubes [12, 13]. Due to the asymmetric distortion, the angles between the C-C nearest neighbor bonds measured on the STM image deviate from $\pi/3$.

Apart from the phenomena discussed above, there are other factors specific to the imaging of carbon nanosystems but here we can only give references to these phenomena. These factors include the effect of the different electronic structures of the nanosystem and its support [14], the effect of lattice imperfections and substituted atoms [15], the second tunneling gap [16] (between the nanosystem and the support), point contact effects [17], the lateral spreading of the charge along a nanotube while tunneling [18], etc. All these effects make the simulation of the imaging mechanism an important tool in interpreting STM images and scanning tunneling spectroscopy (STS) curves of carbon nanosystems.

3. STM theory

3.1. PERTURBATIC TREATMENT

Treating the tip-sample coupling as a first-order perturbation the total tip-sample current is given as

$$I_t = \frac{2\pi c}{\hbar} \int_{-\infty}^{+\infty} dE\, [f_t(E) - f_s(E)] \sum_{\alpha,\beta} |\langle\alpha|v|\beta\rangle|^2 \delta(E-E_\alpha)\delta(E-E_\beta)\,. \quad (3)$$

where $f_t(E)$ and $f_s(E)$ are the occupation numbers of the tip and sample, respectively and $\langle\alpha|v|\beta\rangle$ is the coupling matrix element of an $|\alpha\rangle$ tip state to a $|\beta\rangle$ sample state. This equation is the well-known starting expression of the perturbatic theories of elastic tunnel processes.

3.2. TERSOFF-HAMANN THEORY

In this simplified model [19] only one atom at the tip apex is taken into account, with an s-wave orbital. Both the tunneling matrix elements and the tip density of states (DOS) are taken to be constant in the eU_t energy window. With these assumptions the tunnel current is directly proportional to the LDOS integrated between the Fermi level of the tip and sample, that is

$$I_t(\vec{r}, U_t) \propto \int_{E_F^s - eU_t}^{E_F^s} dE \; \rho_{LDOS}(\vec{r}, E) \; , \tag{4}$$

with

$$\rho_{LDOS}(\vec{r}, E) = \sum_{\beta} \mid \psi_\beta(\vec{r}) \mid^2 \delta(E_\beta - E) \; , \tag{5}$$

where ψ_β and E_β are the electron wavefunction and eigenvalue of state β, respectively. We then approximate the constant current images as isosurfaces of $I_t(\vec{r}, U_t)$. Using this approximation Rubio et al. simulated [20] STM images of single wall nanotubes by determining the $\rho_{LDOS}(\vec{r}, E)$ values from *ab-initio* calculations.

For an infinitesimally small bias Eq. 4. becomes even simpler:

$$I_t(\vec{r}, U_t) \propto rho_{LDOS}(\vec{r}, E_F^s) \; , \tag{6}$$

which means that the tunnel current at tip position $\vec{r}$ is simply proportional to the sample LDOS on the Fermi level at the center of the tip.

3.3. TIGHT-BINDING STM THEORY

To go beyond the simple Tersoff-Hamann result one has to make some justificated assumptions for the $\langle\alpha|v|\beta\rangle$ tunneling matrix elements and for the tip DOS. A simplest formalism to accomplish this is the *tight-binding* theory. Assuming just one orbital per atom, the tight-binding expression of the tunnel current takes the following form [21]

$$I_t(U_t) = (2\pi)^2 \, \frac{e}{h} \int_{E_F^s - eU_t}^{E_F^s} dE \sum_{i,i' \in t} \sum_{j,j' \in s} v_{ij} v^*_{i'j'} n^t_{ii'}(E^t_F - E^s_F + eU_t + E) n^s_{jj'}(E) \tag{7}$$

where the E_F's are the Fermi levels of the tip (t) and sample (s), v_{ij} is a tight-binding coupling between the orbitals located on sites i of the tip and j of the sample, and $n^s_{jj'}(E) = (-1/\pi)$ Im $G^s_{jj'}(E)$ with $G^s_{jj'}$ a Green's function element of the sample. A similar notation, $n^t_{II'}(E)$, is defined on the tip side.

For the applications illustrated below, the tip was treated as a single s-atom (i), like in Tersoff-Hamann theory, with a Gaussian DOS. On the nanotube, there is one π orbital per C atom (j). The tip–sample coupling terms were Slater-Koster sp hopping interactions decaying exponentially with the distance d_{ij}:

$$v_{ij} = v_0 \, w_{ij} \, e^{-d_{ij}/\lambda} \cos\theta_{ij} \tag{8}$$

where θ_{ij} is the angle between the π orbital on site j and the ij direction, and w_{ij} is a cutting weight factor

$$w_{ij} = e^{-ad_{ij}^2} / \sum_{j'} e^{-ad_{ij'}^2} \tag{9}$$

introduced for convergence reasons. The parameters are $\lambda = 0.85$ Å, $a = 0.6$ Å^{-2}.

4. Wave packet dynamical simulation of the STM image

For a correct interpretation of some effects mentioned in Sec. 2.1. it is necessary to go beyond the first-order perturbation theory. This is the case, e.g. for *point contact imaging* [17] when, because of the small tip-sample separation, the conduction mechanism switches from tunneling to ballistic flow. Point contact imaging is accomplished experimentally by applying relatively large (more than $1\,nA$) tunnel currents with moderate (less than $1\,V$) biases. This is because to maintain the small $R_{tunnel} = U_t/I_t$ value the feedback loop have to push the tip so close to the sample that the potential barrier gets punctured. Atomic resolution is still possible [22] in point contact mode and the switching from tunneling to point contact causes characteristic changes [17] in STS spectra.

In the wave packet dynamical method the current density is determined by calculating the scattering of wave packets (WPs) incident on the barrier potential. The initial WPs are constructed from the stationary states of the reservoir (tip or support bulk) from which the WPs are arriving. The total tunnel current at a given STM bias is calculated as a statistical average of the tunnel currents for WPs of different allowed incident energies and directions ($\vec{k}$ vectors) weighted according to the band structure of the two reservoirs.

The tunneling probability for a given initial WP is determined from the time dependent wave function $\psi(\vec{r},t)$ computed from the time dependent Schrödinger equation by the *split operator Fourier transform method* [14].

In this method the time evolution operator $\exp(-i\mathbf{H}\Delta t)$ is approximated (in Hartree atomic units) by the symmetrical unitary product

$$\exp(-i\mathbf{H}\Delta t) = \exp(-i\mathbf{K}/2\Delta t)\exp(-i\mathbf{V}\Delta t)\exp(-i\mathbf{K}/2\Delta t) \quad (10)$$

where Δt is the simulation time step. While the effect of the potential energy propagator $\exp(-i\mathbf{V}\Delta t)$ is a simple multiplication with $\exp(-iV(\vec{r})\Delta t)$ for local potentials, the kinetic energy propagator $\exp(-i\mathbf{K}/2\Delta t)$ is applied in k space by multiplying the $\varphi(\vec{k},t)$ momentum space wave function by $\exp\left(i|\vec{k}|^2/4\Delta t\right)$. To utilise this formula it is necessary to calculate the $\varphi(\vec{k},t)$ momentum space wave function by fast Fourier transform (FFT) of $\psi(\vec{r},t)$. Finally we have to return back to real space by inverse FFT.

5. Simulation results

5.1. GRAPHENE SHEET

As a first application of the tight-binding formalism STM image of a single sheet of graphite (graphene sheet) was calculated, see Fig. 4

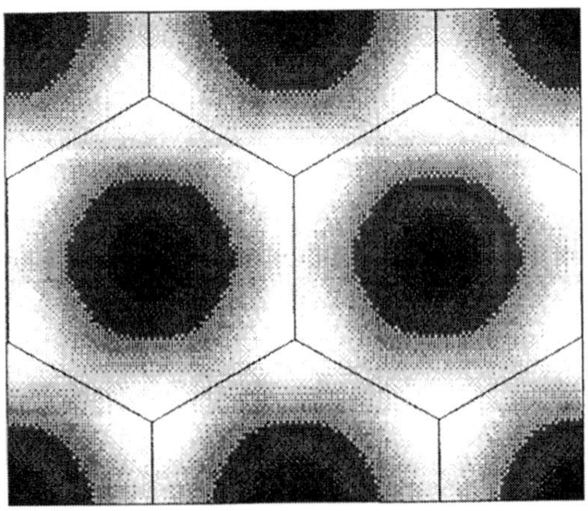

Figure 4. STM current map at constant height (0.5 nm) above a graphene sheet. The tip potential is 0.25 V. See the text for details.

We can see corrugation valleys that correspond to the center of the hexagons of the honeycomb structure (overlaid on the image by black lines). Around the hexagonal holes, there is a network of protruding C-C bonds. The hexagonal atomic lattice does not show up in the image, instead a triangular lattice made of the valleys is seen.

5.2. BERNAL GRAPHITE

When coupling two or more graphene layers with the Bernal graphite stacking, the two atoms per unit cell become unequivalent: atom A has a neighbor directly beneath whereas atom B does not. In a small interval around the Fermi level, the LDOS on site A is much smaller than that on B [23]. As a result, the STM current at small bias is larger when the tip is above an atom B, which therefore appears as a protrusion in the constant-current image. This interpretation is considered as the ad-hoc explanation of the fact that only every other two atoms are seen in the experimental STM images of graphite [24].

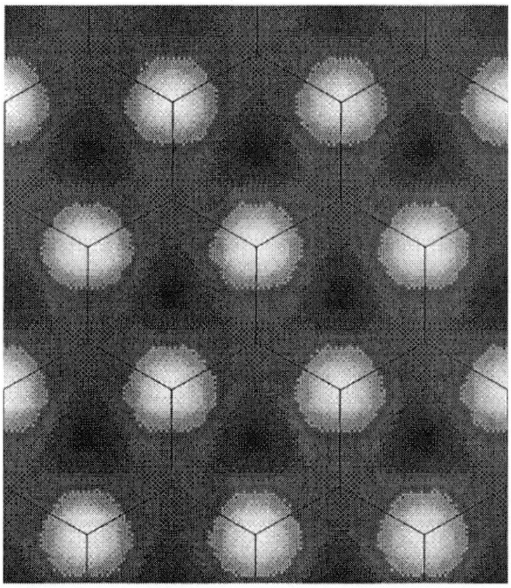

Figure 5. STM current map at constant height (0.5 nm) above a multilayer graphite surface. The tip potential is 0.25 V, which corresponds to a current ratio $I_B/I_A = 2$. The B atoms are clearly resolved, whereas the A atoms do not come out. See the text for details.

As can be seen in Fig. 5. there are no marked local maxima of the current at the locations of the A atoms, only the B atoms are seen (as white features). By increasing the bias, the differences between the densities of states on A and B sites become less important and the asymmetry washes out gradually.

5.3. *SINGLE WALL NANOTUBES*

Fig. 6. shows the STM images computed by the tight-binding formalism for four single-wall nanotubes with diameter around 1.4 nm.

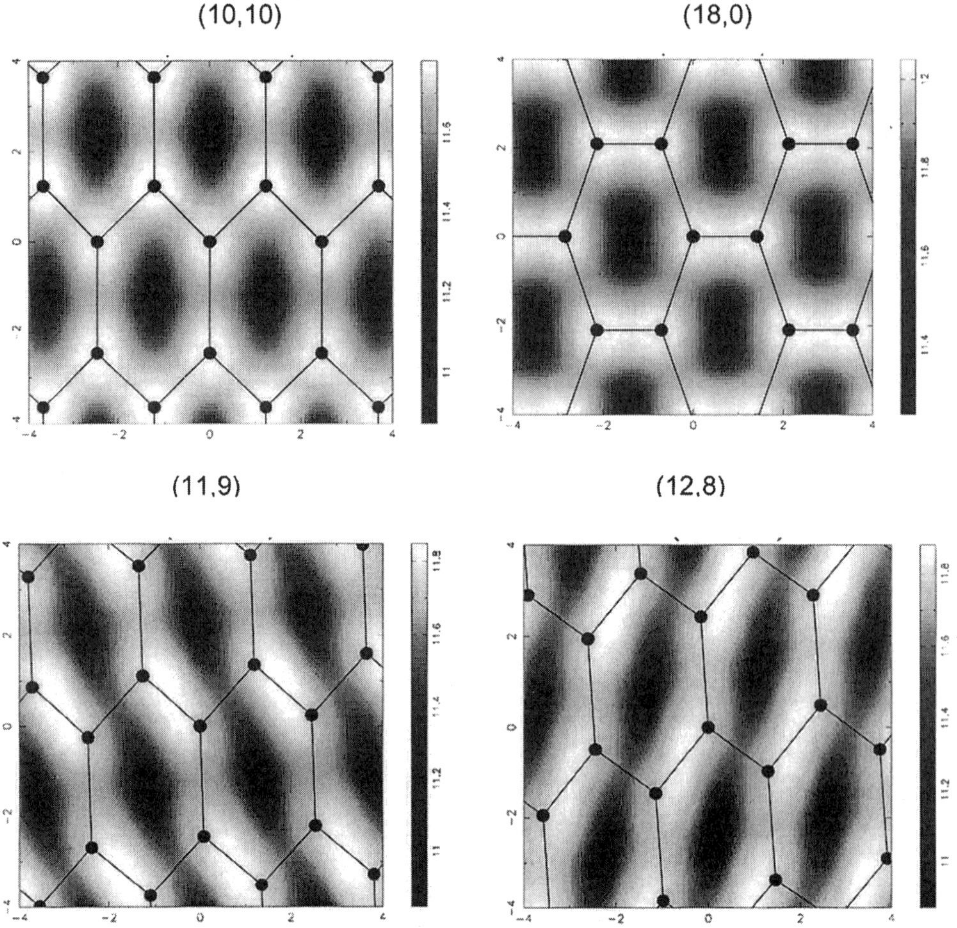

Figure 6. Gray-scale representation of the axial distance at constant current of the STM tip apex above the topmost part of four single-wall carbon nanotubes. The tube axis is parallel to the horizontal direction. The vertical distance of the tip to the atom located at the center of each image was set to 0.5 nm. For the two metallic, non-chiral (10,10) and (18,0) nanotubes, the tip potential was 0.3 V. The chiral (11,9) and (12,8) nanotubes are semiconductors, with a band gap of ~0.6 eV. For both of them, the tip potential was 0.5 V. All coordinates in the figure are in Å.

In agreement with the interpretation of graphene images summarized above, the centers of the honeycomb hexagons correspond to sharp dips of corrugation. In these two-dimensional maps of the radial distance of the tip, the nanotube axis is along the horizontal, x direction. Only the topmost parts of the nanotubes have been imaged. The curvature of the nanotube causes a distortion of the images in the form of an inflation of the y (see Sec. 2.1.) as discussed in [21], here by a factor of 1.7. This distortion is clearly visible on the honeycomb lattice that was superimposed on the images. It is also responsible for the elongate shape of the corrugation deeps at the center of the hexagons.

In the (18,0) zig-zag nanotube illustrated in Fig. 6, the largest protrusions are found on the bonds parallel to the axis. These protruding bonds form a triangular lattice of oblate humps. This resembles the triangular lattice formed by every other two atoms in multilayered graphite. In the (10,10) armchair nanotube, the largest protrusions are realized on the atoms. Here all the bonds look the same and the image has the honeycomb symmetry. In the case of chiral (11,9) and (12,8) nanotubes, one third of the bonds protrudes more than the others, like with the zig-zag geometry, but not always the ones closest to the axial direction. This bond anisotropy again destroys the hexagonal symmetry, as often observed experimentally [25]. In the chiral nanotubes, the protruding bonds form stripes that spiral around the nanotube [26]. The elongate holes at the center of the hexagons are no longer aligned with the circumference, as indeed often observed in the experimental images [27–29].

Fig. 7. shows snapshots of the time dependent WP probability density for a Gaussian WP incident from the tip bulk onto the tip-nanotube-support tunnel junction. In this calculation focussing on geometrical effects the tunnel junction is modeled by a simple 2D jellium potential [14]. The geometrical and material parameters of the carbon nanotube and the tip are the same as in [14]. The carbon nanotube is modeled by a cylinder of $0.5\,nm$ radius floating above the support at a distance of $0.335\,nm$. The STM tip is taken as a hyperbolic cylinder of $0.5\,nm$ apex radius and $15°$ aperture angle. The effective surface of these objects is assumed to be $0.071\,nm$ outside their geometric surface (defined as a smooth surface matching the nuclear skeleton of the surface atoms). An infinitesimally small bias is used and all the electrodes are assumed to have the same, free electron metallic DOS. The STM constant current loop is simulated by finding for each X_{apex} lateral tip displacement that Z_{apex} vertical tip displacement which yields a constant WP tunneling probability.

With the help of this WP dynamical calculation we are in position to verify the heuristic assumptions made in the derivation of the geometric line cut in Sec. 2.1. It can be seen in Fig. 7. that the majority of the tunnel current is indeed flowing in a narrow channel around the line section connecting the closest points of the tip and the sample, as was sketched intuitively in Fig. 2. For $X_{apex} = 1.6\,nm$ we can observe the switching of the tunneling point from the side to the apex of the tip.

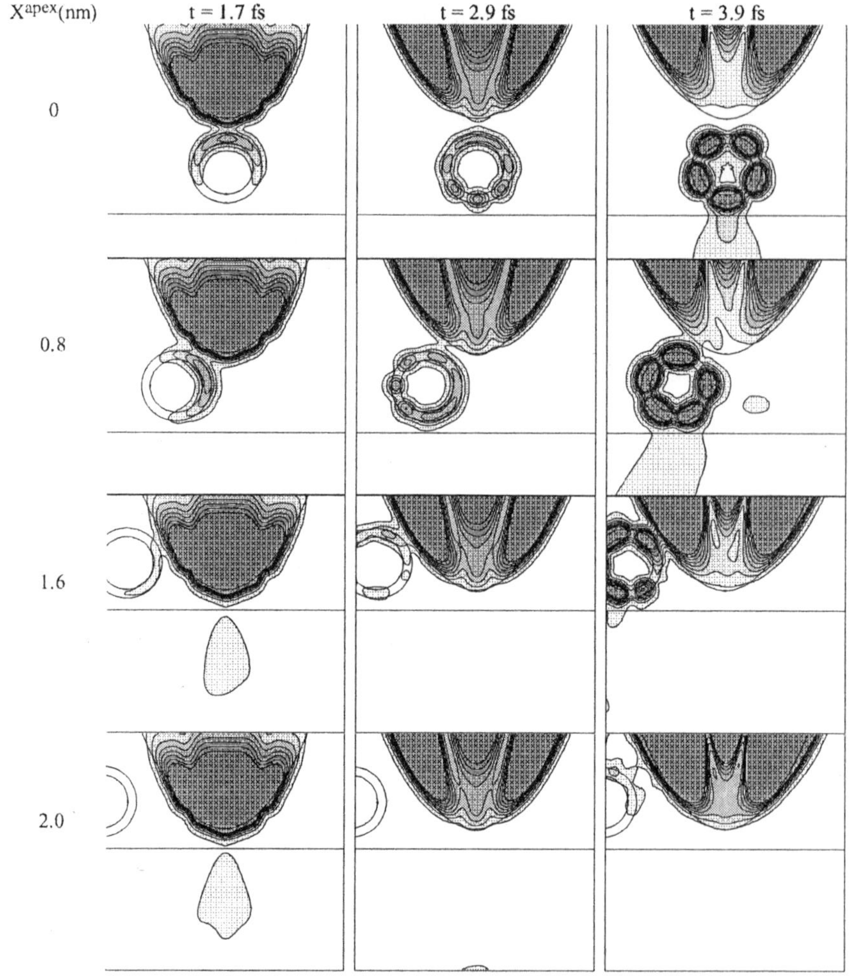

Figure 7. Probability density of the scattered wave packet for selected time instants and X_{apex} lateral tip displacements (0.0, 0.8, 1.6, and 2.0 nm). Size of the presentation window is 3.84 nm. Contour lines are drawn on sqare root scale. Each frame is normalized to its maximum density. Maximum density values are 4.2, 0.6, and 0.07 nm^{-2} for 1.7, 2.9, and 3.9 fs, respectively. Density in the tube region becomes appreciable on the lower right frame because of the renormalization.

The curve drawn by the tip apex point when the simulated tip is scanned over the nanotube while keeping the tunneling probability at a constant value is called a *quantum line cut* [14]. The good matching of the so defined quantum line cut (drawn by crosses in Fig. 8.) with the geometric line cut defined in Sec. 2.1. justifies the constant tip-sample separation ap-

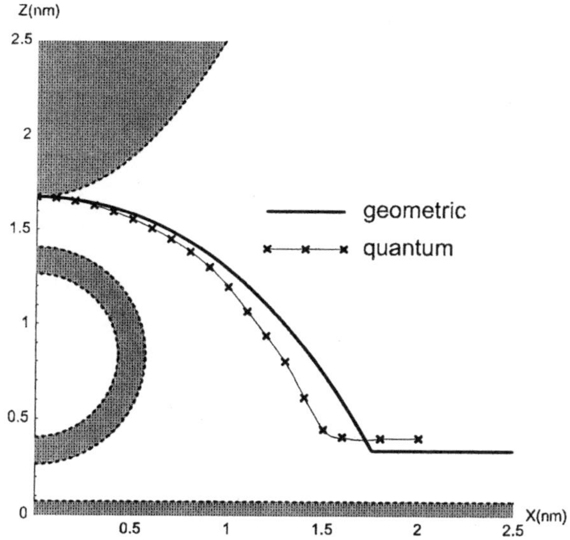

Figure 8. Comparison of geometric and quantum line cut. Thick solid line is the geometric line cut. Crosses show calculated points of the quantum line cut. A thin solid line connecting the crosses was drawn to guide the eye.

proximation. As seen in Fig. 8., however, the tip-sample separation is larger above the support plane than over the nanotube. This is because the WP has to tunnel through two tunnel resistances (tip-nanotube and nanotube-support) when the tip is above the nanotube but it has to tunnel through only one tunnel resistance when the tip is above the support plane. This causes a larger tunnel resistance above the nanotube than above the support plane which makes the feedback loop to retract the tip somewhat when the tunneling point is switched from the nanotube to the support plane. The different electronic structure of the nanotube and the support not included in this model causes a further increase of the tunnel gap above the support [14] as compared to that above the nanotube.

6. Summary and outlook

Scanning tunneling microscopy is an extremely attractive tool for imaging individual carbon nanostructures and to perform simultaneously spectroscopic measurements on the imaged nano-object. Because the imaging of supported 3D objects differs strongly from that of the flat, homogeneous surfaces, cautious image analysis and simulation are necessary to correctly interpret the experimental images.

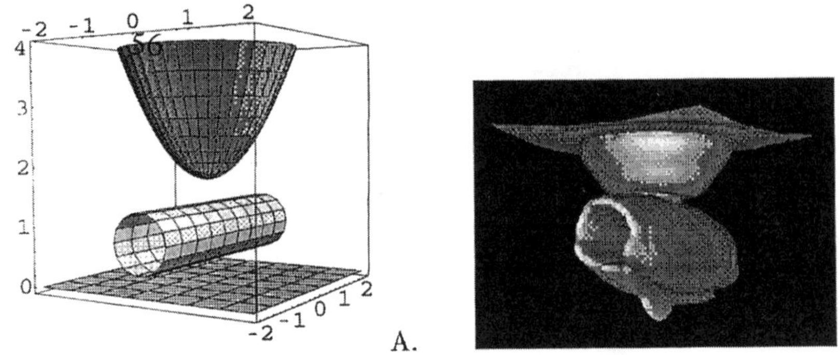

Figure 9. A) Geometric model of STM tip, nanotube and support surface. All dimensions are in nm. B) Snapshot (at $t = 3.5\,fs$) of the constant probability density surface of a wave packet tunneling through the tunnel junction shown in (A).

Recent developments in computer power and better algorithms also make it possible to use more realistic models in the simulations. These enhancements extend the sphere of the phenomena which can be interpreted by these simulations. As an example, in Fig. 9(b). shows a snapshot from the wave packet development calculated in three dimensions (3D) for the STM junction model of Fig. 9(a). Note that while the WP tunnels into the support surface (see the lower protrusion in the figure) it simultaneously spreads along the tube. Details are published elsewhere [18].

The tight-binding and wave packet dynamical STM simulation methods are simple enough to be utilized in the near future for more complex nanosystems like coiled nanotubes, Y-connections, nanotube crossings, etc.

Acknowledgments

This work has been partly funded by the interuniversity research project on reduced dimensionality systems (PAI P4/10) of the Belgian Office for Scientific, Cultural, and Technical affairs, by the Hungarian OTKA Grant T 30435, and by EUS Centre of Excellence ICAI-CT-2000-70029. LPB gratefully acknowledges financial support from the Belgian FNRS.

References

1. Kroto, H.W., Heath, J.R., O'Brien, S.C., Curl, R.F., and Smalley, R.E. (1985) C60: Buckminsterfullerene, *Nature* **318**, 162-163.
2. Scharff P. Fundamental Properties and Applications of Fullerene and Carbon Nanotube Systems, *This Volume.*
3. Dresselhaus, M. S., Dresselhaus, G., Ecklund P. C. (1996) *Science of Fullerenes and Carbon Nanotubes*, Academic Press, San Diego.

4. Iijima, S. (1991) Helical microtubules of graphitic carbon, *Nature* **354**, 56-58.
5. Smith, B.W., Monthioux, M., and Luzzi, D.E. (1998) Encapsulated C60 in carbon nanotubes, *Nature* **396**, 323-324.
6. Ouyang, M., Huang, J.-L., Cheung, Ch.L., Lieber, Ch.M. (2001) Atomically resolved single-walled carbon nanotube intramolecular junctions, *Science* **291**, 97-100.
7. Nagy, P., Ehlich, R., Biró, L.P., Gyulai, J. (2000) Y-branching of single wall carbon nanotubes, *Appl. Phys. A* **70**, 481-483.
8. Binnig, G. and Rohrer, H. (1982) Scanning tunneling microscopy *Helv. Phys. Acta* **55**, 726-735.
9. Wiesendanger, R. (1994) *Scanning probe microscopy and spectroscopy*, Cambridge University Press, Cambridge.
10. Lucas, A.A., Morawitz, H., Henry, G.R., Vigneron, J.P., Lambin, Ph., Cutler, P.H., and Feuchtwang, T.W. (1988) Scattering theoretic approach to elastic one-electron tunneling through localized barriers: Applications to scanning tunneling microscopy, *Phys. Rev. B* **37**, 10708-10720.
11. Ge, M. and Sattler, K. (1994) STM of single-shell nanotubes of carbon, *Appl. Phys. Lett.* **65**, 2284-2286.
12. Venema, L.C., Meunier, V., Lambin, Ph., and Dekker, C. (2000) Atomic structure of carbon nanotubes from scanning tunneling microscopy, *Phys. Rev. B* **61**, 2991-2996.
13. Kim, Ph., Odom, T.W., Huang, J., and Lieber, C.M. (2000) STM study of single-walled carbon nanotubes, *Carbon* **38**, 1741-1744.
14. Márk, G.I., Biró, L.P., and Gyulai, J. (1998) Simulation of STM images of three-dimensional surfaces and comparison with experimental data: carbon nanotubes, *Phys. Rev. B* **58**, 12645-12648.
15. Clauss, W., Bergeron, D.J., Freitag, M., Kane, C.L., Mele, E.J., and Johnson, A.T. (1999) Electron backscattering on single-wall carbon nanotubes observed by STM, *Europhys. Lett.* **47**, 601-607.
16. Biró, L.P., Gyulai, J., Lambin, Ph., Nagy, J.B., Lazarescu, S., Márk, G.I., Fonseca, A., Surján, P.R., Szekeres, Zs., Thiry, P.A. Lucas, A.A. (1998) Scanning tunnelling microscopy (STM) imaging of carbon nanotubes, *Carbon* **36**, 689.
17. Márk, G.I., Biró, L.P., Gyulai, J., Thiry, P.A., Lucas, A.A., and Lambin, Ph. (2000) Simulation of scanning tunneling spectroscopy of supported carbon nanotubes, *Phys. Rev. B* **62**, 2797-2805.
18. Márk, G.I., Biró, L.P., Koós, A., Osváth, Z., Gyulai, J., Benito, A.M., Thiry, P.A., and Lambin, Ph. (in press) Charge spreading effects during 3D tunneling through a supported carbon nanotube, in *Electronic Properties of Novel Materials – Molecular Nanostructures*, March 2001, Kirchberg, Austria, AIP Conference Proceedings, edited by Kuzmany, H., Fink, J., Mehring, M., Roth, S.
19. Tersoff, J., Hamann, D.R. (1985) Theory of the scanning tunneling microscope, *Phys. Rev. B* **31**, 805-813.
20. Rubio, A., Sanchez-Portal, D., Artacho, E., Ordejon, P., and Soler, J.M. (1999) Electronic states in a finite carbon nanotube: A one-dimensional quantum box, *Phys. Rev. Lett.* **82**, 3520-3523; Rubio, A. (1999) Spectroscopic properties and STM images of carbon nanotubes, *Appl. Phys. A* **68**, 275-282.
21. Meunier, V. and Lambin, Ph. (1998) Tight-binding computation of the STM image of carbon nanotubes, *Phys. Rev. Lett.* **81**, 5888-5891.
22. Agrait, N., Rodrigo, J.G., and Vieira, S. (1992) On the transition from tunneling regime to point contact: graphite, *Utramicroscopy* **42-44**, 177-183.
23. Charlier, J.C., Michenaud, J.P., and Lambin, Ph. (1992) Tight-binding density of electronic states of pregraphitic carbon, *Phys. Rev. B* **46**, 4540-4543.
24. Tománek, D. and Louie, S.G. (1988) First-principles calculations of highly asymmetric structure in scanning-tunneling-microscopy images of graphite, *Phys. Rev. B* **37**, 8327-8336.
25. Venema, L.C., Wildöer, J.W.G., Dekker, C., Rinzler, A.G., and Smalley, R.E. (1998) STM atomic resolution images of single-wall carbon nanotubes, *Appl. Phys. A* **66**,

S153-S155.
26. Kane, C.L. and Mele, E.J. (1999) Broken symmetry in STM images of carbon nanotubes, *Phys. Rev. B* **59**, R12759-R12762.
27. Wildöer, J.W.G., Venema, L.C., Rinzler, A.G., Smalley, R.E., and Dekker, C. (1998) Electronic structure of atomically resolved carbon nanotubes, *Nature* **391**, 59-62.
28. Odom, T.W., Huang, J.L., Kim, Ph., and Lieber, Ch.M. (1998) Atomic structure and electronic properties of single-walled carbon nanotubes, *Nature* **391**, 62-64.
29. Hassanien, A., Tokumoto, M., Kumazawa, Y., Kataura, H., Maniwa, Y., Suzuki, S., and Achiba, Y. (1998) Atomic structure and electronic properties of single-wall nanotubes probed by STM at room temperature, *Appl. Phys. Lett.* **73**, 3839-3841.

CARBON NANOTUBES UNDER INTERNAL PRESSURE

B.A. GALANOV.
[1]Institute for Problems in Materials Science,
National Academy of Science, Kiev, Ukraine

S.B. GALANOV.
[2]Ukrainian Research and Design Institute of Steel Construction,
Kiev, Ukraine

Y. GOGOTSI
[3]Department of Materials Engineering, Drexel University,
Philadelphia, PA 19104
gogotsi@drexel.edu

Abstract

A considered application of carbon nanotubes is nanopiping in nanofluidic devices. The use of nanotubes for fluid transport requires large-diameter tubes that can sustain prescribed loading without failure. Two models of the stress-strain state of long multiwall carbon nanotubes, subjected to internal pressure, are described. Cylindrical nanotubes having a Russian doll structure have been considered. It is assumed that the deformations are linear elastic and negligible along the tube axis (in comparison with the radial deformations). This assumption is not restrictive for potential applications of nanotubes, where their deformations must be small and reversible. The distance between the layers is small in comparison to the radii of curvature of graphite layers. In the case of several carbon layers, a discrete model (DM) is proposed. The solutions of DM equations, with corresponding boundary conditions, determine the stresses between the layers, the forces in the layers, and the deformation of the layers. For the case of thick walls built of numerous carbon layers, a continuous model (CM) is proposed. The main CM equation is the Euler's differential equation with corresponding boundary conditions. Its solution defines the continuous distribution of the stresses and strains across the wall thickness of the tube.

Notations

a_0 - distance between two adjacent layers before deformation ($a_0 = const$) (Fig. 4)

E. Buzaneva and P. Scharff (eds.), Frontiers of Multifunctional Nanosystems, 59–74.

a_n - distance between the layers (n) and (n-1) after deformation (defined through experiments for DM);

R_0 - inner radius of the tube before deformation (Fig. 2)

R_i - radius of layer i before deformation;

i - layer number ($i = \overline{0,n}$,Fig. 4)

R_n - external radius of the tube before deformation (Fig. 4)

p_{ext} - pressure outside the tube (external);

p_{int} - pressure inside the tube (internal);

σ_i - radial stress between the layers ($i-1$)and i ($\sigma_i = const$);

E_1 - tangential (hoop)Young's modulus;

E_2 - radial Young's modulus;

T_i - tangential (hoop) stress in layer i, measured in $Pa \cdot m$ $(\sigma_{\tau i} = T_i / a_0)$;

$\varepsilon_i = \sigma_{\tau i} / E_1$ - tangential (hoop) deformation of layer i;

$(n+1)$ - number of layers;

$\sigma(r)$ - distribution of radial stresses (for CM);

$\sigma_\tau(r)$ - distribution of tangential (hoop) stresses (for CM).

$\varepsilon_{ri} = \sigma_i / E_2$ - radial deformation of spacing between layers (i -1) and (i);

$\varepsilon_{rn} = (a_n - a_0) / a_0$ - relative change of distance between the layers (n) and (n-1) after deformation (defined through experiments for DM);

T_z - stress $(Pa \cdot m)$ in the layers in the tube axis direction (for DM of closed tube);

σ_z - stress in the tube wall in the tube axis direction (for CM of closed tube);

1. Introduction

Carbon nanotubes (Dresselhaus, 1996; Ebbesen, 1997; Harris, 1999; Iijima, 1994) are among the most exciting new materials being investigated because of their potential uses in new technologies and devices. In particular, multi-wall carbon nanotubes are of interest to the growing micro- and nanofluidic industry, which is a new branch of microengineering with great promise (Burns, 1998; Delamarche, 1997; Jo, 2000; Wuu, 1999). Use of nanopipes may be required to enable the development of functional nanomachines or other nanofluidic devices. The use of nanotubes would provide cylindrical channels similar to pipes used in the macro-world. They can also act as

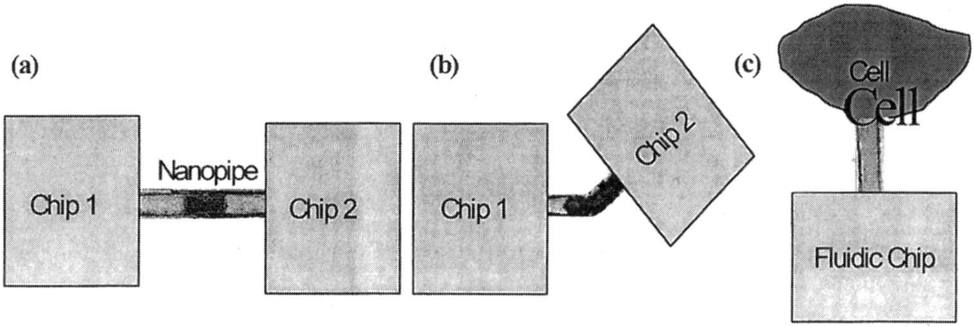

Figure 1. Schematic showing the possible use of carbon nanopipes for connecting microfluidic chips (a,b) and probing or injecting small objects such as cells (c).

interconnects (Fig.1) between microfluidic chips or between a chip (e.g., a drug delivery system) and the subject (e.g., a cell). Nanopipes that act like tiny straws could deliver medicines, slowly and over time, to a person's bloodstream or to a highly specific location in the body (Schulz, 2000).

We have discovered that hydrothermal catalytic synthesis can produce multi-wall carbon nanotubes with a high aspect ratio, a high degree of graphitization, and wide internal channels (Gogotsi, 2001a; Gogotsi, 2001c). Fig. 2 shows typical hydrothermal nanotubes. The thin ones, which were produced by a hydrothermal treatment of single-wall tubes (Gogotsi, 2001a; Gogotsi, 2001c), but could also be synthesized from ethylene glycol solutions, have less than 20 carbon layers in the wall (Fig. 2a) and are similar to nanotubes produced by arc discharge or plasma-assisted CVD (Harris, 1999). Larger nanotubes (Fig. 2b,c), which were produced by catalytic decomposition of ethylene glycol or polyethylene mixtures with water in presence of a nickel catalyst, have large internal diameters (50% to 90% of the outer diameter). We occasionally refer to them as "carbon nanopipes" or "micropipes" (for diameters larger than 100 nm) because they look like pipes used in the macro world, but are orders of magnitude smaller. Hydrothermal nanopipes have graphitic walls (Fig. 2a,c) and a high continuity of graphene planes. Those

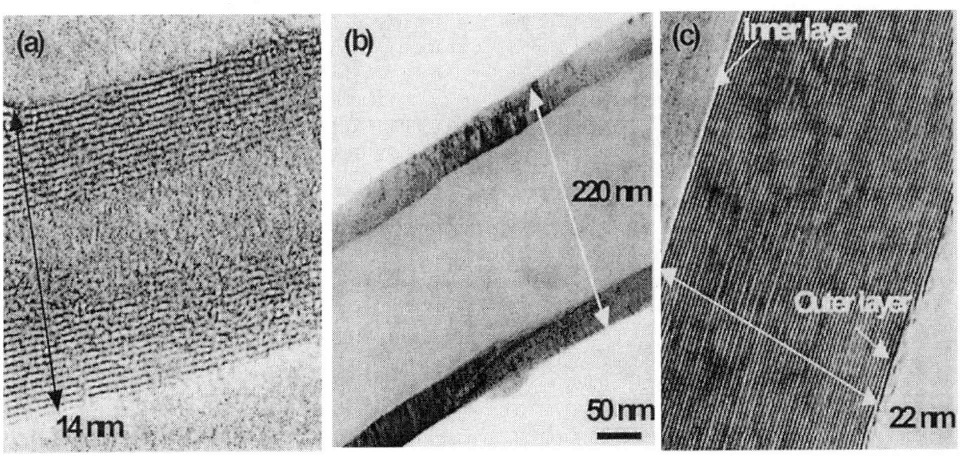

Figure 2. TEM micrographs of multiwall carbon nanotubes produced by hydrothermal synthesis and ranging from 14 nm·(**a**) to 220 nm (**b**) in diameter. **c** is a high-resolution image of a tube wall. The number of layers in the wall can vary from just a few in thin tubes (14 layers in **a**) to dozens in micro-pipes (65 layers in **c**).

tubes that are capped (Figure. 3) appear to be gas-tight and can maintain hundreds of atmospheres of internal pressure in the vacuum of a transmission electron microscope (10^{-8} *torr*), even at temperatures above 500°C, thus, demonstrating a very high strength. Some of the closed-end tubes contained a high-pressure encapsulated aqueous fluid (Gogotsi, 2001b), which displayed clearly segregated liquid and gas phases separated by well-defined curved interfaces (Figure. 3).

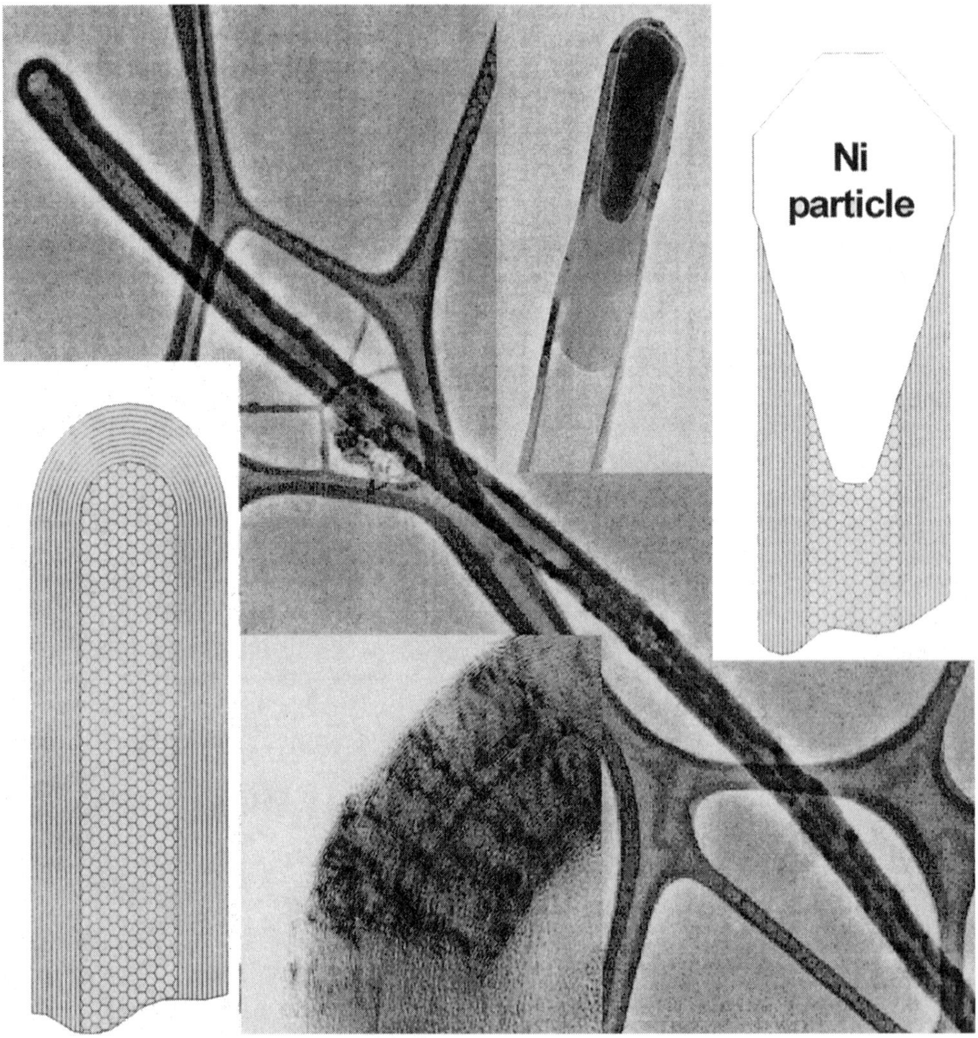

Figure 3. TEM micrographs and schematic pictures of tube end termination by a carbon dome or a catalyst particle, which results in trapping of aqueous fluid (liquid and gas) inside the hydrothermally grown nanotubes.

If we consider that a closed tube was sealed at 750°C and the hydrothermal fluid was captured at this temperature, the pressure of the tube contents at room temperature may be up to ~30 MPa. The demonstrated ability of the closed hydrothermal nanotubes to act as miniature pressure vessels and maintain the pressurized fluid, even when heated under the high vacuum of a TEM column (10^{-8} Torr), confirms their high strength and wall perfection. This is in agreement with the high strength of multiwall nanotubes reported in the literature (Wong, 1997). However, only strength in the longitudinal direction has been measured (Ruoff, 1995; Wong, 1997) and it was shown to decrease drastically with increasing tube diameter. Interesting atomistic analysis conducted for single-wall tubes (Yakobson, 2000) does not provide a sufficient theoretical basis for understanding mechanical behavior of multi-wall tubes. The use of nanotubes for fluid transport requires large-diameter tubes that can sustain a prescribed loading without failure. Relatively high pressures may be required for the liquid transport to overcome strong interactions with the tube walls, which increases with decreasing tube diameter. No internal pressure experiments have been reported for nanotubes so far. Stresses in macroscopic cylindrical shells have been thoroughly investigated (Ugural, 1999) and long cylindrical conduits, such as gas pipelines, have been analyzed in numerous publications. However, the applicability of a classical approach to anisotropic nano-shells, such as nanotubes, is questionable. The objective of this work was to perform computational analysis of stresses and strains in nanotubes of various wall thickness.

2. Model Equations

As shown in Fig. 4, carbon nanotubes range in diameter from a few nanometers to

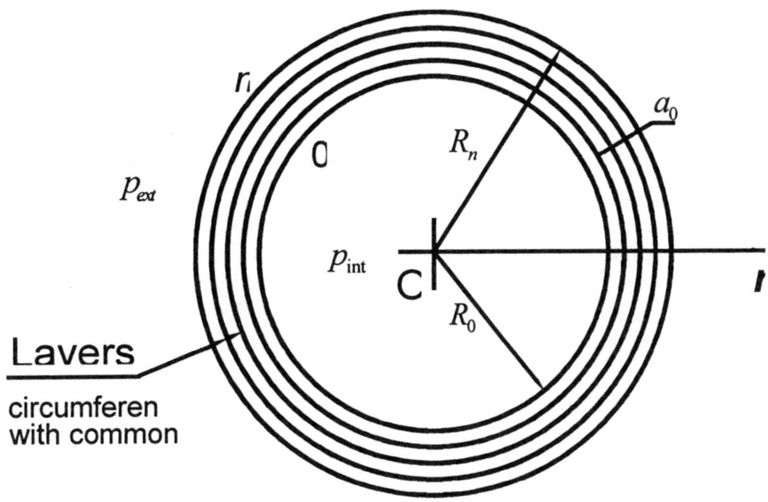

Figure 4. Idealized geometric model of a carbon nanotube depicting a cross-section normal to the tube axis.

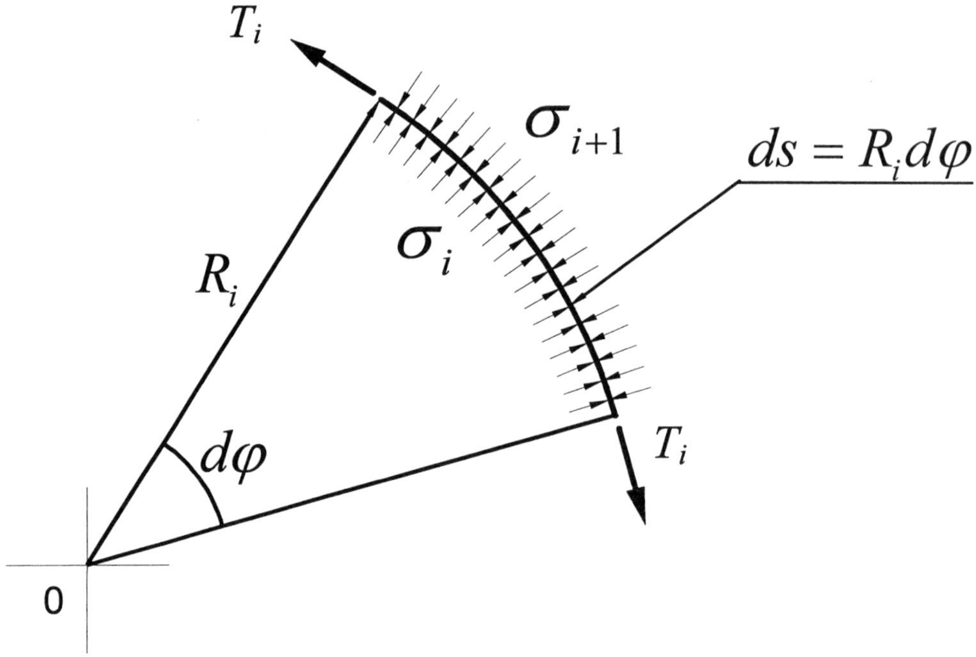

Figure 5. The loading scheme of a layer element $ds = R_i d\varphi$.

hundreds of nanometers. For the thinnest tubes (nanopipes) with few graphene layers in the wall (Fig. 2a), continuum solid mechanics is expected to break down and a discrete model must be used. However, the tubes shown in Fig. 2c have thick walls with more than 50 graphene layers in the wall. Although an exact dimensional limit for the continuum mechanics breakdown is not known, we believe that the tubes with the walls of that thickness can be described using continuum mechanics and a different model has been developed for the micropipes.

These models have been developed for nanotubes built of co-axial graphene cylinders (Russian doll structure). It is well known that many nanotubes, including some nanotubes produced by hydrothermal synthesis (Libera, 2001), have a herring-bone structure. Also, some nanotubes are supposed to have a scroll structure. The continuous model should be applicable to the scroll tubes having thick walls. However, herring-bone tubes will require separate analysis and are not covered in this study.

2.1. *DISCRETE MODEL (DM)*

The geometric scheme of a non-deformed cross-section, normal to the tube axis, is shown in Fig. 4. Cylindrical nanotubes having a Russian doll structure have been considered. It is assumed that the distance between the layers, a_0 (about 0.34 nm), is small with regard to the radii of curvature, R_i, of the graphene layers. The layers are assumed to have no thickness. This helps us to avoid an ambiguous estimate of the layer cross-section, that resulted in unreasonably high strength values assigned to nanotubes by some researchers. Because of this, the layer is considered as a membrane shell (Gere, 1997) and the stress, T_i, (see Fig. 5) is measured in $Pa \cdot m$ instead of Pa. From the equilibrium equation of layer element, ds, (see Fig. 5) and Hook's law, the stress, T_i, and the deformation, ε_i, for the layer i can be obtained from:

$$T_i = R_i(\sigma_{i+1} - \sigma_i),$$
$$\varepsilon_i = \frac{T_i}{E_1 a_0}. \tag{1}$$

As a result, the deformative increment of radius R_i is

$$\Delta R_i = \frac{T_i R_i}{E_1 a_0}$$

and the distance between the layers i and $(i+1)$ after deformation is changed by the amount:

$$\Delta a_i = \Delta R_{i+1} - \Delta R_i = \frac{1}{E_1 a_0}(T_{i+1}R_{i+1} - T_i R_i)$$

On the other hand, under the stress σ_{i+1}, this distance is changed by the quantity:

$$\Delta a_i = \frac{\sigma_{i+1} a_0}{E_2}$$

Therefore, the compatibility condition of the deformation must be satisfied

$$\frac{\sigma_{i+1} a_0}{E_2} = \frac{1}{E_1 a_0}(T_{i+1}R_{i+1} - T_i R_i) \tag{2}$$

This system of difference equations for the compatibility of the layer deformations can be written by inserting equation (1) into equation (2) in the following form:

$$\frac{E_1 a_0^2}{E_2}\sigma_{i+1} = R_{i+1}^2(\sigma_{i+2} - \sigma_{i+1}) - R_i^2(\sigma_{i+1} - \sigma_i), \tag{3}$$
$$i = \overline{0,(n-1)}$$

with the boundary conditions of two types:

$$\begin{cases} \sigma_{n+1} = -p_{ext} \\ \sigma_n = E_2 \dfrac{a_n - a_0}{a_0} \end{cases} \tag{4}$$

and

$$\begin{cases} \sigma_{n+1} = -p_{ext} \\ \sigma_0 = -p_{\text{int}} \end{cases} \tag{5}$$

where a_n is the known distance between the layers (n) and (n-1) after deformation (defined through experiments).

For the boundary conditions (4) and (5), equation (3) is presented respectively in the following form:

$$\begin{cases} \dfrac{a_0^2 E_1}{E_2}\sigma_n = R_n^2(-p_{ext} - \sigma_n) - R_{n-1}^2(\sigma_n - \sigma_{n-1}), \sigma_n = E_2 \dfrac{a_n - a_0}{a_0}, \\ \dots\dots\dots\dots\dots\dots\dots\dots\dots\dots\dots\dots, \\ \dfrac{a_0^2 E_1}{E_2}\sigma_2 = R_2^2(\sigma_3 - \sigma_2) - R_1^2(\sigma_2 - \sigma_1), \\ \dfrac{a_0^2 E_1}{E_2}\sigma_1 = R_1^2(\sigma_2 - \sigma_1) - R_0^2(\sigma_1 + p_{\text{int}}) \end{cases} \tag{6}$$

and

$$\begin{cases} \dfrac{a_0^2 E_1}{E_2}\sigma_n = R_n^2(-p_{ext} - \sigma_n) - R_{n-1}^2(\sigma_n - \sigma_{n-1}), \\ \dots\dots\dots\dots\dots\dots\dots\dots\dots\dots, \\ \dfrac{a_0^2 E_1}{E_2}\sigma_2 = R_2^2(\sigma_3 - \sigma_2) - R_1^2(\sigma_2 - \sigma_1), \\ \dfrac{a_0^2 E_1}{E_2}\sigma_1 = R_1^2(\sigma_2 - \sigma_1) - R_0^2(\sigma_1 + p_{\text{int}}) \end{cases} \tag{7}$$

The linear system of equations (6) has the unknowns ($\sigma_1, \sigma_2, ..., \sigma_{n-1}, p_{\text{int}}$) and can be solved by the consecutive elimination of unknowns beginning with the first equation. The linear system (7) has the unknowns ($\sigma_1, \sigma_2, ..., \sigma_n$).

2.2. *CONTINIOUS MODEL (CM)*

This model is applied for cases when the number of layers is large. Then the continuos equation analogous to equation (3) ($a_0 \to 0$) can be written as a differential equation in the following form:

$$\frac{d^2\sigma}{dr^2} + \frac{2}{r}\frac{d\sigma}{dr} - \frac{E_1}{E_2}\frac{\sigma}{r^2} = 0 \tag{8}$$

with the corresponding boundary conditions (4),(5):

$$\begin{cases} \sigma(r = R_n) = \sigma_n \\ \dfrac{d\sigma}{dr}(r = R_n) = \sigma'_n \end{cases} \tag{9}$$

$$\begin{cases} \sigma(r = R_n) = -p_{ext} \\ \sigma(r = R_0) = -p_{\text{int}} \end{cases} \tag{10}$$

where

$$\sigma'_n \approx \frac{\sigma_n - \sigma_{n-1}}{a_0}, \sigma_n = E_2 \frac{a_n - a_0}{a_0}, \sigma_{n-1} = (1 + \frac{R_n^2}{R_{n-1}^2} + \frac{E_1 a_0^2}{E_2 R_{n-1}^2})\sigma_n + p_{ext}\frac{R_n^2}{R_{n-1}^2}$$

The stress σ_{n-1} was obtained from the first equation (6).
The solution of the linear boundary problem (8), (9) or (10) determines the continuos distributions of the stresses $\sigma(r)$ and $\sigma_\tau(r) = r\dfrac{d\sigma}{dr}$ ($T_i = \sigma_\tau(R_i)a_0$).
Equation (8) is Euler's equation (Smirnov, 1958). By means of a variable substitution of $r = e^t$ equation (8) transforms to form:

$$\frac{d^2\sigma}{dt^2} + \frac{d\sigma}{dt} - \frac{E_1}{E_2}\sigma = 0$$

with the general solution

$$\sigma(t) = C_1 e^{\alpha t} + C_2 e^{\beta t}$$

where $\alpha = 0.5(-1 + \sqrt{1 + 4E_1E_2^{-1}})$, $\beta = -0.5(1 + \sqrt{1 + 4E_1E_2^{-1}})$ and C_1, C_2 are the arbitrary constants. Therefore equation (8) has the general solution:

$$\sigma(r) = C_1 r^{\alpha} + C_2 r^{\beta}, R_0 \le r \le R_n \qquad (11)$$

and the constants C_1, C_2 are determined from boundary conditions (9) or (10).

For a closed tube, the stresses T_z, σ_z in the tube axis direction can be estimated using the following equations:

$$T_z = \frac{-p_{ext} R_n^2 + p_{int} R_0^2}{(n+1)(R_n + R_0)},$$

$$\sigma_z = \frac{-p_{ext} R_n^2 + p_{int} R_0^2}{R_n^2 - R_0^2}$$

for DM and CM respectively.

It follows from equations (6)-(10) that the stress distribution in the tube wall depends on E_1 / E_2 ratio only.

3. Numerical Implementation

In this section, we consider the numerical implementation of the models with boundary conditions (4) and (9). System (6) was preliminarily reduced to the recurrent formula

$$\sigma_i = (1 + \frac{R_{i+1}^2}{R_i^2} + \frac{E_1 a_0^2}{E_2 R_i^2})\sigma_{i+1} - \frac{R_{i+1}^2}{R_i^2}\sigma_{i+2},$$

$$i = (n-1), (n-2), \ldots, 0, \quad \sigma_n = E_2 \varepsilon_{rn}, \sigma_{n+1} = -p_{ext}, \qquad (12)$$

$$\varepsilon_{rn} = \frac{a_n - a_0}{a_0}$$

and it was solved for the following initial data taken for graphite (Dresselhaus, 1996):

$$E_1 = 1060 GPa, E_2 = 36.5 GPa, R_0 = 30 nm,$$
$$n = 30, a_0 = 0.3355 nm, p_{ext} = 0, \varepsilon_{rn} = -1.3295 \cdot 10^{-5} \qquad (13)$$

We used elastic moduli of graphite, which are well known and are considered to be similar to that of nanotubes. Both slightly higher and lower values have been reported for nanotubes (see (Harris, 1999) for a review) from simulations and experimental measurements. They depend on nanotube chirality and diameter. However, since the methods of mechanical testing of nanotubes provide qualitative rather than quantitative information, more work will be required to determine the exact value of the Young's modulus for carbon nanotubes. The interplanar distance in (002) direction was taken as the theoretical value for graphite. We obtained values slightly below 0.34 nm for the hydrothermal nanotubes with the best wall ordering, but it can be above 0.34 nm for tubes with disordered walls. A typical tube radius and wall thickness for hydrothermal nanopipes (Fig. 2c) has been selected. Simultaneously, the corresponding boundary value problem (8), (9) was solved as well. The boundary value problems were considered for the dimensionless variable $\rho = r/R_n$. The solution (11) of CM has form

$$\sigma(\rho) = C_1\rho^{\alpha} + C_2\rho^{\beta}, \rho_0 \le \rho \le 1\,,\; \rho_0 = R_0/R_n \tag{14}$$

where $\alpha = 4.912, \beta = -5.912, C_1 = \dfrac{R_n\sigma_n' - \beta\sigma_n}{\alpha-\beta}, C_2 = \dfrac{\alpha\sigma_n - \sigma_n' R_n}{\alpha - \beta}$.

The constants C_1, C_2 are defined from the boundary conditions corresponding to (9):

$$\begin{cases} C_1 + C_2 = \sigma_1 \\ \alpha\, C_1 + \beta\, C_2 = \sigma_n' R_n \end{cases}$$

The results of the calculations are shown in Fig. 6 and Fig. 7. For the internal pressure p_{int}, we have: $p_{\text{int}} = -\sigma(\rho_0) = -C_1\rho_0^{\alpha} - C_2\rho_0^{\beta} = 30.13MPa$ for CM and $p_{\text{int}} = -\sigma_0 = 30MPa$ for DM. Since the models are linear, the results for any $\overset{*}{\varepsilon}_{rn}$ or $\overset{*}{p}_{\text{int}}$ value can be obtained from corresponding data shown in Fig. 6 and Fig. 7 by theirs multiplication by factor $\left(\dfrac{\overset{*}{\varepsilon}_{rn}}{\varepsilon_{rn}}\right)$ or $\left(\dfrac{\overset{*}{p}_{\text{int}}}{p_{\text{int}}}\right)$. For this reason, as appears from (12), (13) and (14)

$$p_{\text{int}} = k(-\varepsilon_{rn}) \tag{15}$$

for an arbitrary ε_{rn} (since $p_{ext} = 0$), $k = 2257GPa$ for DM and $k = 2267GPa$ for CM. The factor *k* is an internal pressure for $\varepsilon_{rn} = -1\cdot 10^{-5}$. The graphs in Fig. 6 and Fig. 7 for ε_{ri} and ε_r can be used to calculate the change in interplanar spacing a_0 at any position across the tube wall for a given internal pressure. We attempted such measurements in high-resolution TEM for tubes having an estimated internal pressure of 30 *MPa* and placed in a high vacuum (10^{-8} *torr*) of a TEM column (Libera, 2001). However, as expected from the model, the changes in the interplanar spacing were far below the measurement accuracy.

The plot of k vs. n is presented in Fig. 8. It can be used to calculate by equation (15) the internal pressure in the tube with $R_0 = 30nm$ and wall thickness $t = na_0\ (3 \le n \le 30)$ for an arbitrary deformation ε_{rn}. If the tube is closed then

$$T_z = 1.243\cdot 10^{-11} GPa\cdot m\ \left(\sigma_z = T_z / a_0 = 3.705\cdot 10^{-2} GPa\right),$$

$$\sigma_z = 3.845\cdot 10^{-2} GPa\,.$$

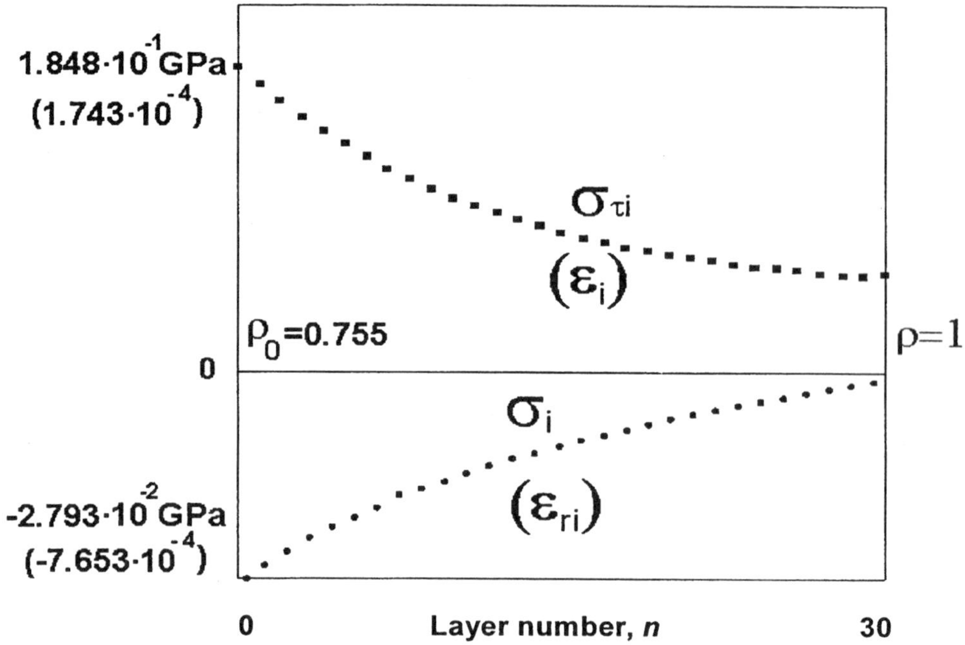

Figure 6. The dependencies of stresses σ_i , $\sigma_{\tau i} = T_i / a_0$ and deformations $\varepsilon_{ri} = \sigma_i / E_2$, on $\rho\,(\rho_0 \leq \rho \leq 1)$ for DM and $p_{\text{int}} = 3.0 \cdot 10^{-2} GPa$. ρ_0=0.755 corresponds to the inner tube wall and ρ=1 corresponds to the outermost graphene layer. Dots show stress values $\sigma_{\tau i}$ in each layer (i) and stress values σ_i between the layers (i-1) – (i) of the tube wall but also the corresponding deformations ε_i and ε_{ri}.

calculations can be easily done for any set of values (13) available from direct experimental measurements.

The calculations showed that the results of DM and CM practically agree. The distinction between the DM and CM is essential if the interplanar distance a_0 is commensurable with the radii of curvature of the layers. This would be the case for small diameter nanotubes. However, since tubes with the internal diameter smaller than 4 nm could not be filled with liquids (Ugarte, 1996), nanotubes of that size are of limited interest for the fluidic applications considered here. However, thin tubes may be important for gas storage (Gadd, 1997) or making nanowires by metal encapsulation (Harris, 1999) and the DM model should be used to analyze them.

We considered the case of a defect-free nanotube. Dislocations, disclinations, stacking faults, incommensurate layers and other defects have been observed in real nanotubes. All these defects are expected to decrease the moduli and strength of

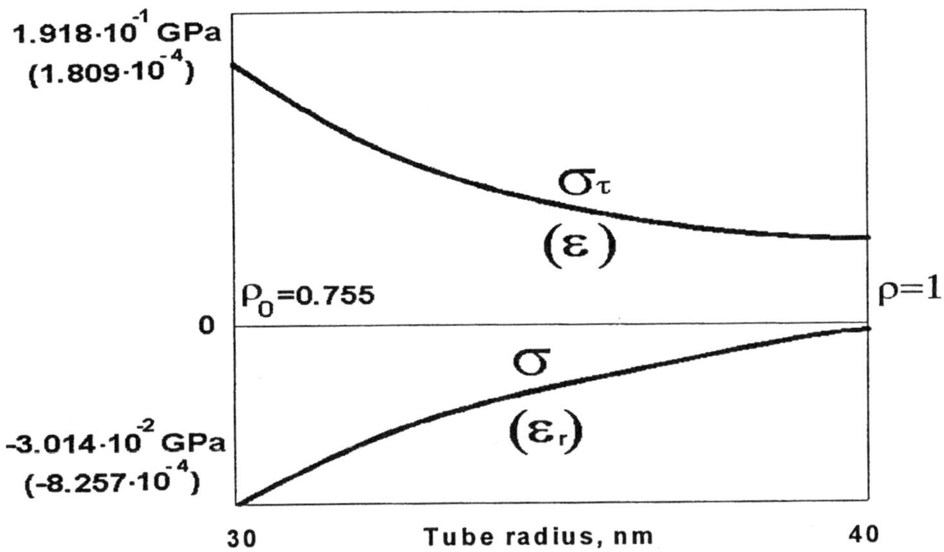

Figure 7. The dependencies of stresses σ , $\sigma_\tau = \rho \dfrac{d\sigma}{d\rho} = \rho(\alpha C_1 \rho^{\alpha-1} + \beta C_2 \rho^{\beta-1})$ and deformations $\varepsilon_r = \sigma / E_2, \varepsilon = \sigma_\tau / E_1$ on ρ ($\rho_0 \le \rho \le 1$) for CM and $p_{\text{int}} = 3.013 \cdot 10^{-2} GPa \quad (T_i = \sigma_\tau(\rho_i) a_0, \rho_i = R_i / R_n)$. ρ_0=0.755 corresponds to the inner tube wall and ρ=1 corresponds to the outermost graphene layer.

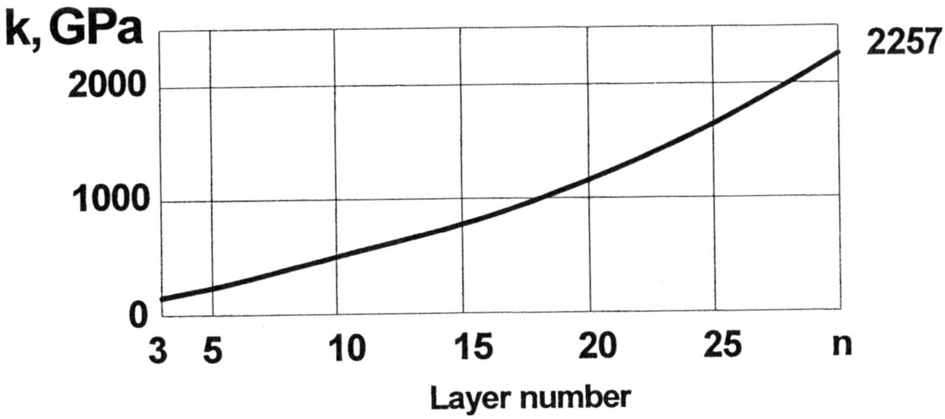

Figure 8. Plot of dependence of k vs. the layer number n (see formula (15)). Other initial parameters have the same values as in equation (13): $E_1 = 1060GPa, E_2 = 36.5GPa, R_0 = 30nm, a_0 = 0.3355nm, p_{ext} = 0$.

nanotubes. However, since even for very thick graphite whiskers (Bacon, 1960) the Young's modulus of 800 GPa was measured in the longitudinal direction, we expect only minor changes in the values calculated using our models. However, the calculations can be easily done for any set of values (13) available from direct experimental measurements.

Finally, if the tube experiences a high and irreversible deformation, a nonlinear model must be used.

4. Conclusions

- Two models of stress-strain state of long multiwall carbon nanotubes subjected to internal pressure have been described.
- In the case of several carbon layers, a discrete model (DM) is proposed. The solutions of DM equations with corresponding boundary conditions determine the stresses between the layers, the forces in the layers and the deformation of the layers.
- For the case of thick walls built of numerous carbon layers, a continuous model (CM) is proposed. The main CM equation is the Euler's differential equation with corresponding boundary conditions. Its solution defines the continuous distribution of the stresses and strains across the wall thickness of the tube.
- For tubes of large radius (about 30 nm) and having numerous carbon layers, both models provide similar results.

5. Acknowledgements

The work done at Drexel University was supported by NSF grant CTS-0196006. B.A.G. was supported by an international supplement to NSF grant DMR-9874955. Y.G. is thankful to his former student Mr. J.A. Libera for TEM micrographs.

References

1. Bacon, R. 1960. Growth, structure and properties of graphite whiskers. *J. Appl. Phys.* **31**, 283-290.
2. Burns, M. A., B. N. Johnson, S. N. Brahmasandra, K. Handique, J. R. Webster, M. Krishnan, T. S.Sammarco, P. M. Man, D. Jones, D. Heldsinger, C. H. Mastrangelo & D. T. Burke. 1998. An integrated nanoliter DNA analysis device. *Science* **282**, 484-487.
3. Delamarche, E., A. Bernard, H. Schmid, B. Michel & H. Biebuyck. 1997. Patterned delivery of immunoglobulins to surfaces using microfluidic networks. *Science* **276**, 779-781.
4. Dresselhaus, M. S., G. Dresselhaus & P. C. Eklund. 1996. *Science of fullerenes and carbon nanotubes* (Academic Press).
5. Ebbesen, T. W. 1997. *Carbon nanotubes: Preparation and properties* (CRC Press, Boca Raton).
6. Gadd, G. E., M. Blackford, D. Moricca, N. Webb, P. J. Evans, A. M. Smith, G. Jacobsen, S. Leung, A. Day & Q. Hua. 1997. The world's smallest gas cylinders. *Science* **277**, 933-936.
7. Gere, J. M. & S. P. Timoshenko. 1997. *Mechanics of materials* (PWS Publishing, Boston).
8. Gogotsi, Y., J. A. Libera, A. F. Koster van Groos & M. Yoshimura. 2001a. Hydrothermal Synthesis of Carbon Nanotubes, in *Joint sixth international symposium on hydrothermal reactions (ISHR) & fourth international conference on solvo-thermal reactions (ICSTR)* (Kochi, Japan) in press.
9. Gogotsi, Y., J. A. Libera, A. Güvenç-Yazicioglu & C. M. Megaridis. 2001b. *In-situ* multiphase fluid experiments in hydrothermal carbon nanotubes. *Appl. Phys. Lett.* **79**, 1021-1023.
10. Gogotsi, Y., J. A. Libera & M. Yoshimura. 2001c. Formation of Nanostructured Carbons under Hydrothermal Conditions, in *Applications of fullerenes* (eds. Osawa, E. & Nakazawa, T.) (Kluwer, Dordrecht).
11. Harris, P. J. F. 1999. *Carbon nanotubes and related structures* (Cambridge University Press, Cambridge).
12. Iijima, S. 1994. Carbon nanotubes. *MRS Bull.* **19**, 43-49.
13. Jo, B.-H., V. Lerberghe, L. M. Motsegood, M. Kathleen & D. J. Beebe. 2000. Three-dimensional micro-channels fabrication in polydimethylsiloxane (PDMS) elastomer. *J. Microelectromechanical Systems* **9**, 76-81.
14. Libera, J. A. & Y. Gogotsi. 2001. Hydrothermal synthesis of graphite tubes using Ni catalyst. *Carbon* **39**, 1307-1318.
15. Ruoff, R. S. & D. C. Lorents. 1995. Mechanical and thermal properties of carbon nanotubes. *Carbon* **33**, 925-930.
16. Schulz, W. 2000. Crafting a national nanotechnology effort. *C&EN* **2000**, 39-42.
17. Smirnov, V. I. 1958. *A course of higher mathematics* (Fizmatgiz, Moscow).
18. Ugarte, D., A. Chatelain & W. A. DeHeer. 1996. Nanocapillarity and chemistry in carbon nanotubes. *Science* **274**, 1897-1899.
19. Ugural, A. C. 1999. *Stresses in plates and shells* (McGraw-Hill, Boston).

20. Wong, E. W., P. E. Sheehan & C. M. Lieber. 1997. Nanobeam mechanics: Elasticity, strength, and toughness of nanorods and nanotubes. *Science* **277**, 1971-1974.

21. Wuu, D.-S., R.-H. Horng, C.-C. Chan & Y.-S. Lee. 1999. Plasma-deposited amorphous silicon carbide films for micromachined fluidic channels. *Appl. Surf. Sci.* **144-145**, 708-712.

22. Yakobson, B. I., G. Samsonidze & G. G. Samsonidze. 2000. Atomistic theory of mechanical relaxation in fullerene nanotubes. *Carbon* **38**, 1675-1680.

RECOGNITION TEMPLATES FOR BIOMATERIALS WITH ENGINEERED BIOREACTIVITY

BUDDY D. RATNER, HUAIQIU SHI, JANET WANG, SHENG PAN
University of Washington Engineered Biomaterials (UWEB)
University of Washington
Department of Bioengineering, Box 351720
Seattle, Washington 98195 USA

1. Introduction

Biology ubiquitously uses molecular recognition and specificity to do its work. In contrast, the biomaterials widely used in medicine today non-specifically adsorb complex protein mixtures to their surfaces. These ill-defined protein layers are "read" by the cells (neutrophils, macrophages) that interrogate the implanted materials – the protein coat may be the trigger of subsequent biological reactions. Given the heterogeneity in composition (there are 200+ proteins in blood) and organization (proteins are adsorbed with no control of orientation or conformation) of adsorbed protein layers, it is no surprise that biological reactions, particularly *in vivo*, are poorly controlled. If fact, implanted synthetic materials are almost universally seen by the body as foreign objects and walled off into a tough, avascular, collagenous bag[1].

Medical implants are widely used clinically and save lives and improve the quality of life for millions (Table 1). Medical implants are primarily fabricated from synthetic materials (polymers, metals, ceramics, glasses). These implants are walled off from the body within 3 weeks of implantation as described above. This implant encapsulation can impede the healing and performance of many devices. Capsule formation around implants may be a manifestation of the lack of recognition by the body of the non-specifically adsorbed proteins coating the implant[1,2].

In contrast to synthetic materials, biological tissue and normal wounds heal by a different process. Normal wound healing and integration are characterized by the absence of a collagenous capsule, new blood vessel formation and reconstruction of damaged tissues. Obviously, this normal reconstruction should be preferable to the aberrant, capsular healing noted with modern, "biocompatible" biomaterials. Is it possible to convert synthetic biomaterials to materials that will heal in the body by controlling the non-specific protein adsorption and thus use the surfaces of materials to deliver correct healing signals to the body? The approach of increasing the recognition and specificity of biomaterial surfaces is being pursued in our research group[2]. With appropriate recognition structures on biomaterials, we may be able turn on normal healing rather having the biomaterial targeted as foreign.

E. Buzaneva and P. Scharff (eds.), Frontiers of Multifunctional Nanosystems, 75–78.

Table 1. *Numbers of Medical Devices/year Worldwide*

Device	Number/year
Intraocular lens	7,000,000
contact lenses	75,000,000
vascular grafts	400,000
hip and knee prostheses	1,000,000
Catheters	300,000,000
heart valves	200,000
stents (cardiovascular)	>2,000,000
breast implants	300,000
dental implants	500,000
Pacemakers	200,000
renal dialyzers	25,000,000
left ventricular assist devices	100,000

Molecular recognition, a process that functions at nano size scales, implies both chemical and shape complementarity to provide "lock and key" specificity. Our work focuses on new types of planar recognition surfaces for protein and biomolecule recognition. Flat surfaces permit molecular template/imprint ideas to be readily incorporated in diagnostics, biosensors and biomaterials. Two ideas will be discussed in this paper: an imprinting approach and a recognition surface made by randomly distributing organic functional groups that contribute to the recognition site.

2. Molecular imprinting to create protein recognition cavities

The first idea uses molecular-scale imprinting to create recognition sites for proteins[3,4]. Imprinting has been found to be an intriguing approach to preparing artificial recognition sites for low molecular weight compounds[5]. However, only limited success has been achieved with proteins because of the labile and macromolecular nature of proteins. In these experiments, tailored protein-binding cavities have been created by an imprinting approach based on RF-plasma deposition of organic thin films (Figure 1). The structure of this nanopit surface has been confirmed by a range of methods including transmission electron microscopy, atomic force microscopy (AFM), electron spectroscopy for chemical analysis (ESCA), time of flight secondary ion mass spectrometry (Tof-SIMS), contact angle methods and light microscopy. All evidence, consistent with Figure 1, suggests a polysaccharide-like surface with imprinted "nanocavities" complementary to the shape and chemistry of protein molecules. This surface allows only the template protein molecules to fill the cavities and to bind strongly. The bound protein in its "pit" is prevented from exchange with dissolved protein because of strong non-covalent interactions and steric hindrance. The non-template protein that weakly adsorbs on the surface is displaceable. Binary protein mixture competitive adsorption experiments show a highly selective adsorption of template proteins to their imprints. The recognition specificity is demonstrated by the fact that lysozyme imprints have a 26 fold increase in adsorption of lysozyme (MW

14,600) versus ribonuclease A (MW 13,700). Recent experiments have shown that recognition surfaces for a large protein (fibronectin) can also be created by this method[6]. The fibronectin protein bound into the recognition cavities retains biological activity as suggested by cell attachment studies.

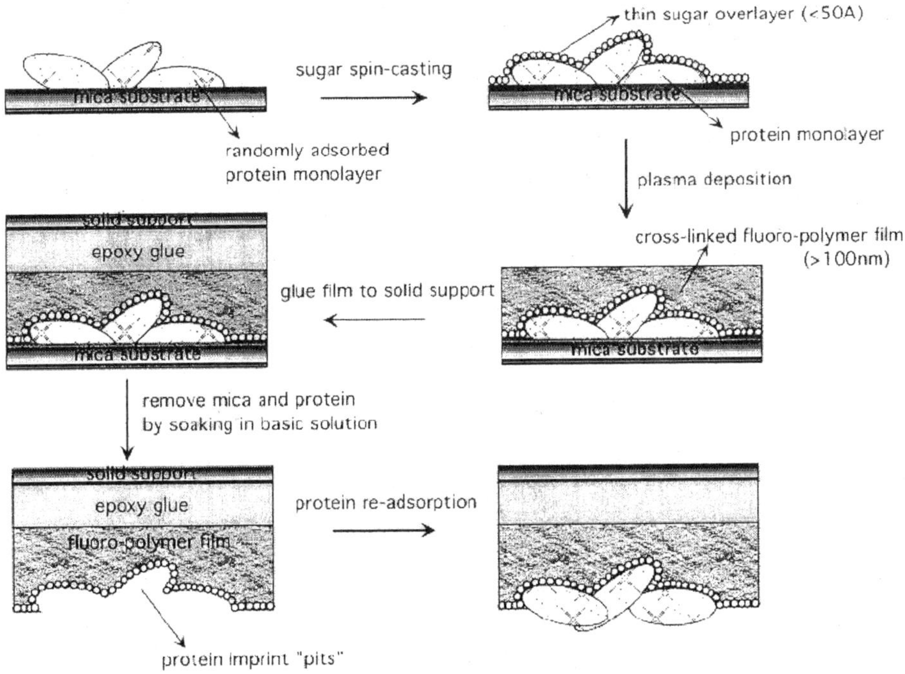

Figure 1. A schematic representation of the process used to create protein imprints.

3. Randomized arrays of molecules to mimic recognition motifs

The second concept takes advantage of random mixtures of complex chemical elements to stochastically form recognition sites[7]. Receptor sites can be envisioned as an array of chemical groups immobilized in a specific geometry. Evidence from these experiments suggests that such sites can be synthesized by randomly "scattering" amino acid units on a surface. In this experiment, amino acid terminated n-alkane thiols that were self assembled on gold were shown to generate cell-recognizable receptor sites[8,9]. The head groups were arginine (R), glycine (G) and aspartic acid (D), and an RDG tripeptide. These surfaces were characterized by ESCA, Tof-SIMS, AFM and infrared reflection-absorption spectroscopy (IRAS). The randomized R-G-D surfaces showed endothelial cell attachment as good or better than the RGD peptide. The RGD-like activity of the surface could be blocked by exposing the endothelial cells to soluble RGD peptide thus saturating surface receptors.

Other approaches exist to create recognition surfaces including immobilization of intact protein molecules. The approaches described here provide new options. In particular, imprint surfaces may provide enhanced affinity for specific proteins important for the healing of medical devices without using expensive, fragile proteins.

4. Acknowledgements

This work was supported by University of Washington Engineered Biomaterials (UWEB, an NSF Engineering Research Center, NSF EEC-9529161) and NESAC/BIO (an NIH National Resource for surface analysis, NIH RR-01296).

References

1. Ratner,BD (1993): New ideas in biomaterials science -- a path to engineered biomaterials. J. Biomed. Mater. Res. 27, 837-850.
2. Ratner,BD (1997): The engineering of biomaterials exhibiting recognition and specificity. J. Mol. Rec. 9, 617-625.
3. Shi,H; Tsai,W-B; Ferrari,S; Ratner,BD (1999): Template imprinted nanostructural surfaces for protein recognition. Nature 398, 593-597.
4. Shi,H; Ratner,BD (1999): Template recognition of protein imprinted polymer surfaces. J. Biomed. Mater. Res. 50, 1-11
5. Shi,H; Ratner,BD (1999): Recognition templates for biomaterials with engineered bioreactivity, Current Opinions in Solid State & Materials Science 4, 395-402
6. Wang, J. (2001) Protein template-imprinting to enhance specific protein adsorption and cell adhesion. Master's Thesis, University of Washington, Seattle, Washington
7. Jozefowicz,M; Jozefonvicz,J (1997): Randomness and biospecificity: random copolymers are capable of biospecific molecular recognition in living systems. Biomaterials 18(24), 1633-1644.
8. Pan, S., (1999) Molecular Engineering and Characterization of Self-Assembled Biorecognition Surfaces" Ph.D. Thesis, University of Washington, Seattle, Washington
9. Pan, S., Castner, D and Ratner, BD, J. Biomed. Mater. Res., submitted

LAYER-BY-LAYER METHOD FOR IMMOBILIZATION OF PROTEIN MOLECULES ON BIOCHIP SURFACE

G.K. ZHAVNERKO[a],

[a]*Institute of Chemistry of New Materials, National Academy of Sciences, Minsk 220141, Belarus*

SUN-JU YI[b], SOO-MI KWEON[b], and KWON-SOO HA[b]

[b]*Biomolecule Research Team, Korea Basic Science Institute, Yeoeun-dong 52, Yusung-Ku, Taejon 305-333, South Korea*

Abstract

The article presents a Layer-by-Layer method to produce new biochip surfaces on gold or mica for protein binding. The chip surfaces were modified by two polyelectrolytes, poly(sodium 4-styrenesulfonate) and poly(diallyldimethylammonium chloride), which produce negatively and positively charged surfaces, respectively. Then, the modified surfaces were used for the binding of several proteins, such as bovine serum albumine, apo-transferrine, tissue transglutaminase and SAG (sensitive to apoptosis gene) protein. The proteins were self-assembled from aqueous solution on the modified surfaces and adsorption process has been examined by Surface Plasmon Resonance and Atomic Force Microscopy. Our results suggested that negatively charged surface was preferable for the binding of four proteins, even though two surfaces could be used.

Keywords: *self-assembled monolayers, protein film, AFM, SPR.*

E. Buzaneva and P. Scharff (eds.), Frontiers of Multifunctional Nanosystems, 79–90.

1. Introduction

Adsorption of protein molecules on solid surfaces plays a key role in surface science, in particular, for structural researches of biomolecules by Scanning Probe Microscopy [1-5]. Immobilization of protein molecules on solid surfaces also extremely important because they are used as receptors in sensor devices for medical diagnosis [6-7] and environment sensing [8-10].

Surface engineering for protein bindings is a quickly growing research area. The surface has been modified by various methods, including Langmuir-Blodgett transfer [11-14], covalent binding [15-18], and spontaneous adsorption from solution [19-21]. One of promising methods for spontaneous organization of polyelectrolyte molecules on solid surfaces is the self-assembling method by electrostatic Layer-by-Layer (LbL) adsorption of oppositely charged polymers [22-26]. LbL method also gives a chance to produce monomolecular protein layers on solid surface [27-34]. Investigation of protein binding to structurally well-defined organic surfaces is important for optimization of recipient layer properties, e.g., optimization of protein orientation, its stability, activity and layer capacity.

Doubtless, there are many bifunctional reagents for surface modification for protein bindings, including antibody-antigen pairs [35-36], biotin-avidin [37-40], or N-hydroxysuccinimide group-lysine residues of protein [41-42] as coupling units between different molecular layers. However, as a rule, these reagents are expensive, and also require the special storage conditions and labor-consuming techniques of a surface treatment. Thus, we have chosen a combined method of LbL and LB to reduce the price and to simplify the process of proteins immobilization on solid surface. The LbL method allows to change "easily" the properties of an inorganic surface. The method also creates "friendly" environment (hydrocarbon, charged, etc.) as well as forms non-specific binding monolayers of proteins. The LB method can supply an opportunity to work with protein at molecular level on a aqueous surface, to change the protein conformation (and, as a result, to open the reactionary centers of molecules) and to vary the concentration of protein at an experimenter request. Thus, combining two methods, we shall have an opportunity to design complex biological systems in aqueous environment.

In present article we have undertaken the attempt to find a simple way of qualitative protein monolayer fabrication on a solid substrate. The aim of this work is to demonstrate in comparison the effects of protein-immobilizing surfaces on the quality of monolayer films formed from different protein molecules. The processes of gold surface modification have been examined using Surface Plasmon Resonance (SPR) and Atomic Force Microscopy (AFM) methods. Gold surface was modified with thiol-containing molecules firstly. Subsequent polymeric sub-layers have been prepared by alternative adsorption of polyelectrolytes. Layer-by-Layer method was used to prepare polymeric sub-layer and to make protein monolayers on biochip surface. Monomolecular films from different proteins at modified gold surface have been self-assembled from aqueous solutions. The influence of inorganic Co^{2+} ions on protein monolayer morphology has been also studied.

2. Experimental

2.1. *MATERIALS*

Positively charged poly(diallyldimethylammonium chloride) (PDDA), M_w~100000, and poly(sodium 4-styrenesulfonate) (PSS), M_w~70000 were purchased from Aldrich. PSS was dialyzed against Milli-Q water for two days and then lyophilized before use. Bovine serum albumine (BSA), apo-transferrine (TF), and tissue transglutaminase (tTG) were purchased from Sigma. SAG (Sensitive to apoptosis gene) protein was prepared from *E. coli* expressing human SAG. Thioacids, 3-mercaptopropanoic acid, 11-mercaptoundecanoic acid (11-UDA) and 16-mercaptohexadecanoic acid were purchased from Aldrich. L-Cysteine ethyl ester hydrochloride ($HSCH_2CH(NH_2)CO_2C_2H_5{\cdot}HCl$ (CstEst) m_p~123-124° C was synthesized in accordance with [43] from L-Cysteine (Aldrich).

2.2. *SUBSTRATE PREPARATION AND MODIFICATION*

Gold-coated (450Å) slide glasses were prepared by sputtering gold at a rate of 140Å/min on glass plate. Thin (30Å) titanium sub-layer was used to improve gold film adhesion to glass. Then, the slide glass was cleaned by exposure to H_2O:NH_4OH:H_2O_2 mixture (7:4:1, v:v:v) at 50-60° C during 5-10 minutes. Muscovite mica for AFM observation was cleaved with adhesive tape immediately prior to use.

Gold surfaces on slide glass or fresh cleaned mica were modified by alternative adsorption of 1 mg/ml PDDA or PSS solutions. The treated surfaces were rinsed with pure water 3-5 times between each deposition and finally dried with nitrogen gas. The outermost layer becomes «negatively» or «positively» charged after last treatment with PSS and PDDA or Co^{2+} ions, respectively.

Protein monolayers were formed on the modified gold or mica surfaces by incubation with various proteins in Phosphate buffer contained 25 mM potassium dihydrogen phosphate, pH 7.0, containing 60 mM disodium hydrogen phosphate.

The plates with gold spots were held for 12–16 h in 1 mM alcoholic solution of the thiol compounds. After extensive rinsing with alcohol to remove unbound thiols, the surfaces were blown dry with N_2.

2.3. *INSTRUMENTATION*

Surface modification with polyelectrolytes or protein binding on the modified surfaces has been measured by a SPR system (self-assembled). Monochromatic light from a 633nm HeNe laser was focused on a gold-coated slide glass and the totally reflected light was analyzed by a linear CCD array. Index matching between a glass right angle prism and the slide glass was achieved by a microscope immersion oil. The resulting data were processed by a fitting program (self-developed).

AFM images were obtained with a Nanoscope IIIa (Digital Instruments, USA. The device was equipped with a <D> calibrated using the manufacturer's grating. AFM images were obtained by both contact mode (CM) and tapping mode (TM). 100 mm and 200 mm nanoprobe cantilevers (spring constants of 0.12 and 0.38 N/m) with oxide-sharpened Si_3N_4 integral tips were used for CM regime and tapping silicon cantilevers with resonance frequency 280-320 kHz were used for TM regime. The applied force was varied over a wide range from several nN up to hundreds of nN in contact mode. Film thickness was estimated by measuring a depth of an artificial hole that was scratched preliminary in CM regime. Film roughness was measured on images obtained in less destructive TM regime. Contact angles were measured in air on static drops on surface modified. Measurements were repeated on at least four drops.

3. Results and discussion

3.1. *ADSORPTION OF PROTEIN ON MONOLAYER FILM FROM BIFUNCTIONAL REAGENTS*

Among various bifunctional reagents that is suitable for gold surface modification we have chosen a set of thioacids and cysteine ester derivative for surface functionality with carboxylic and amino groups, consequently. Fig. 1 shows the components of multilayer biochip for protein binding on gold surface.

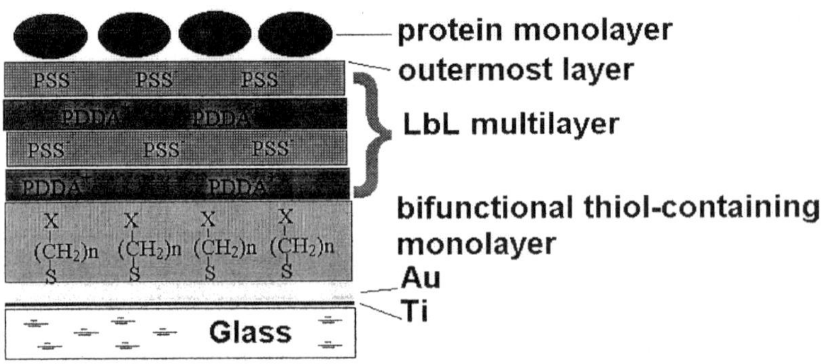

Figure 1. Schematic showing of multilayer system constructed on glass plate.

First of all, we used the hydrophobization procedure for surface activation before treatment with sulfur-containing reagents. It has allowed us to store plated without special care before use. The SPR signal shift to the higher angle was observed by the subsequent treatment of gold surface with 1 mM thioacids or CystEst derivative for 12 hours. 11-UDA provided most reproducible results among thioacids studied. Thus, 11-UDA was used to modify gold surface in further experiments.

Any or little shift was recorded at subsequent adsorption of protein's investigated on gold surface activated with 11-UDA. This was also rightly for TG and BSA molecules on the covering from cysteine ester derivative. AFM observations have shown presence of protein aggregates on separate places of such surface only (Figure. 2).

On the contrary, it was established that two proteins, namely TF and SAG, formed an extended monomolecular film with excellent quality on gold surface modified with cysteine ester derivative (see Figure. 3 as an example).

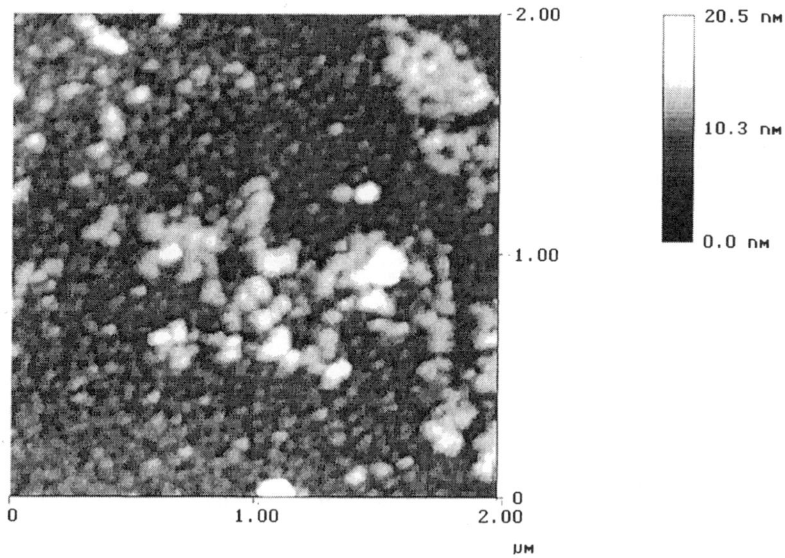

Figure 2. AFM image of partial adsorption of Transglutaminase molecules on Au/CstEst surface.

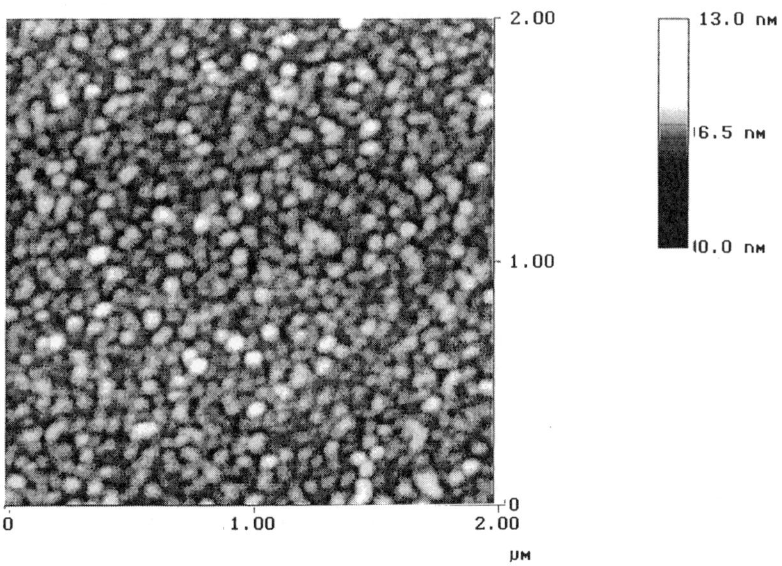

Figure 3. AFM images (tapping mode) of monolayer covering from TF on gold surface modified with CstEst derivative (0.2 mg/ml, KH_2PO_4/Na_2HPO_4 buffer).

3.2. *SURFACE MODIFICATION WITH POLYELECTROLYTE LAYERS*

Polyelectrolyte films on modified surface of gold or mica were prepared by LbL method. First, Au/UDA surface was treated with 1 mM NaOH (in water) to provide negative charged surface. Then, the surfaces were modified by consecutive adsorption

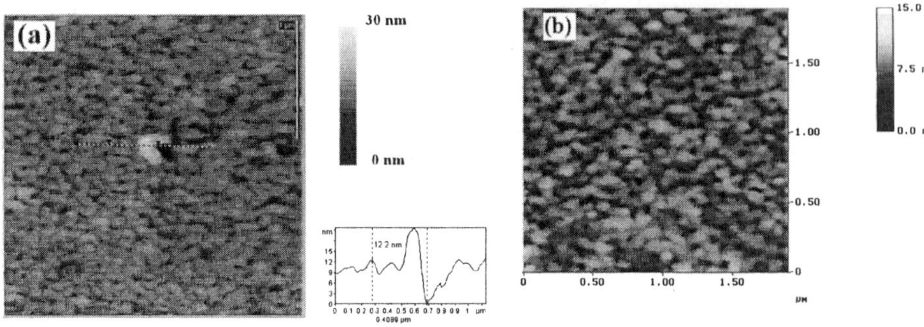

Figure 4. (a) High resolved AFM image and (b) film morphology and section analysis of 3 (PDDA/PSS) layers adsorbed on Au/UDA surface.

of opposite charged polyelectrolytes (PDDA and PSS). The steady growth was observed by SPR system. It was desirable to use saline and aqueous solutions for PDDA and PSS, respectively, to form a maximum homogeneous and smooth film surface (Figuure. 4).

The deposition of PSS monolayers from salt solution, which is widely used in practice, results in a rough non-uniform film. The feature of our PDDA/PSS multilayer system showed a nonlinear shift of SPR signal depending on the number of deposited layers (Figure. 5).

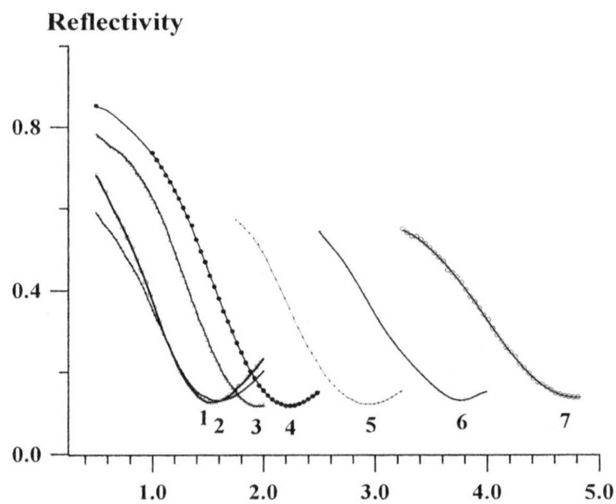

Figure 5. SPR signal shift depending on the number of polyelectrolyte layers: 2, 4, 6 - PDDA, 3, 5, 7 – PSS monolayers on 1 - clean gold surface.

The explanation of this effect is not a subject of given article and will be published further. Emphasize only, that such nonlinear response of SPR system on growing LbL film was not due to the aggregated structure of multilayer film. Indeed, thickness of a monolayer was about 2 nm per monolayer and the film roughness was practically constant (~ 0.5 nm) and did not depend of layer's number. For example, total film thickness was about 12 nm for six polyelectrolyte monolayers (3 (PDDA/PSS) layers), in accordance with AFM data of section analysis via scratched artificial defect in the film (Fig. 4a). The polymeric molecules on a surface had the characteristic toroid form (Fig. 4b), and the difference of heights between a wall and a hole in torus ring was the same 2 nm. It allowed to correlate the observable ring structures with polymeric molecules. Polyelectrolyte film with the same quality can be formed on a mica surface too. The growth of polyelectrolyte multimolecular film on mica was started from adsorption of PDDA molecules because a mica is negatively charged [44] in water and we observed formation practically quite similar film and its morphology was identical to represented on Fig. 4a.

TABLE 1: *Comparison properties of protein used and monolayers form them on modified surface*

Protein		**BSA**	**TG**	**TF**	**SAG**
Molecular Mass, kDa		67.5	80,0	80,0	16,6
pI		4.9	5.17	5.3-5.6	5.13
Roughness of modified surface (nm)	Mica/Co^{2+}	0.3±0.1	0.46±0.2	1.6±0.1	0.9±0.2
	Mica/$(PDDA)_2/(PSS)_2/Co^{2+}$	1.4±0.1	2.8±0.2	2.1±0.1	2.7±0.3
	Mica/PDDA/PSS/PDDA	1.1±0.2	2.7±0.4	0.7±0.2	7.6±2.2
	Mica/$(PDDA)_2/(PSS)_2$	1.8±0.5	2.6±0.5	1.65±0.2	1.77±0.2
Thickness of modified surface (nm)	Mica/Co^{2+}	1.0±0.1	2.8±0.2	2.6±0.2	5.1±0.2
	Mica/$(PDDA)_2/(PSS)_2/Co^{2+}$	8.5±0.4	10±0.2	7.0±0.2	9.1±0.4
	Mica/PDDA/PSS/PDDA	6.0±1.0	6.4±1.2	6.3±1.2	9.0±0.4
	Mica/$(PDDA)_2/(PSS)_2$	7.1±0.3	10.8±0.6	9.1±0.3	8.8±0.3

Positive and negative charged surfaces were produced from 3 monolayers (PDDA/PSS/PDDA) and 4 monolayers (2(PDDA/PSS)), respectively. These two variants of surface modification with polyelectrolyte were chosen for two reasons: (i) for elimination of possible film defects; (ii) for the optimum response of SPR system on protein molecules adsorption on polyelectrolyte surface (because of large SPR shift at more thick LbL film).

3.3. *ADSORPTION OF PROTEINS ON A SURFACE MODIFIED WITH POLYELECTROLYES*

Adsorption of all proteins investigated was carried out simultaneously on the identically prepared surfaces in an identical conditions. Table 1 shows the parameters of a roughness of a surface with adsorbed protein molecules (measured in TM regime) and the total "LbL+protein" film thickness measured on depth of artificial defect (measured in CM regime). The measurements of roughness were made on films formed on mica to exclude the contribution to a roughness the granular structure of gold surface. The morphology of a surface together with film roughness in 1-3 nm exceeded similar value of LbL layer in 2-3 times, testifying about protein molecules adsorption on a surface. In particular, globular structure, that is given on Fig. 6, is typical enough and was received for each of proteins investigated after it adsorption on negatively charged PSS outermost layer.

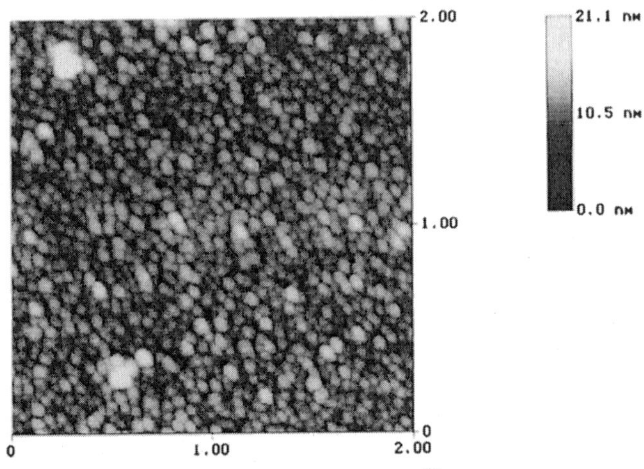

Figure 6. AFM image (tapping mode) of TF monolayer (from 1 mg/ml PBS solution) on mica/PDDA/PSS/PDDA/PSS surface.

The coverings, formed on such surface, represented extremely uniform and extended films. This result was unexpected enough because in the conditions of our experiment (phosphate buffered saline solution) all proteins, having pI in an interval ~5-6 (Table 1), were charged negatively and, hence, should not cooperate with the surface bearing the same charge. Nevertheless, besides ours, the similar observations were made by F.Caruso et. all at studying of IgG adsorption on PSS [37] and by J.Buijs et. all for IgG on negatively charged polystyrene latex [45]. On the contrary, on positively charged PDDA outermost layer the formation of high-quality monomolecular film was found for BSA and TG molecules only.

On the contrary, the film morphology and significant degree of surface roughness for SAG protein (Table 1) suggested the possible aggregation of the protein on PDDA sub-layer. As opposed to that, TF molecules formed a smooth film (Fig. 7) with an insignificant degree of a roughness of a surface (Table 1), that, obviously, was stipula-

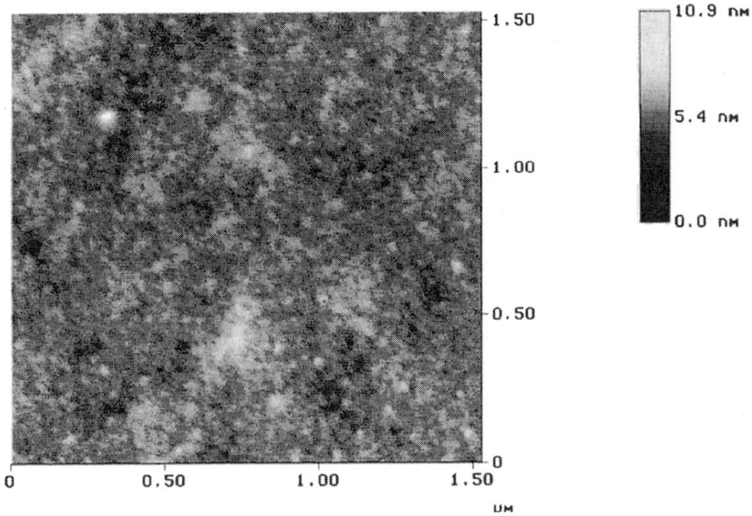

Figure 7. AFM images (tapping mode) of TF molecules (from 1 mg/ml PBS solution) on mica/PDDA/PSS/PDDA surface

ted denaturation of this protein on PDDA sub-layer. We believe that not only presence of a positive charge on a surface was resulted in TF denaturation on the PDDA surface. Really, it was possible to form an uniform monomolecular film with globular structure from TF molecules (Fig. 8) as well as from BSA ones at recharge of negatively charged 2(PDDA/PSS) surfaces with cobalt ions. Our measurement of contact angle on polyelectrolyte film gave the value of θ_{PDDA}~60°, while θ_{PSS}~45° in conditions of our experiments and even decreases with increase of number of layers in multilayer system.

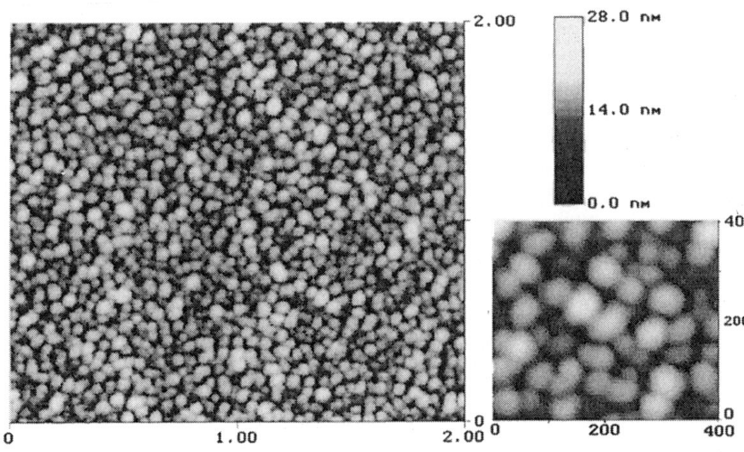

Figure 8. High-resolved AFM images (tapping mode) of Transferrin monolayer on mica/PDDA/PSS/PDDA/PSS /Co^{2+} surface.

Hence, PSS provided more hydrophilic surface than PDDA. Possible explanation of TF and SAG molecules behaviour on PDDA surface is next. Large globular TF molecules are more surface active than smaller SAG proteins, since TF was more denaturable, has a larger surface area and more surface binding domains. Therefore, TF protein adsorption may be increased by the hydrophobicity elevation of the surface and rearrangements of TF molecules at PDDA surfaces. At the same time, that protein is stable on more hydrophilic surfaces.

We have also tried, for comparison, to form protein's films on a surface of mica after it recharge with cobalt ions. In this case, as it was seen from surface morphology and films roughness, a monolayer from TF molecules was only stable, while other proteins were denatured, apparently, on such kind of an inorganic surface (see roughness values in Table 1).

Thus, it is not ascertain yet a common rule for monolayer covering's formation from globular proteins on a surface. It is obvious only, that negative PSS surface is more preferable in formation of monomolecular films from protein molecules than positive surface. Indeed, physical adsorption processes involve not only hydrophobic or electrostatic interactions between the protein and the solid phase, but the interaction of protein with a surface is also result of various types of interactions due to a combination of hydrophobic, dipole-dipole, and ion-dipole interactions. Hence, as it follows from experimental data discussed, we hardly can consider as only possible the variant of monolayer stabilization on PSS surface in result of hydrophobic interactions between protein and surface, as for TF molecules on PDDA surface. Probably, the interactions like ion-dipole are dominant in the case of interaction of proteins with PSS sub-layer, take into account the large number of amide groups in the structure of proteins.

We also believe, that the polymeric surface constructed from polyelectrolytes is also preferable for adsorption of biological objects in comparison with inorganic surface (for example, such as mica charged with Ni^{2+}) with aim to study samples structure by AFM method.

4. Conclusions

The conditions of quality layer adsorption should be studied for each separately taken protein. The conditions of stable protein layers formation have been established for BSA, TF, TG and SAG molecules. In particular, gold surface modification with cysteine ester derivative was enough for quality protein monolayer film formation from TF and SAG proteins. It was also found that polymeric surfaces from polyelectrolytes are preferable before an inorganic or composite "polymer-metal ions charged" surfaces at quality monomolecular protein film formation. From different polymeric surfaces charged with Co^{2+} cations the best quality of protein monolayer was obtained for BSA and TF molecules.

Results obtained suggest that LbL method is well suited for protein biological receptor layer formation on a polyelectrolyte surface for immunosensing applications.

References

1. Rabe, J. P. (1992), Molecules at interfaces: STM in materials and life sciences, *Ultamicroscopy* **41**, 42-44.
2. Hansma, H.G. and Hoh, J.H. (1994) Biomolecular imaging with the atomic force microscope, *Annu. Rev. Biophys. Biomol. Struct.* **23**, 115-139.
3. Bustamante C. and Keller, D. (1992), Scanning force microscopy in biology, *Physics Today*, 32-38.
4. Chen X., Davies M.C., Roberts C.J., Tendler S.J.B., Williams P.M., Davies J., Dawkes A.C., and Edwards J.C., (1997) Recognition of Protein Adsorption onto Polymer Surfaces by Scanning Force Microscopy and Probe-Surface Adhesion Measurements with Protein-Coated Probes, *Langmuir* **13**, 4106-4111.
5. Tanigawa, M. and Okada, T. (1998), Atomic force microscopy of supercoiled DNA structure on mica, *Analytica Chimica Acta* **365,** 19-25.
6. Williams, D.F.(1992), *Medical and Dental Materials*; In *Materials Science and Technology* 14, (Cahn, R.W., Haasen, P., Kramer, E.J., Eds.); pp.1-27 VCH: Weinheim, Germany.
7. Davis, S.S., Hunneyball, I.M., Illum, L., Radcliffe, J. H., Smith, A., and Wilson, C.G. (1985), Recent advances in the use of microspheres for targeted therapy, *Drugs Exp. Clin. Res.* **11**, 633-640.
8. Vukusic, P.S., Bryan-Brown, G.P., and Sambles, J.R. (1992), Surface-plasmon resonance on gratings as a novel means for gas sensing, *Sens. Actuators B* **8,** 155-160.
9. Caruso, F., Rodda, E., and Furlong, D.N. (1996), Orientational aspects of antibody immobilization and immunological activity on quartz crystal microbalance electrodes, *J. Colloid Interface Sci.* **178,** 104-115.
10. Mayo, C.S. and Hallock, R.B., (1989) Immunoassay based on surface plasmon oscillations, *J. Immunol. Methods* **120,** 105-114.
11. Barraud, A., Perrot, H., Billard, V., Martelet, C., and Therasse, J. (1993), Study of immunoglobulin G thin layers obtained by the Langmuir-Blodgett method: application to immunosensors, *Biosens. Biolectron.* **8,** 39-48.
12. Tronin, A., Dubrovsky, T., De Nitti, C., Gussoni, A., Erokhin, V., and Nicolini, C. (1994) Langmuir-Blodgett-films of immunoglobulines IgG – ellipsometric study of the deposition process and of immunological activity, *Thin Solid Films* **238,** 127-132.
13. Calvert T.L. and Leckband D.(1997), Two-Dimensional Protein Crystallization at Solid-Liquid Interfaces, *Langmuir* **13,** 6737-6745.
14. Yeung, C. and Leckband D. (1997), Molecular Level Characterization of Microenvironmental Influences on the Properties of Immobilized Proteins, *Langmuir* **13**, 6746-6754.
15. Lin, J.N., Herron, J., Andrade, J.D., and Brizgys, M. (1988), Characterization of immobilized antibodies on silica surfaces, *IEEE Trans Biomed. Eng.* **35**, 466-471.
16. Leggett, G.J., Roberts, C.J., Williams, P.M., Davies, M.C., Jackson, D.E., and Tendler, S.J.B. (1993), Approaches to the immobilization of proteins at surfaces for analysis by scanning-tunneling-microscopy, *Langmuir* **9**, 2356-2362.
17. Millot M.C., Martin F., Bousauet D., Sebille B., and Levu Y. (1995), A reactive macromolecular matrix for protein immobilization on a gold surface. Application in surface plasmon resonance, *Sensors and Actuators B* **29**, 268-273.
18. Mirsky V.M., Riepl M., and Wolfbeis O.S. (1997), Capacitive monitoring of protein immobilization and antigen-antibody reactions on monomolecular alkylthiol films on gold electrodes, *Biosensors and Bioelectronics* **12,** 977-989.
19. Green, R.J., Davies, M.C., Roberts, C.J., and Tendler, S.J.B. (1999), Competitive protein adsorption as observed by surface plasmon resonance, *Biomaterials* **20**, 385-391.
20. Caruso, F., Furlong, D.N., and Kingshott, P. (1997), Characterization of ferritin adsorption onto gold, *Journal of Colloid and Interface Science* **186,** 129–140.
21. Blomberg, E., Claesson, P., Froberg, J., and Tilton, R. (1994), Interaction between adsorbed layers of lysozyme studied with the surface force technique, *Langmuir* **10**, 2325-2334.
22. Decher, G.,, and Hong, J.D. (1991), Buildup of ultrathin multilayer films by a self-assembly process .2. Consecutive adsorption of anionic and cationic bipolar amphiphiles and polyelectrolytes on charged surfaces, *Ber. Bunsen-Ges. Phys. Chem.* **95**, 1430-1434.
23. Decher, G., Hong, J.D., and Schmitt, J. (1992), Buildup of ultrathin multilayer films by a self-assembly process .3. Consecutively alternating adsorption of anionic and cationic polyelectrolytes on charged surfaces, *Thin Solid Films* **210/211**, 831-835.

24. Lvov, Y., Decher, G., and Mohwald, H. (1993), Assembly, structural charaterization, and thermal-behavior of layer-by-layer deposited ultrathin films of poly(vinyl sulfate) and poly(allylamine), *Langmuir* **9**, 481-486.
25. Decher, G., Lvov, Y., and Schmitt, J. (1994), Proof of mutilayer structural organization in self-assembled polycation polyanion molecular films, *Thin Solid Films* **244**, 772-777.
26. Sukhorukov, G. B., Schmitt, J., and Decher, G. (1996), Reversible swelling of polyanion/polycation multilayer films in solutions of different ionic strength, *Ber. Bunsen-Ges. Phys. Chem.* **100**, 948-953.
27. Blomberg, E., Claesson, P., Froberg, J., and Tilton, R. (1994), Interaction between adsorbed layers of lysozyme studied with the surface force technique, *Langmuir* **10**, 2325-2334.
28. Lvov, Y.; Ariga, K.; Ichinose, I.; and Kunitake, T. (1995), Assembly of multicomponent protein films by means of electrostatic layer-by-layer adsorption, *J. Am. Chem. Soc.* **117**, 6117-6123.
29. Onda, M., Lvov, Y., Ariga, K., and Kunitake, T. (1996), Sequential actions of glucose oxidase and peroxidase in molecular films assembled by layer-by-layer alternate adsorption, *Biotech. Bioeng.* **51**, 163-167.
30. Lvov, Yu., Ariga, K., Ichinose, I., and Kunitake, T. (1996), Molecular film assembly via layer-by-layer adsorption of oppositely charged macromolecules (linear polymer, protein and clay) and concanavalin A and glycigen, *Thin Solid Films* **284/285**, 797-801.
31. Caruso, F., Niikura, K., Furlong, D.N., and Okahata Y. (1997), Assembly of Alternating Polyelectrolyte and Protein Multilayer Films for Immunosensing, *Langmuir* **13**, 3427-3433.
32. Houska, M. and Brynda, E. (1997), Interactions of Proteins with Polyelectrolytes at Solid/ Liquid Interfaces:Sequential Adsorption of Albumin and Heparin, *Journal of Colloid and Interface Science* **188**, 243–250.
33. Brynda, E. and Houska, M. (1998), Preparation of organized protein multilayers, *Macromol. Rapid Commun.* **19**, 173–176.
34. Caruso, F., Furlong, D.N., Ariga, K., Ichinose, I., and Kunitake, T. (1998), Characterization of Polyelectrolyte-Protein Multilayer Films by Atomic Force Microscopy, Scanning Electron Microscopy, and Fourier Transform Infrared Reflection-Absorption Spectroscopy, *Langmuir* **14**, 4559-4565.
35. Sadana, A., and Beelaram, A.M. (1995), Antigen-antibody diffusion-limited binding kinetics of biosensors: a fractal analysis, *Biosensors and Bioelectronics* **10**, 301-316.
36. Hinterdorfer, P., Baumgartner, W., Gruber, H.J., Schilcher, K., and Schindler, H. (1996), Detection and localization of individual antibody-antigen recognition events by atomic force microscopy, *Proc. Natl. Acad. Sci. U.S.A.* **93**, 3477-3481.
37. Lee, G.U., Kidwell D.A., and Colton, R.J. (1994), Sensing discrete streptavidin biotin interactions with atomic-force microscopy, *Langmuir* **10**, 354-357.
38. Pierce, M., Stuart, J., Pungor, A., Dryden P., and Hlady V. (1994), Adhesion force measurements using an atomic-force microscope upgraded with a linear position-sensitive detector, *Langmuir* **10**, 3217-3221.
39. Florin, E.-L., Moy, V.T., and Gaub, H.E. (1994), Adhesion forces between individual ligand-receptor pairs, *Science* **264**, 415-417.
40. Moy, V.T., Florin, E.-L., and Gaub, H.E. (1994), Intermolecular forces and energies between ligands and receptors, *Science* **266**, 257-259.
41. Patel, N., Davies, M.C., Hartshorne, M., Heaton, R.J., Roberts, C.J., Tendler. S.J.B., and Williams P.M.(1997), Immobilization of Protein Molecules onto Homogeneous and Mixed Carboxylate-Terminated Self-Assembled Monolayers, *Langmuir* **13**, 6485-6490.
42. Patel, N., Davies, M.C., Heaton, R.J., Roberts, C.J., Tendler, S.J.B., and Williams, P.M. (1998), A scanning probe microscopy study of the physisorption and chemisorption of protein molecules onto carboxylate terminated self-assembled monolayers, *Appl. Phys. A* **66**, S569–S574.
43. Cherbuliez, E. and Plattner, P. (1929), Determination of the amino acids resulting from thehydrolysis of proteins. II. Acetyl esters of some amino acids, *Helv. Chim. Acta* **12**, 317-329.
44. Butt, H.J., Downing, K.H., and Hansma, P.K. (1990), Imaging the membrane protein bacteriorodopsin with the atomic force microscope, *Biophys. J.* **58**, 1473-1480.
45. Buijs, J., Lichtenbelt, J.W.Th., Norde, W., and Lyklema, J. (1995), Adsorption of monoclonal IgGs and their F(ab')(2) fragments onto polymeric surfaces, *Colloids Surf., B: Biointer.* **5**, 11-23.

ENZYME ELECTRODES WITH ENZYME IMMOBILISED BY SOLGEL TECHNIQUE

MALGORZATA PRZYBYT, BEATA BIALKOWSKA
Technical University of £ódŸ, Department of Food Chemistry and Biotechnology, Institute of General Food Chemistry, Stefanowskiego 4/10, 90-924 £ódŸ, Poland,
mprzybyt@snack.p.lodz.pl

Abstract

In this work, the preliminary results of immobilisation of enzymes in silica gel on oxygen electrode are reported. As the model enzyme glucose oxidase was used. The influence of the composition of the casting solution (gel precursor, pH of the enzyme solution, sol-water ratio) on the electrode response was investigated. In addition, the addition of -aminopropyltrieethoxy silane to the casting solution was checked. The best electrode with stable signal was obtained when the formed gel was not very rigid (buffer, pH 6 or 7 and high water-sol ratio). For the optimal composition of the casting solution, some properties of the glucose electrode were investigated (stability, influence of pH and temperature). The method of sol-gel entrapment was also used to obtain the electrodes for disaccharides by co-immobilisation of invertase, lactase and maltase with glucose oxidase. In addition, other oxidases were immobilised with this method. For electrode with immobilised tyrosinase some additional measurements were done like influence of pH, temperature, addition of organic solvents, NaCl and SDS. Also, the broad spectrum of different substrates of this enzyme was tested.

Keywords : *enzyme electrode, glucose electrode, polyphenol biosensor, sol-gel technique, enzyme immobilisation*

1. Introduction

A biosensor can be defined as a device in which biological sensing element is closely connected with a transducer, which converts any observed change into a measurable signal. In any biosensor the most important is the connection of biological material with the transducer which can be achieved by immobilisation of the first. The method of immobilisation can change and improve the properties of the biosensor, therefore there is

E. Buzaneva and P. Scharff (eds.), Frontiers of Multifunctional Nanosystems, 91–108.

continuous search for new and better methods. There are many methods of immobilisation among them the entrapment in gel lattice (organic or inorganic) is very popular because of its simplicity. Such entrapment in inorganic gel can be achieved using the solgel technique.

The method of sol-gel entrapment of biomolecules is a very promising technique of immobilisation for biosensors construction, because of its simplicity, low temperature of the process, large amount and low leakage of entrapped material. Since 1990 when Braun at al. [1] reported for the first one the entrapment of proteins in silica gel, the sol-gel process has become an attractive way of immobilisation of biological material in biosensor construction and has been reviewed some times [2,3].

Sol-gel technique provides a simple method to obtain the glassy materials by hydrolysis (acidic or basic) and condensation of metal alkoxides. Typically, the silica gels are used starting from tetramethoxysilane [4-7] or tetraethoxysilane [8-10]. After hydrolysis (typically acidic) the gel is obtained quickly by changing the pH. But also, other precursors and gels are used like alumina [11-13] or vanadia [14], because of their greater electrical conductivity, which is important in case of application in electrochemical methods of measurement [3]. For immobilisation of biomolecules, the conditions of precursor hydrolysis and condensation like pH and organic solvent content require stringent control to avoid denaturation of proteins. Therefore, the gels are obtained using mild conditions of hydrolysis, without adding alcohols and with high water-sol ratio [2,3]. When the sol-gel technique is used to construct the optical biosensors, the conditions of the process must aloud to obtain transparent gels. The solgel technique is very flexible and allows obtaining the final gel in form of blocks of different shapes, fibres, thin films and granules [3]. The properties of silica gels can be modified by hydrolysing and condensation of organosilicon derivatives like 3-aminopropyltriethoxysilane, 3-glycidoxyprpyltrimethoxy-silane, 2-(3,4-epxycyclohexyl)-ethyltrimethoxysilane [3,15,16] or others. The resulting gel has better structure, porosity and regular distribution of immobilised biomaterial as compared with conventional one [16]. The use of organosilicon derivatives allows also incorporating ionogenic or redox-active groups into the gel structure [3]. Very promising composite materials for biosensor construction can be obtained by mixing silica sol with redox polymers [17], colloidal gold [18] or graphite powder [19,20]. Such composites have good electrical properties [3] and can be used in screen-printing approach [21].

The sol-gel technique has been applied in construction of biosensors of different types : conductometric [4], amperometric [6-8,10-23], photometric [8,10,24] and fluorometric [5,9]. The simple enzyme electrode can be obtained by immobilisation of oxidase on the surface of oxygen electrode. The concentration of oxidase substrate is proportional to oxygen depletion measured by the electrode. This is the concept of so called first generation enzyme electrode. The sol-gel process was used by Tatsu et al. to immobilise glucose oxidase in such electrode and the resulting biosensor was used in flow injection analyser [25].

In this work, the preliminary results of immobilisation of enzymes in silica gel obtained by sol-gel process on oxygen electrode are reported and some properties of obtained electrodes are evaluated.

2. Experimental

2.1 *REAGENTS*

As gel precursors tetramethoxysilane (**TMOS**) (99+%, Aldrich Chemie GmbH, Germany) or tetraethoxysilane (**TEOS**) (99+%, Aldrich Chemie GmbH, Germany) were used. To some gels 3-aminopropyltriethoxysilane (**APTES**) (Merck, Germany) was added. As the model enzyme glucose oxidase (**GOD**) from *Aspergillus niger* (solution, 5370 U/ml, Serva) was used. Other enzymes used in experiments were : catalase from bovine liver (solution, 155000 U/ml, Serva), invertase from bakers yeast (solid, 500 U/mg, Sigma), -galactosidase (lactase) from *Escherichia coli* (solid, 388 U/mg, Sigma), -galactosidase (lactase) from *Aspergillus oryzae* (solid, 3,8 U/mg, Sigma), -glucosidase (maltase) from bakers yeast (solid, 3,5 U/mg, Sigma), -glucosidase from almonds (solid, 6,6 U/mg, Sigma), mutarotase from porcine kidney (suspension, 25000 U/ml, Sigma), ascorbate oxidase from cucurbita species (solid, 146 U/mg, Sigma), phenolase from potato (solid, 570 U/mg, Sigma), laccase from *Rhus vernicifera* (solid, 180 U/mg, Sigma), tyrosinase from mushroom (solid, 3000 U/mg, Sigma), alcohol oxidase from *Candida boidinii* (solid, 0,4 U/mg, Sigma), galactose oxidase from *Dactylium dendroides* (solid, 16 U/mg, Sigma), choline oxidase from *Alcaligenes* species (solid, 14 U/mg, Sigma), cholinesterase, butyryl (pseudocholinesterase) from horse serum (solid, 8,7 U/mg, Sigma). The enzymes were used with no further purification, if solid they were dissolved in 0,05 M phosphate buffer, pH 6. All other reagents were, if possible, of analytical grade. Double-distilled water was used throughout. At pH 5,5 and lower 0,05 M acetate buffers were used, for higher pH 0,05 M phosphate ones were used.

2.2 *APPARATUS*

The measurements of dissolved oxygen were done using oxygen electrode CTN-920.S (MES-EKO, Wroc³aw, Poland) connected with microcomputer oxygen meter CO-551 (Elmetron, Zabrze, Poland). The temperature was maintained by the thermostat U1 (MLW, Medingen, Germany). The constant stirring rate of the solution was maintained by magnetic stirrer BMM 21 (DHN Wigo, Piastów, Poland).

2.3 *IMMOBILISATION OF THE ENZYMES AND ELECTRODE PREPARATION*

The stock solutions of sol were prepared as follows : **SOL I** - 4,5 ml **TEOS**, 1,4 ml H_2O and 0,1 ml 0,1 M HCl [8]; **SOL II** - 5 ml of **TMOS**, 1,0 H_2O and 0,05 ml 0,1 M HCl [4] were stirred vigorously at room temperature until the transparent homogenous solution was obtained (**SOL I** - 3 hours, **SOL II** - 15 minutes) and stored in refrigerator. **SOL I** was stable for 1 week, **SOL II** for 6 weeks.

A casting solution was prepared by mixing 200 ìl of the sol solution with enzyme solution and buffer with pH ranging from 5 to 9 (the total volume of enzyme and buffer ranging from 200 to 800 ìl). 20 ìl of this mixture was dropped immediately after mixing on the surface of oxygen electrode covered with nylon mesh. The mesh was used as a mechanical holder of gel layer because of the poor adhesion of it to the teflon membrane of oxygen electrode. To some of casting solutions, **APTES** was added. The compositions of casting solutions and gelation times are given in Table 1. When the gel was formed, the resulted enzyme electrode was dipped in 100 ml of phosphate buffer, pH 7, at 25 °C unless stated others.

4. Measurement procedure

The measurement procedure was similar to reported earlier [26]. The response of the electrode was measured using microcomputer oxygen meter and expressed as % of saturation with oxygen from air. The measurements were done in phosphate buffer, pH 7, at 25 °C in stirred solution (stirring rate 500 rpm). When the electrode response become stable 0,1 ml of 0,1 M substrate solution (if not stated others) was added. The solutions of saccharides in water (glucose, galactose, lactose, maltose, saccharose, cellobiose) were made one day before use to allow the mutarotation. All other substrate solutions were made freshly every day. Subsequent addition of substrate was made when the response reached steady state. The typical response of glucose electrode is given on Figure 1. As the result, the calibration curves were obtained (the dependence of decrease of % O_2 against concentration of substrate) and the slopes of the linear part of them (sensitivity) were calculated. The examples of such calibration curves are given on Figure 2.

5. Results

5.1 *GLUCOSE ELECTRODE*

As a model enzyme glucose oxidase was used because of its high activity, selectivity, stability and very well known properties. This enzyme catalyses reaction :

$$\beta - D - glu\cos e + O_2 \xrightarrow{GOD} gluconic\_acid + H_2O_2 \qquad (1)$$

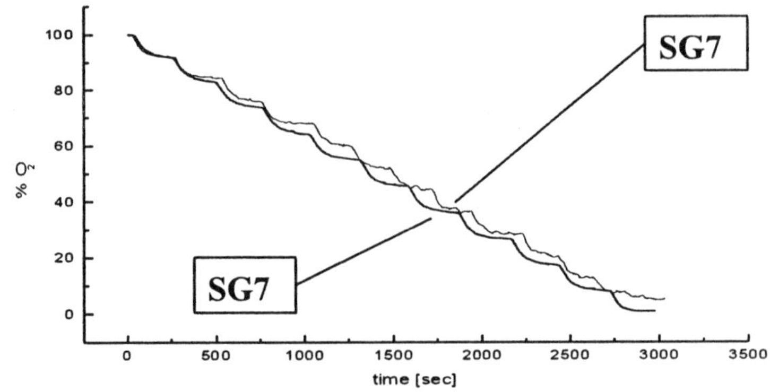

Figure 1. Examples of response of glucose electrodes : SG70 - 200 μl SOL I + 180 μl buffer, pH 7 +20 μl GOD ; SG74 - 200 μl SOL I + 750 μl buffer, pH 7 + 50 μl GOD

The depletion of oxygen concentration in diffiusional layer measured by the electrode is thus the measure of glucose concentration.

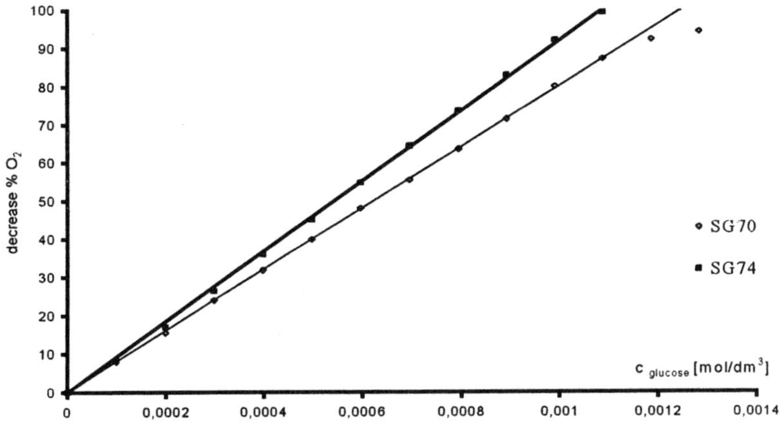

Figure 2. Examples of the calibration curves of glucose electrode : **SG70** - 200 μl **SOL I** + 180 μl buffer, pH 7 + 20 μl **GOD** ; **SG74** - 200 μl **SOL I** + 750 μl buffer, pH 7 + 50 μl **GOD**

TABLE 1. *Dependence of gelling time on composition*

Sol	pH of the buffer	sol-buffer ratio	volume of added **APTES**	gelling time [min]
TMOS	5	1 : 1	none	3
	6			4
	7			1
	8			0,5
	9			0,25
	5	1 : 4	None	70
	6			30
	7			1,5
	8			8
	9			7
	6	1 : 1	2,5μl	> 0,1
TEOS	5	1 : 1	None	9
	6			17
	7			4
	8			1,5
	9			1
	7	1 : 1,5	None	2,5
		1 : 2		3
		1 : 3		3
	5	1 : 4	None	30
	6			40
	7			4
	8			1,5
	9			2
	6	1 : 1	2,5μl	1
			5μl	0,5
			10μl	0,5
			20μl	0,5
	7	1 : 1	5μl	0,5

In preliminary investigations, the response of glucose electrode and its sensitivity was measured for gels prepared from different precursors, with different pH of the buffer and different sol-buffer ratio. For some compositions of the casting solution, also the influence of **APTES** addition was tested. The details of tested compositions are given in Table 1. The casting solutions with gelation times longer than 30 minutes and shorter than 0,25 minute cannot be used to prepare the glucose electrode because of technical problems. For very long gelation times the sol is drying and cracking before the gel is formed. For very short times, it is impossible to take the desired volume from the casting solution and drop it on the electrode because the gel is formed in the pipette. When **TMOS** was used as a precursor, some precipitation of silica could be observed for sol-buffer ratio 1:4. The more silica was precipitated the less pH of the buffer was.

For some tested compositions of the casting solution the response of the electrode is not stable but is varying around some value ("noisy" response). Such noisy response can be observed for example for gel formed from **TEOS** , pH 7, sol-water ratio 1:1 (line **SG70** on Figure 1). The electrodes with noisy response are not suitable to measure the glucose content in solution. Probably, such unstable response is caused by the poor adhesion of gel layer to teflon membrane and its unhomogenity. The adhesion could be improved by addition of surfactant to the casting solution [7,29]. Thus, to some casting solutions, surfactant Triton X-100 was added but no improvement of the electrode response was observed. The typical calibration curves for glucose were linear with some deviations from linearity for higher concentrations of glucose (Figure 2). The reciprocal plots (not shown) of the data indicates that the response of electrode is controlled by diffusion of substrates to electrode surface. The deviations from linearity are caused by the lack of oxygen in diffusion layer near the electrode surface [27]. In Table 2 analytical properties of glucose electrodes are summarised.

The typical response time of the electrodes (time interval from addition of substrate until the signal become stable) is varying from 3 to 5 minutes with some tendency to be lower for higher water content in gel indicating the lower diffiusional resistance of gel layer.

The results of investigations of influence of the composition of the casting solution on gelling time and glucose electrode response indicate that **TEOS** is better gel precursor than **TMOS,** because for the latter the response of the electrodes is noisy and there are some problems with gel formation (precipitation of silica for high buffer-sol ratio). For **TEOS** better electrodes with higher sensitivity are obtained with high buffer-sol ratio (Table 2) especially for basic pH. The lower response of electrodes with low sol-buffer ratio is probably caused by greater diffiusional resistance of the gel, which is in this case denser [22]. The addition of **APTES** caused the shortening of the gelation time but does not improve the properties of the electrodes. In contrary, when the high amount of **APTES** is added the response of the electrode become unstable.

For further experiments with glucose oxidase the gels of following composition were chosen : **SG60** - buffer pH 6, sol-buffer ratio 1:1, **SG61** - buffer pH 6, sol-buffer ratio 1:1 + 2,5 l **APTES** added, **SG74** buffer pH 7, sol-buffer ratio 1:4. For these gels the dependence of electrode response on pH (Figure 3) and temperature (Figure 4) was investigated. For all tested gels the response of the electrode shows the maximum at pH = 7, with sharp loss of the sensitivity for greater values. Similar results were obtained by other authors [6].

TABLE 2. *Analytical parameters of glucose electrodes*

Sol	buffer pH	sol-buffer ratio	volume of added APTES	linear range [mol/dm^3]	sensitivity [%/mol/dm^{-3}]	r
TMOS	6*	1 : 1	none	0 - 0,0006	88240	0,999
	7*			0 - 0,0008	80410	0,999
	8*			0 - 0,0006	123600	0,997
	9*			0 - 0,0008	42140	0,998
	7	1 : 4	none	0 - 0,0011	79220	0,999
	8			0 - 0,0007	64520	0,997
	9			0 - 0,0009	102960	0,999
TEOS	5	1 : 1	none	0 - 0,0008	88330	0,998
	6			0 - 0,0011	86190	0,999
	7*			0 - 0,0009	88190	0,998
	8*			0 - 0,0007	63520	0,997
	9*			0 - 0,0007	77180	0,999
	7	1 : 1,5	none	0 - 0,001	95080	0,999
		1 : 2		0 - 0,001	86020	0,997
		1 : 3		0 - 0,001	87000	0,999
	5	1 : 4	None	0 - 0,001	92760	0,999
	6			0 - 0,001	90880	0,999
	7			0 - 0,001	92260	0,999
	8			0 - 0,0009	99840	0,999
	9			0 - 0,001	100110	0,999
	6	1 : 1	2,5 μl	0 - 0,0009	103500	0,997
			5 μl	0 - 0,001	86440	0,999
			10 μl*	0 - 0,0007	81990	0,999
			20 μl*	0 - 0,001	83880	0,999
	7*	1 : 1	5 μl	0 - 0,0008	117190	0,999

*Compositions of gel for which the response of the glucose electrode is not stable in steady state but is varying about some value ("noisy" response)

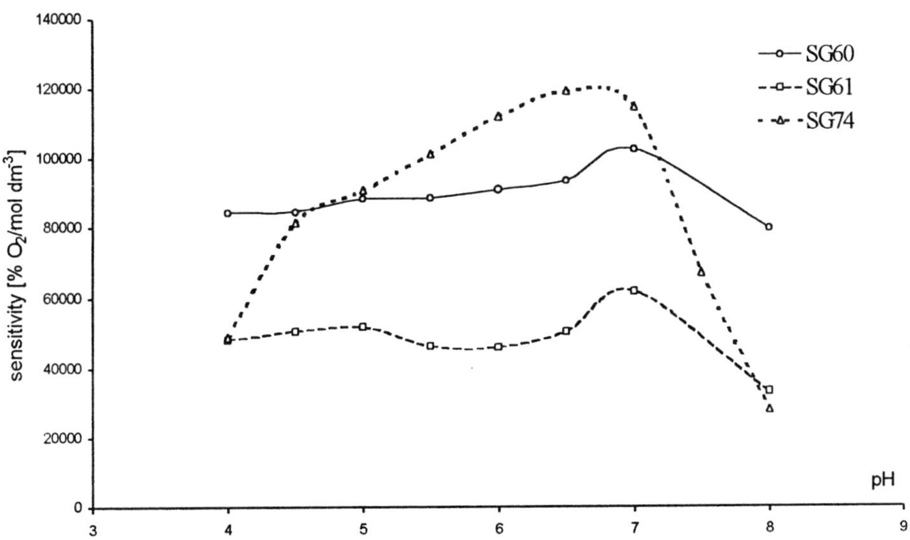

Figure 3. The dependence of glucose electrode sensitivity on pH. (description of legend key in text)

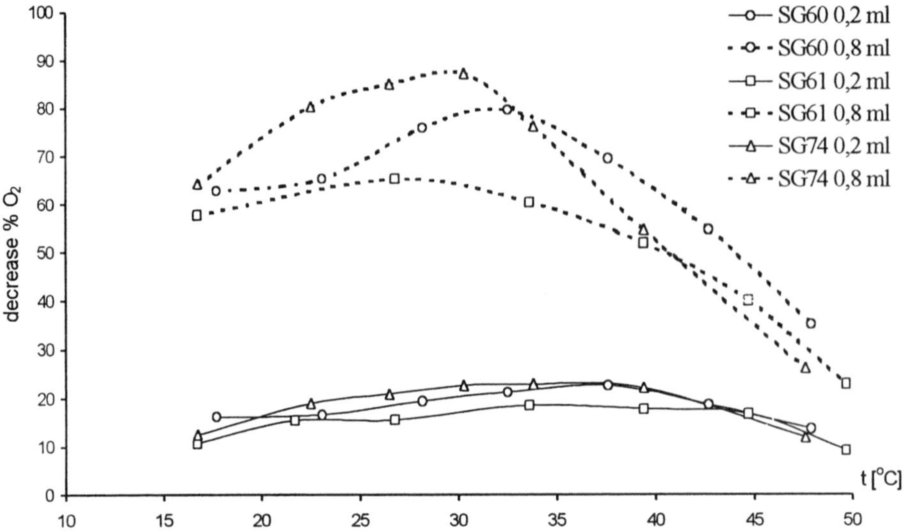

Figure 4. The dependence of glucose electrode response on temperature for 0,2 ml and 0,8 ml of added 0,1 M glucose solution. (description of legend key in text)

With increasing temperature the response of electrodes is growing up to ⊇ 35C but for the temperatures higher than 30C the calibrations curves became very unlinear (not shown) and because of that the maximum response is obtained for different temperatures for low and high glucose concentration (Figure 4). The glucose oxidase is known as a very stable enzyme [28] but the obtained results indicate that in silica environment its stability is poor. Also, the stability of the electrodes were tested. For this purpose, they were stored for 10 days in 50 ml of buffer, pH 7, at refrigerator (4C). After this storage, the responses of electrodes were very low (for **SG60** only 7,7 % of the initial sensitivity) and unstable. Similar results for GOD immobilised in silica gel were obtained by other authors [6]. For **SG61** and **SG70** the response was so unstable and noisy that, the calibration curves could not be red. In storage buffer, the traces of **GOD** activity could be found. The results indicate that there is leakage of enzyme from the gel. Also the gel by itself is not stable and some cracking of it could be observed.

Glucose oxidase was co-immobilised with catalase to obtain the glucose electrodes with broader range of calibrations curves. The upper limit of glucose concentration that can be assayed by glucose electrode is limited by the solubility of oxygen in solution. By addition of catalase the limit could be extended twice, because the oxygen consumed in reaction 1 is in a half recovered in the reaction :

$$H_2O_2 \xrightarrow{\text{catalase}} H_2O + \tfrac{1}{2}O_2 \qquad (2)$$

To casting solutions **SG60** and **SG74** catalase was added and the ratio of activities, catalase to **GOD** was 14,4:1. As the result, the linear parts of the calibration curves were extended to 0,0017 mol/dm^3 for **SG60** and 0,0013 mol/dm^3 for **SG74**. The sensitivity decreased 0,505 fold for **SG60** and 0,608 for **SG74**.

5.2 *ELECTRODES FOR DISACCHARIDES*

By co-immobilisation of **GOD** with suitable hydrolases (and mutarotase optionally) the electrodes for disaccharides (maltose, lactose, saccharose and cellobiose) could be obtained. The hydrolase hydrolysis disaccharide to glucose and other monosaccharide in reaction :

$$disaccharide + H_2O \xrightarrow{hydrolase} glu\cos e + other\_monosaccharide \quad (3)$$

Glucose produced in reaction 3 is then consumed by **GOD** in reaction 1. Because **GOD** oxidases only □-D-glucose and during hydrolysis of some disaccharides (e.g. saccharose, maltose) □-D-glucose is produced, mutarotase is added [30]. Mutarotase increases the rate of reaction of mutarotation of glucose :

$$\alpha - D - glu\cos e \xleftrightarrow{mutarotase} \beta - D - glu\cos e \quad (4)$$

and thus, the response of the electrode for disaccharide is increasing.

Table 3. *Analytical parameters of electrodes for disaccharides at pH 7*

Composition	Enzyme*	Activity [U]**	substrate	Sensitivity [%/mol/dm^{-3}]	linear range [mol/dm^3]	r
SG60	Invertase Invertase + mutarotase	50 50 + 12,5	saccharose	50346 61676	0 - 0,0005 0 - 0,001	0,999 0,999
SG74	Invertase Invertase + mutarotase	50 50 + 12,5		36074 107308	0 - 0,0008 0 - 0,0009	0,999 0,999
SG60	Maltase Maltase + mutarotase	4 4 + 12,5	maltose	no response		
SG74	Maltase Maltase + mutarotase	4 4 + 12,5		3017,4 2368,5	0 - 0,007 0 - 0,005	0,998 0,987
SG74	Lactase from *E. coli* Lactase from *E. coli* + mutarotase	13,5 13,5 + 12,5	lactose***	27907 45381	0 - 0,0008 0 - 0,001	0,999 0,999
SG74	Lactase from *A. oryzae* Lactase from *A. oryzae* + mutarotase	1,9 1,9 + 12,5	lactose***	22836 21981	0 - 0,0008 0 - 0,0008	0,999 0,999
SG74	□-glucosidase □-glucosidase + mutarotase	2,64 2,64 + 12,5	cellobiose***	21686 25687	0 - 0,0008 0 - 0,0008	0,998 0,995

* in each case the activity of added **GOD** was 5,4 U per electrode
** activity of enzyme immobilised on electrode
*** for this saccharides the compositions **SG60** was not tested

For hydrolases only the compositions of casting solutions **SG60** and **SG74** were tested (not for all enzymes - see Table 3), because the addition of **APTES** (**SG61**) does not change the properties of electrode significantly. Typically, the response of electrodes was lower for sol-buffer ratio 1:1 or in case of maltase, there was no response. The reason is that the gel is denser and denaturation of the enzymes during the gelling process occurs by high concentration of ethanol, which is produced during hydrolysis of **TEOS**. Some results (calibra tion curves) for different disaccharides are presented on Figure 5 and some analytical parameters of them are collected in Table 3.

Except maltase, the improvement of the electrode response by addition of mutarotase could be seen especially for maltase and lactase from *Escherichia coli*. For lactase from *Aspergillus oryzae* this effect was not observed but for this enzyme the calibration was made at pH very different from its optimum pH, which is 4,5 and due to that its activity was low. The calibration curves for disaccharides are not linear. The double reciprocal plot of the data (not shown) indicates that the electrode response is controlled by the kinetic of the enzymatic reaction [27]. From these plots, the apparent Michaelis constants for enzymes could be calculated using Lineweaver-Burk equation [22]. They are 18,6 ; 5,39 ; 2,47 and 16,6 mM for invertase, lactase from *Escherichia coli*, □-glucosidase and maltase, respectively.

5.3 *ELECTRODES WITH OTHER OXIDASES*

The enzymes tested were : galactose oxidase, ascorbate oxidase, alcohol oxidase, three types of polyphenolase (laccase, tyrosinase from mushroom and phenolase from potato) and choline oxidase. Choline oxidase was also co-immobilised with pseudocholinesterase (Figure 8) to assay choline esters due to the course of reactions :

$$butyrylcholine + H_2O \xrightarrow{pseudocholinesterase} choline + butyryl\_acid \qquad (5)$$

$$choline + 2O_2 + H_2O \xrightarrow{choline\_oxidase} betaine + 2H_2O_2 \qquad (6)$$

Such electrode is very promising to assay pesticides by inhibition of esterase [31]. For all enzymes except alcohol oxidase, the tested compositions of casting solution were **SG60** and **SG74**. Some examples of the calibration curves are shown on Figures 6, 7 and 8, and analytical parameters of calibration curves are collected in Table 4.

For all enzymes, the response of the electrodes was higher for sol-buffer ratio 1:4. For some enzymes, the precipitation of the protein could be observed during formation of gel with sol-buffer ratio 1:1 causing the very low response or even none. It is especially characteristic for choline oxidase and esterase. The ratio of sensitivities for sol **SG74** and **SG60** is 51,5 and 96,8 for choline oxidase and co-immobilised choline oxidase and esterase, respectively (Figure 8). The reason is that in sol **SG60** the concentration of ethanol is much higher as compared with **SG74**. It could be also seen that for the same enzyme the response for the substrate with greater molecule are much lower for the denser gel (compare the results with tyrosinase for catechol and (+)catechine - Figure 7). For sol composition **SG60** the ratio of sensitivities for

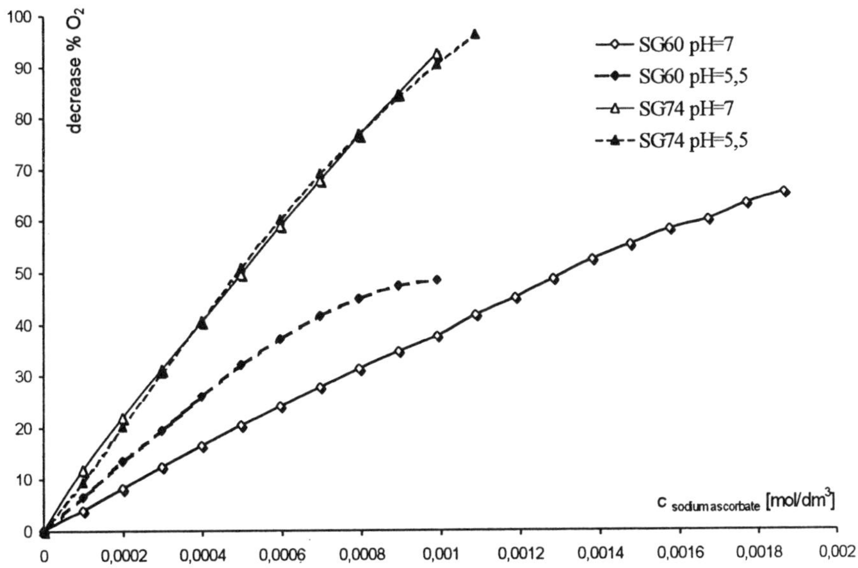

Figure 5. Examples of calibration curves for disaccharides with casting solution composition **SG74** (details about added activities of enzymes in Table 3) at pH 7

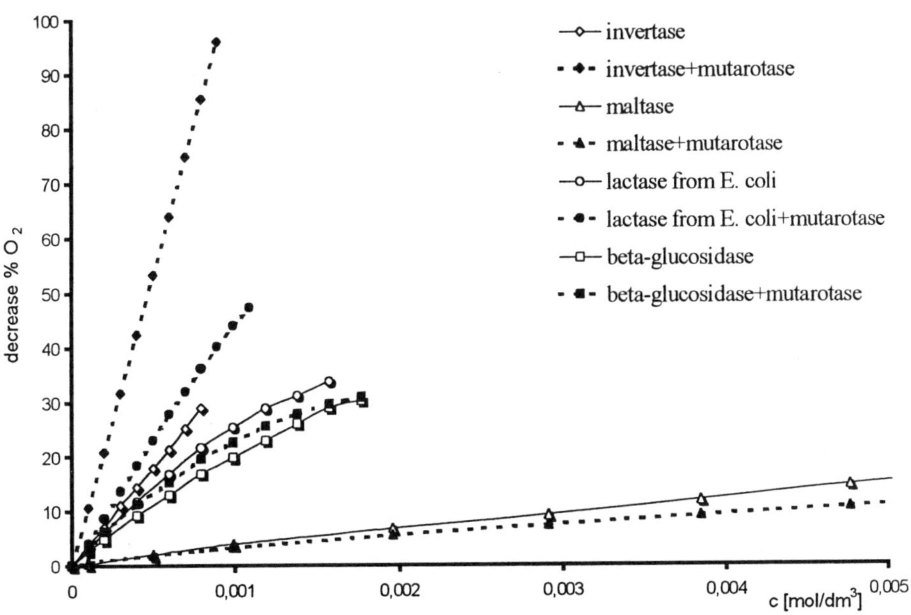

Figure 6. Calibration curves of electrode with ascorbate oxidase for vitamin C

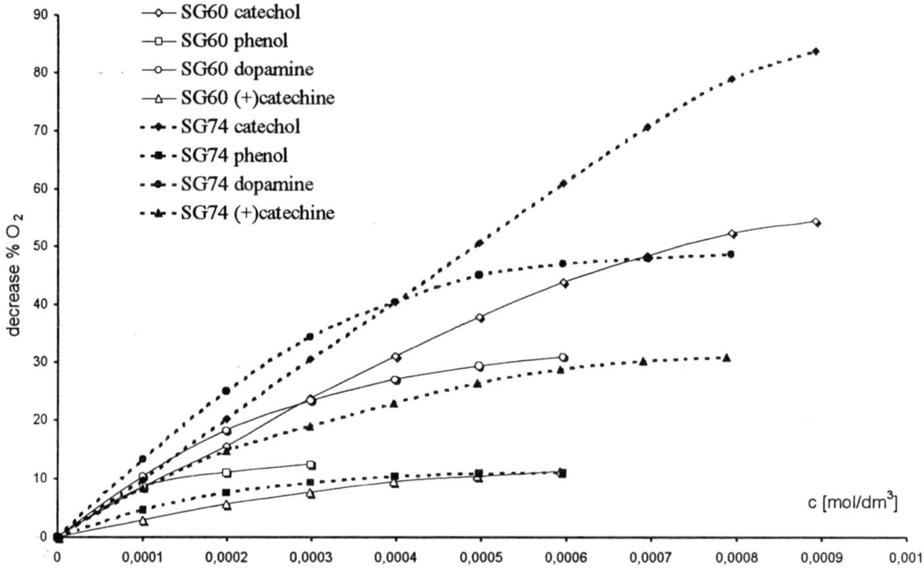

Figure 7. Calibration curves of electrode with tyrosinase for different substrates at pH 7 (4 U of enzyme immobilised on electrode)

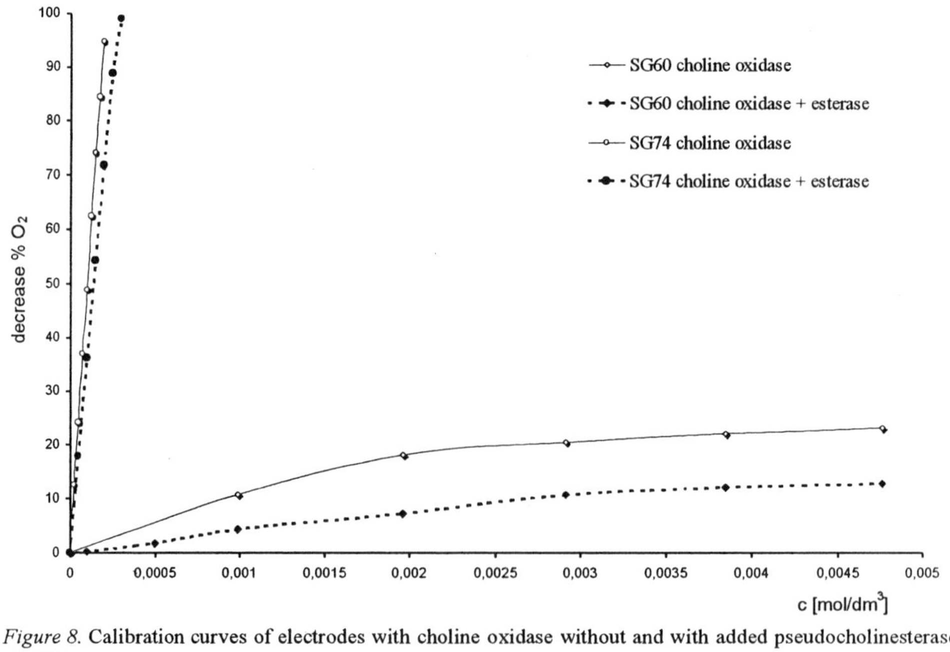

Figure 8. Calibration curves of electrodes with choline oxidase without and with added pseudocholinesterase at pH 7

(+)catechine and catechol is 0,336 and for **SG74** 0,744. It is caused by greater diffiusional resistivity of denser gel. For all tested oxidases, the calibration curves are not linear and their linear ranges are sometimes very narrow. The reason of such non-linearity is that the response of the electrode is controlled by the kinetics of enzyme reaction not by the diffusion of the substrate and oxygen [27]. For some enzymes, there is also another reason of non-linearity. Polyphenolases, especially tyrosinase, and ascorbate oxidase belong to the group of the enzymes that undergo suicide inactivation by the substrates [32,33]. The higher is the concentration of the substrate and the longer is the time of enzyme contact with it the lower is the activity of the enzyme. This effect is especially significant for tyrosinase and orto-diphenols. It causes that during prolonged time of measurements and for higher concentrations of substrate, the response of enzyme electrode with immobilised tyrosinase begins to grow [34,35]. This effect was observed in presented results for all tested polyphenolases and for ascorbate oxidase at pH 5,5. To diminish the effect of suicide inactivation the activity of immobilised tyrosinase must be very high. For sol composition **SG60** the response of the electrode begun to grow at catechol concentration 0,5 mM when 1,6 U of tyrosinase is immobilised and at 0,9 mM for 4 U. Also the sensitivity of the electrode is in first case lower and linear range is more narrow.

5.4 *SOME PROPERTIES OF ELECTRODE WITH TYROSINASE*

From three tested polyphenol oxidases the highest sensitivity of electrode was obtained for tyrosinase (Figure 7, Table 4). The enzyme electrodes with immobilised tyrosinase could be very usefull to assay phenolic compounds in enviromental analysis [36] and in assay of polyphenols in food and drinks [37, 38]. Thus, the properties of tyrosinase electrode were examined in some details. Tyrosinase oxidases phenol and orto-diphenols differently substituted in the benzoic ring in course of reactions :but with different rates.

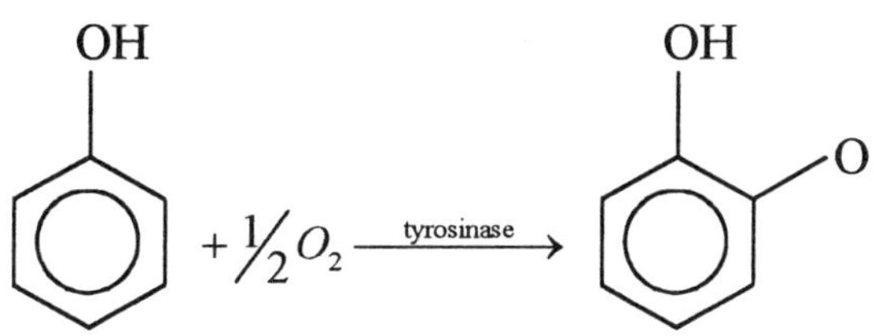

(7)

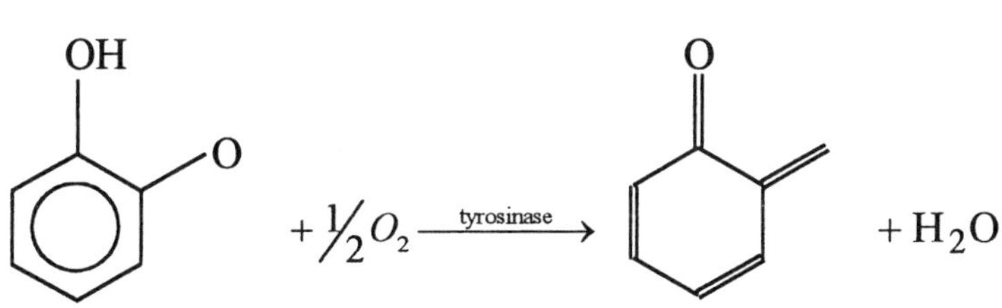

(8)

TABLE 4. *Analytical parameters of some electrodes with immobilised oxidases*

Composition	enzyme	activity [U]*	substrate	pH of assay	linear range [mol/dm^3]	sensitivity [%/mol/dm^{-3}]	r
SG60	galactose oxidase	2,5	galactose	7	0 - 0,0012	5614,9	0,997
			lactose	6	0 - 0,005	1143,6	0953
SG74			galactose	7	0 - 0,004	16745	0,994
			galactose	6	0 - 0,006	6858,5	0,992
			lactose	6	0 - 0,006	791,15	0,998
SG74	alcohol oxidase	0,2	ethanol	7	0 - 0,0016	11880	0,985
			methanol	7	0 - 0,0004	42810	0,976
SG60	ascorbate oxidase	4	sodium ascorbate	7	0 - 0,0014	38363	0,999
				5,5	0 - 0,0006	63568	0,999
SG74				7	0 - 0,0007	99356	0,998
				5,5	0 - 0,0007	10051	0,999
SG60	choline oxidase	2	choline	7	0 - 0,002	9529,8	0,994
SG74				7	0 - 0,000175	490562	0,999
SG60	choline oxidase + pseudocholinesterase	2 + 6	butyryl choline chloride	7	0 - 0,003	3704,2	0,998
SG74				7	0 - 0,00025	358494	0,999
SG60	tyrosinase	1,6	catechol	7	0 - 0,0003	69275	0,996
		4	catechol	7	0 - 0,0004	78457	0,999
			dopamine	7	0 - 0,0002	93830	0,997
			(+)catechine	7	0 - 0,0002	26373	0,996
SG74		4	catechol	7	0 - 0,0007	101755	0,999
			dopamine	7	0 - 0,0002	126361	0,999
			(+)catechine	7	0 - 0,0003	75739	0,997
SG74	phenolase	4	catechol	7	0 - 0,0004	39218	0,989
SG74	laccase	4	catechol	7	0 - 0,002	15674	0,979

* activity of enzyme immobilised on electrode

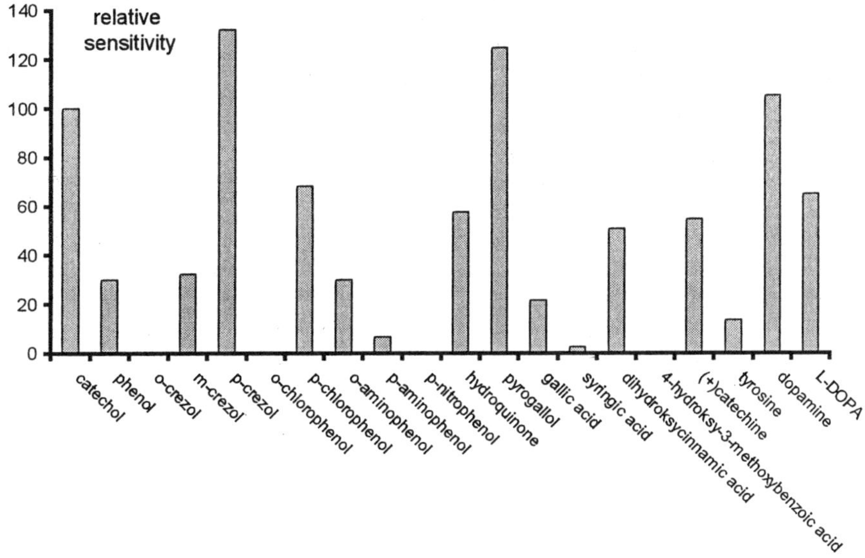

Figure 9. Relative sensitivity of electrode with tyrosinase at pH 7 and temperature 25C (the sensitivity for catechol is taken as 100 %)

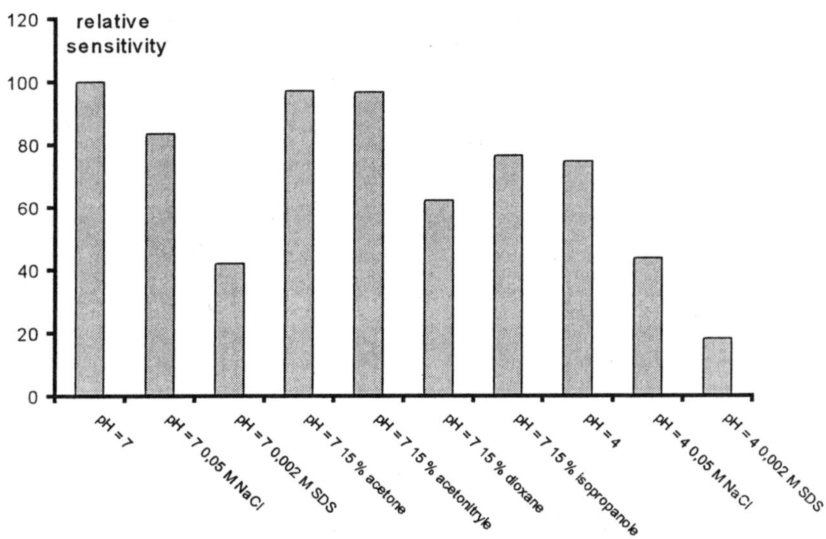

Figure 10. The influence of organic solvents, SDS and NaCl on tyrosinase electrode response at 25C expressed as relative sensitivity (the sensitivity for catechol in phosphate buffer, pH 7, without additivities, is taken as 100 %)

For this experiments only the composition of casting solution **SG74** with 4 U of tyrosinase was chosen. As can be seen from Figure 7 the electrode response is different

For this experiments only the composition of casting solution **SG74** with 4 U of tyrosinase was chosen. As can be seen from Figure 7 the electrode response is different for different substrates. It is caused by differencies of enzyme reaction rate. The electrode was tested at last for 20 different substrates important for different aplications and the selectivity spectrum is shown on Figure 9. For many of tested substrates the linear range of calibrations curves was very narrow and the suicide inhibition by the substrate was observed for very low cncentrations of them. Such results indicate that there could be some problems in practical use of this electrode in analysis of real samples.

The activity of tyrosinase (polyphenol oxidase) is known to be influenced by addition of organic solvents [39], sodiumdodecyl sulphate (**SDS**) [40] and chloride anions [41]. Therefore the influence of four organic solvents (acetone, acetonitryle, dioxane and isopropanole), **SDS** and NaCl at pH 7 and 4 (only for **SDS** and NaCl) was tested using catechole as the substrate. The results expressed as relative sensitivity are presented on Figure 10. As can be seen acetone and acetonitrile do not change thesensitivity of the electrode but the others are inhibiting the activity of tyrosinase. **SDS** and NaCl cause the lowering of sensitivity of the electrode at both tested pH. The results for **SDS** are rather suprising because **SDS** is known to activate polyphenol oxidase [40]. The lowering of sensitivity caused by NaCl is in good agreement with properties of the enzyme which is inhibited by chloride anions [41].

The optimum pH for the tyrosinase electrode is 7, at this value the sensitivity shows a sharp maximum (results not shown). The influence of temperature is rather strange. At temperatures ranging from 20 to 35C the response of the electrode is lowering and the inhibitory effect of catechol is more pronounced the higher is the temperature.

The obtained results indicate that tyrosinase in silicagel enviroment is rather unstable and very sensitive to influence of enviroment in which the electrode is used. Also the stability of the electrode is very poor.

6. Conclusions

- The technique of immobilisation by sol-gel transition can be applied to different enzymes to obtain enzyme electrodes.
- The better gel precursor for immobilisation of enzymes on oxygen electrode is **TEOS** than **TMOS**.
- The ratio of water to sol during gel formation must be high to avoid the denaturation of the enzyme by ethanol during immobilisation, to diminish the diffiusional resistance of the gel layer and thus, to obtain electrodes with high sensitivity.
- The addition of **APTES** does not improve significantly the properties of the electrodes.
- The wet silica gel in water is rather unstable and some leakage of the enzyme during storage is observed.

- The calibration curves are not linear. The non-linearity is caused by different reasons depending on the enzyme type and immobilised amount of the enzyme.
- The technique of immobilisation of the enzymes by sol-gel transition needs improvement to increase the stability of the gel and to avoid the leakage of the enzyme.

7. Acknowledgement

The author is grateful to the Committee for Scientific Research (Poland) for a financial support through grant no. 7 T08E 008 16.

References

1. S. Braun, S. Rappoport, R. Zusman, D. Avnir, M. Ottolenghi, Mater. Lett., 10 (1990), 1.
2. B.C. Dave, B. Dunn, J.S. Valentine, J.I. Zink, Anal. Chem., 66 (1994), 1120 A.
3. M.M. Collinson, A.R. Howells, Anal. Chem., 72 (2000), 702 A.
4. W.-Y. Lee, S.-R. Kim, T.-H. Kim, K.S. Lee, M.-C. Shin, J.-K. Park, Anal. Chim. Acta, 404 (2000), 195.
5. S. de Marcos, J. Galindo, J.F. Sierra, J. Galbon, J.R. Castillo, Sens. Actuators B, 57 (1999), 227.
6. T. Yao, K. Takashima, Biosens. Bioelectr., 13 (1998), 67.
7. S.L. Chut, J. Li, S.N. Tan, Analyst, 122 (1997), 1431.
8. U. Narang, P.N. Prasad, F.V. Bright, K. Ramanathan, N.D. Kumar, B.D. Malhotra, M.N. Kamalasanan, S. Chandra, Anal. Chem., 66 (1994), 3139.
9. X. Wu, M.M.F. Choi, D. Xiao, Analyst, 125 (2000), 157.
10. A. Kumar, R. Malhotra, B.D. Malhotra, S.K. Grover, Anal. Cim. Acta, 414 (2000), 43.
11. Z. Liu, B. Liu, J. Kong, J. Deng, Anal. Chim. Acta, 392 (1999), 135,
12. Z. Liu, J. Deng, D. Li, Anal. Chim. Acta, 407 (2000), 87.
13. Z. Liu, B. Liu, J. Kong, J. Deng, Anal. Chem. 72 (2000), 4707.
14. O. Lev, V. Glezer, J. Am. Chem. Soc., 115 (1993), 2533.
15. P.C. Pandley, S. Upadhyay, H.C. Pathak, Electroanalysis, 11 (1999), 59.
16. P.C. Pandley, S. Upadhyay, H.C. Pathak, Sens. Actuators B, 60 (1999), 83.
17. T.-M. Park, E.I. Iwuoha, M.R. Smyth, R. Freaney, A.J. McShane, Talanta, 44 (1997), 973.
18. S. Bharathi, O. Lev, Anal. Commun., 35 (1998), 29.
19. S. Samphath, O. Lev, Anal. Chem., 68 (1996), 2015.
20. L. Coche-Guorente, S. Cosnier, P.Labbo, Chem. Mater., 9 (1997), 1348.
21. J. Wang, P.V.A. Pamidi, D.S. Park, Anal. Chem., 68 (1996), 2705.
22. J. Li, L.S. Chia, N.K. Goh, S.N. Tan, Anal. Chim. Acta, 362 (1998), 203.
23. J. Wang, P.V. Pamidi, K.R. Rogers, Anal. Chem., 70 (1998), 1171.
24. J.W. Aylott, D.J. Richardson, D.A. Russell, Analyst, 122 (1997), 77.
25. Y. Tatsu, K. Yamashita, M. Yamaguchi, M. Yamamura, H. Yamamoto, S. Yoshikawa, Chem. Lett., (1992), 1615.
26. M. Przybyt, H. Sugier, Anal. Chem. (Warsaw), 34 (1989), 63.
27. F. Scheller, F. Schubert, *Biosensors*, Elsevier, Amsterdam, 1992, p. 55.
28. R. Wilson, A.P.F. Turner, Biosens. Bioelectr., 7 (1992), 165.
    O. Lev, Analusis, 20 (1992), 543.
29. M. Filipiak, K. Fludra, E. Gościmińska, Biosens. Bioelectr., 11 (1996), 355.
30. M. Bernabei, S. Chiavarini, C. Çremisini, G. Palleschi, Biosens. Bioelectr., 8 (1993), 265.

31. F.G. Conovas, J. Tudela, C.M. Madrid, R. Varon, F.G. Carmona, J.A. Lozano, Biochim. Biophys. Acta, 912 (1987), 417.
32. K. Tokuyama, C.R. Dawson, Biochim. Biophys. Acta, 56 (1962), 427.
33. J. Zhang, B. Li, G. Xu, G. Cheng, S. Dong, Analyst, 124 (1999), 699.
34. G.A. Rivas, V.M. Solis, Electroanalysis, 6 (1994), 1136.
35. P. Annerfjord, J. Emnous, G. Marko-Varga, L. Gorton, F. Ortega, E. Domonguez, Biosens. Bioelectr., 10 (1995) 607.
36. B.R. Eggins, C. Hickey, S.A. Toft, D. Min Zhou, Anal. Chim. Acta, 347 (1997) 281.
37 L. Campanella, G. Favero, M. Pastorino, M. Tomassetti, Biosens. Bioelectr., 14 (1999) 179.
38. R. Batra, M.N. Gupta, Biotechnol. Lett., 16 (1994) 1059.
39. B.M. Moore, W.H. Flurkey, J. Biol. Chem., 265 (1990) 4982.
40. P. Wesche-Ebeling, M.W. Montgomery, J. Food Science, 55 (1990) 1320.

TEMPLATE-DIRECTED LATTICES OF NANOSTRUCTURES: PREPARATION AND PHYSICAL PROPERTIES

SERGEI G. ROMANOV
Institute of Materials Science and Department of Electrical and Information Engineering, University of Wuppertal, Gauss-str. 20, 42097 Wuppertal, Germany
A.F. Ioffe Physical Technical Institute, 194021, Polytekhnicheskaya str., 26, St. Petersburg, Russia

Abstract:

An introduction to the template method of preparation of regular 3-dimensional arrays of nanostructures in an internal space of opals is given. Application of all-dielectric opal-based materials as photonic crystals is discussed. Different opaline materials have been examined by means of the optical diffraction and the light emission and correlations of observed phenomena with the material structure have been demonstrated. Novel approach towards engineering of opaline materials via heterostructuring is presented. Recent progress in the field of nanostructured superconductors in opals is discussed. Striking enhancement of the critical magnetic field, which is followed the reduction of the superconductor volume fraction, has been shown. Outlook of using opal-based superconducting lattices as materials for quantum storage and processing of information is given.

Key words: opal/photonic crystal/photonic bandgap/optical diffraction /photoluminescence/ emission rate/emission anisotropy/photonic bandgap heterostructure/ superconductor nanoparticles/lattice of weak links and loops/critical magnetic field/magnetoresistance/topological magnetic vertices/quantum information

E. Buzaneva and P. Scharff (eds.), Frontiers of Multifunctional Nanosystems, 109–138.

1. Introduction

Design of novel solids by assembling nanosize particles is within the main stream of the modern material science. First of all, this is because the properties of individual nanoparticles differ from that of their bulk parent materials. The imitation of naturally built atomic or molecular crystals by using nanoparticles in highly ordered arrays allows to expand these properties towards macroscopic objects. Moreover, if nanophase materials possess a structural inhomogeneity, which matches the characteristic length of particular physical phenomenon (e.g. de Broglie wavelength, exciton radius, coherence length, radiaton wavelength), then collective effects take a lion share in ensemble physical appearance.

There are several basic methods in use for preparation of nanostructures ensembles. The straightforward subtractive method is based on top-down approach, where a chunk of material is divided down to pieces of a microscopic size (Figure. 1). Usually, nanopatterning is applied to thin films and results in formation of 1- or 2-dimensional ensembles of nanostructures. This method features highly ordered arrays with well-controlled number of nanostructures as well as flexibility in sizing and shaping of each individual nanostructures and ensemble as a whole. It is limited in terms of dimensionality, number of nanostructures in array and aspect ratio. High cost and time consuming are also its drawbacks.

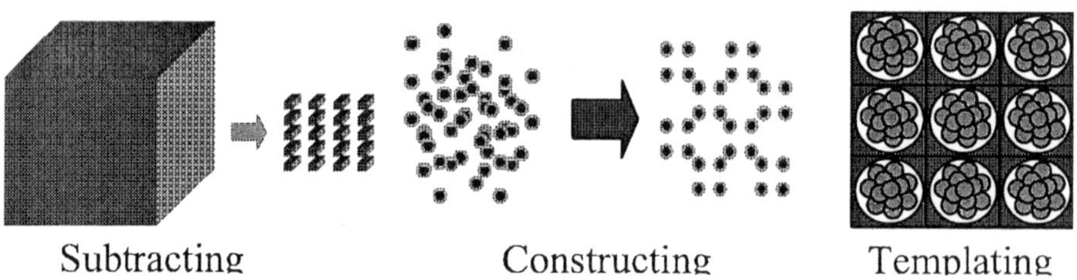

Figure 1. Methods of nanostructure ensemble preparation

Another method is constructive, where the desired nanostructures are formed from individual atoms or molecules (Figure.1). Typical products of this method are dense statistically ordered arrays with intrinsic disorder made of, generally, unlimited number of nanostructures. Well-known examples of this bottom-up approach are self-organised quantum dots and 3-dimensional colloidal crystals. The spacing and ensemble symmetry cannot be voluntarily controlled because they result from internal interaction. In contrast to the subtractive approach, this method is of moderate cost and time consumption.Templating is another method, which is based on the formation of nanoparticles under the influence of a predefined topological relief. If there exists a material with a system of interconnected voids inside, this free space can be infilled with a "guest" material. Examples of porous templates are zeolites, carbon nanotubes, asbestos, porous silicates, opals and so on. In the "host-guest" composite the guest can fill voids either completely or partially providing that the guest synthesis does not destroy the host. In any case, the size of guest particles is restricted by voids, the host dictates the spatial arrangement of guest particles and the their shape are forced to follow the void geometry. In atom-size voids, like in zeolites, the guest-guest interaction in dense arrays can overtake the topology of the template, and, likely, the overall configuration will be the superposition of topological and interactive potentials. Templating is flexible in terms of a combination of various host and guest materials. Low cost and fast growth are its attractive feature.

1. Templating was first applied by Watson [1] in 1968 to stabilise indium nanograins in voids of a porous glass. In early 70s V.N. Bogomolov suggested to exploit ordered templates to achieve crystalline ordering of nanoparticles in all three dimensions [2]. Since that, numerous experiments have been performed using template-directed synthesis of nanoparticles, unfortunately, there were no comprehensive reviews devoted to this method so far. My vision of this area was made public in [3] but it is restricted to a small fraction of the field. One of particularly fast progressing sort of templated materials relates to use of artificial opals as hosts for a vide variety of guest materials. Opal use as a template was launched in early 80s by Y. A. Kumzerov and V.N. Bogomolov in an attempt to prepare nanophase superconductors [4]. The renewed interest in colloidal opal-like crystals was targeted with 3-dimensional photonic crystals and peculiar heat/electron transport materials [5]. The first systematic overview of the properties of opal-based materials with regard to superconductor, semiconductor and dielectric infill materials has been outlined in 1991 [6], further developments of semiconductor infilled opals for the light scattering and electron transport were highlighted in Refs. [7,8]. The aim of the present paper is to up-date this concept in the light of our latest experiments with opal-based photonic crystals (PhC) and nanostructured superconductors.

2. Opal and its modifications

Opal is the face centred cubic (FCC) package of identical spheres (Figure.2). Details of the structure, composition, etc. of opals can be found elsewhere [9]. Silicon dioxide, nowadays, is not the only material for spheres, but different polymers are successfully employed as well. Increasing the variety of materials opens new possibilities for the modification and functionalisation of opal properties, e.g. polymers can be readily doped with light emitting molecules, or prepared as core-shell particles of various compositions themselves or can be reduced to hollow nanocapsules [10]. Another modern tendency is to employ vulnerability of polymer beads in the race for opaline packages made from non-spherical particles.

Recently opals have been synthesised as thin films [9] consisting from one to hundreds layers. Basic advantages of this modification as compared with bulky opals are the greatly reduced crystallisation time, the dramatic decrease of the density of intrinsic packing defects and the improvement of mechanical stability due to the substrate support. Important advantage of a thin film opal over its bulky counterpart is the orientation of the [111] growth direction normal to the substrate. In our studies we relay nowadays on the use of thin film opals as a promising template, the crystallinity of

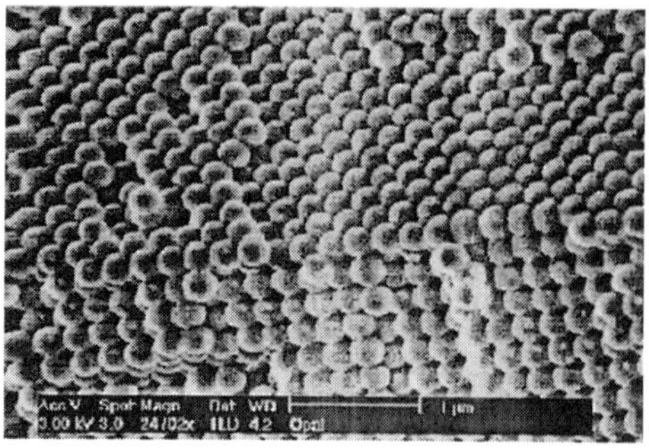

Figure 2 SEM image of direct opal

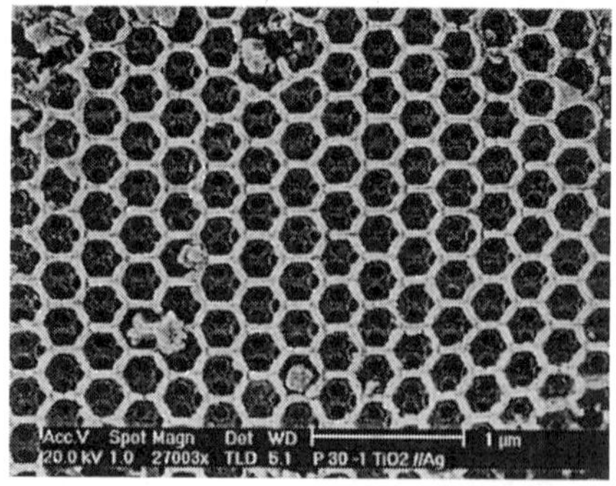

Figure 3. SEM image of TiO_2 inverted opal

whose package allows to improve significantly the visibility of structure-induced effects in physical properties of opaline materials.

The straightforward crystallisation of opaline film has been made by casting a suspension of PMMA beads, prepared by a modified emulsion polymerisation [11], on hydrophilic substrates. By drying the suspension in an atmosphere of controlled humidity (98%), opaline films covering areas of several cm2 can be prepared. These thin-film opals consist of about 20-50 monolayers of beads, and possess monocrystalline domains of size from 10 μm to 1000μm. SEM inspection shows that the long-range order is preserved over many crystallites in the area of several square millimetres. In general, thin film PMMA-based opals combine the advantage of a cheap and quick synthesis with easy machinability and post-deposition processing by e.g. e-beam- or photolithography. More sophisticated methods of opal crystallisation [12,13,14,15,16] offer better long range ordering, but experiments to be discussed below were performed with opals obtain through gravity crystallisation only. Cracks, which divide the opaline film on separate crystallites, pose the main issue for further improvement of the film quality. A similarity of the crystallite lattice orientation over the film allows to avoid averaging of optical characteristics.

Achieving a high RI contrast is a crucial problem for the development of a PhC. To enhance the RI contrast of colloidal PhCs, the filling of the opaline template with a semiconductor and subsequent eliminating of a template has been performed (Fig. 3). The problem with PhCs for the visible is a limited choice of semiconductors, whose RI exceeds 2.8 and whose electronic bandgap is above ~2 eV. Promising inverted opaline structures were synthesised from TiO_2 [17] and Si [18]. Obviously, Si inverted opals can be used at frequencies below 1 eV. TiO_2 so far was synthesised in the anatase form of 2.3 RI and filling fraction of about 11%, which specifications are far below the requirements sufficient for reaching the omnidirectional bandgap. Our attempt to reach the visible light was to implement the SnS_2, which combines a 2.1 eV-wide electronic gap (2.4 eV indirect gap) with a bulk RI of 3.4 at 1.8 eV. SnS_2 replicas are prepared by chemical vapour deposition from SnCl4 vapour and H_2S gas at room temperature and ambient pressure in the voids of PMMA opaline template. The PMMA beads are subsequently dissolved in an organic solvent (THF), leaving behind the semiconductor replica on the glass substrate [19]. The replica is uniform over an area of cm2 thus copying the template ordering. A crucial point is the density of synthesised semiconductor because this factor limits the effective RI. So far RIs of SnS2 replicas between 1.9 and 2.8 have been realised experimentally. In parallel we are using TiO2 replicas as well (Figure.3), because their chemical stability allows to use them as carriers for secondary synthesised "guest" materials.

Preparation of superconductor-infilled opals is still far behind the technology for PhC applications. Opals used for infilling with superconductors (SC) are bulk opals made from silica spheres. Voids of a template were completely infilled with In by high-pressure and high-temperature injection [11]. Presently, we are developing the method of electrochemical infilling of thin opaline films with metals, which is a way to prepare structurally perfect samples without applying mechanical stress.

3. Opal-Based Photonic Crystals

3-dimensional photonic bandgap (PBG) materials working in the visible have been given increasing attention in recent years due to their ability to confine electromagnetic (EM) waves in all three dimensions and to control the emission rate of PhC-embedded light sources. Since the first proposal made in 1972 by V. P. Bykov [20] and in the late 80s - by E. Yablonovich and S. John [21], PhCs have been gaining much interest both for fundamental and applied research. These crystals are periodic dielectric structures that are designed to form the energy band structure for photons, which either allow or forbid the propagation of electromagnetic waves of certain frequency ranges in the same manner as the periodic potential does for electrons in atomic crystals. The absence of propagating EM modes in photonic crystals gives rise to distinct optical phenomena such as the localisation of photons, the inhibition of spontaneous emission, the ultra-refraction, etc. Indeed, the most interesting applications of PhCs for optoelectronics are expected in the near-IR and the visible parts of the spectrum. Naturally, to exploit the PhC in its full capacity, it should be able to cope with EM waves of arbitrarily mode structure, what, certainly, calls for 3-dimensional PBG structures. It is instructive to mention that the omnidirectional PBG means the presence of a frequency range, where the density of optical modes is reduced for all directions simultaneously. The complete PBG is the omnidirectional PBG with zero density of optical modes in the bandgap that requires an infinitely large PhC.

To obtain a 3-dimensional PBG in the visible, a periodic ensemble of scatterers with a half-wavelength lattice parameter has to be realised. Although considerable progress has been made in recent years to microfabricate structures in the sub-μm regime, this process still poses a challenge. Without characterising particular techniques of nanoscale 3-dimensional patterning of bulk materials, it is worth to list competitors: nanolithography patterning and subsequent layer-by-layer assembling of 2-dimensional patterns, e.g. by wafer fusion technique, x-ray lithography, focused ion beam etching, electrochemical etching, autocloning, 4-beam mixing using photopolymers, self-assembly, etc.[22,23,24,25,26,27].

Self-assembling of opals aims with the preparation of ordered 3-dimensional arrays of scatterers from arbitrarily large number of monodisperse spheres. Schematics of two basic modifications of opal-based PhCs is shown in Figure.2. Direct opals constructed from spheres represent the most widely used configuration because, which performs as an incomplete PhC. Spheres of a direct opal can be modified by adding e.g. a laser dye [28] or coating with a shell prior the opal crystallisation. These procedures provide with a certain flexibility the modification of the PBG and emission properties of opals, but, except the hypothetic so far case of metal core-shells, there is no chance to reach an omnidirectional PBG with direct opals due to symmetry considerations. Moreover, the RI of opal spheres is usually about 1.5 and lower, because it is difficult to self-assemble spheres made from semiconductors. The filling fraction of high-RI component in direct opal is on excess of 74% that is far above the optimum value, which maximises the bandwidth of the PBG for a given RI contrast. Correspondingly, experiments with direct opals are usually playing around the Bragg gap, which is the lowest energy bandgap in the PBG diagram of a direct opal.

Theory predicts, that the increase of the RI contrast leads to broadening of all stop-bands and to the development of the omnidirectional bandgap between the 8th and 9th bands in the FCC lattice of air spheres occupying 70-80% in the dielectric environment of the RI on excess of 2.9 [29, 30]. This gap is believed to be very susceptible to a disorder, therefore a good margin by RI contrast has to be kept. Practically, this assumes the application of semiconductors with the RI larger than three. The inverse opal or the opal replica satisfies the conditions for opening the omnidirectional PBG.

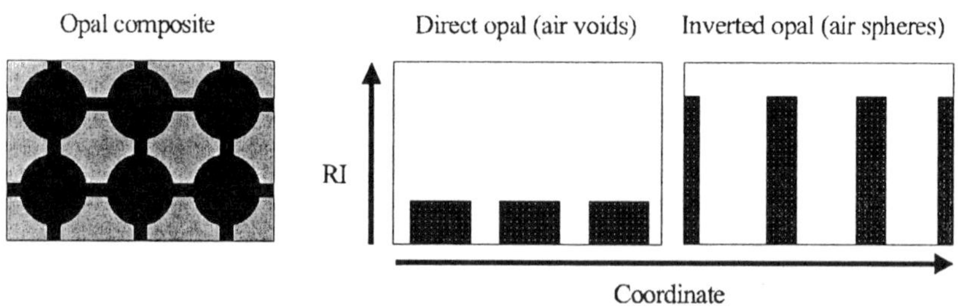

Figure 4. Left panel: schematics of opaline photonic crystals. Middle and right panels: schematics of the RI distribution over the space in the direct and inverted opals, correspondingly.

4. Optical characterisation

4.1 *OPTICAL DIFFRACTION FROM THIN OPALINE FILMS*

Advantages of thin film opals are readily seen in optical reflectance/transmission spectra. In contrast with bulky opals, where the application of single domain spectroscopy with a microscope resolution is necessary for observation of, e.g., the Fabry-Pérot oscillations [31], thin film opals allow to observe these and other fine features from a large area. Moreover, averaging over the azimuth can be avoided due to the similarity of lattice orientations in separate crystallites.

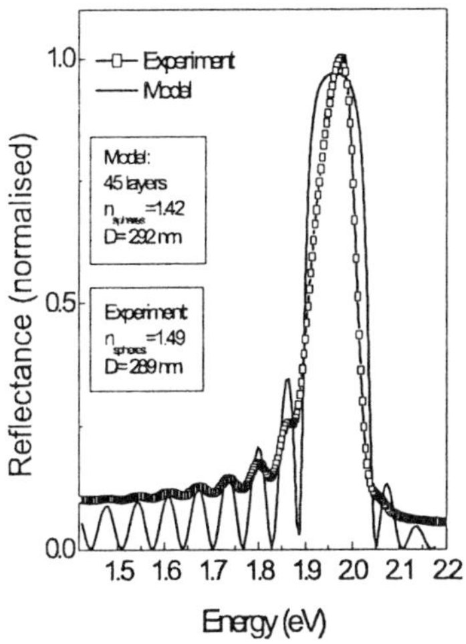

Figure 5. Fabry-Pérot oscillations in the experiment and in the theory

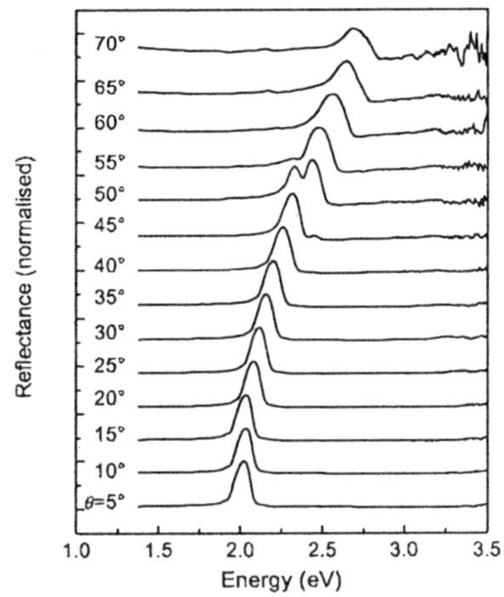

Figure 6. Angular dispersion of PMMA opaline film

Diffraction has been measured in the Bragg configuration, when the angle of the light incidence θ_{inc} is equal to the angle of reflection θ_{ref}, under the illumination with the collimated white light. The scattered light was collected within the ~10o solid angle along the direction of the specular reflection.

Reflectance fringes from a film consisting of 45 layers of *D*=292 nm beads indicate the thickness homogeneity (Figure 5). These Fabry-Pérot oscillations occur due to the interference of the light reflected from opposite surfaces of the opal film. Quantitatively, with increasing the spot area from 1 mm^2 to 1 cm^2, the magnitude of Fabry-Pérot oscillations reduces approximately by factor of three, which can occur only with well-preserved film thickness. Remarkably, the Fabry-Pérot oscillations exist over the wide frequency range from the low energy side of the resonance but vanishes rapidly from the other side. This evolution can be explained by a rapid variation of the dielectric constant of the PhC in the vicinity of the PBG. The reflectance spectra were calculated using a 3-dimensional transfer matrix method and theoretical curves fit nicely the experiment.

The diffraction resonance can be identified with the resonance frequency, the full width at the half-maximum (FWHM) and the depth of the transmission minimum. Figure. 6 shows the set of reflectance spectra of PMMA opaline film taken with unpolarised light at observation angles from $\theta = 5°$ to 70°. The FWHM of the Bragg resonance is $\Delta E/E_0 \approx 5\%$. This value is in a good agreement with the calculated 6.3% FWHM of the [111] gap [32]. The most interesting feature is the splitting of the Bragg resonance for external angles between $\theta \approx 45°$ and $55°$ (Figure. 6). Branching of the stop-band [33] is a co.existence of Bragg peaks from (111) and (200) planes. In accord with the Laue conditions, the diffraction resonance occurs when the end of the vector from the lattice point touches the boundary of the Brillouin zone (see, inset of Figure. 7). However, the peak positions in the vicinity of the crossing point do not follow exactly the Bragg law. The reason is the mixing of closely separated bands near U point, where the (200) gap is only open in the vicinity of the U point and closed for all other directions [33].

SnS_2 and TiO_2 replicas demonstrate optical features similar to their PMMA templates because they preserve an ordering of the template. The direct consequence of the increased RI contrast is the broadening of the PBG (Figure. 9). For example, FWHM of the Bragg peak up to 15% in the in the SnS_2 inverted opal, whose carcass has the effective RO of 2.2 [35].

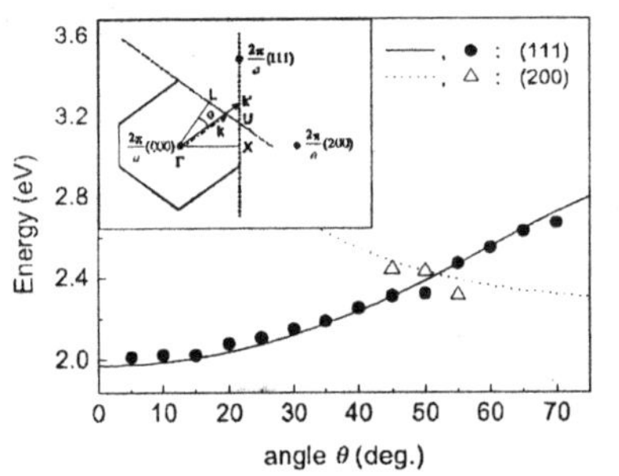

Figure 7. Band branching. Curves calculated dispersion of (111) and (200) bands. Insert – Diffraction conditions in the extended Brillouin zone.

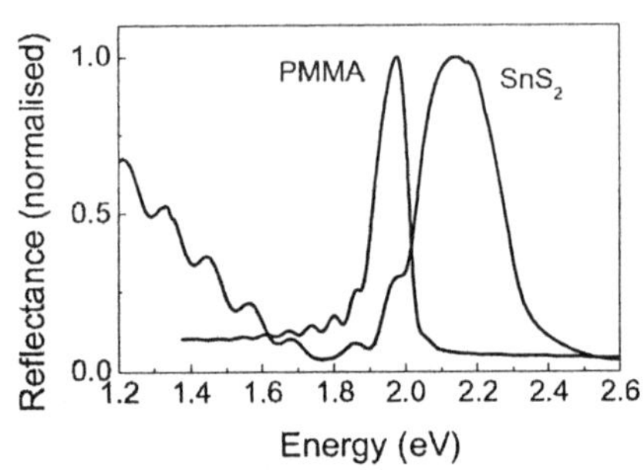

Figure 8. Comparison of the diffraction resonances in PMMA template and SnS_2 replica.

The spectral position of the resonance is defined by the Bragg condition with the Snell law incorporated, $\lambda^2 = 4d^2(n_{eff}^2 - \sin^2\theta)$, where d is the plane spacing in a given direction. Experimental values correspond well to that calculated using the Bragg law, if the RI is given by the effective medium approximation

$$n_{eff}^2 = \sum_i n_i^2 f_i, \qquad (1)$$

where f_i is the volume fraction of the i-component and n_i - its RI.

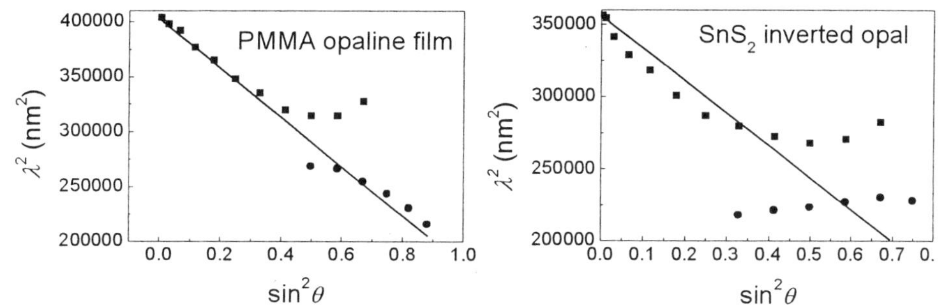

Figure 9. Deviation from the Bragg law (straight line) at branching angles.

As Figure. 9 shows, deviation from the Bragg law occurs in both PMMA and SnS_2 opals, but it is more pronounced in SnS_2 replica. PBGs of both opals calculated by the plane wave method are shown in Figure. 10, where letters are for high-symmetry points in the Brillouin zone of opal. One can see the quick changing the energy of the Bragg gap the *k*-vector.

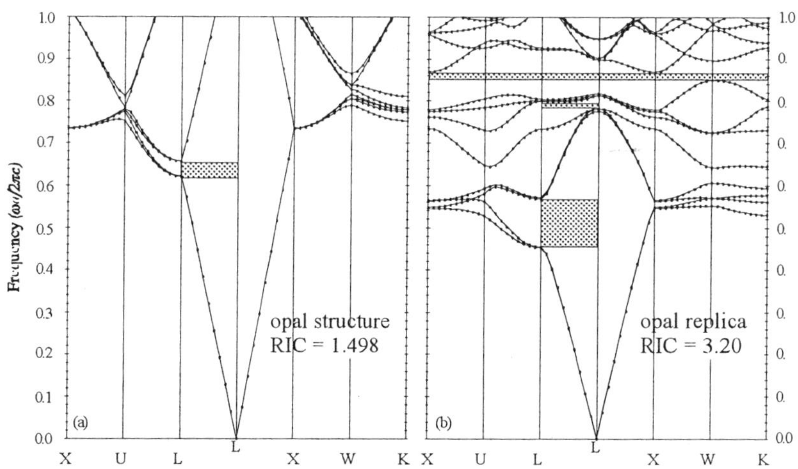

Figure 10. PBG structure for the direct opal (left panel) and for the inverted opal (right panel)

Polarisation-dependent measurements of the angular dispersion of the Bragg resonance demonstrated that the band branching occurs only for s-polarised (E-vector perpendicular to the plane of incidence) EM waves. p-polarised EM waves do not show band branching and follow the Bragg law over the whole angular range. The probable explanation comes from the fact, that p-polarised EM wave samples the opal only in-depth, because the reflectance from the air-opal interface is suppressed near the Bruster angle. For p-polarised waves the diffraction takes place on the (111) plane stack, whereas for s-polarised waves the azimuthal arrangement of scatterers is essential (insert, Figure. 12). It is obvious, that the 60° periodicity observed for the branching resonance follows the periodicity of changing the in-plane scattering conditions.

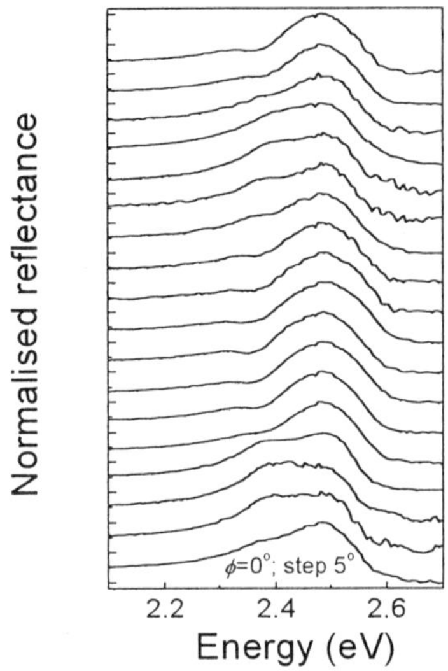

Figure 11. Dependence of the Bragg resonance for s-polarised EM waves at q=50° upon changing the azimuthal rotation of the opal.

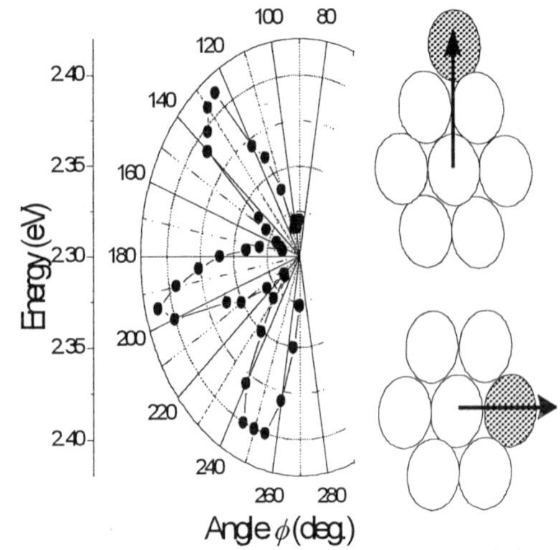

Figure 12. Energy position of the low-frequency peak in

4.2 *SPONTANEOUS EMISSION FROM LIGHT SOURCES EMBEDDED IN OPALINE FILMS*

The major goal of PhCs is to employ a suppression of the spontaneous emission (SE) in the spectral range of a PBG for improved performance of PhC-embedded light sources [4,15]. In order to enhance a SE rate, an emitter should be coupled to localised modes of a PhC, as it follows from the Purcell effect [36]. Ideally, such a localised mode is an in-gap mode associated with a single lattice defect hosted by otherwise perfect PhC with a complete PBG. Lasing by this mechanism was first demonstrated with the 2-dimensional PhC [37]. If there exists an EM wave with the frequency and the k-vector in the stop band of a non-omnidirectional PhC, it should be the mode of the defect state, firstly, and it has to be localised because its magnitude decays exponentially at the distance away from the defect, secondly. Of course, this is an anisotropied localisation. Non-controllable defects in opals give rise to defect modes, which are continuously spread over frequencies and fill the PBG-related minimum in the density of states.

Studies of SE processes require embedding of efficient light source in a PhC. In the low-field regime the photoluminescence (PL) intensity is a linear function of excitation power. Correspondingly, deviations from this linearity can be treated as a manifestation of sample-related alterations of the emission process. In what follows, the example to be discussed is thin film latex opals impregnated with CdTe nanocrystals. But it worth to mention that similar effect was observed in many other combinations of emitter/opal, namely, dye loaded direct and inverted opals, CdTe loaded replicas, etc.

Samples were prepared from suspensions of 240 nm diameter polystyrene spheres by slow sedimentation on glass slides. These opals were infiltrated with CdTe nanocrystals (~3 nm size) in the form of aqueous colloidal solutions [38]. PL spectra were excited by the 351 nm line of a continuous wave (cw) Ar-ion laser focused on a 0.1 mm diameter spot. The excitation power was kept low to avoid degradation of the CdTe emission. PL spectra were collected from the film face opposite to that exposed to laser beam, thus allowing the emission to traverse the film. Emission spectra were measured within 5° fraction of the solid angle along the different directions, which were labelled by the angle θ between the [111] axis and the detection direction, in order to trace the correlation between the spectrum and the angular-dependent stop band.
CdTe nanocrystals under focused laser beam show very broad PL spectrum without sharp features (Figure. 13). The reduction of the emission intensity is clearly seen in the stop band for all plots at different θ-angles. The transmission spectrum can be obtained from PL data by dividing the spectrum obtained at $\theta = 0°$ to that at 80°. We assume that this high angle PL spectrum takes into account all characteristics of the CdTe luminescence in the opal environment but not the stop band. The dip in the relative PL spectrum is centred at 2.2 eV and its FWHM is $\Delta E / E_B = 0.066$(Figure. 14). For comparison, the transmission spectrum shows a dip of relative width $\Delta E / E_B \approx 0.067$ centred at 2.2 eV, which is the frequency of the Bragg resonance. In the stop band the suppression for external light exceeds three times, whereas for the internal light source two-fold suppression is applied.

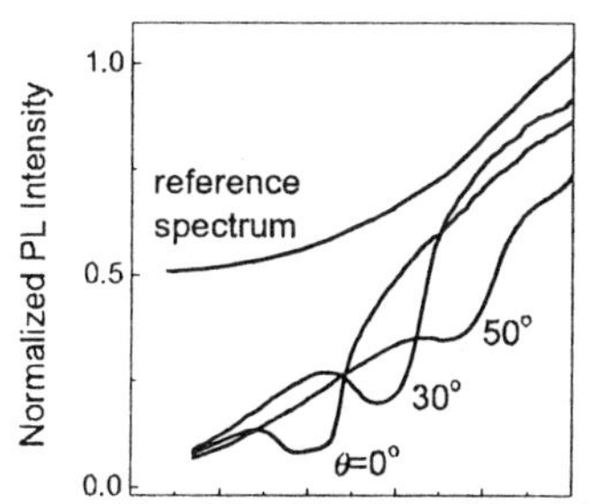

Figure 13. PL spectra of CdTe-opal at different angles of the light collection. Upper curve is the reference spectrum collected at 80°, where no stop band features are observed in emission and reflectance.

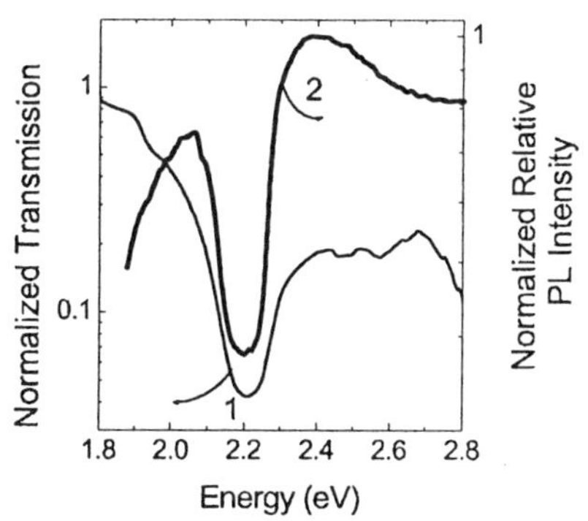

Figure 14. Relative PL spectrum (thick line) and transmission spectrum (thin line) of CdTe-opal for $\theta = 0°$.

In order to visualise small pumping-dependent non-linear changes in the PL spectrum shape, PL spectra, obtained at different levels of the excitation power, were divided by each other. The ratio PL spectrum shows the peak, whose central frequency is exactly the Bragg resonance frequency as it is seen from the comparison with the relative PL spectrum (Figure.14, θ=0°). Its FWHM is $\Delta E/E_B$=0.059, which is very close to the FWHM of the stop band. The frequency of the Bragg peak increases when the angle of detection increases, which effect corresponds to the shift from ΔL direction along the LX branch of the opal PBG (Figure.10), moreover both angular dependences of the ratio peak position and the stop band are similar (Figure.15). Conclusion to draw is that the observed peak in the ratio spectrum is the PBG related effect.

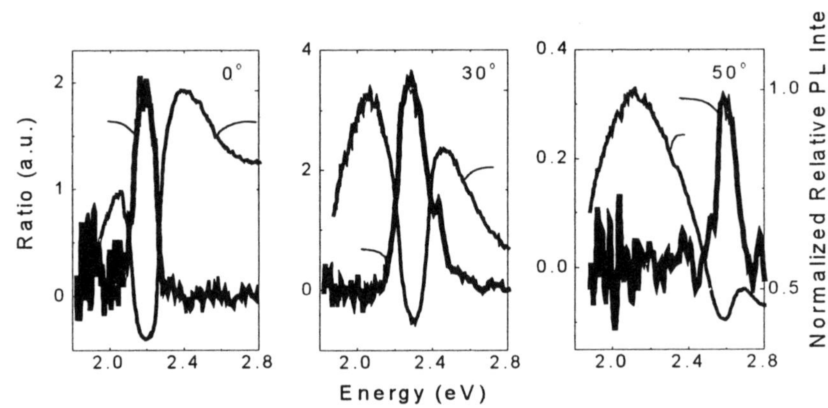

Figure 15. Ratio PL spectra (thick curves) after subtraction of the smooth background super-imposed with corresponding relative PL spectra for different angles of the light collection (shown in panels). Excitation power ratio is 3 mW/0.3 mW for all ratio spectra.

The dependence of the ratio spectrum upon the absolute level of the excitation power for the fixed angle is shown in Figure. 16 for 10-fold power increase increment. At the lowest level of excitation the ratio shows the minimum at the Bragg frequency. For the harder excitation this minimum is replaced with the maximum, and the magnitude of the ratio peak increases with further increase of the base power.

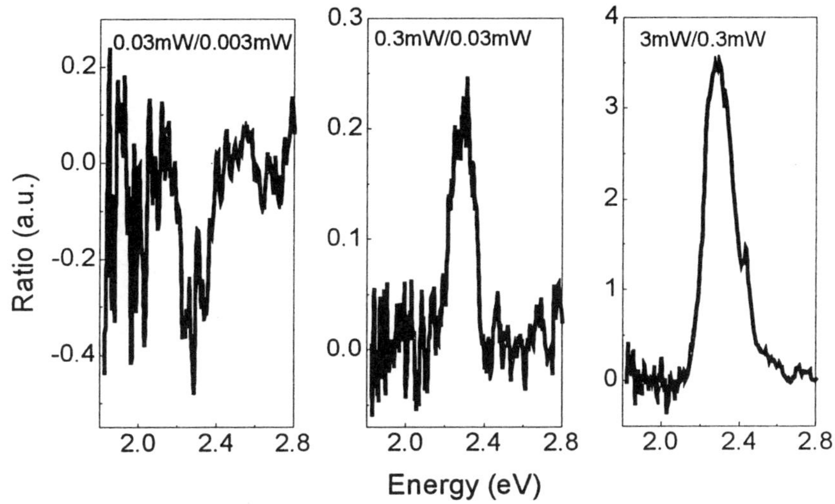

Figure 16. Development of the SE feature, as it is visualised by the ratio PL spectrum, for θ=30° with increase of the optical gain. Ratio increments are shown in panels.

Emission intensity of the light source embedded in PC is the function of both the electronic and photonic energy band structures. Under the weak field excitation, the power-dependent changes in the electronic subsystem of CdTe-opal is merely the linear increase of the number of excited states. As a PhC concerns, emission intensity becomes, in particular, the function of the density of photon states (DOSP) and the rate of a SE [39]. The aperture of the collecting optics defines the fraction of the solid angle, over which the DOSP is integrated. If this fraction along the direction of the diffraction resonance were less than the Bragg cone, then the density of photon eigenstates of the perfect opal would be zero. However, lattice defects produce optical modes, which fills the gap in the DOSP. Because the DOSP does not change with changing the excitation power, the ratio spectrum shows the spectrum of the SE rate, or, in other words, compares the emission rate inside and outside the stop band.

Changing of the SE rate minimum to the maximum along the increase of the excitation power can be understood as the interplay between the superlinear increase of the emission intensity coupled to localised modes and the excitation-independent suppression of the emission rate due to the minimum in the DOSP. At the low excitation power the higher emission rate cannot compensate the decreased number of photon states in the stop band and the minimum of the ratio spectrum is observed, but this superlinearity takes over at higher excitation.

Summarising, the increase of the SE rate at stop band frequencies opens a way for designing efficient PBG-based light sources.

4.3 *DIRECTIONALITY OF EMISSION*

Anisotropy is one of the most prominent features of the emission from an incomplete PhC. The emission intensity at a given frequency varies with changing the emission collection angle due to the anisotropy of the stop-band. In a periodic media the energy transported by EM waves is governed by the topology of dispersion surfaces, which are isofrequency cross-sections of the PBG in the k-space (Figure.17, left panel). In fact, the average Pointing vector is parallel to the group velocity vector and so it is perpendicular to dispersion surfaces. In the homogeneous medium the dispersion surface looks like a ring. For the energy range of the Bragg stop band, dispersion surfaces contain some discontinuities, which are forbidden directions, because of gap opening along these directions. These surfaces were calculated for SnS_2 inverted opal for all k-directions in the LUX□· plane using the plane wave method (Figure.17, right panel). The configuration of dispersion surface changes with the frequency in accord with the PBG diagram. Remarkable result that the angular diagrams of emission intensity (Figure.18) are in a good agreement with the topology of calculated dispersion surfaces, having pronounced dips in the directions of partial gaps and remaining close to the standard $\cos\theta$ like intensity distribution within the passbands.

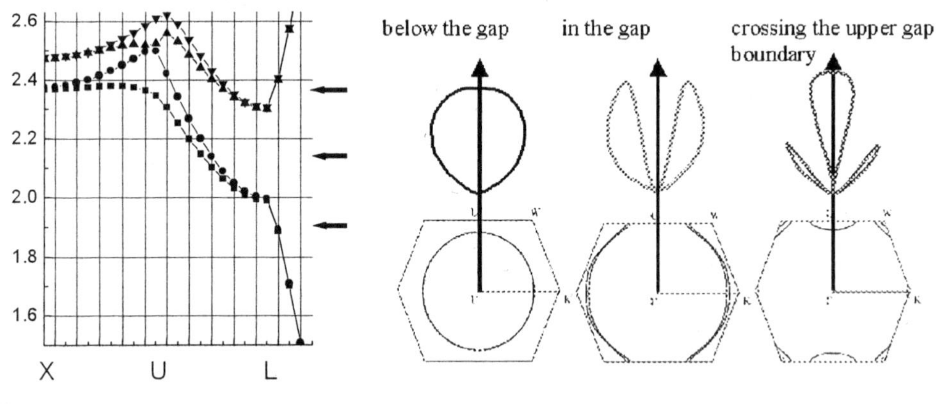

Figure 17. Left panel – PBG structure of SnS_2 inverted opal in the vicinity of the Bragg gap. Right panel – dispersion surface calculated for this PBG at frequencies indicated by arrows in the left panel and schematics of directionality diagrams of the emission at these frequencies.

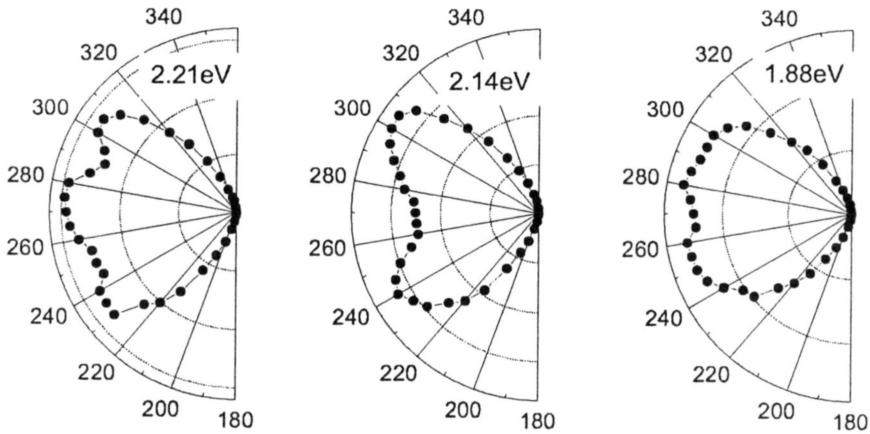

Figure 18. Experimental directionality diagrams for the sample, which PBG and dispersion surfaced are depicted in Figure. 17.

4.4 *EMISSION FROM HIGH-RI CONTRAST OPALINE REPLICA*

Experimental set-up for PL measurements was the same as that described in the section III.2. The PMMA and SnS_2 samples were loaded with organic dyes. A Perylene dye was introduced in the SnS_2 replica by dipping it in the solvent and following drying. A Coumarin dye was embedded in PMMA due course of bead synthesis. Different dyes were used to match different spectral regions of the stop band of these PhCs.

In the case of low-RI PhCs the PL spectrum can be closely reconstructed using the transmission spectrum of PhC and the PL spectrum of an unstructured material of a same composition. Figure. 19c shows reconstructed PL spectra obtained by multiplying transmission spectra of dye/PMMA opal (Figure.19b) and the PL spectrum of unstructured dye/PMMA film (Figure. 19a). Because of the good resembles of calculated (Figure.19c) and measured (Figure. 19a) PL spectra, one can deduce that the emission intensity reduction at stop band is governed by the DOSP spectrum, latter is represented by the transmission spectrum within the aperture cone along the direction of emission collection. Here the DOSP is, as it was mentioned earlier, is taken within the aperture of collecting optics.

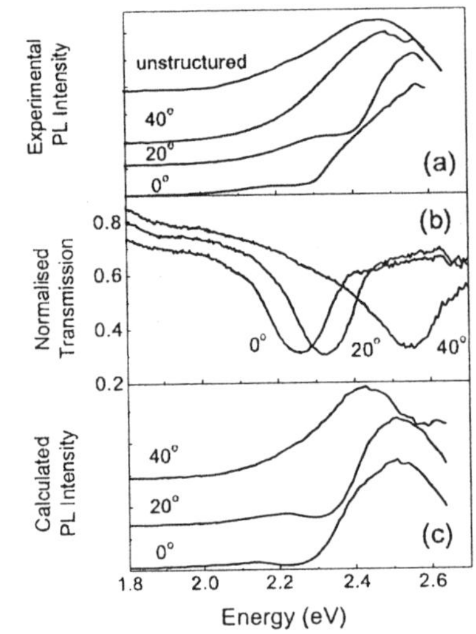

Figure 19. (a) Experimental PL spectra of Coumarin/PMMA opal obtained at different angles and similar spectrum from unstructured sample. (b) Transmission spectra at the same angles. (c) Reconstructed PL spectra.

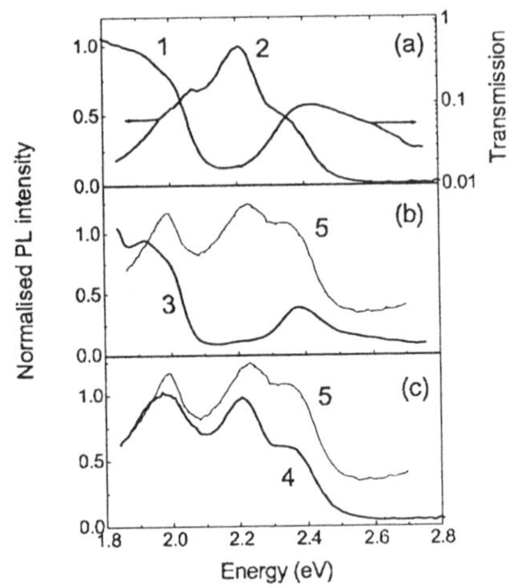

Figure 20. (a) Transmission spectrum of Perylene/SnS_2-replica at θ=0° (1) and PL spectrum of unstructured Perylene/SnS_2 film (2). (b) – Calculated (3) and experimental (5) PL spectra for the full stop band depth. (c) The same as (b) for the 1/10 of the stop band depth.

In contrast, a similar procedure applied to the SnS_2 replica leads to a different result (Figure. 20). If the transmission spectrum were applied in its full depth, the product of the reference emission and opal transmission would be entirely dominated by the bandgap (Figure.20b). However, with the artificially reduced transmission dip it is possible to reproduce satisfactorily the experimentally observed PL spectrum (Figure .20c). This can be understood as the consequence of the weak PBG effect upon the emission from near surface volume and difference in coupling of externally and internally generated light with PhC defect modes.

The PL spectrum of Perylene/SnS_2 replica contains three bands at 2.05, 2.21 and 2.36 eV, whereas the reference spectrum, which is the same dye on the surface of unstructured SnS_2 film, contains practically one band at 2.21 eV (Figure. 20a). PBG suppresses strongly the 2.21 eV band, which matches the stop band, thus increases relatively the PL intensity at the sides of the stop band. Figure. 21 shows that 2.05 eV band exists for all angles , whereas the simulated spectrum cannot reproduce this band for higher angles. According to the theoretical consideration of the dipole emission at the edges of the PBG [26], the emission enhancement takes place at edges of both complete and incomplete gaps. The persistence of 2.05 eV, apparently, is the manifestation of the emission enhancement at the PBG edge. Thus, unlike a low RI contrast PMMA opal, the PL spectrum of high RI contrast PhC is affected non only by the DOSP spectrum, but also by another factor, which is likely the enhancement of a SE rate. Angular independence of this feature is, probably, the result of the formation of a deep minimum in the DOSP spectrum, which is integrated over the whole solid angle.

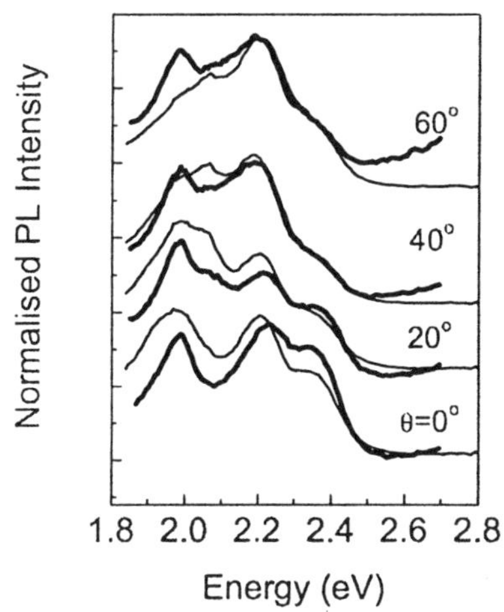

Figure 21. Comparison of experimental (thick lines) and reconstructed (thin lines) spectra of Perylene/SnS_2 replica

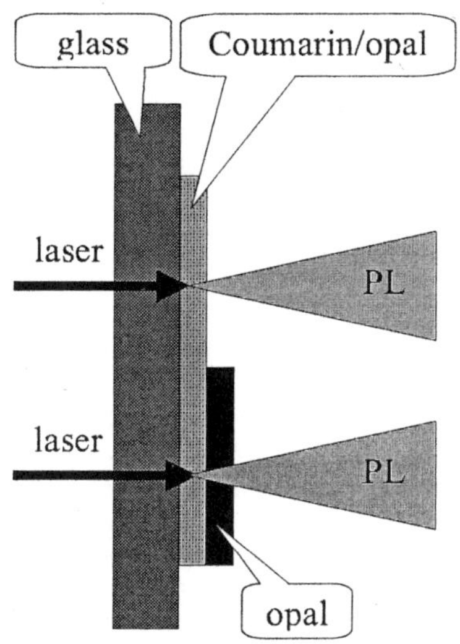

Figure 22. Layout of the experiment, which allows to compare emission from single and double layer photonic crystals.

Summarising the emission section it is reasonable to emphasise the following material aspects of designing the emitter-PhC system. The RIC has to be high to ensure the strong interaction between the light source and the structure. The light source should be incorporated either in the body of the PhC or occupy only a small fraction of its internal volume. Opaline light sources where functions of a PhC and a light source are separated, allow independent tuning of the PBG and the emission band. For example, these are inverted opals loaded with dye molecules [40], rare earth ions [41] or semiconductor nanocrystals. The width of an emission band has to be much narrower than the PBG width, i.e significantly less than 5% of the midband frequency in order to avoid the re-distribution of the emission intensity towards the direction of the allowed propagation. Most organic dyes do not satisfy this demand, but rare earth ions and semiconductor quantum dots do. The emitting component must have a long lifetime under strong laser pumping or electrical excitation. The number of intrinsic defects in an opal should be minimised in order to allow the fabrication of intentional defect, in which mode the light source can couple its emission.

4.5 *PHOTONIC CRYSTAL HETEROSTRUCTURES*

The idea of PhC heterostructures has been suggested as the method to engineer the DOSP spectrum, in particular, to realise the operation of PhCs via artificial defect modes. In the case of opals, the heterostructuring has been inspired by the experiment with double PBG structure obtained by inhomogeneous loading of bulk opal with GaP [42]. In the first instance, this approach can be used to construct the plane defect in the body of PhC by fabricating 3-layer structure. The well-known alternative is the superlattice approach suggested for 1-dimensional diffraction gratings [43]. Recently, the engineering of the DOSP profile was demonstrated with colloidal multiple layered PhCs [44]. In general, the difference between the PhC superlattice and the PhC heterostructure is that the former leads to the creation of the band of propagating modes in the stop band frequency range whereas the latter results in the localised states. This makes heterostructures more suitable for enhancement of the emission as compared with superlattice.

An interface between different colloidal crystals can be obtained by interrupting the colloidal crystallisation after acquiring a certain thickness of the layer and crystallising the next layer from modified spheres, preferably of the same size in order to use colloidal epitaxy for avoiding interface defects. Next interruption and then resumption of the opal crystallisation from spheres identical to that in the first layer is a way to 3-layer structures. The trick leading to defect-look alike PBG is to keep the thickness of the intermediate layer within the wavelength scale. Otherwise, an independent PBG structure will be formed in the intermediate layer. Advances of this structure are numerous, e.g. the defect band can be formed within the stop band, light emitting species can be added exclusively in this layer to avoid coupling of the light source to optical modes of the free space and the emission output can be canalised explicitly through this defect band. Further development of PhC heterostructures assumes the use of inverted opals as heterostructure components and engineering of the heterostructure interface to use its properties for waveguiding.

So far, only two-layer structures have been examined. Figure. 22 shows the layout for the simplest experiment: one dye-loaded opaline film topped with another made of the same but unloaded spheres. The transmission of 2-layer structure is about 1 order of magnitude lower at the frequency of the Bragg resonance than that in 1-layer structure. Taking into account that the excitation of localised modes inside a PhC can be done more readily by the internal light source than by external one, it is reasonable to expect better isolation of the emitting layer from coupling of its emission with modes of a free space. Being excited in the back window configuration, as shown in Figure. 22, the sample shows a PL spectrum either affected by additional PhC or not. Comparing them one can see the better-pronounced dip in relative PL spectrum of two-layer structure (Figure.24). The meaning of this observation is following. If there are many defect states in the stop band, than they can couple together thus reducing the localisation strength. All emitters of this frequency contribute to the output signal, correspondingly, the light emitted in the depth of PhC is transported along the coupled defect states to the sample surface and it becomes suppressed only slightly. Adding another layer of the passive PhC suppresses further the emission intensity in the stop band due to improved efficiency of the isolation of the light source from the free space. In the spirit of localisation, the quality factor of the light source embedded in the bottom layer appears higher because escaping of the energy with stop band frequencies via the second layer becomes more difficult as compared with the single layer structure.

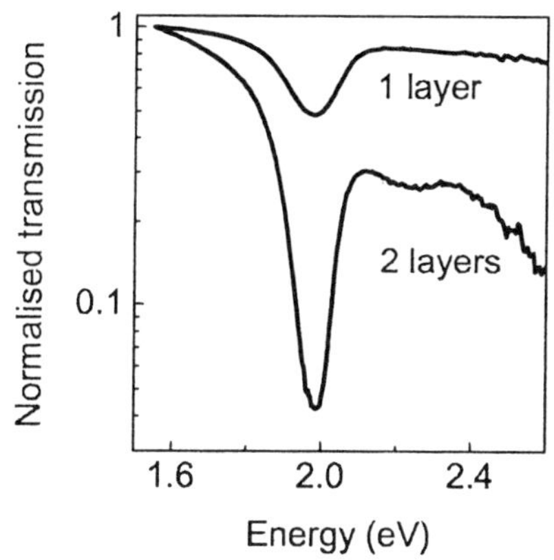

Figure 23. Transmission of 1 and 2 layer opaline films

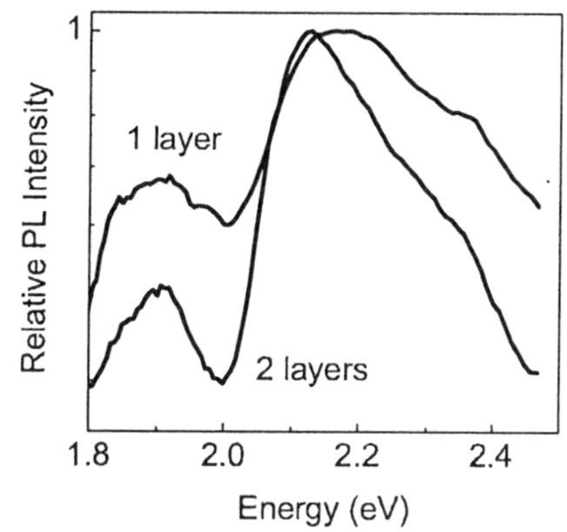

Figure 24. Relative PL intensity obtained from the experiment Fig. 22.

Another representative heterostructure has been prepared by sedimentation of 240 nm latex spheres on the TiO_2 inverted opal, prepared on the base of template assembled from 275 nm spheres. The light source was formed in the TiO_2 replica by impregnating it with CdTe nanocrystallites. This experiment was done to prove the feasibility of preparation procedure and to study the effect upon the emission from the heterogeneous structure possessing of two different bandgaps. The cover latex opal was made thick to acquire its own symmetry regardless the properties of the interface. It worth to note here, that optical properties of interface can be the particularly important issue regarding the coupling of eigenmodes of one PhC with that of another. Two PBG structures co-exist in this sample. Moreover, due to the different RI contrast, namely, 1.58 and 1.5 for TiO_2 and Latex, respectively, and due to different dielectric fractions, which are, correspondingly, 0.2 and 0.8, the angular dispersions of stop bands are different in this heterostructure.

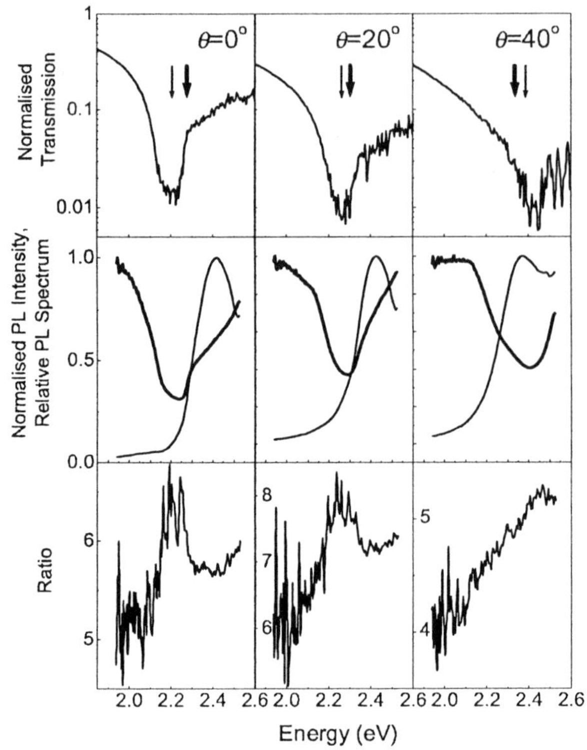

Figure 25. Upper raw – Transmission spectra of Latex-CdTe/TiO_2 hetero photonic crystal at three angles with regard to [111] axis. Thin arrow indicate the position of the central frequency of the Bragg resonance for Latex opaline film, thick arrows – for TiO_2 opaline replica. Middle raw – PL spectra (thin lines) and relative PL spectra (thick lines) obtained as usually with the help of PL spectra at 70°. Bottom raw – ratio spectra obtained for the excitation power decrease from 1 to 0.1mW.

Stop bands of heterostructure layers overlap in the wide range of angles around [111] (Figure. 25). Correspondingly, the emission at stop band frequencies, which is coupled to defect modes of TiO_2 replica, can be additionally reflected by the stop-band of the coating layer. This additional degree of isolation co-operates in the enhancement of the SE rate. The corresponding profile of the SE rate, which is given by ratio spectra, shows the enhancement in the spectral range where stop bands of both films overlap essentially. Amazingly, the enhancement appears sensitive to the stop band of the cover layer, whereas the emitter is situated in another PhC. One possible reason could be the quality of the TiO_2 layer. In this particular example, the ordering of the TiO_2 replica was disturbed because of the possible inhomogeneity of CdTe nanoparticles distribution. The likely reason of the correlation of the SE emission enhancement with the stop band of the cover PhC is the mismatch of the PhC eigenmodes at the PhC interface, which slows down the escape of the emission from active PhC layer. Coupling of EM wave through the interface and related physical phenomena emission are very important issues to be addressed in further experiments.

5. 3-dimensional lattices of weakly coupled superconducting nanograins

Early studies of superconductors (SC) in a confined geometry have been performed in two different directions. First of all, this was the search for the size-induced enhancement of the superconductivity, which brought such results as a huge increase of the critical magnetic field in ensembles of nanosize grains and a moderate increase of the temperature of the superconducting transition. The opal is known as the template for assembling the superconductor nanoparticles since early 80s [37]. Opal-based nanocomposites possess a periodicity, which is comparable or less than a coherence length and a magnetic penetration depth for metallic SCs. Later, lithographically defined 2-dimensional arrays of SC nanostructures absorbed the attention and pre-defined the achievements in this area [45]. The lattice of grains gives rise to the multiple connectivity that is the precursor for the magnetic flux quantization. Oscillations of the critical temperature (*Tc*) vs magnetic field (*H*), which are known as Little-Parks oscillations [46], is the prominent example of magnetic field quantization on closed microsize loops. In the array of submicron loops the contour currents, which encircle loop circumferences, which are, in fact, topological excitations, substitute the magnetic vortices [47]. Generally, these contour currents form a particular interference pattern resulting from interplay of a transport current, an external magnetic field and symmetry of the array [46].

The behavior of 3-dimensional regularly nanostructured SCs is poorely studied [48,49], mostly because of difficulties with extending the array of nanostructures in the third dimension. 3-dimensional arrays can be prepared either by stacking patterned 2-dimensionally layers or by assembling nanograins. The former method results rather in layered structures, whereas the latter encounters the problem of uncontrollable intergrain weak links. With templated arrays the problem of intergrain contacts is lifted due to reversing the conductive component from a lattice of grains to its replica.

In opal-based nanocomposites the size and spacing of nanoparticles as well as the geometry of constrictions are identical throughout the array. Small deviations unavoidable in real structures can be considered similarly to defects in atomic crystals because of the high concentration, which is typically 1012 cm^{-3}, of nearly identical nanostructures. This justifies the assumption of identity of the Cooper pair condensate in each grain as well as the identity of the intergrain coupling throughout the array. Schematics of the SC-opal composite and the spatial variation of the SC order parameter are shown in Figure.26. It is seen, that intergrain bridges play the role of weak links and the lattice arrangement of grains gives rise to the lattice of loops, which are able to quantize the magnetic flux.

Important advantage of the opal template is the control over the filling fraction of SC in composite. In disordered composites, prepared by e.g. co-evaporation with a dielectric, the filling fraction of a SC has to be on excess of ~60% in order to approach the percolation threshold. In opal-based composites, the SC filling fraction can be varied from 26% to below 10% without loosing the homogeneity of the current distribution. Most of experiments so far were made of In-opal composite. It was found that SC properties of In depends dramatically on the In volume fraction (Figure. 27). With about 20 vol.% of In the "anti-dot" modification of SC component can be identified, which is characterised by the high critical current (*Ic*), hyperbolic (Josephson-type) current-voltage characteristic (IVC), the critical temperature (*Tc*) in the range of the bulk SC and slightly enhanced critical magnetic field (*Hc*) [14,50,51]. Oppositely, when the volume fraction is reduced below 10%, the "wireframe" modification of SC is formed, where *Tc* is enhanced, *Hc* is raised by more than 10 times, but the *Ic* is decreased dramatically and IVC shows a parabolic curve. In what follows the properties of one representative example of "wireframe" In-opal will be underlined.

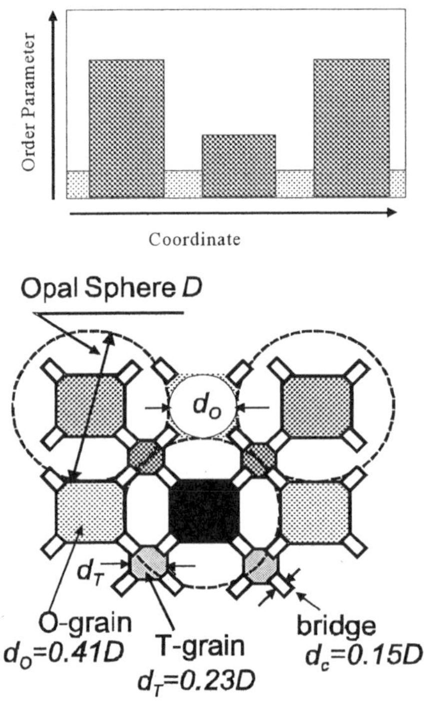

Figure 26. Schematics of SC-opal composite and the diagram of the order parameter along the current path.

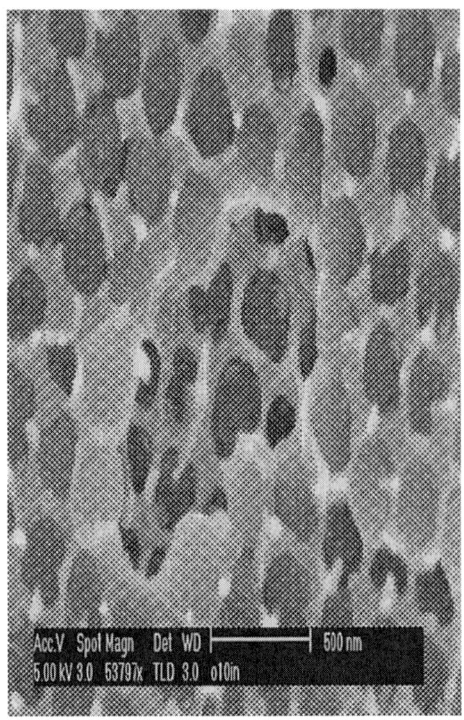

Figure 27. SEM images of "anti-dot" (top) and "wireframe" (bottom) In-opal.

5.1 *SUPERCONDUCTIVITY IN "WIREFRAME" IN-OPAL*

Two types of alternated voids can be distinguished in an FCC package of spheres: O-voids with characteristic size *dO=0.41D* connected each with T-voids of *dT=0.23D* via windows of minimum characteristic diameter *db=0.15D*. Because O-voids are 8-fold co-ordinated and T-voids are 4-fold co-ordinated, there are two identical roads interconnecting O-voids through two different T-voids [47]. The porosity of the ideal FCC package of spheres is about 26%. In order to reduce the volume fraction of SC, additional dielectric material was deposited on the inner surface of opal voids. In the record sample constructed from spheres of $D = 234 \pm 10$ nm the free volume was initially reduced down to 13% by coating with amorphous silica and then by depositing of 26 monolayers of TiO_2. Details of TiO2 deposition are given elsewhere [52]. Effectively, the overall reduction of the void size was about two times as compared with ideal package for the same lattice parameter, i.e. $dO \approx 60$ nm, $dT \approx 25$ nm and $db \approx 7$ nm apply. Voids of a
template were completely infilled with In by high-pressure and high-temperature injection [46]. SEM of the infilled opal with the reduced In fraction is shown in Figure.27.

5.1.1 *Resistive superconducting transition*

The *R(T)* curve for "wireframe" In-opal (Figure. 28) resembles that of strongly doped semiconductor with overall ratio $R_{4K}/R_{300K} \approx 0.72$. Its resistive SC transition shows the onset at $T<4.6$K followed by the peak centred at $T= 4.24$K and then the gradual decrease down to $T_{c0}(R=0) \approx 3.7$K. T_{c0} exceeds sufficiently that of the bulk In $T_{c\ bulk} = 3.41$ K in correspondence with the reduced dimensionality of the SC condensate. The effective diameter d of the SC path can be estimated using the empirical expression $T_{c0} = 3.41 + 5.1/d[\text{nm}]$ [53]. Values of d as 18 and 6 nm apply, if $T_{c0}=3.7$ K or $T_{fluct}=4.25$ K are used, respectively. These are measures of the SC condensate dimension in the In wireframe, which, probably, reflect the modulation of In wire cross-section.
Using the Drude formulation of the resistivity $\rho = \hbar k_F / N_0 e^2 l_{eff}$ one can extract the mean free path and after solving the multiple-resistor circuit for the opal lattice - the coherence length $\xi(T)$. Having the coherence length of bulk, the Ginzburg-Landau coherence length at $T=0$ for a given mean free path is $\xi(0) \approx 21$ nm. This value is essentially smaller than that of the bulk material, but exceeds the characteristic diameter of the SC path. As the result, the order parameter of SC does not change across the current paths. At the same time the penetration depth in In-opal is on excess of 750nm, that suggests very weak screening of the magnetic field by the wireframe.

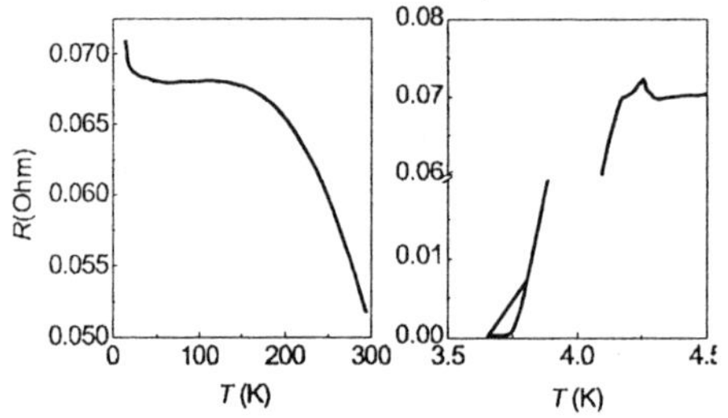

Figure 28. R(T) of "wireframe" In-opal (left) and its resistive superconducting transition (right)

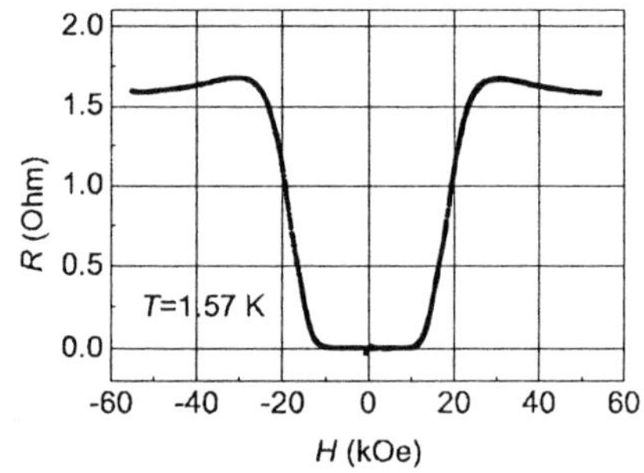

Figure 29. Magnetoresistance of In-opal.

5.2 *CRITICAL MAGNETIC FIELD*

Magnetoresistance curve measured at T=1.37K is shown in the Fig. 29. The onset of the resistance $H_c(R=0)$ appears at 35 times higher magnetic field than that of the bulk In, which is only 280 Oe at T=0. The SC transition is spread broadly by a field. The resistance approaches the resistive anomaly R_N at 37kOe and then the negative magnetoresistance spreads to ~50kOe. Another feature is a kink at approximately the midpoint of the transition. The first derivative dR/dH was used to reveal the precise values of characteristic fields (Fig.30). Likely, that the SC transition is composed of two overlapping transitions at fields H_1 and H_2, (H_1 =15.6kOe and H_2 =18.7 kOe at T=1.37K) shown by derivative peaks. $H_c(T)$ dependences are both fit the standard expression $H_c=H_c(0)(1-(T/T_c)^2)$ (Fig. 31). Correspondingly, the similarity of mechanisms of these transitions is the probable conclusion. These two transitions, naturally, can be related to T- and O-grains, respectively, and the small difference can be explained by the coupling of these grains through bridges. Thus the high value of H_c is the direct consequence of the confinement of the SC condensate at nanosize grains.

5.2.1 *Magnetoresitance of In-opal*

The magnetoresistance $R(H)$ increases nearly linearly with the field $R[Ohm] \propto (0.16-0.18)H[kOe]$ in the temperature range from 3.735 to 1.57 K (Figures.29,30). Moreover, the prefactor increases with the bias current. This sort of $R(H)$ is typical for the resistivity due to the dissipative motion of flux lines under the Lorenz force. Remarkably, there is the asymmetry of $R(H)$ curves for the increase and decrease of the magnetic field, which is ~15% of H_c magnitude. E.g., the penetration field at T=3.39K is by ~0.5 kOe larger than the depenetration field.

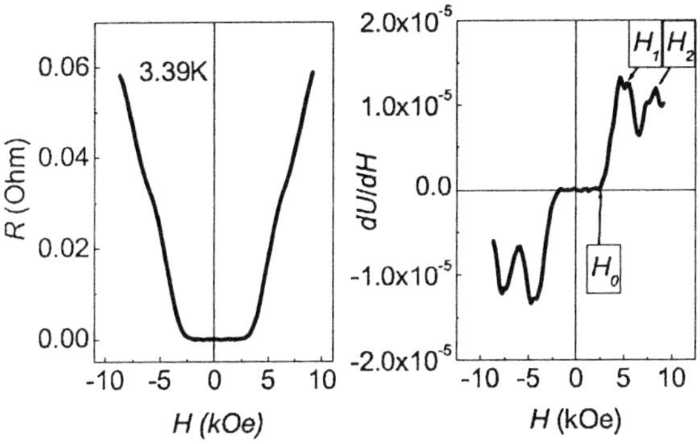

Figure 30. R(H) (left) and its first derivative (right).

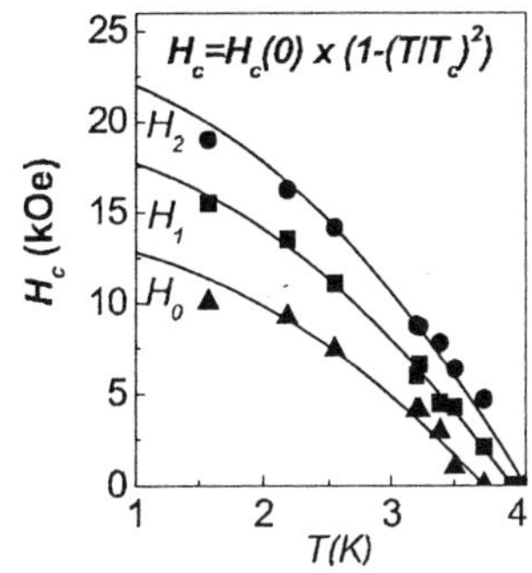

Figure 31. H-T phase boundary diagram of In-opal

The magnetic penetration length in In-opal exceeds considerably the coherence length and approaches the grain size that means the absence of magnetic vortices in the body of grains. Under the applied external field, the supercurrent, which is induced to screen the interior of a SC, must flow along the loop circumference. As the result, this supercurrent generates the magnetic field in loops so that to fulfill the fluxoid quantization conditions. If the external field corresponds to the non-integer number of flux quanta per loop, the direction of the screening current is to add or subtract the local field from the external field depending on whether the captured flux is above or below the half-quanta, correspondingly. Circulating currents are conveniently described as topological magnetic vortices. The magnetic field associated with the bias current is included in the balance of circulating currents as well. In order to change the configuration of a lattice of circulating currents, the whole system should transit from one discrete state to another. This is the way to generate the collective effect because the ensemble of fixed radius loops in In-opal undergoes the strict hierarchy, i.e. any transition in the individual loop should be multiplied by the enormous number of identical loops in the In-opal sample. Thus a macroscopic value of the magnetic field should separate the states of the lattice with and without flux quantum in loops. This is, likely, the reason for the magnetoresistance asymmetry.

The presence of the lattice defects destroys the strict discreteness of ensemble states. In this case the individual flux line, which is associated with the circulating current in the loop with a defect, can change its state at the arbitrary field value. Once it happens, the dissipation of the energy takes place in analogy with the viscous motion of the magnetic vortices in an inhomogeneous SC under the Lorentz force at a current above the depinning threshold. Certainly, this dissipative motion depends on the density of defects and the interaction between circulating currents.

5.3 *3-DIMENSIONAL LATTICE OF SUPERCONDUCTING LOOPS AS A MATERIAL FOR QUANTUM COMPUTING*

SC state is described with a complex order parameter, which magnitude is a number of Cooper pairs. For SCs the phase is identical for all particles of a superfluid. Alternatively, the discontinuity of the phase is provided by Josephson junctions. The current in Josephson junctions is determined by the phase difference between two SCs. Magnetic vortices in SCs are topological excitations, they also can be characterised by a local singularity of the phase distribution. The number of Cooper pairs and the phase are non-commuting quantum variables. Quantum computers are devices proposed to store information in quantum variables and to process the information by making these variables interacting while preserving the quantum coherence [54,55]. Typically, the system under consideration is a two-states quantum system. States of such system is called qubits. SC circuits represent a particular solid-state solution towards the realisation of qubits. SC persistent current qubits were designed as the loop with two Josephson junctions [56]. Two different quantum states in such circuit are associated with circulating currents generated in the loop due to the phase shift of the order parameter in weak links. Third junction is used to control the states of qubits. Another sort of qubits are fluxons trapped in the long annular Josephson junction, which constitute well-defined quantum states of the junction [57]. Long Josephson junction provides a double-well potential where trapped fluxons possess a relatively large decoherence time.

There is a clear analogy of SC-opal composite with the requirements designed for SC materials capable of quantum computation (Figure.32). The smallest loop in SC-opal includes 4 weak links, which provide the slip of the phase of the order parameter between grains. Correspondingly, each weak link generates a current, which encircles the loop. Due to the nature of this process, such a current can be described as the particular quantum state. Totally, there are 4 such different states in a loop. Moreover, the condition of the flux quantization maintains a coherent feedback between these states.

As it was described in the previous section, there is a hierarchy of magnetic stable states in such systems accordingly to the each scaling level of its lattice structure. By virtue of that, the total number of all accessible magnetic states is considerably increased. In the loop lattice the actual distribution of the current depends on the lattice symmetry, but it is also changes with changing the external magnetic field and the bias current due to the mutual cancellation of screening currents in common arms of adjacent loops. The field around a single vortex in SC-opal is less effectively screened and the repulsive interaction is of longer range as compared with the bulk type II superconductor. Thus, loops within the cluster of penetration length dimensions are strongly coupled. Correspondingly, the behavior of the network as a whole is essentially different from that of a linear collection of loops with the smallest areas. In the case of distributed Josephson media in opals the information can be store and processed differently from counting Cooper pairs or fluxons. It is quite similar to recording of 3D hologram, when the information is memorized in the whole storage medium but not in its separate part. [58].

Experimental evidences of the existence of multiple quantum states in the SC-opal are numerous, but indirect. E.g., Figure.33 shows the IVC of "anti-dot" opal. The resistive part of this IVC spreads over the wide voltage range and its hyperbolic shape resembles that of the Josephson junction. Remarkable features are the "anomalous" hysteresis, where the critical current of the resistive to SC state transition exceeds that from SC to resistive state, and the irreversibility of the IVC at its each point in the resistive state.

One of the hypothesis explaining this behaviour relates the irreversibility with the necessity for the system to make the transition from one quantum state to another, which states are macroscopic because of the discreteness of the flux quantization in the loop lattice. Correspondingly, the specificity of the response of the SC-opal upon changes in the bias current is the topologically non-equal processes of penetration and depenetration of the magnetic flux in the loop lattice. Studies of the microwave response of In-opal support the view on the resistive state of IVC as mainly the dissipation due to changes of the configuration of topological vortices [36].

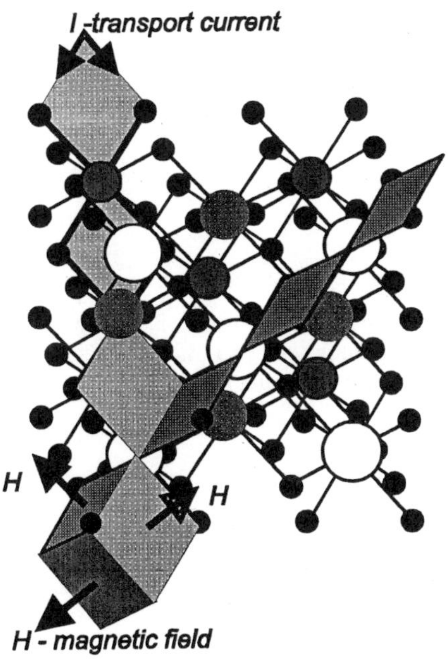

Figure 32. Schematics of the lattice of interconnected SC grains in opal

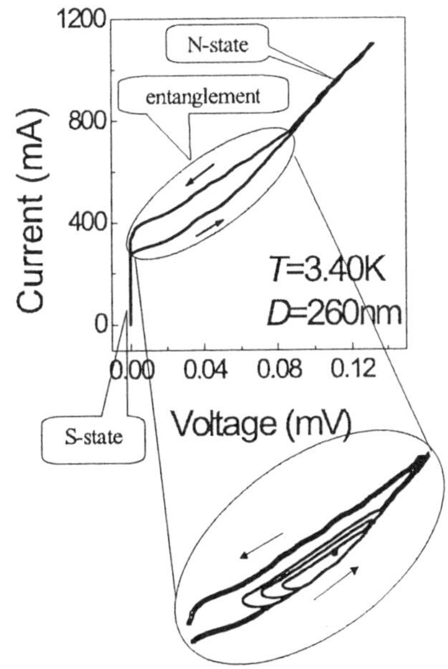

Figure 33. IVC of In-opal and its irreversability

6. Conclusions

State-of-art opal-based 3-dimensional thin film PhCs possessing of the nearly faultless crystalline packages have demonstrated a variety of diffraction phenomena. Qualitative criteria of the crystallinity could be (i) matching the calculated bandgap width and the experimental value of the FWHM of the diffraction resonance, (ii) the in-plane orientation dependence of the Bragg resonance and (iii) the Fabry-Perot oscillations. Omnidirectionality of all-dielectric PhCs is also closed achieved with inverted opals made from high RI semiconductors. However, further progress is dependent upon the improvement of the functionality of these materials. In particular, in the area of PBG-enhanced light sources the necessity of elaborating either complex combinations of different PhCs or incorporating the intentional defects in single crystal PBG materials are widely realised. Our approach towards PhC heterostructures is designed within this main stream in attempt to bring novel physics and application prospects.

Remarkable progress in understanding of the physics of regular arrays of SC nanograins has been achieved with fabrication of “wireframe” SC-opal composites. This is the first time, when enhancement of critical magnetic field was obtained in the regular structure with homogeneous distribution of the supercurrent in the nanocomposite. From the other hand, 3-dimensional lattices of weakly couples SC grains in opal represent the SC circuits, which offer considerable opportunities to build the SC information media with multiple quantum distinguishable states.

7. Acknowledgments

I gratefully acknowledge my colleagues C.M. Sotomayor Torres, T. Maka, D.N. Chigrin, R. Zentel, M. Mueller, A. Rogach, N. Gaponik, A. V. Fokin, D. Cassagne, D.V. Shamshur, V.G. Soloviev and A.V. Chernyev for their valuable contributions.

References

1. N.K. Hinley, J. H. P. Watson, Superconducting Metals in Porous Glass as Granular Superconductors *Phys. Rev.* **183**, 525, (1969)
2. V.N. Bogomolov, Liquids in ultrathin channels, *Soviet Physics Uspehi* **21**, 77 (1978),
3. S.G. Romanov and C.M. Sotomayor Torres,Three-dimensional lattices of nanostructures: the template approach, in “Handbook of Nanostructured Materials and Technology”, Ed. H.S. Nalwa, v.4, ch.4, pp.231-323, Acad. Press, 2000
4. V.N. Bogomolov, V.V. Zhuravlev, A.I. Zadorozhnii, E.V. Kolla, Yu.A. Kumzerov, Voltage-current characteristics of a regular system of weakly coupled superconducting particles. *JETP Lett*, **36**, 365 (1982)
5. Y. Xia, B. Gates, Y. Yin, Yu Lu Monodispersed Colloidal Spheres: Old Materials with New Applications *Adv. Mater.* **12**, 693, 2000

6. V.N. Bogomolov, Yu.A. Kumzerov, S.G. Romanov. Fabrication of three-dimensional superlattices of nanostructures. in "*Physics of Nanostructures*" eds.J.H.Davies and A.R.Long. IOP Publishing, Bristol, Philadelphia, p.317, 1992
7. S.G. Romanov, 3-Dimensional Photonic Crystals at Optical Wavelengths (review), *J.Nonlinear Optical Physics & Materials*, **7**, 181-200, (1998)
8. S. G. Romanov, Concept Of Templated Lattices Of Semiconductor Nanostructures, *Journal of Porous Materials*, 7 (1/3), 153-157 (2000)
9. V.G.Balakirev, V.N.Bogomolov, V.V.Zhuravlev, Y.A.Kumzerov, V.P.Petranovskii, S.G.Romanov, LA.Samoilovich. Three-dimensional superlattices in the opals. *Crystallography Reports*, **38**, 348-353 (1993)
10. Special issue on Photonic Crystals *Adv. Mater.*, **13**, no.6, 369 (2000)
11. J.W. Goodwin, J. Hearn, C.C. Ho, R. H Ottewill, Sudies on the preparation and characterisation of monodisperse polystyrene latices, Colloid Polym. Sci. 252, 464 (1974)
A. van Blaaderen, R. Ruel, P. Wiltzius, Template-directed colloidal crystallization, *Nature* **385**, 321 (1997)
B. Gates, D. Qin, Y. Xia, Assembly of nanoparticles into opaline structures over large areas, *Adv. Mat.* **11**, 466 (1999)
12. P.A. Kralchevsky, N.D. Denkov. V.N. Paunov, O.D. Velev, I.B. Ivanov, H. Yoshimura, K. Nagayama, Formation of two-dimensional colloid crystals in liquid films under the action of capillary forces, *J. Phys.: Condens. Matter* **6**, A395 (1994)
13. R. Amos, J.G. Rarity, P.R. Tapster and T.J. Shepherd, Fabrication of large-area face-centered-cubic hard-sphere colloidal crystals by shear alignment, *Phys. Rev. E* **61**, 2929 (2000)
14. JEGJ Wijnhoven, W.L. Vos, Preparation of photonic crystals made of air spheres in titania, *Science* **281**, 802 (1998)
A. Blanco, E. Chomski, S. Grabtchak, M. Ibisate, S. John, S.W. Leonard, C. Lopez, F. Meseguer, H. Míguez, J.P. Mondia, G.A. Ozin, O. Toader, H.M. van Driel, Large-scale synthesis of a silicon photonic crystal with a complete three-dimensional bandgap near 1.5 micrometres, *Nature* **405**, 437 (2000)
15. M. Müller, R. Zentel, T. Maka, S.G. Romanov, C.M. Sotomayor Torres, Photonic crystal films with high refractive index contrast, *Adv. Mat.* **12**, 1499 (2000)
16. V.P. Bykov, Spontaneous emission in a periodic structure, *Sov. Phys. JETP* **35**, 269 (1972).
17. E. Yablonovich, Inhibited spontaneous emission in solid-state physics and electronics, *Phys Rev. Lett* **58**, 2059 (1987); S. John, Strong localization of photons in certain disordered dielectric superlattices, *Phys Rev. Lett* **58**, 2486 (1987)
18. S. Noda, N. Yamamoto, M. Imrada, H. Kobayashi, M. Okato, Alignment and stacking of semiconductor photonic bandgaps by wafer-fusion, *J. Lightwave Technol.* **17**, 1948 (1999)
B. Cuisin, A. Chelnokov, J.M. Lourtioz, D. Decanini Y. Chen, Fabrication of three-dimensional photonic structures with submicrometer resolution by x-ray lithography, *J. Vac. Sci. Technol. B* **18**, 3535 (2000)
19. K. Wang, A. Chelnokov, S. Rowson, P. Garoche, J.M. Lourtioz, Focused-ion-beam etching in macroporous silicon to realize three-dimensional photonic crystals, *J. Phys. D: Appl. Phys.* **33**, L119 (2000)
20. S.Y. Lin, J.G. Fleming, A three dimensional optical photonic crystal, *J. Lightwave Technol.* **17**, 1944 (2000)
21. M. Champbell, D.N. Sharp, M.T. Harrison, R.G. Denning, A.J. Turberfield, Fabrication of photonic crystals for the visible spectrum by holographic lithography, *Nature* **404**, 53 (2000)
22. E. Kuramochi, M. Notomi, T. Tamamura, T. Kawashima, S. Kawakami, J. Takahashi, C. Takahashi, Drilled alternating-layer structure for three-dimensional photonic crystals with a full band gap, *J. Vac. Sci. Technol.* **B 18** , 3510 (2000)
23. M. Mueller , R. Zentel, T. Maka , S. G. Romanov , C. M. Sotomayor Torres, Dye-Containing Polymer Beads as Photonic Crystals, *Chem. Materials*, **12**, 2508-2512 (2000)
24. H.S. Sözüer, J.W. Haus, R. Inguva, Photonic bands: Convergence problems with the plane-wave method, *Phys. Rev. B* **45**, 13962 (1992)
25. K. Busch, S. John, Photonic band gap formation in certain self-organizing systems, *Phys. Rev. E* **58**, 3896 (1998)
26. Yu. A. Vlasov, M. Deutsch, D.J. Norris, Single-domain spectroscopy of self-assembled photonic crystals, *Appl. Phys. Lett.* **76**, 1627 (2000)
A. Reynolds, F. López-Tejeira, D. Cassagne, F.J. García-Vidal, C. Jouanin, and J. Sánchez-Dehesa, Spectral properties of opal-based photonic crystals having a SiO2 matrix, *Phys. Rev. B.* **60**, 11422 (1999)
27. H.M. van Driel, W.L. Vos, Multiple Bragg wave coupling in photonic band-gap crystals, *Phys. Rev. B.* **62**, 9872 (2000)

28. S.G. Romanov, T. Maka, C.M. Sotomayor Torres, M. Müller, R. Zentel, D. Cassagne, J. Manzanares-Martinez and C. Jouanin, Diffraction of light from thin-film polymethylmethacrylate opaline photonic crystals, *Phys. Rev. E* **63**, 056603 (2001)
29. M. Müller , R. Zentel, T. Maka , S. G. Romanov , C. M. Sotomayor Torres, Photonic Crystal Films with High Refractive Index Contrast, *Adv. Materials* **12**, 1499-1503 (2000)
30. E.M. Purcell, Spontaneous emission probabilities at radio frequencies, Phys.Rev. 69, 681 (1946)
31. O.J. Painter, A. Husain, A. Sherer, J. D. O'Brien, I. Kim, P.D. Dapkus, Room temperature photonic crystal defect lasers at near infrared wavelengths in InGaAsP, *J. Lightwave Technology* **17**, 2082-2088 (1999)
A. L. Rogach, L. Katsikas, A. Kornowski, D. Su, A. EychmÜller, H. Weller, Synthesis and characterization of thiol-stabilized CdTe nanocrystals, *Ber. Bunsenges. Phys. Chem.*, **100**, 1772-1778. (1996)
32. T. Suzuki, P.K.L. Yu, Emission Power Of An Electric Dipole In The Photonic Band Structure Of The Fcc Lattice, *J. Opt. Soc. Am. B*, **12**, 570 (1995)
33. M. Megens, JEGJ Wijnhoven, A. Lagendijk, W.L. Vos, Light sources inside photonic crystals, *J. Opt. Soc. Am. B* **16** 1403 (1999)
34. S.G. Romanov, A.V. Fokin, R.M. de la Rue, Eu3+ Emission in an Anisotropic Photonic Bandgap Environment, *Appl. Phys. Lett.* **76**, 1656 (2000)
35. S. G. Romanov, H. M. Yates, M. E. Pemble, R. M De La Rue, Opal-Based Photonic Crystal with Double Photonic Bandgap Structure, *J. Phys.: Cond. Matter* **12**, 8221-8229 (2000)
36. P. St. J. Russel Bragg Resonance of light in optical superlattices, *Phys. Rev. Lett.* **56**, 596 (1986)
37. P. Jiang, G. N. Ostojic, R. Narat, D. M. Mittleman, V. L. Colvin, The Fabrication and Bandgap Engineering of Photonic Multilayers, *Adv. Mater.*, **13**, 389, (2001)
38. V.V. Moschalkov, V. Bruyndoncx, L. Nan Look, M.J. Van Bael, Y. Bruynseraede, A. Tonomura, Quantization and confinement phenomena in nanostructured superconductors, in "Handbook of Nanostructures and Nanotechnology", ed. H.S. Nalwa, Acad. Press, 2000, v.3, 451.
39. W.A. Little, R.D. Parks, Observation of Quantum Periodicity in the Transition Temperature of a Superconducting Cylinder, *Phys. Rev. Lett.*, **9**, 9 (1962)
40. H.S.J. van der Zant, M. N. Webster, J. Romijn, J. E. Mooij, Vortices in two-dimensional superconducting weakly coupled wire networks, *Phys. Rev. B*, **50**, 340 (1994)
41. V..N.Bogomolov, Y.A.Kumzerov, S.G.Romanov, V.V.Zhuravlev. Josephson properties of the three-dimensional regular lattice of the weakly coupled nanoparticles. *Physica C: Superconductivity*, **208**, 371 (1993)
42. E.V. Charnaya, C. ein, K.j. Lin, C. S. Wur, Yu. A. Kumzerov, Superconductivity of gallium in various confined geometries, *Phys. Rev. B*, **58**, 467 (1998)
43. S.G.Romanov, A.V.Fokin, K.Babamuratov, Resistance anomalies and negative magnetoresistance in a regular 3D lattice of superconducting nanostructures. *JETP Lett.*, **58**, 824, (1993)
44. S.G. Romanov, D.V. Shamshur. Suppression of the superconductivity in 3-dimensional lattice of weakly coupled indium nanoparticles in opal. *Solid State Physics*, **42**, no. 4, 581-588 (2000) (*Fizika tverdogo tela*, **42**, no. 4, pp.5-12 (2000) in Russian))
45. V.V. Tretyakov, S.G. Romanov, A.V. Fokin and V.I. Alperovich. Electron Quantitative Microanalysis of Composition of Opals-Based Nanostructured Materials, *Mikrochimica Acta*, No.S15, pp.211-217 (1998)
46. H.T. Vogel, M.M. Garland, *J.Appl.Phys.*, **38**, 5116 (1967)
47. S. Lloyd, A potentially realizable quantum computer, *Science*, **261**, 1589 (1993)
48. D.P. DiVincenzo, Quantum computation, *Science*, **269**, 225 (1995)
49. T.P. Orlando, J.E. Mooij, L. Tan, C:H: van der Wal, L.S. Leditov, S. Lloyd, J.J. Mazo, Superconducting persistent-current qubit, *Phys. Rev. B*, **60**, 15398 (1999)
50. A.Franz, A. Wallraff, .A. V. Ustinov, Measurements of critical-current diffraction patterns in annular Josephson junctions, *Phys. Rev. B* **62**, 119 (2000)
51. S. Ktitorov and Yu.I.Kuzmin, private communication

TEMPLATES FOR METAL NANOWIRE SELF-ASSEMBLY

Mathias Brust*, Yun-Ping Liu, Thomas O. Hutchinson,
Christopher J. Kiely
Centre for Nanoscale Science, Department of Chemistry, The University of Liverpool, Liverpool L69 7ZD, United Kingdom, E-mail: M.Brust@liv.ac.uk

The miniaturization of electronic devices is the mayor economical driving force in nanotechnology, and consequently, traditional lithographic fabrication techniques are currently being pushed towards their size resolution limits. This is generating a need for research into alternative fabrication strategies, which can provide smaller structures. At present, these structures are predominantly of fundamental interest since size dependent phenomena such as Coulomb blockade effects[1] and conductance quantisation[2] are beginning to play an important role and can be studied.

Two very promising approaches to smaller structures and devices are nano-manipulation using scanning probe microscopes such as atomic force (AFM) and scanning tunnelling microscopes (STM), and self-assembly.[3,4] While the former can be applied quite readily to the fabrication and imaging of surface confined structures with atomic or molecular precision, the latter could, in principle, be exploited for the inexpensive mass fabrication of self-assembled nano-objects and devices. This would resemble the chemical synthesis of a large number of precisely defined molecules, only on a somewhat larger size scale. The possibility of such *chemical nanofabrication* is demonstrated by biological systems, which are, without exception, entirely constructed in this way. Artificial self-assembling systems are of course substantially different in that they are lacking organizational complexity and tend to develop towards the formation of thermodynamically stable structures. The controlled fabrication of even very simple nanostructures by self-assembly is at present still challenging, and the future development of a technology based on self-assembly would first require a thorough understanding of the physicochemical factors that govern self-organization processes. Some important factors include the size, shape and chemical nature of the building blocks, the chemical nature and topography of the substrate and the properties of the medium in which the processes are taking place.

In the context of ongoing work on nanostructure self-organisation from gold nanoparticles we present here a brief overview of our recent results on the use of templates for the directed self-assembly of linear structures. Examples of two different systems are given to illustrate the importance of particle reactivity in one case, and that of substrate topography in the other.

E. Buzaneva and P. Scharff (eds.), Frontiers of Multifunctional Nanosystems, 139–146.

Gold nanoparticles have been chosen as building blocks since they can be readily prepared with controllable chemical reactivity over a range of different sizes. The particles used in this investigation were prepared via a two-phase liquid/liquid (water/toluene) reduction route using sodium borohydride as a reducing agent and tetraoctylammonium bromide as a phase transfer catalyst.[5-8] They were stabilised in toluene solution by two different strategies, i.e. ligand- and charge stabilisation. The preparations were carried out either in the presence of the stabilising ligand dodecanethiol giving a particle core size of 1.5 to 3 nm, or in its absence leading to particles of 4 to 7 nm core diameter charge-stabilised by the adsorption of bromide and quarternary ammonium ions. The larger particles are quite reactive due to the comparatively weak adsorption of the bromide ions. They can, for example, be linked covalently to each other or to surfaces via dithiol molecules and are also readily transformed into ligand-stabilised particles by addition of thiols.[9,10] After reaction with thiols such particles are analogous to the smaller particles prepared in the presence of thiols. Thiol-capped particles of both sizes can be isolated as pure, solid products, which readily re-dissolve in various organic solvents (toluene, chloroform, pentane, etc.) and are stable for many months due to the presence of a compact self-assembled monolayer (SAM) of thiol molecules on their surface.

One of the most interesting features of both the ligand- and the charge-stabilised particles used in this work is their ability to form well-ordered, two-dimensional superlattices by spontaneous self-organisation.[11,12] Such structures form simply by allowing a solution of particles in an organic solvent to evaporate on a suitable substrate. A typical example of hexagonal ordering of thiol-capped gold nanoparticles obtained in this way is shown in Figure 1. The factors that govern this two-dimensional crystallisation process are manifold and still subject of current research. From molecular dynamics simulations it has been concluded that the interlocking of bundles of ligand molecules contributes significantly to the formation of ordered structures.[13] The nature of the substrate also plays an important role. On amorphous carbon, for example, polydisperse thiol-capped gold particles form size segregated, monodisperse superlattices coexisting with ordered arrays of a bimodal size distribution[14], while the same particles show fractal growth patterns without any order on oxide covered silicon surfaces.[15] Very recent work by Lin et al. has demonstrated that unusually large, ordered rafts of hexagonal superlattices can be obtained on flat silicon nitride substrates if the evaporating solvent is "contaminated" by excess dodecanethiol molecules.[16] The extremely slow de-wetting in the presence of the non-volatile thiol has been given as an explanation for this phenomenon. It is, however, also possible that entropy driven crystallisation[17] is responsible for this effect. As the solvent evaporates thiol molecules and metal particles are competing for solvent volume and the total entropy of the system increases temporarily by crystallisation of the particles, which frees up extra volume for the thiol molecules remaining in solution. Similar effects may explain the spontaneous alignment of nanoparticles along topographical features as will be explained later.

The intriguing property of gold (and various other) nanoparticles to self-organise into regular arrays suggests their application as building blocks also for other and possibly more complex nanostructures. The first example given in this communication is the use of ropes of single wall carbon nanotubes (SWNTs) as a linear

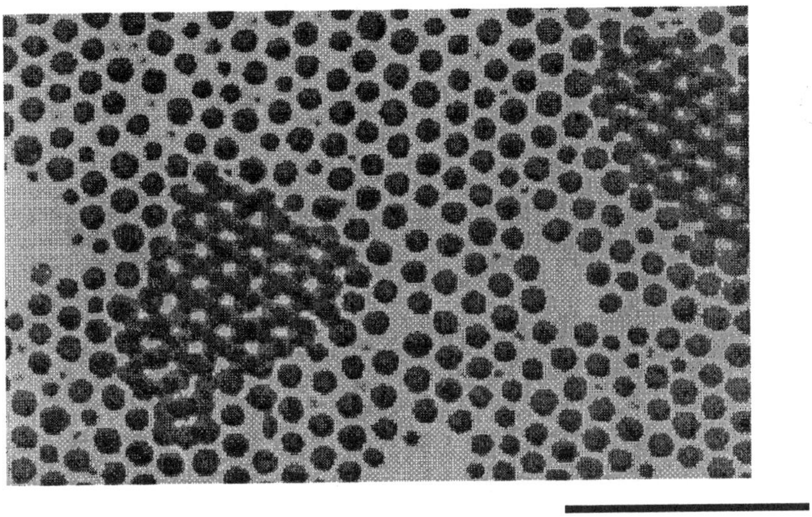

Figure 1 Transmission electron micrograph showing hexagonal ordering in a monolayer raft of dodecanethiol-capped gold nanoparticles deposited onto amorphous carbon by solvent evaporation. Note the onset of ordered bilayer formation (dark lines).

template for the formation of gold nanowires. We have previously reported the interaction of gold nanoparticles stabilised by tetraoctylammonium bromide with C_{60} fullerene in toluene solution.[18] After addition of fullerene the particles aggregate immediately, which is indicated by a sudden colour change from wine red to deep blue, and finally precipitate quantitatively as a black powder consisting of agglomerates of gold particles individually surrounded by a shell of fullerene molecules. The aggregation is probably caused by an electron transfer step from the particles to the fullerene molecules with subsequent electrostatic binding facilitated by the relatively low dielectric constant of the solvent. The adsorption of the fullerene is weak and after repeated washing with toluene the material decomposes to give gold micro-grains surrounded by a thin layer of gold bromide.

Carbon nanotubes react with the same type of gold particles in a similar way resulting in the decoration of SWNT ropes by the particles as shown in Figure 2.[19] Fitzmaurice and co-workers have first published this decoration phenomenon and also obtained continuous tubular gold structures after sintering the particles by a heat treatment.[20] Thiol-capped gold nanoparticles, however, do not interact measurably with SWNTs suspended in toluene presumably because electron transfer is impeded by the passivating ligand shell. This illustrates the importance of the chemical reactivity of the building blocks as a parameter that can be addressed to control nanostructure self-assembly.

A very different kind of templated self-assembly of line structures can be obtained by evaporating a solution of thiol-capped gold nanoparticles on a substrate with linear topographical features. Such features can be easily obtained by so-called

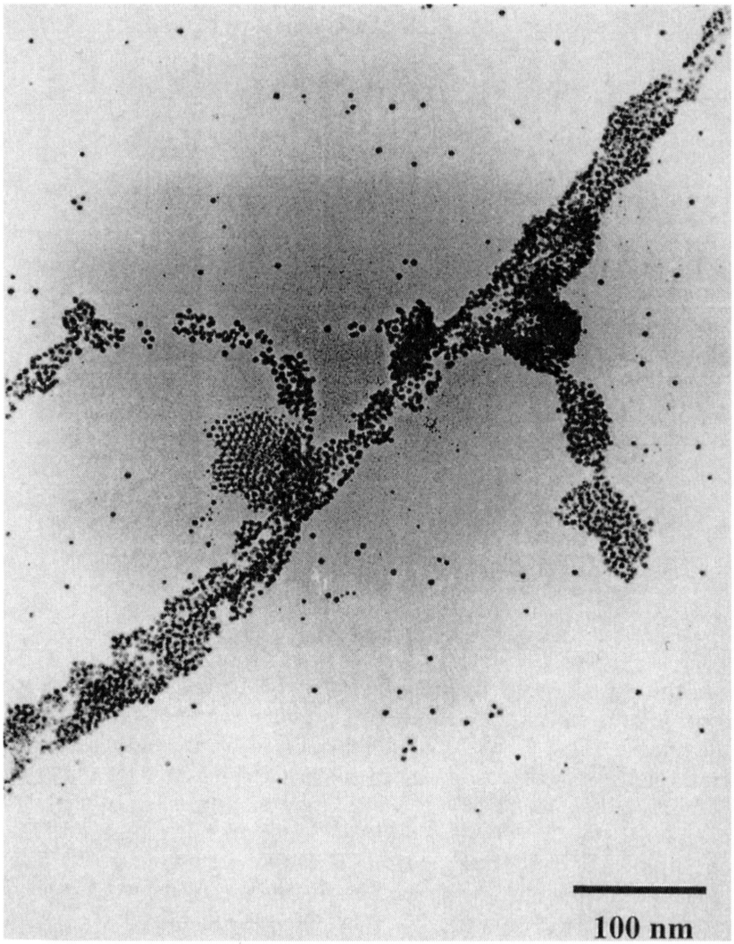

Figure 2 Transmission electron micrograph of a rope of single wall carbon nanotubes decorated by tetraoctylammonium bromide stabilised gold nanoparticles.

friction transfer of polytetrafluoroethylene (PTFE) lines onto flat glass or silicon surfaces.[21] The transfer of lines of ca. 10 nm width and 10 to 15 nm height with an average spacing of ca. 100 nm is achieved simply by rubbing a block of PTFE unidirectionally across the heated (200-250 °C) substrate.[22] An atomic force microscopy (AFM) image of PTFE lines on a glass microscope slide is shown in Figure 3. Substrates treated in this way can be used directly for the assembly of lines of thiol-capped gold nanoparticles by drop casting and solvent evaporation. Preliminary results suggest that the particles preferentially align along the PTFE lines as the solvent evaporates. Since the particles used so far were in the size range between 2 and 6 nm their alignment in close contact with the larger PTFE features has been somewhat difficult to show unambiguously by AFM. Very recent scanning electron microscopy (SEM) data, however, indicate that this is indeed the case. A clearer demonstration of

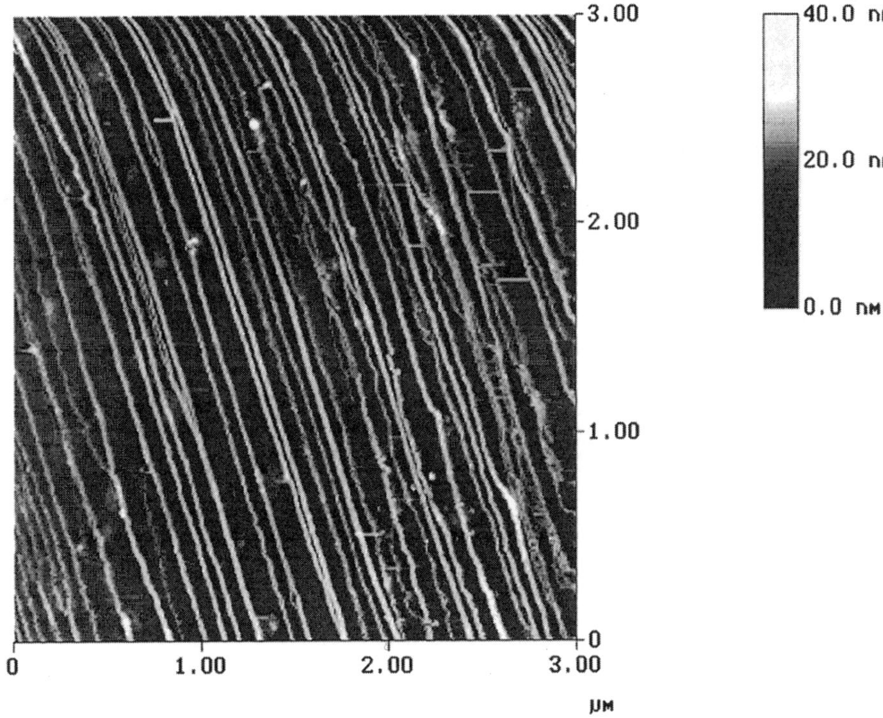

Figure 3 Atomic force microscopy (AFM) image of friction transferred PTFE lines on glass.

the influence of substrate topography on nanostructure self-assembly is achieved by depositing the particles on a carbon replica of the friction-transferred features.[22] Such replicas are prepared by vapour deposition of a thin film (ca. 30 - 50 nm) of amorphous carbon onto the substrate containing the PTFE lines followed by immersion in water, which causes the carbon film to lift off. Figure 4 shows a transmission electron microscopy (TEM) image of lines of thiol-capped gold nanoparticles self-assembled on such a substrate. The line spacing is comparable to that of the PTFE lines shown in Figure 3. Similar to gold particles decorating the surface of SWNT ropes the lines of individual particles assembled on the corrugated carbon thin film can be sintered into continuous metallic structures by heat treatment or by exposure to the electron beam in the transmission electron microscope. Such a sintering sequence is shown Figure 5. By this method gold nanowires of only 3 to 4 nm thickness and up to 100 nm length can be obtained.[22]

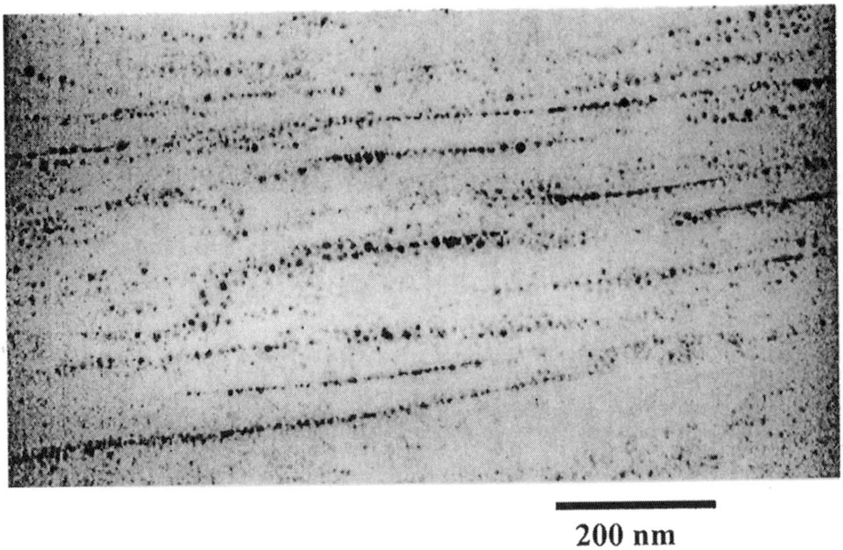

Figure 4 Transmission electron micrograph showing self-assembled lines of thiol-capped gold nanoparticles on a carbon thin film replica of friction transferred PTFE lines.

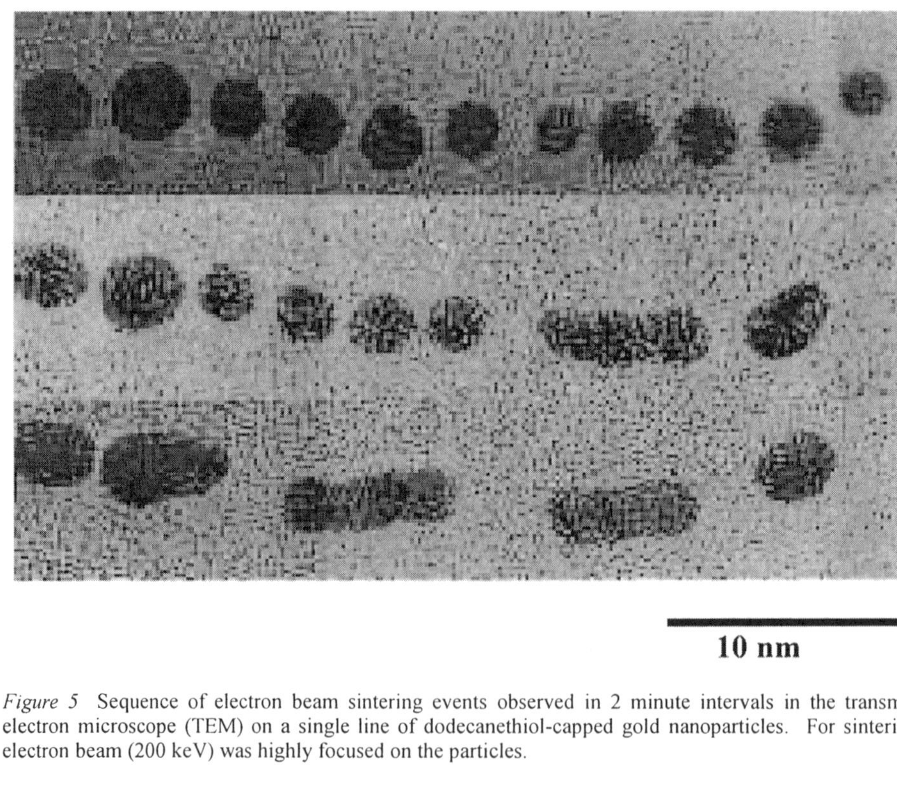

Figure 5 Sequence of electron beam sintering events observed in 2 minute intervals in the transmission electron microscope (TEM) on a single line of dodecanethiol-capped gold nanoparticles. For sintering the electron beam (200 keV) was highly focused on the particles.

This second example of templated nanostructure self-assembly emphasises the role of substrate topography as opposed to the chemical and electrostatic interactions that cause the decoration of SWNT ropes with gold particles. The alignment of chemically inert particles along topographical features can be explained in close analogy to the aforementioned entropy driven crystallisation of superlattices in the presence of a soluble "contaminant". It has been shown that chemically non-interacting microspheres deposited from suspension onto silicon substrates can align preferentially along the corner sites of micromachined linear features, i.e. precisely were the wall meets the floor.[23] This effect is only observed in the presence of an excess of smaller particles, which remain in suspension. The reason for such directional self-assembly is the total increase of entropy of the system by liberating solvent volume for the majority species of suspended smaller particles. This is achieved by maximising the total contact area between the large particles and the walls of the container, i.e. by maximal volume (solvent) exclusion. Effectively, the topographical features and the presence of the smaller particles generate an "entropic force field" which directs the self-assembly process.[23] The same argument can be evoked to account for the alignment of nanoparticles shown in Figure 4, although here no "contaminant" molecules or smaller particles have been added deliberately.[22] The particle preparations are, however, polydisperse, which can generate "entropic force fields" that cause the larger particles to precipitate along surface features or, on flat substrates, to crystallise in superlattices. We are currently studying this effect in detail by deliberately changing the size distributions and by adding molecular contaminants such as octanol to the solutions of nanoparticles. Preliminary results on flat substrates show clearly that both in the presence of excess smaller particles, and after addition of a molecular contaminant, the formation of ordered superlattices is significantly enhanced.

In conclusion, two different cases of linear alignment of gold nanoparticles have been presented and the importance of chemical interactions, surface topography and thermodynamics of self-organisation processes has been qualitatively discussed. This field of research is clearly still in its infancy and we are only beginning to understand in detail some of the parameters that govern the self-assembly of nanometre-scale systems. Its ultimate future goal is to control these processes in a similar way, chemists are able to control and predict the formation of molecules.

Acknowledgements

This work was supported by the EPSRC (GR/N65165/01) and by a NATO Linkage Grant (PST.CLG.976545). MB acknowledges the award of an EPSRC Advanced Research Fellowship (AF/100171).

References

1. R.P Andres, T. Bein, M. Dorogi, S. Feng, J.I. Henderson, C.B. Kubiak, W. Mahoney, R.G. Osifchin, R.
2. Reifenberger, Science 272 (1996) 1323.

3. S. Ciraci, A. Buldum, I.P. Batra, J. Phys. Condens. Mat. 13 (2001) 537.
4. P. Moriarty, Rep. Prog. Phys. 64 (2001) 297.
5. M. Li, H. Schnablegger, S. Mann, Nature 402 (1999) 393.
6. M. Brust, M. Walker, D. Bethell, D.J. Schiffrin, R. Whyman, J. Chem. Soc., Chem. Commun. (1994) 801.
7. M. Brust, J. Fink, D. Bethell, D.J. Schiffrin and C.J. Kiely, J. Chem. Soc., Chem. Commun. (1995) 1655.
8. M. Brust, D. Bethell, D.J. Schiffrin, C.J. Kiely, Adv. Mater 7 (1995) 795.
9. A.C. Templeton, W.P. Wuelfing, R.W. Murray, Acc. Chem. Res. 33 (2000) 27.
10. D. Bethell, M. Brust, D.J. Schiffrin, C.J. Kiely, J. Electroanal. Chem. 409 (1996) 137.
11. M. Brust, D. Bethell, C. J. Kiely, D. J. Schiffrin, Langmuir 14 (1998) 5425.
12. R.L. Whetten, J.T. Khoury, M. Alvarez, S. Murthy, I. Vezmar, Z.L. Wang, P.W. Stephens, C.L. Cleveland, W.D. Luedtke,U. Landman, Adv. Mater. 8 (1996) 428.
13. J. Fink, C.J. Kiely, D. Bethell, D.J. Schiffrin, Chem. Mater. 10 (1998) 922.
14. W.D. Luedtke, U. Landman, J. Phys. Chem.100 (1996) 13323.
15. C.J. Kiely, J. Fink, M. Brust, D. Bethell, D.J. Schiffrin, Nature 396 (1998) 444.
16. M.D.R. Taylor, P. Moriarty, M. Brust, Chem. Phys. Lett. in press.
17. X.M. Lin, H.M. Jaeger, C.M. Sorensen, K.J. Klabunde, J. Phys. Chem. B 105 (2001) 3353.
18. P.D. Kaplan, J.L. Rouke, A.G. Yodh, D.J. Pine, Phys. Rev. Lett. 72 (1994) 582.
19. M. Brust, C.J. Kiely, D. Bethell, D.J. Schiffrin, J. Am. Chem. Soc. 120 (1998) 12367.
20. M. Brust, C.J. Kiely, Colloid. Surface. A in press.
21. S. Fullam, D. Cottell, H. Rensmo, D. Fitzmaurice, Adv. Mater. 12 (2000) 1430.
22. J.C. Wittmann, P. Smith, Nature 352 (1991) 414.
23. T.O. Hutchinson, Y.-P. Liu, C. Kiely, C.J. Kiely, M. Brust, Adv. Mater. in press.
24. A.D. Dinsmore, A.G. Yodh, Langmuir 15 (1999) 314.

LAYER-BY-LAYER ASSEMBLY OF NANOTUBES AND NANOFILMS FROM NANOPARTICLE AND POLYMER BLOCKS FOR ELECTRONIC APPLICATIONS

N. I. KOVTYUKHOVA*[1, 2], B. R. MARTIN[1], J. K. N. MBINDYO[1], T. E. MALLOUK*[1], M. CABASSI[3], P. SMITH[3], B. RAZAVI[3], T. S. MAYER[3]
[1]Department of Chemistry, The Pennsylvania State University, University Park, PA 16802
[2]Institute of Surface Chemistry, N.A.S.U., 17, General Naumov Str., 03680 Kyiv, Ukraine
[3]Department of Electrical Engineering, The Pennsylvania State University, University Park, PA 16802

1. Introduction

The near-future trends in computing, electronics, optoelectronics have put forward the necessity in radically new concepts of fabricating ultrahigh density electronic circuits. A perspective approach to this problem is constructing the circuits from the ground up using nanoscale There have been numerous reports of molecule- and polymer-based devices that can function as rectifiers, transistors, and switches [1,2]. Some elaborate logic circuits based on molecular diodes [3], and nanoscale crossbar arrays that contain bistable memory elements [4] have recently been proposed. Obviously, current rectification is needed at each crossing point in order to isolate individual memory elements from each other. Real challenge to practical use of these devices is the problem of making the appropriate connections between nanoscale devices, and their integration into the larger scale circuits.

Our approach to this wiring problem involves membrane replication methods [5] to make nanowires of precisely controlled geometric parameters, and connection chemistry [6]. The wires are made by electrochemically growing a series of metals in the cylindrical pores of an alumina or polycarbonate membrane. One can compose nanowires of precisely defined series of "stripes", which can subsequently be derivatized with appropriate SAMs. Interactions between complementary molecules (hydrophobic, hydrophilic, anionic, cationic, single-stranded DNA) on specific stripes programs the adhesion of nanowires to each other and to complementary patterns on planar surfaces. For example, deterministic assembly of nanowires into cross- or T-shaped pairs has been recently realized [7]. This technique, and other assembly methods such as microfluidic alignment [8], hold some promise for constructing logic gates and cross-point memories from sub-lithographic components. However, to date nanowires that incorporate the two-terminal nonlinear devices needed for these circuits - diodes and switches - have not been available.

E. Buzaneva and P. Scharff (eds.), Frontiers of Multifunctional Nanosystems, 147–162.

On the other hand, wet layer-by-layer assembly has been shown to be an experimentally simple but powerful route to functional nanoscale devices with thickness controlled to nanometer precision [9]. This method involves the spontaneous adsorption of pre-formed colloidal particles and/or organic macromolecules as monolayers [10-21]. Many research efforts have now been focused on making multilayer nanoparticle/polymer films with structure- and sequence-dependent electronic functions. These include Coulomb blockade devices [22], rectifying diodes [18-20,23], electroluminescent devices [24], photon harvesting and charge separation assemblies [25a], and multicomponent energy/electron-transfer cascades that mimic some of the functions of natural photosynthetic assemblies [25b]. Because it is an adsorption technique, the layer-by-layer assembly method can be equally applied to planar or curved surfaces.

In this work we summarize our recent research [26, 27] aimed at developing a new synthetic route to nanowire diodes that combines advantages of both template replication and layer-by-layer assembly techniques. Two different diode structures have been realized: (i) semiconductor/polymer films sandwiched between two segments of a nanowire, i.e., an "in-wire" diode; and (ii) semiconductor/polymer films that cover the walls of a nanowire, such that contacting the wire with another crossing wire gives a rectifying junction. The synthesis of the first structure involves the assembly of a multilayer TiO_2(or ZnO)/polymer film on top of a metal nanowire inside the pores of an alumina or polycarbonate membrane, followed by electrochemical or chemical growth of a second metal segment. Our previous work using planar substrates has shown that multilayer TiO_2/polymer and ZnO/polymer films can be easily grown on suitably primed (Au or Pt) and bare (Ag) metal surfaces, and that these films act as diodes when contacted with an appropriate top metal [17, 18, 23a].

Two different strategies were used to make the second structure, in which the semiconductor/polymer film covers the walls of a metal nanowire. The first one is based on the concept of sequential tubular synthesis, which was introduced by Martin for fabrication of rod-shaped concentric nanocomposite particles [28]. This method exploits surface chemistry of the membrane pore walls. We extend this approach to layer-by-layer assembly of semiconductor/polymer tubule shells stuck to the walls, inside which metal nanowire core is electrochemically or chemically grown. Recently we found that densely-packed TiO_2/polymer films could be prepared on planar Al substrates bearing thin surface layer of native oxide [17]. The second strategy exploits reactivity of metal rod surface, and involves layer-by-layer growth of the semiconductor/polymer shell around the metal nanowire core. This method follows from our recent experiments that show that rod-shaped nanoparticles can be used as a substrate for layer-by-layer assembly of smooth, conformal films of insulating organic polyelectrolytes [29].

2. Experimental Section

Whatman Anopore disks (Al_2O_3-membranes) and OSMONICS Poretics, Polycarbonate (PC) membranes were used as templates for nanowire synthesis. Aqueous TiO_2 and ethanolic ZnO colloids, the emeraldine base form of polyaniline (PAN) and aqueous poly (sodium 4-styrenesulfonate) (PSS) were used for preparing semoconductor/polymer films.

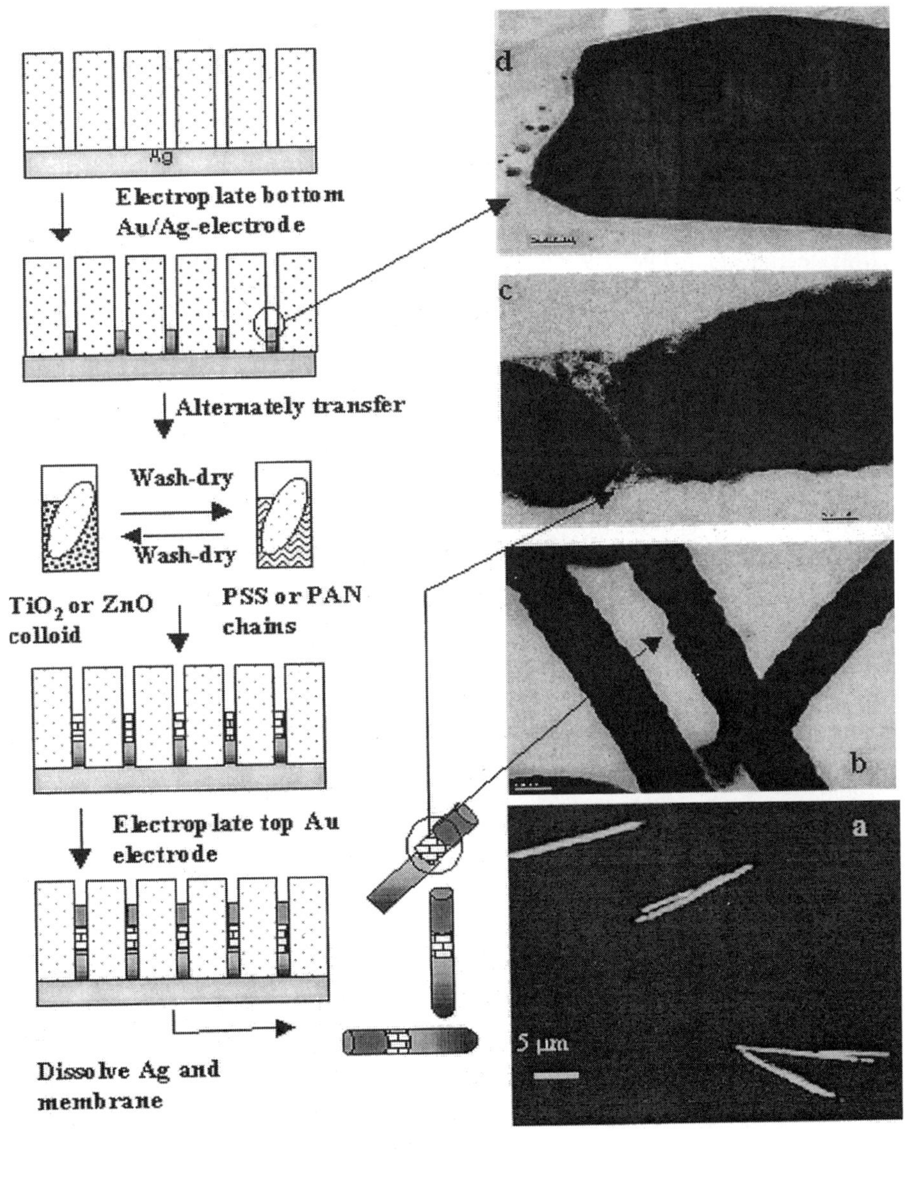

Figure. 1. (Left) Scheme for the preparation of in-wire semiconductor/polymer thin-film devices. (Right) (a) The optical micrograph of an Au/Ag/$(TiO_2/PSS)_{10}TiO_2$/Au nanowires prepared in an OTS-treated alumina membrane. The two Au ends appear darker than the central silver stripe; (b) TEM image of an Au/Ag/$(ZnO/PAN)_9TiO_2$/Au nanowires prepared in a PC membrane. Nanoparticle junction is seen between the two metal; (c) Ag/$(TiO_2/PSS)_{10}TiO_2$/Au nanoparticle junction (end of the bottom Ag-electrode on the right). The image was taken after focusing the electron beam for several seconds; (d) TEM image of the top-end of an Au nanowire prepared using an alumina membrane that had been treated with OTS.

Device structures, in which the semiconductor/polymer films are sandwiched between two segments of a nanowire, were synthesized according to the scheme shown in Fig.1 (for details see [26,27]). This process involves four steps: evaporating silver on the membrane backside and electroplating the bottom metal electrode [6]; layer-by-layer assembly of the multilayer film; electroplating the top metal electrode; and dissolving the template and silver backing. Layer-by-layer assembly of (TiO_2 or ZnO)/(PAN or PSS) multilayer film onto the exposed tips of the metal nanowires was achieved by subjecting the metal-filled membrane to the same procedure as that used for planar substrates [17,18]. However adsorption and washing times were increased to 15-30 min. Control experiments on planar metal substrates showed that no priming procedure was needed for the silver bottom electrode, but that priming of the gold electrode surface with mercaptoethylamine (MEA) should precede the deposition of multilayer (TiO_2 or ZnO)/Polymer films.

Device structures, in which the semiconductor/polymer film covers the walls of a rod-shaped electrode. For preparing these structures two strategies were applied (for details see [26,27]).

1. Layer-by-layer synthesis of multilayer TiO_2/PSS or ZnO/PSS tubules on the pore walls followed by electrochemical or chemical plating of Au-rods inside the tubules is shown in Fig.2. A membrane was successively immersed in aqueous TiO_2 or ethanolic ZnO and PSS stock solutions. After each immersing step, the membrane was placed in a suction filtering unit and washed with the appropriate solvent under vacuum followed by drying in Ar stream. Au-rods inside the TiO_2/PSS tubules were grown using electrochemical or electroless [30] technique.
2. Layer-by-layer deposition of ZnO/PSS or TiO_2/PSS films on the surface of metal nanowires was performed as follows. Gold rods were grown electrochemically and transferred into ethanolic solution by dissolving the template membrane and Ag-backing [6]. Then the gold surface was derivatized with MEA. The layer-by-layer assembly of multilayer ZnO/PSS or PSS/TiO_2 film on the primed surface of the nanowire was conducted as presented in Fig.3. Ethanolic ZnO or aqueous TiO_2 and aqueous PSS were alternately added to the concentrated (by centrifuging) suspension of nanowires in the appropriate solvent. Each adsorption cycle was followed by a centrifugation/washing cycle with the appropriate solvent, which was then replaced by the solvent used in the next adsorption step.

In an alternate method, a gold nanowire array attached to a silver base was used. Such a brush-like substrate was prepared by dissolving the Au-filled template and leaving the Ag-backing untouched. Then this "brush" was subjected to the conventional layer-by-layer assembly procedure (see scheme in Fig. 4).

Nanowires were aligned for electrical measurements using electrofluidic technique [31].

3. Results and Discussion

3.1. *NANOWIRES CONTAINING IN-WIRE DIODE JUNCTION*

TEM images of gold nanowires grown in alumina membranes (not presented) show that the two ends of the wire differ in their topography: the base end is bulging and rounded, while the growing end has an apparent hollow in the middle (cup-shaped). This difference can be understood in terms of adsorption of metal ions on the wall, and preferential wetting of the polar alumina surface by the reduced metal. It was recently shown

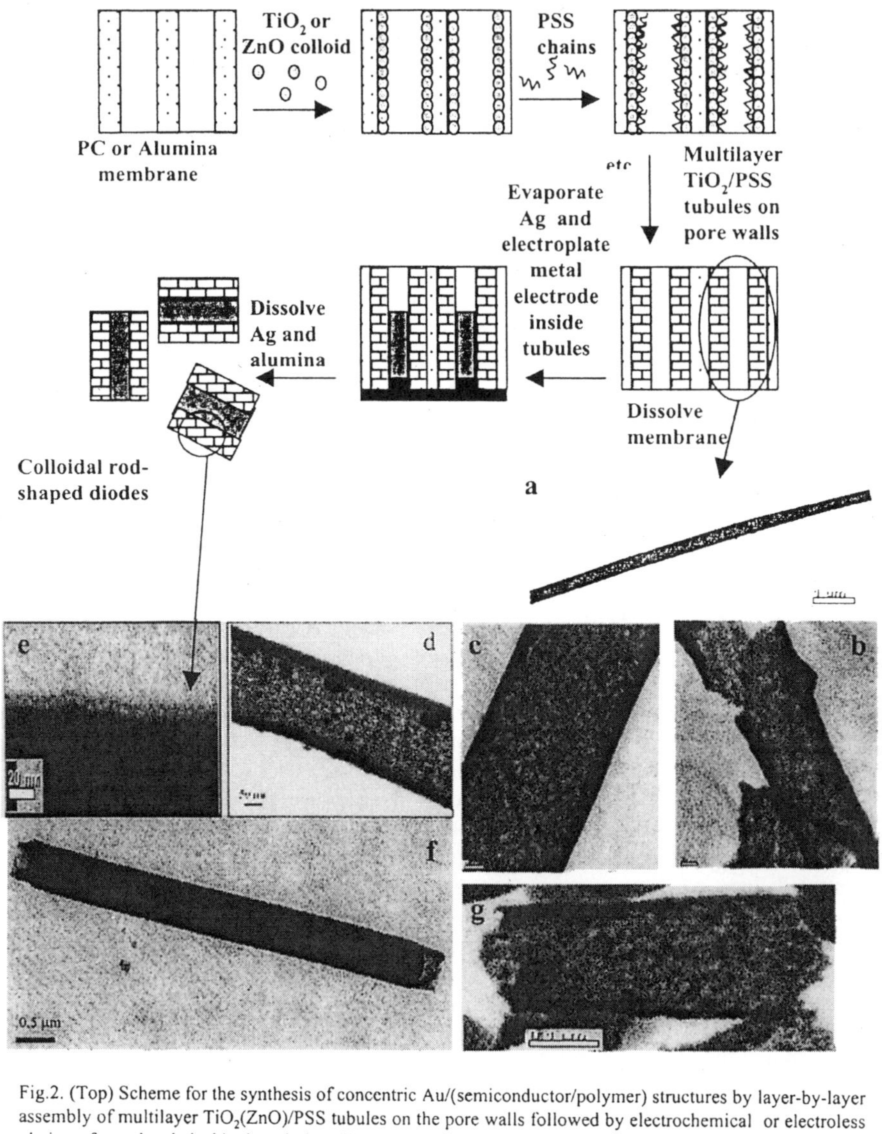

Fig.2. (Top) Scheme for the synthesis of concentric Au/(semiconductor/polymer) structures by layer-by-layer assembly of multilayer TiO_2(ZnO)/PSS tubules on the pore walls followed by electrochemical or electroless plating of metal rods inside the tubules. (Bottom) TEM images of (a-d) - $(TiO_2/PSS)_5TiO_2$ tubules prepared in 200 nm diameter pores of alumina (a,c,d) and PC (b) membrane; (d) $(TiO_2/PSS)_9TiO_2$ tubule activated for electroless Au depositoin by successive adsorption of $SnCl_3^-$ ions and $Ag(NH_3)_2$; (e,f) Low and high resolution TEM images of Au nanowires grown electrochemically inside $(TiO_2/PSS)_9TiO_2$ tubules; (g) (ZnO/PSS)ZnO tubules prepared in 100 nm diameter pores of PC membrane. (b,c,d)-scalebar 50 nm.

that Au(III) complexes are taken up by hydrous aluminum oxide by formation of Au-O-Al bonds. These adsorbed Au(III) ions were spontaneously reduced to the metallic state without added reducing agents [32]. If the same reaction takes place in alumina membranes, it will promote metal growth on the wall leading to a cup-shaped tip.

In order to probe the topography of multilayer TiO_2/polymer films sandwiched between Au and Ag nanowire segments, we prepared Au/(TiO_2/PAN)$_6$/Ag junctions and dissolved the top Ag segments in the nitric acid. The remaining Au wires with (TiO_2/PAN)$_6$ films deposited at their ends were imaged by TEM (Figure 5). Control experiments showed that the ellipsometric thickness of a multilayer TiO_2/PAN film grown on a planar Au/MEA substrate did not decrease after immersion in 6 M HNO_3 for 30 min, indicating that the film is stable in the medium used to etch away the Ag segment. A TEM image of an Au/(TiO_2/PAN)$_6$ nanowire (Figure 5a) shows the profile of the cup-shaped tip. After focusing the electron beam on this tip for some time (typically tens of seconds), a nanoparticle film appears at the tip (Figure 5b), perhaps because of beam-induced melting of the metal. Comparing figures 5a and 5b, one can see that the upper contour line of the film is very close to that of the cup-shaped tip before melting. Apparently, the multilayer film grows on both the cup bottom and cup walls and thus retains the cup shape after the thin walls have melted. This explanation is consistent with observed film height of ~100 nm (Figure 5b), which approximates the cup depth rather than the (TiO_2/PAN)$_6$ film thickness. The ellipsometric thickness of a (TiO_2/PAN)$_6$ film on a planar Au(MEA) substrate is about 10 nm.

The tip shape (which may influence the uniformity and continuity of nanoparticle junctions between metal segments) can be altered by passivation of the pore walls by alkylsilane derivatives, such as OTS, ethyltriethoxysilane or chlorotrimethylsilane. Transmission IR spectra of an alumina membrane treated with OTS showed new intense bands at 2960, 2920, 2850 cm^{-1} (C-H), 1470 cm^{-1} (C-H), and 1100 cm^{-1} (Si-O). Membranes derivatized this way were hydrophobic, confirming the adsorption of OTS on the pore walls and faces of the membrane. TEM images of Au nanowires grown from OTS-treated alumina showed a tendency for the previously cup-shaped end to become pointed (Fig. 1d). Thus, the derivatization of the wall of the membrane to some extent allows one to smooth the surface of the rod ends.

Figure 1 shows the reaction scheme for in-wire layer-by-layer assembly and dissolution of the template and silver backing. The optical micrograph in Figure 1a shows one of the suspended Au/Ag/(TiO_2/PSS)$_{10}$$TiO_2$/Au nanowires, in which a silver "stripe" can be clearly seen between two gold ends. In a TEM image of Au/Ag/(ZnO/PAN)$_9$ZnO/Au nanowires prepared in 70-nm pores of the PC membrane one can see the metal/nanoparticle-film/metal junction (Fig.1b). However a TEM image (recorded within the first several seconds) of Au/(TiO_2/PAN)$_{10}$/Ag/Au nanowire, which was prepared in 200-nm pores of the alumina membrane, shows no visible signs of a metal/film/metal junction. In the alumina membrane, the metal ions used in the second electrodeposition step fill voids between nanoparticle film and pore walls, so that the top electrode follows the tip surface of the bottom electrode resulting in apparently featureless nanowire walls. This process is promoted by the strong affi-

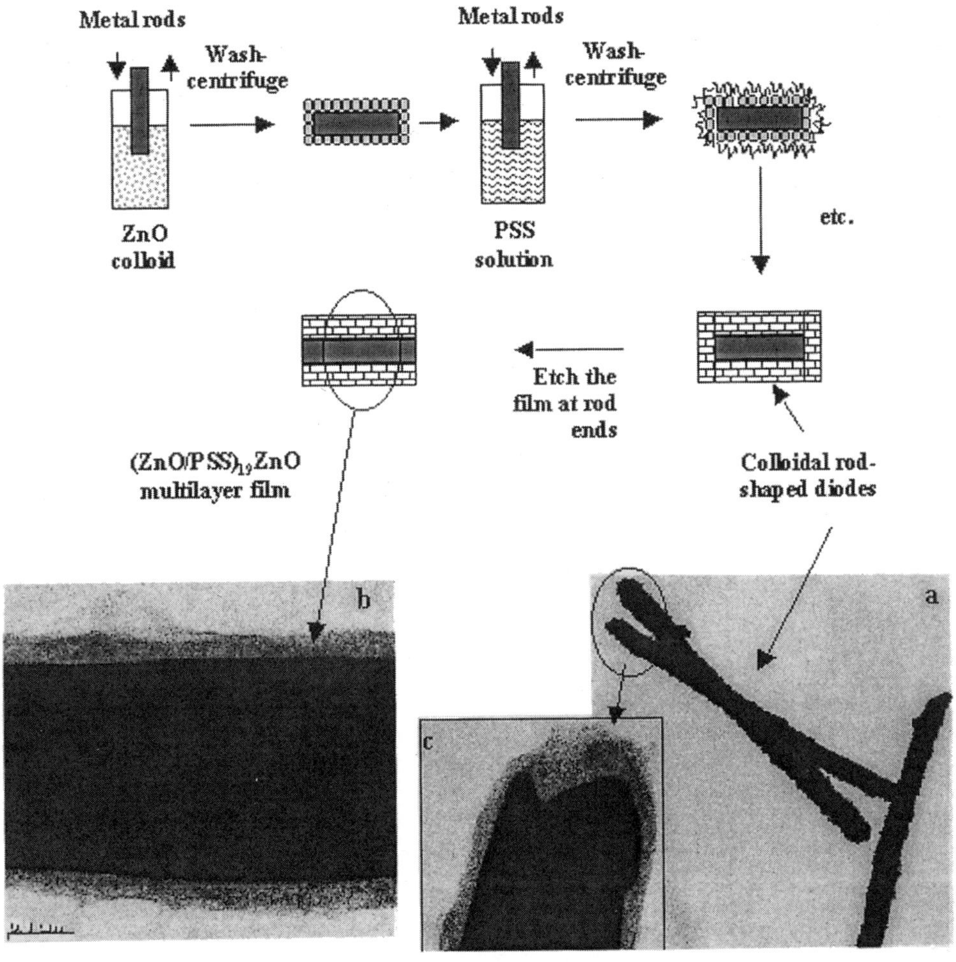

Figure.3. Scheme for layer-by-layer assembly of ZnO/PSS film on the surface of template-grown Au-rods. (a-c) Low and high resolution TEM images of Au nanowires that were grown in alumina membrane, then released with HNO_3 and NaOH and subjected to 20 ZnO/PSS adsorption cycles. One can see the nanoparticle film completely covering the walls (b) and ends (c) of the rods (scale bar 50 nm).

nity of the metal ions to the alumina pore walls. During TEM imaging, focusing the electron beam on this featureless nanowire for several seconds leads to beam-induced metal melting in the vicinity of the Au/film/Ag junction (Fig. 1c). A break appears in the wire, and particles of 5-10 nm diameter, which adhere to both metal ends, are observed. Apparently, the TiO_2 nanoparticles are present between the two electroplated metal segments. Optical and TEM data confirm that a multilayer TiO_2/polymer film can be assembled layer by layer on the tip of a metal rod in the membrane, and that this film does not prevent electroplating the second metal segment on the top of the film.

Interestingly, for Au/Ag/(TiO_2/PSS)$_{10}$$TiO_2$/Au nanowires grown in OTS-treated alumina membranes, a very thin (< 5 nm) film can be seen on the walls of the bottom metal segments. This suggests that the OTS SAM deposited on the pore walls adheres to the nanowire surface after the membrane is dissolved away. In principle, this effect could be used to prepare ultrathin insulating films around the metal nanowires.

3.2. *DIODE DEVICES IN WHICH THE SEMICONDUCTOR/POLYMER FILM COVERS THE WALLS OF A NANOWIRE*

3.2.1. *Multilayer TiO_2/polymer tubules self-assembled on the pore walls.*

TEM images of $(TiO_2/PSS)_9TiO_2$ tubules grown on the alumina membrane pore walls followed by membrane dissolution are shown in Figure 2. The tubules as prepared are typically unbroken with an outside diameter of about 200 nm (the diameter of the pores) and length ranging from 5 m to 15 m (Fig. 2a). The tubule ends are always open and their length is less than the thickness of the template membrane (~70 m). This im plies that the tubules are broken during the membrane dissolution or subsequent handling. A higher resolution TEM image (Figure 2c) shows uniform tubule walls of densely packed TiO_2 particles of ~5 nm diameter (the particle diameter in the stock TiO_2 solution). The thickness of the walls increases with the number of adsorption cycles and can be estimated as 40 nm for a ten-layer TiO_2/PSS tubule. This value is consistent with the ellipsometric thickness (37 nm) of ten-layer TiO_2/PSS films grown on planar substrates. Fig. 2b shows TiO_2/PSS tubules of the same composition but grown inside 200-nm pores of a PC membrane. The latter tubules also have densely packed walls of approximately the same thickness, but unlike those grown in alumina they are broken and never longer than 1-1.5 m. A four-layer ZnO/PSS tubule, prepared inside the 100-nm pores of a PC membrane, has ~ 10-nm thick walls, in reasonable agreement with ellipsometric thickness (~ 3 nm per ZnO/PSS) of planar ZnO/PSS films (Fig.2g). The outer surface of the tubules is quite smooth and its shape appears to follow the pore walls. One can see round-ended branches on some of the tubule walls, which reflect structure of the alumina membrane pores in their branched part. These observations suggest strong adsorption of the first TiO_2 and ZnO particle layer on both alumina and polycarbonate surfaces. Previously we showed that TiO_2 colloidal particles readily form a well-packed monoparticle layer on a planar Al/Al_2O_3 substrate [17]. However breaking of the tubules released from PC membranes is observed in all cases and may be due to the swelling of PC membranes in solutions used to prepare the tubules and dissolve the membrane.

The FTIR spectrum (not shown) of an alumina membrane subjected to ten TiO_2/PSS adsorption cycles shows new band maxima at 2935, 2841, 2347, 1637, 1497, 1450, 1412, 1127 cm^{-1}. These peaks are observed in the spectrum of PSS, and confirms the presence of PSS in the membrane. Apparently the adsorption of PSS polyanions onto the positively charged (at pH < 6) TiO_2-terminated surface is driven by electrostatic interactions.

Gold deposition within the $(TiO_2/PSS)_9TiO_2$ tubules could be done either chemically or electrochemically. In the case of electroless deposition, the tubules were first primed by adsorption of $SnCl_3^-$ ions and then treated with $Ag(NH_3)_2^+$. It is known that treatment of Sn-sensitized surfaces with $Ag(NH_3)_2^+$ leads to reduction of the silver cations and formation of Ag aggregates, which provide a nucleation site for the electroless deposition

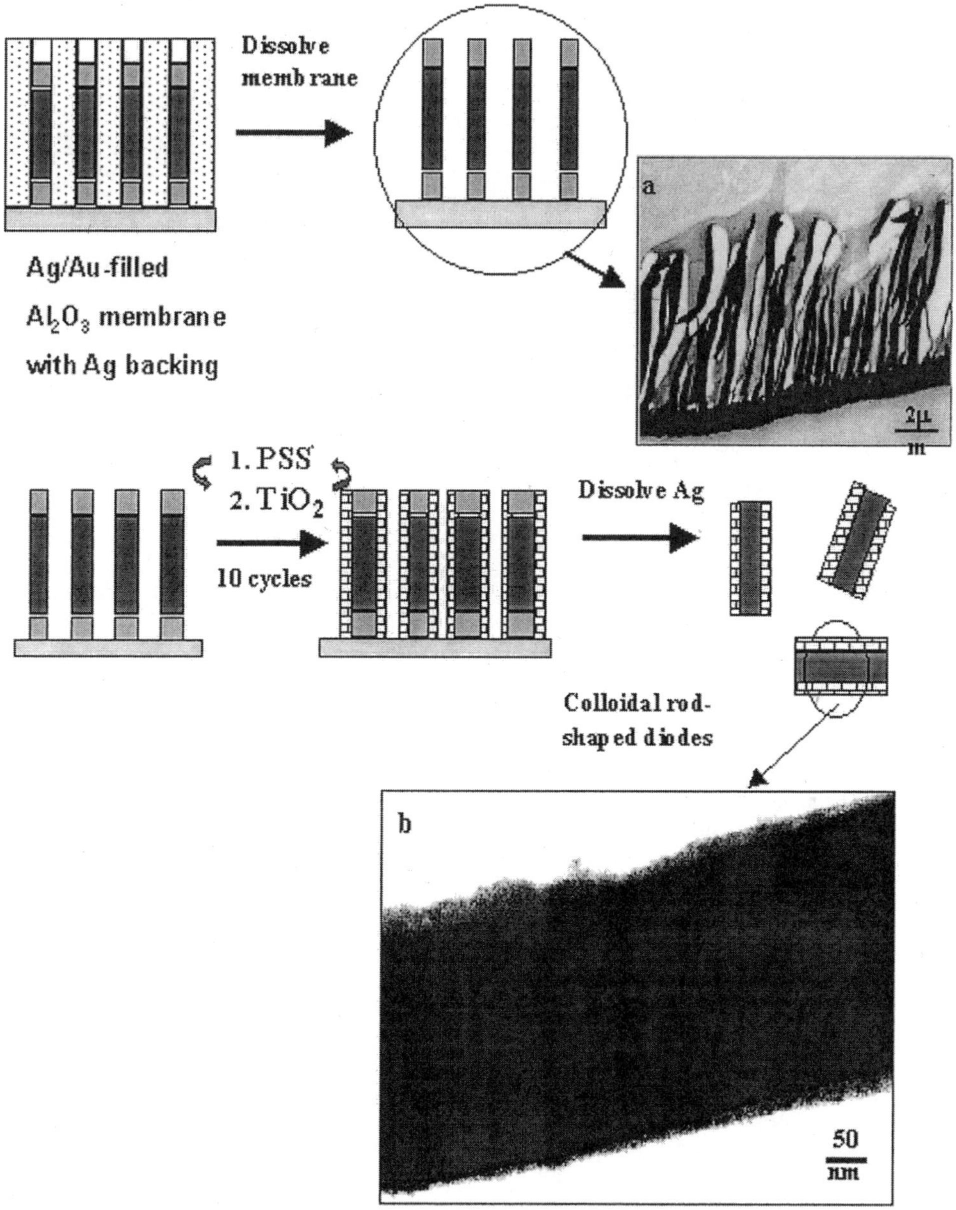

Figure 4. Layer by layer adsorption scheme for the preparation of concentric devices by "brush" technique. (a) A cross-section TEM image of a brush-like substrate prepared by dissolving alumina membrane and leaving Ag-backing untouched; (b) A TEM image of a $Au(MEA)(PSS/TiO_2)_{10}$ nanowire prepared by the "brush" technique.

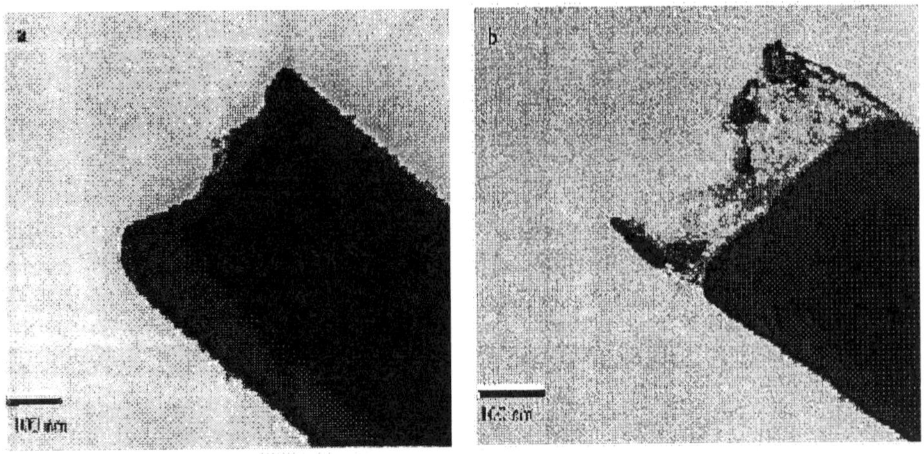

Figure5. (a,b) TEM images of an $Au/(TiO_2/PAN)_6/Ag$ nanowire taken after dissolving the top Ag segment with HNO_3: (a) the image taken immediately after focusing the electron beam; (b) the image taken after focusing the electron beam on the nanowire tip for tens of seconds.

of Au [30]. Indeed, evenly distributed 5-50 nm diameter metallic nanoparticles are clearly seen by TEM on the inner surface of the TiO_2/PSS tubules (Figure 2d).

Typical TEM images of Au-filled $(TiO_2/PSS)_9TiO_2$ tubules (Figure 2e,f) shows solid metal rods with walls bearing a thin nanoparticle film. The film looks featureless and rather smooth. Its roughness, which is determined by the roughness of the pore walls, is comparable to the average size of the TiO_2 colloidal particles, ~ 5 nm. The Au-filled TiO_2/PSS tubules appeared to break within 5 - 50 nm of the ends of the gold rods (Fig.2f). This leaves the rod ends accessible for further derivatization by wet chemical or vapor phase techniques. The length of the coated nanowires prepared this way was in the range of 3 – 5 m for electrochemical plating (Fig.2f) and 2 - 12 m for electroless plating.

3.2.2. *Layer-by-layer assembly on the surface of template-grown Au-rods.*

An important advantage of this strategy, compared to the one described above, is its applicability to a wide range of nanoparticle/polymer films which might be destroyed under the conditions used to dissolve the template membrane. This method also allows one to prepare nanowires with ends accessible for further functionalization, by starting with a sacrificial metal (e.g. Ag or Cu) at the tips of the wires.

Figure 3 shows schematics of the method and TEM images of gold nanowires that were electrochemically grown and released from the alumina membrane, and then subjected to 20 ZnO/PSS adsorption cycles. One can see that the nanoparticle film completely covers the walls and ends of the nanowires (Fig.3b,c). The film thickness is approximately 40 nm, which is less than expected, since on planar Au substrates the average thickness per ZnO/PSS adsorption cycle is 3 nm per ZnO/PSS. This difference tends to increase with the number of adsorption cycles. Although the film is relatively uniform, its surface is rather rough (Fig.3a); in contrast, similar films grown on planar Au are smooth by AFM. It is possible that aggregates are formed from suspended ZnO particles and PSS chains in solution (due to incomplete removal of reagents during the washing-centrifuging stage) and that these aggregates precipitate on the nanowire surface.

An alternate way to deposit these films is to use a nanowire array attached to a silver base (see scheme in Fig.4). Recently we successfully used such a "brush" for preparing polyelectrolyte-coated gold nanowires [29]. Figures 4 a, b show TEM images of the "brush" of an Au(PSS/TiO_2)$_{10}$ nanowire isolated from this "brush" after dissolution of the Ag backing. Dissolving the sacrificial Ag layer on the ends of the Au nanowires leaves them uncovered, but unfortunately, some uncovered areas of the nanowire walls are also observed. It is interesting to note that the same film growth procedure on a planar Au-substrate resulted in high quality ten-layer PSS/TiO_2 film with a thickness of ~ 37 nm. An experiment with nitric acid (used for dissolving the Ag backing) showed that this film was not significantly changed. The poor quality of the (PSS/TiO_2)$_{10}$ film on the nanowire walls probably results from the fact that some nanowires in the "brush" can contact each other, thereby preventing the access of nanoparticles and polymer chains to the surface.

3.3. *ELECTRICAL PROPERTIES*

Optical micrographs of test structures used to characterize the electrical properties of in-rod devices and concentric nanowire/nanoparticle devices are shown in Figures 6 a and b, respectively.

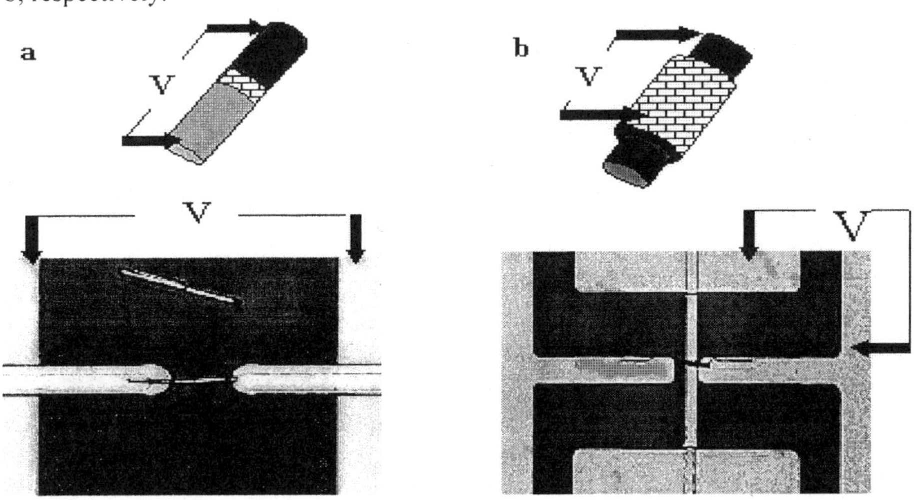

Figure 6. Optical micrographs of test structures for measuring the electrical properties of (a) in-wire devices and (b) devices in which the semiconductor/polymer film covers the walls of the nanowire. In (a), a Au/Ag(TiO_2/PSS)$_{10}$$TiO_2$/Au nanowire was aligned between two gold pads and 175 nm thick gold contacts were evaporated on the top of the ends of the nanowire. In (b), an Au nanowire bearing a ZnO/PSS multilayer film on the walls was aligned between two gold pads. Electrical contact was made by evaporation of 30 nm thick Ag-stripe at right angles to the wire, followed by evaporation of 175 nm thick gold contacts on the ends.

A typical I-V curve for the Au(MEA)/$(ZnO/PSS)_{19}$ZnO/Ag structure (Figure 7a) with the ZnO/PSS film deposited on the walls of an Au nanowire reveals the expected current rectifying behavior with turn-on potentials at ~ -1.25 V and ~ 2.02 V for forward (Ag "-") and reverse (Ag "+") biases respectively. The difference between numerical values of the turn-on potentials is in reasonable agreement with the difference between the work functions of Ag (~4.6 eV) and Au (~5.3 eV). The operating voltage is also close to the ZnO bandgap (3.28 eV, as estimated from the UV-visible spectrum of starting ZnO sol), and both branches of this curve can be successfully linearized using the Schottky equation (ln(I) V) (Fig.7b). These observations indicate that rectification is determined by charge injection at the metal/ZnO/PSS-film interface rather than by a tunneling mechanism, although the I-V curve shows quite satisfactory linearity in Fowler-Nordheim $\{\ln(I/V^2)\ \ 1/V\}$ coordinates as well.

Typical I-V characteristics of Au/$(TiO_2/PSS)_9TiO_2$/Ag structures prepared as in-rod devices (the TiO_2/PSS film is sandwiched between bottom silver and top gold segments of the nanowire), and devices in which the TiO_2/PSS film covers the Au nanowire walls, are shown in Figures 7c.1and 2, respectively. For comparison purposes, the I-V characteristic of the same structure grown on a planar Au substrate and top-contacted with evaporated Ag is also shown (Fig.7c.3). (This planar device was prepared by the layer-by-layer self-assembly of a ten-layer PSS/TiO_2 film on an Au-substrate primed with MEA). One can see that all three curves are rather similar in shape, showing an abrupt current increase at turn-on potentials.

Although for all three devices, the differences between the numerical values of the "forward" and "reverse" turn-on potentials are close (~ 0.35-0.5 eV), they are less than that expected for the difference in work functions between Ag (~4.6 eV) and Au (~5.3 eV). These values also differ for different devices prepared the same way. The charge transport in these TiO_2/PSS thin-film devices cannot be described by either the Schottky or a Fowler-Nordheim equations. In some cases the I-V curve shape suggests breakdown characteristic of insulators, but such "breakdowns" are quite reproducible in 4-6 successive voltage sweeps, and this is not usually observed for breakdown in insulators. The comparison between the I-V curves of the Au/$(TiO_2/PSS)_9TiO_2$/Ag structures and bare Au/TiO_2/Ag (Fig.7c.4) and Au/PSS/Ag (Fig.7c.5) suggests that the PSS plays a role in the abrupt transition from a non-conducting to a conducting state. The I-V curve of bare Au/PSS/Ag device shows breakdown at relatively low potentials (~ +0.26 V and –0.46 V) for both forward and reverse bias modes. It is interesting to note that no "breakdowns" were observed when PAN or PAH was used instead of PSS in planar Au(TiO_2/polymer)Ag devices. Thus electrical properties of PSS do not allow to consider it as the best candidate for diode preparation despite the fact that among all the polymers investigated by us (PSS, PAN, PAH, PAA, PDDA, PEI) PSS provides the most regular growth of multilayer TiO_2/polymer films.

An in-wire Au(MEA)/$(TiO_2/PAN)_3TiO_2$/Pt device (Fig.8a) shows current rectifying behaviour with turn-on potentials at ~ -0.2 and ~ +0.9 V for forward and reverse bias, respectively. The difference in turn-on potentials (0.7 V) is higher than that expected from the difference in work functions of Pt (~5.6 eV) and Au (~5.3 eV), while the turn-on potentials are lower than barrier heights at the metal/film interfaces that might be expected from the TiO_2 conduction band edge potential (~ 4.0 eV). The I-V characteristic of a planar Au(MEA)/$(TiO_2/PAN)_9TiO_2$/Pt device shows a difference in

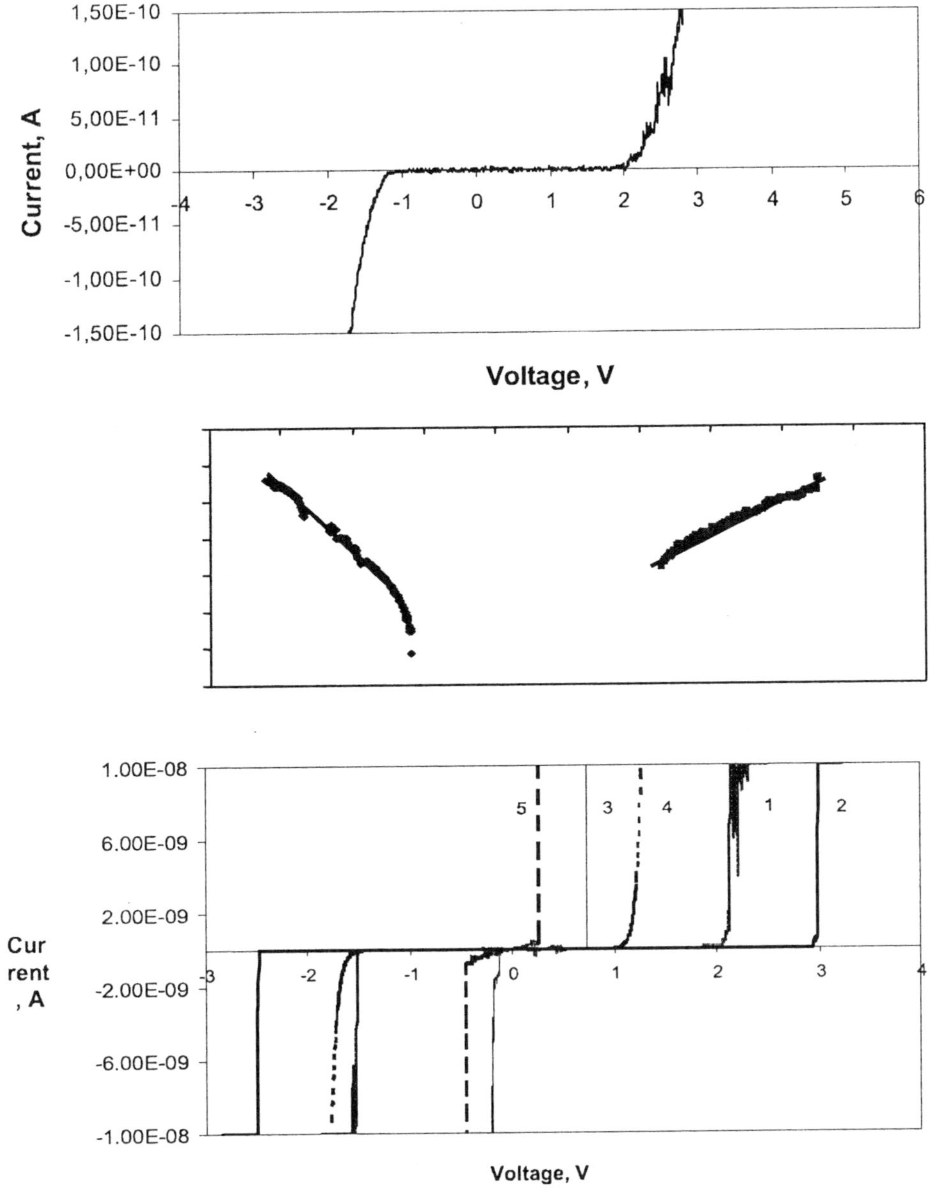

Figure 7. *I*-V characteristics of (a) a concentric Au(MEA)/$(ZnO/PSS)_{19}$ZnO/Ag device. The data are plotted in Schottky coordinates in (b).

(c) nanowire and planar Ag$(TiO_2/PSS)_9TiO_2$/Au devices and planar Ag/TiO2/Au and Ag/PSS/Au devices:
(1) an in-wire Ag$(TiO_2/PSS)_9TiO_2$/Au device; (2) a concentric Au/$(TiO_2/PSS)_9TiO_2$/Ag device;
(3) a planar Au/$(TiO_2/PSS)_9TiO_2$ /Ag, device area ~ 1 mm^2; (4) a planar Au/TiO_2/Ag device, area ~ 1 mm^2;
(5) a planar Au/PSS/Ag device, area ~ 1 mm^2. All I-V curves were measured in two sweeps starting from 0 volts.

turn-on potentials as high as 1.12 V (Fig.4b-2). Comparing both I-V curves to those for bare Au/TiO_2/Pt (Fig.8b-1) and Au/PAN/Pt (Fig.8b-3) planar devices suggests that the TiO_2 particles responsible for the rectifying behavior, because PAN shows insulating behavior in the voltage range of –5 V to +3 V. It should be noted that the described above I-V curves for PAN-containing devices were measured in a single positive-going sweep starting from –7V.

Both Au/(TiO_2/PSS)$_9$$TiO_2$/Ag structures prepared as concentric device (in which the TiO_2/PSS film covers the Au-nanowire walls and is in contact with a silver electrode evaporated top) and in-wire device show similar I-V behaviour (Fig.8c-1,2). When the I-V curves are obtained in single sweeps starting from either negative or positive potentials, enhanced asymmetry is seen compared to curves measured in two sweeps starting from 0 V each. The same tendency is found for planar Au(MEA)/(TiO_2/PSS)$_9$$TiO_2$/Ag devices (Fig.8e). and bare Au/TiO_2/Ag (Fig. 8f-2) and Au/PSS/Ag (Fig.8f-1) structures. One possible explanation for this hysteresis is the localization of injected charge carriers at interface and surface states. Saturation and emptying of surface traps is a slow process that can determine the current flow on this time scale. This idea is supported by the observation of current (see, e.g., Fig.8e-3) immediately after switching the voltage and the subsequent decrease in current by several orders of magnitude. The nature of the traps is not understood at present; they cannot be unambiguously assigned to either TiO_2 particles or PSS chains since both bare Au/TiO_2/Ag (Fig. 8f-2) and Au/PSS/Ag (Fig.8f-1) structures show similar behavior.

It is interesting to note that the hysteresis for an Au(MEA)/(ZnO/PSS)$_{19}$ZnO/Ag structure prepared as a concentric device (Fig.8d), although it exists, is much less pronounced than that of the TiO_2/PSS thin-film devices. One can assume that surface states of the semiconductor particles affect the electrical properties of semiconductor/polymer films. The very different shapes of the I-V curves for TiO_2/PSS and TiO_2/PAN thin-film structures suggest that the polymer component also exerts an influence on the rectifying properties of these devices. Currently, the layer-by-layer synthetic technique does not allow one to control the lateral structure of multilayer films. Due to phase boundaries in the organic and inorganic layers (at least in the case of 3D-particles), the formation of inhomogeneous contacts, in which both the semiconductor and polymer touch the metal surface, is possible. Without knowing the details of the interfacial composition it is difficult to arrive at a complete understanding of the electrical properties. Nevertheless, the observed hysteresis holds some promise for use in memory devices, if persistent high- and low-conductivity states can be attained at the same potentials.

4. Conclusions

We have shown that it is possible to couple layer-by-layer assembly of polymer/colloid films and tubules with the template synthesis of metal nanowires, in order to realize simple devices that act as rectifiers. Both concentric and in-wire semiconductor-containing films give rise to current rectifying effects in "striped" metallic nanowires. Current rectification appears to arise at the interface between the high work function metal (Au or Pt) and the oxide semiconductor nanoparticles. Charge trapping effects that are manifested in the current-voltage curves suggest that semiconductor surface states are important in these devices. The observation of hysteresis in the current-voltage properties of these nanowires is surprising and potentially interesting in terms of switching devices. At present, the reason for this hysteresis is not well understood. Whether it is a consequence of electronic effects, such as filling/emptying of surface trap states, or whether it arises from field-induced reorganization of the thin film structure, is currently unknown. Future work will attempt to address the mechanism of these effects in polyelectrolyte/oxide semiconductor devices.

The availability of in-wire and concentric rectifying films opens up the possibility of constructing logic gates and cross-point devices by means of electrofluidic or chemically driven nanowire assembly. While the device characteristics shown here are poor by the standards of the semiconductor industry, it is important to note that only primitive layer sequences (simple polymer/colloid repeats) were used. It should be possible to make more structured films in which asymmetry is built in (e.g., with electron- and hole-conducting segments) to allow better rectification ratios, lower turn-on voltages, and more complex device functions. Preliminary data show that it is also possible to make in-wire junctions with single layers of self-assembling monolayers. This opens the door to using a wider variety of molecular electronic switches and other devices in self-assembling circuits. These possibilities will be explored in future work.

5. Acknowledgment.

We thank the DARPA Moletronics Program and the Office of Naval Research for support of this work.

References

1. (a) Abruña, H. d.; Denisevich, P.; Umana, M.; Meyer, T. J.; Murray, R. W. *J. Am. Chem. Soc.* 1981, *103*, 1; (b) Denisevich, P.; Willman, K. W.; Murray, R. W. *J. Am. Chem. Soc.* 1981, *103*, 4727; (c) Pickup, P. G.; Kutner, W.; Leidner, C. R.; Murray, R. W. *J. Am. Chem. Soc.* 1984, *106*, 1991.
2. (a) Kittlesen, G. P.; White, H. S.; Wrighton, M. S. *J. Am. Chem. Soc.* 1985, *107*, 7373; (b) Kittlesen, G. P.; Wrighton, M. S. *J. Mol. Electron.* 1986, *2*, 23; (c) Friend, R.; Burroughes, J.; Shimoda, T. *Phys. World* 1999, *12*, 35; (d) Roman, L. S.; Berggren, M.; Inganas, O. *Appl. Phys. Lett.* 1999, *75*, 3557; (e) Chen, J.; Reed, M. A.; Rawlett, A. M.; Tour, J. M., *Science* 1999, *286*, 1550, (f) R. M. Metzger, B. Chen, U. Höpfner, M. V. Lakshmikantham, D. Vuillaume, T. Kawai, X. Wu, H. Thachibana, T. V. Hughes, H. Sakurai, J. W. Baldwin, C. Hosch, M. P. Cava, L. Brehmer, G. J. Ashwell, *J. Am. Chem. Soc.* 1997, *119*, 10455.
3. Ellenbogen, J. C.; Love, J. C., *Proc. IEEE* 2000, *88*, 386.
4. (a) Rueckes, T.; Kim, K.; Joselevich, E.; Tseng, G. Y.; Cheung, C.-L.; Lieber, C. M. *Science* 2000, *289*, 94; (b) Collier, C. P.; Wong, E. W.; Belohradsky, M.; Raymo, F. M.; Stoddart, J. F.; Kuekes, P. J.; Williams, R. S.; Heath, J. R. *Science* 1999, *285*, 391; (c) Collier, C. P.; Mattersteig, G.; Wong, E. W.; Luo, Y.; Beverly, K.; Sampaio, J.; Raymo, F.; Stoddart, J. F.; Heath, J. R. *Science* 2000, *289*, 1172.
5. (a) D.Al-Mawlawi, C.Z.Liu, M. Moskovits, *J.Mater.Res.* 1994, *9*, 1014; (b) M.Nishizava, V.P. Menon, C.R. Martin, *Science* 1995, *268*, 700; (c) V. Menon, C.R. Martin, *Anal. Chem.* 1995, *67*, 1920; (d) C.R. Martin, *Chem. Mater.* 1996, *8*, 1739; (e) L. Wang, K.Yu-Zhang,, A. Metrot, P. Bonhomme, M. Troyon,

Thin Solid Films 1996, *288*, 86; (f) C.R. Martin, R.V. Parthasarathy, *Adv.Mater.* 1995, *7*, 487; (g) B.B. Lakshmi, P.K. Dorhout, C.R. Martin, *Chem. Mater.* 1997, *9*, 857.
6. (a) B.R. Martin, D.J. Dermody, B.D. Reiss, M. Fang, L.A. Lyon, M.J. Natan, T.E. Mallouk, *Adv. Mater.* 1999, *11*, 1021; (b) J.K.N. Mbindyo, B.D. Reiss, B.R. Martin, C.D. Keating, M.J. Natan, T.E. Mallouk, *Adv. Mater.*, in press.
7. B. D. Reiss, J. N. K. Mbindyo, B. R. Martin, S. R. Nicewarner, T. E. Mallouk, M. J. Natan, C. D. Keating, *Mater. Res. Soc. Symp. Proc.*, in press.
8. Huang, Y.; Duan, X.; Wei, Q.; Lieber, C. M. *Science* 2001, *291*, 630.
9. (a) Fendler J.H., Meldrum F. *Adv. Mater.* 1995, *7*, 607; (b) Mallouk T. E., Kim H.-N., Ollivier P. J, Keller S. W., in G. Alberti and T. Bein (eds.), *Comprehensive Supramolecular Chemistry*, vol. 7, Elsevier Science, Oxford, UK, 1996, pp. 189.
10. Iler R.K *J. Colloid Interface Sci.* 1966, *21*,569.
11. Decher G., *Science* 1997, *277*, 1232.
12. Keller S. W., Kim H.-N., Mallouk T. E. *J. Am. Chem. Soc.* 1994, *116*, 8817.
13. Colvin V.L., Golstein A.N., Alivisatos A.P. *J. Am. Chem. Soc.* 1992, *114*, 5221.
14. Fendler J. (1996) Self-assembled nanostructured materials, *Chem. Mater.* 8, 1616-1624.
15. Ollivier P.J., Kovtyukhova N.I., Keller S.W., Mallouk T.E. *Chem. Commun* 1998, 1563.
16. Fang M., Kim H.-N., Saupe G. B., Miwa T., Fujishima A., and Mallouk T. E. *Chem. Mater.* 1999, *11*, 1526.
17. Kovtyukhova N., Ollivier P.J., Chizhik S., Dubravin A., Buzaneva E., Gorchinskiy A., Marchenko A., Smirnova N. *Thin Solid Films* 1999, *337*, 166.
18. Kovtyukhova N.I., Gorchinskiy A.D., Waraksa C.C. *Mater. Sci&Eng.* 2000, *B69-70*, 424.
19. Kleinfeld E. R., Ferguson G. S. *Science* 1994, *265*, 370.
20. Kovtyukhova N., Ollivier P., Martin B., Mallouk T., Buzaneva E., Gorchinskiy A. *Chem. Mater.* 1999, *11*, 771.
21. Schmitt J., Decher G., Dressick W., Brandow S.L., Geer R.E., Shashidhar R., Calvert J.M. *Adv. Mater.* 1997, *9*, 61.
22. Feldheim D. L., Grabar K. C., Natan M. J., Mallouk T. E. *J. Am. Chem. Soc.* 1996, *118*, 7640.
23. (a) Cassagneau T., Fendler J.H., Mallouk T. E. *Langmuir* 2000, *16*, 241; (b) Cassagneau T., Mallouk T. E., Fendler J. H. *J. Am. Chem. Soc.* 1998, *120*, 7848.
24. Gao M., Richter B., Kirstein S., Mohwald H. *J. Phys. Chem.* 1998, *102*, 4096.
25. (a) Kaschak D. M. Mallouk T. E. *J. Am. Chem. Soc.* 1996, *118*, 4222; (b) Kaschak D. M., Lean J. T., Waraksa C. C., Saupe G., Usami H., Mallouk T. E. *J. Am. Chem. Soc.*, 1999, *121*, 3435.
26. N. I. Kovtyukhova, B. R. Martin, J. K. N. Mbindyo, P. A. Smith, B. Razavi, T. S. Mayer, T. E. Mallouk, *J. Phys. Chem. B*, published on web.
27. N. I. Kovtyukhova, B. R. Martin, J. K. N. Mbindyo, T. E. Mallouk, M. Cabassi, T. S. Mayer, *Mater. Sci.&Eng.*, 2001, in press.
28. V.M. Cepak, J.C. Hulteen, G. Che, K.B. Jirage, B.B. Lakshmi, E.R. Fisher, C.R. Martin, *Chem. Mater.* 1997, *9*, 1065.
29. J. S. Yu, J. Y. Kim, S. Lee, J. K. N. Mbindyo, B. R. Martin, and T. E. Mallouk, J. Chem. Soc., Chem. Comm. 2000, 2445.
30. G.O. Mallory, J.B. Hajdu Eds., Electroless Plating: Fundamentals and Applications, American Electroplaters and Surface Finishers Society: Orlando, FL, 1990.
31. P.A. Smith, C. D. Nordquist, T. N. Jackson, T. S. Mayer, B. R. Martin, J. Mbindyo, T. E. Mallouk, *Appl. Phys. Lett.* 2000, *77*, 1399.
32. T. Yokoyama, Y. Matsukado, A. Ichida, Y. Motomura, K. Watanabe, E. Izawa, J. Coll. Interface Sci., 2001, *233*, 112.

NON-THERMAL PLASMA SYNTHESIS OF NANOCARBONS

A.HUCZKO AND H. LANGE
Department of Chemistry, Warsaw University,
02-093 Warsaw, Poland
Y. Q. ZHU, W. K. HSU, H. W. KROTO AND D. R. M. WALTON
School of Chemistry, Physics and Environmental Science,
University of Sussex, Brighton ,BN1 9QJ, UK

Non-thermal plasma application for the synthesis of customized carbon nanotubes (CNTs) is reviewed. Carbonaceous nanostructures were deposited via a dielectric barrier discharge chemical vapor deposition using various mixtures of plasma gases and carbon-bearing reactants. Aligned carbon nanofibers/tubes (20-50 nm in diameter) were generated by hollow cathode glow discharge plasma decomposition of ferrocene in helium atmosphere on an alumina membrane. The average temperature in the reaction zone, determined by the optical emission spectroscopy, was below 200 °C. The as-deposited product consists of bundles of amorphous carbon nanofibers. Heating at 1100 °C under Ar produced well-crystallized carbon nanotubes.

1. Introduction

Thermal (equilibrium) plasma has long been used in powder spheroidization, welding, spraying, cutting, metallurgy, environmental engineering, and high temperature chemistry (Fauchais, 2001). Much of the efforts have focused recently also on the generating and application of stable non-thermal (non-equilibrium) plasmas, particularly in the industrial production of ozone, NO_x and SO_x abatement, and thin film growth (Kogelschatz et al., 1999). Different types of gas discharges have been used for these applications. In particular, the hollow cathode discharge (Stark et al., 1999) has been demonstrated as this is a very promising source of reactive species. Recently Jung et al. (1999) discussed the current status and future trends of advanced non-thermal plasma technology for microelectronics while Shi (1896) reviewed the basic features of plasma polymerization of organic thin films. A variety of techniques for producing polymer thin films with some proposed applications of these coatings in mechanical, electrical and optical industry was presented. Diamond-like carbon films on suitable substrates have attractive properties for electronic, optoelectronic, mechanical and biomedical applications and since the late 1970's numerous methods of diamond layer production have been developed on the basis of hydrocarbon decomposition in glow discharge plasma excited by RF electric field (Holland et al., 1976). The plasma-assisted

E. Buzaneva and P. Scharff (eds.), Frontiers of Multifunctional Nanosystems, 163–172.

deposition of amorphous hard carbon (a-C) and hydrocarbon (a-C:H) coatings is also of interest for wear and optoelectronic applications. Low-pressure arc (Bolt et al., 1996), unbalanced DC magnetron (Anisimova et al., 1998), standard radiofrequency (RF) and electron cyclotron resonance (ECR-RF) (Marechal, et al., 1999), and hollow cathode plasma (Buron et al., 1999) are the commonly used plasma sources.

With the discovery of novel carbon allotropes, fullerenes (Kroto et al., 1985; Krätschmer et al., 1990) and carbon nanotubes (CNTs) (Iijima, 1991; Ebbesen et al., 1993) carbon plasmas have also emerged as an efficient source of these new fascinating nanostructures. Carbon nanotubes have been proposed as new materials in panel displays, scanning electron microscope tips, hydrogen storage, catalyst supports, long ballistics conductors and next-generation molecular electronic devices (Huczko, 2001). They are excellent field emitters and this fact may prove to be very useful in electronics. Indeed, the first prototypes of flat panel displays with CNTs used as cold cathodes have already been reported (Saito et al., 1998).

There are many ways (Ebbesen et al., 1993; Terrones et al., 1998) nowadays to synthesize these nanocarbons but both arc discharge and laser ablation plasmas still remain the basic production techniques. The final choice of the CNT synthesis route does not influence distinctly the morphology of the product. Both, carbon plasmas and CVD produce nanotubes in relatively low yields, random orientation, and tangled distribution (Liu et al., 1998; Journet et al., 1998). It hampers both the fundamental and the applied studies of tubular nanocarbons. To make electronic devices with CNTs more efficient one has to control the growth of vertically well-aligned and ordered carbon nanotubes. It has been successfully achieved by using the chemical vapor deposition (CVD) technique with the catalyst-decorated substrates (Terrones et al., 1997) or the process was carried out by using the floating catalyst method (Huang et al., 1999). However, the CVD systems rarely operate below 600 °C and the growth temperature is further required to be as low as 550 °C for the practical purpose of FED applications, in which the sodalime glasses are often used for the manufacture of the devices. Thus, there have been many reports published recently on other than thermal activation modes of reactants at temperatures below the glass strain point. Tailored nanotubes have been produced using different plasma-assisted CVD techniques and both, catalyst-coated substrate and floating catalyst, approaches have been successfully tried.

Nucleation and growth of CNTs by microwave plasma-enhanced (MPE) CVD at a temperature between 520 °C and 1000 °C have been recently frequently reported (Zhang et al., 2000; Choi et al., 2000; Murakami et al., 2000; Bower et al., 2000; Okai et al., 2000; Wang et al., 2000). At a lower growth temperature the tubes were usually curly, whereas they were straight and self-aligned upward at temperatures above 600 °C.

Liang et al.(2000) reported the deposition on aligned MWCNTs on the Ni-coated silicon substrates by the hollow cathode glow discharge (HCGD) PE CVD using $CH_4/N_2/H_2/Ar$ gas mixtures. The growth conditions included the pressure at 15-25 Torr, the substrate temperature at 500-750 °C, and the concentration of CH_4 at 4.0 ~ 12.0 vol. %.

There have been also many other plasma sources to enhance the growth of carbon nanotubes. Ho et al. (2001) used radio frequency (RF) PE CVD to synthesize dense and well-aligned MWCNTs on the 15-nm thick Ni-coated quartz glass at 650 °C by the decomposition of C_2H_2/NH_3 gases. Vertically well-aligned MWCNTs were also grown on Ni-coated glass substrates using the same reactants by hot filament PE CVD (Han et al., 2001) at low temperatures below 600 °C. Plasma intensity was found to be the critical parameter to determine the growth of MWCNTs (diameter of about 50 nm). Field emission from the tubes was measured using a phosphor anode with an onset electric field of 1.5 V/ìm.

Chung et al. (2001) developed a noble CNT deposition method using a layer-by-layer technique to selectively etch out with CF_4 amorphous carbon and graphite. The vertically aligned CNTs were deposited on the patterned Ni-Co catalyst layer using a RF high density plasma. Silicon, quartz and glass (at low deposition temperature <600 °C) were used as substrates and the MWCNTs were deposited using a CH_4 and H_2 mixture at a pressure of 20 mTorr. The tubes exhibited a turn-on-field of 1.2 V/ìm and the emission current of the layer-by-layer deposited CNTs film was higher than that of conventional plasma deposited tubes.

Aligned hydrogenated amorphous CNTs on AAO substrates were grown by electron cyclotron resonance CVD using the C_2H_2/Ar (Tsai et al., 2000). By adjusting the pore size of anodic alumina, tubes of various diameters could be grown in a range from 230 down to 30 nm.

For the first time, by using the nanocatalytical glow-discharge PE CVD Obraztsov et al. (2000) obtained thin film of MWCNTs oriented perpendicular to the surface. The substrate temperature was ~ 1050 °C, the gas mixture was CH_4:H_2=10:90 and the total gas pressure was ~ 100 Torr. The tubes exhibited excellent field emission current density up to 50 mA/cm^2 at the field of 5 V/ìm and the emission site density of 10^{7}cm^{-2}.

2. Experimental

The formation of nanocarbons was carried out in two different experimental systems. Studies of a dielectric barrier discharge (DBD) PE CVD template deposition were performed in the experimental system presented in Fig. 1. DBD plasma was generated between an Al electrode and Ag-plated ferroelectric ceramics ($BaTiO_3$), which served as a second electrode. The discharge gap was about 200 ìm. A sinusoidal AC voltage of 200-Hz frequency and 800 V amplitude was applied to the electrodes. The runs were carried out under atmospheric pressure and flow conditions. Various mixtures of plasma gases and carbon-bearing reactants (without a catalyst) were tested.

We have also initiated the studies on HCGD plasma synthesis of aligned nanocarbons in the system shown in Fig. 2. Different operational parameters were tried with a catalyst (Fe) introduced as ferrocene vapors. The equipment consisted of a plasma source and a downstream region where substrates are deposited. A substrate and $Fe(C_5H_5)_2$ (20 mg) were placed inside the hollow graphite cathode (40 mm long, 5 mm i.d.) and anode, respectively. A He flow (0.67 kPa; 0.2 cm^3 s^{-1} STP) enabled transport of $Fe(C_5H_5)_2$ (*ca.* 0.4 mg min^{-1}) to the deposition zone. The discharge current and cathode-anode voltage drop were 60 mA and 380 V, respectively.

During discharge, the spectroscopic features of the plasma were monitored optically. The gas temperature in the deposition zone was measured by using the optical emission spectroscopy (OES) technique with the use of a CCD camera.

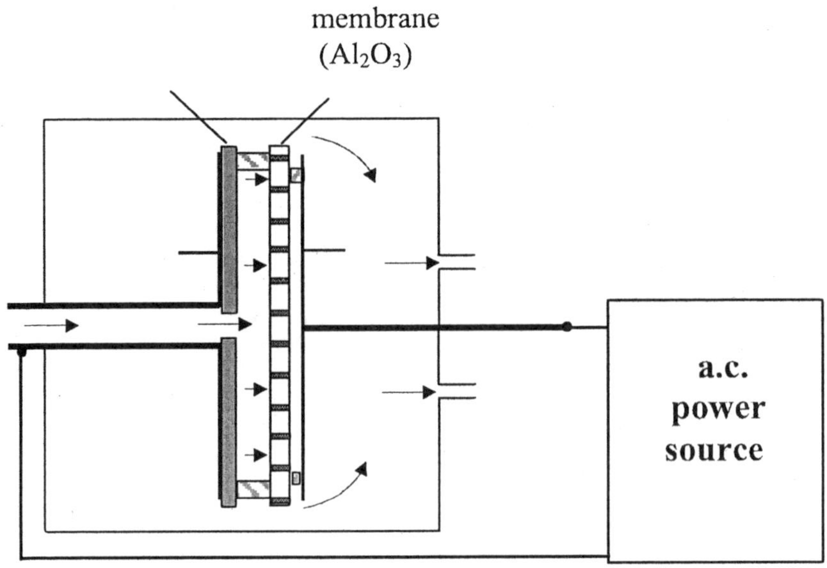

Figure 1. Schematic view of DBD reactor with ferroelectric ceramics and AAO template for the deposition of carbon nanostructures.

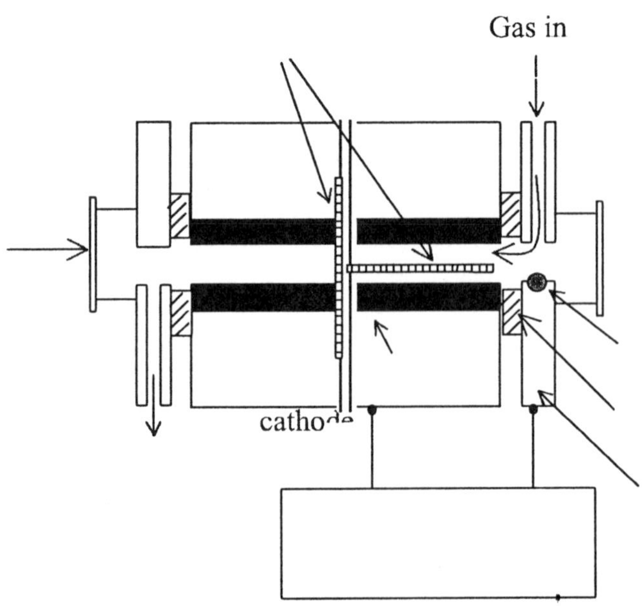

Figure 2. Schematic view of HCGD reactor with AAO template for the deposition of carbon nanostructures.

Anodic aluminum oxide (AAO) membranes 60-100-ìm thick with channels of ~30 nm in diameter were used as substrates. The deposited nanostructures were observed in their as-prepared state using an ultra-high-resolution field-emission SEM (LEO 1530) and TEM (H-7100 and CM200).

3. Results and Discussion

Regarding the DBD system very intriguing nanodeposits were formed, the morphology of which heavily depended on the processed reactants (different hydrocarbons, also with carbon tetrachloride and tiophene addition) – Fig. 3. Due to the relatively high deposition rates, however, the penetration of carbonaceous solids within template channels was in those exploratory experiments rather limited.

After HCGD processing of ferrocene/helium mixture a dark film, deposited on the AAO surface, was examined by scanning electron microscopy. The sample was later subjected to ultrasonic treatment in acetone for 30 min, then transferred to a Cu grid coated with a holey carbon film for transmission electron microscopic investigations. Other as-grown samples were heated to 1100 °C for 5 h under Ar flow. After cooling, they were again treated in acetone in an ultrasonic bath for 3 h, prior to transfer to the Cu grid for TEM examination.

SEM showed that the as-grown black sample consisted of flat carpet-like films (Fig. 4a, top view). The overall size of the film is *ca.* that of the substrate, *i.e.* $10 \times 4 \times 0.06$ mm. The film, *ca.* 3-10 μm thick, comprised relatively well-aligned nanofiber bundles (Fig. 4b, side view). It is noteworthy that these bundles were very homogeneous and contained few nanoparticles. The nanofibers were *ca.* 20-50 nm in diameter and exhibited no direct relationship to the AAO channel structural parameters. At the fiber tips, white particles were apparent, presumably consisting of coalesced Fe particles as a result of $Fe(C_5H_5)_2$ decomposition. TEM showed that the resulting nanofibers were relatively straight and uniform, displaying dimensions similar to those observed by SEM (Fig. 5a). EDX analysis showed that the fibers consisted solely of C and Fe. Thus, the samples were made up of aligned bundles of amorphous carbon nanofibers, with Fe particles embedded therein.

After heat treatment (1100 °C, 5 h), further HRTEM observations showed that the amorphous fibers had been converted into well-crystallized nanotubes (Fig. 5b): uniform, straight, analogous to those produced by arc-discharge methods. In some cases, the nanotubes were buckled and twisted, presumably as a result of 3 h ultrasonic treatment, which caused some of the poorly crystallized tubes to disintegrate. It is interesting to note that we failed to detect any Fe particles or encapsulates in the nanotubes, however large slab-like Fe particles, located in the carbonaceous phases (Fig.5b arrowed), were detected by EDX. This indicates that long-distance migration of Fe and C atoms had occurred during the carbon nanotube graphitization process. At this stage, it is not at all clear how the Fe influences the carbon graphitization and nanotube formation during heat treatment. However a catalytic effect is expected, as it occurred in the $Fe(C_5H_5)_2$ CVD step (temperature > 600 C°) for carbon nanotube production.

The well-known violet CN band was used to determine the plasma temperature, by resolving the rotational structure of the P(0,0) branch, applying the Boltzmann plot method. Our results showed that the average temperature in the discharge zone was *ca.* 190 ± 15 °C.

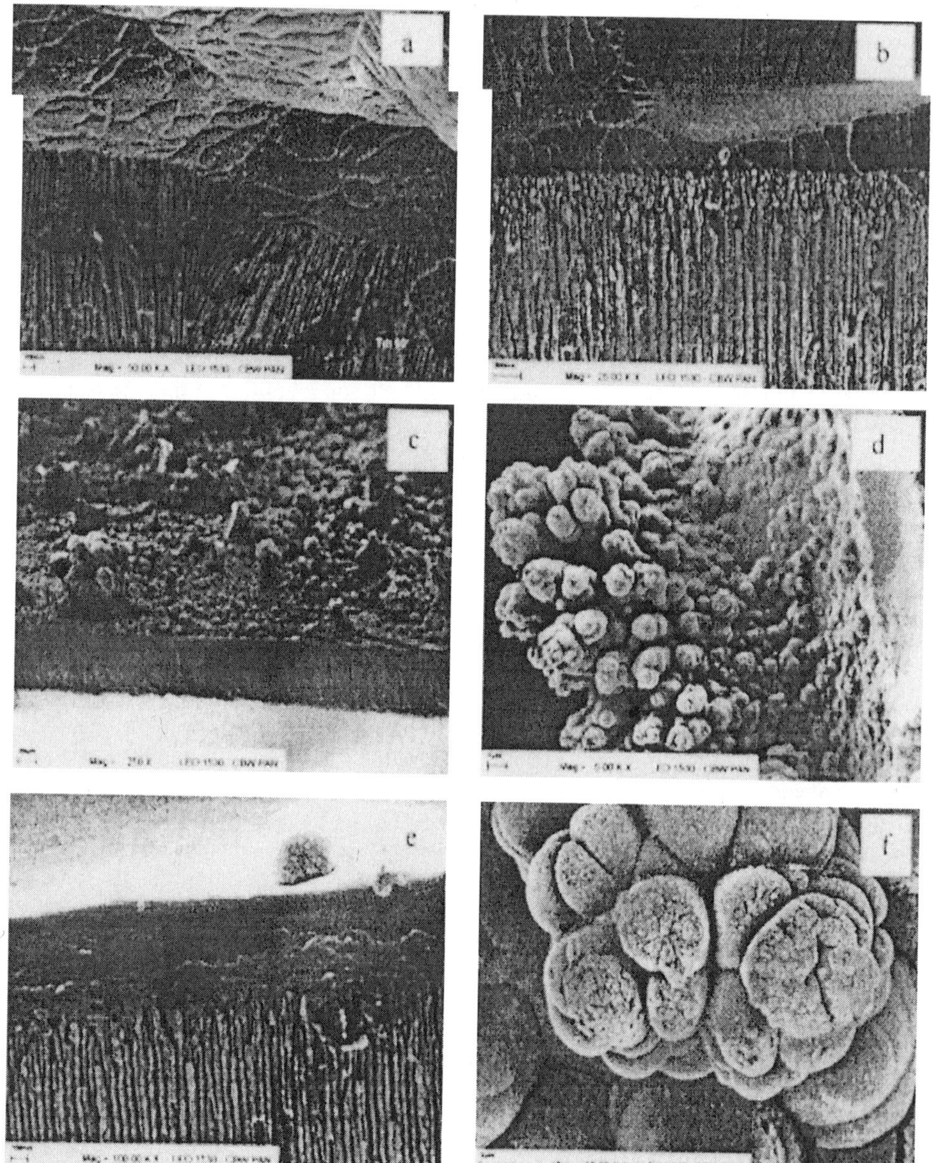

Figure 3. SEM images of AAO template surface and cross-section (DBD plasma); a) starting membrane; b-e) C_6H_6 + NH_3 gas mixture; f) C_6H_6 + CCl_4 +C_4H_4S gas mixture.

The advantages of this single-step approach include: (1) absence of a pre-patterned catalytic substrate; (2) low deposition temperature; (3) high deposition rate, and (4) simple equipment. Emission spectroscopic analysis of the plasma zone revealed the predominance of C_2 radicals, which are believed to constitute the key intermediate phase in nanomaterial formation. In this context, the plasma-generated C_2 species, together with the catalytic Fe, give rise to nanofibers at fairly low 'effective' temperatures (below 200 °C).

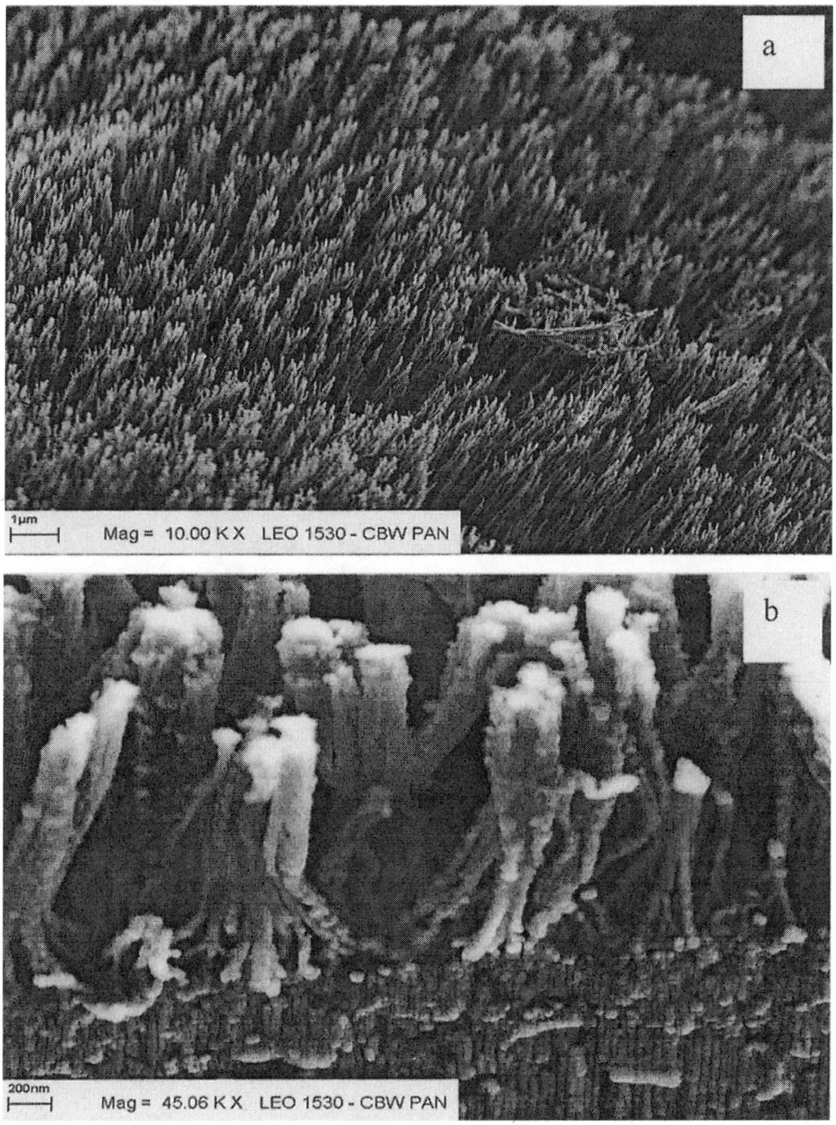

Figure 4. SEM images of nanocarbons on AAO membrane formed in HCGD. Reactants:He/ferrocene vapors.

In summary, we have demonstrated that the HCGD process offers great potential for generating aligned carbon nanofibers at low temperature. Studies are under way to optimize the process since many parameters are involved in CNF and CNT growth. Also, the emission characteristics of the produced tubular nanocarbons will be measured.

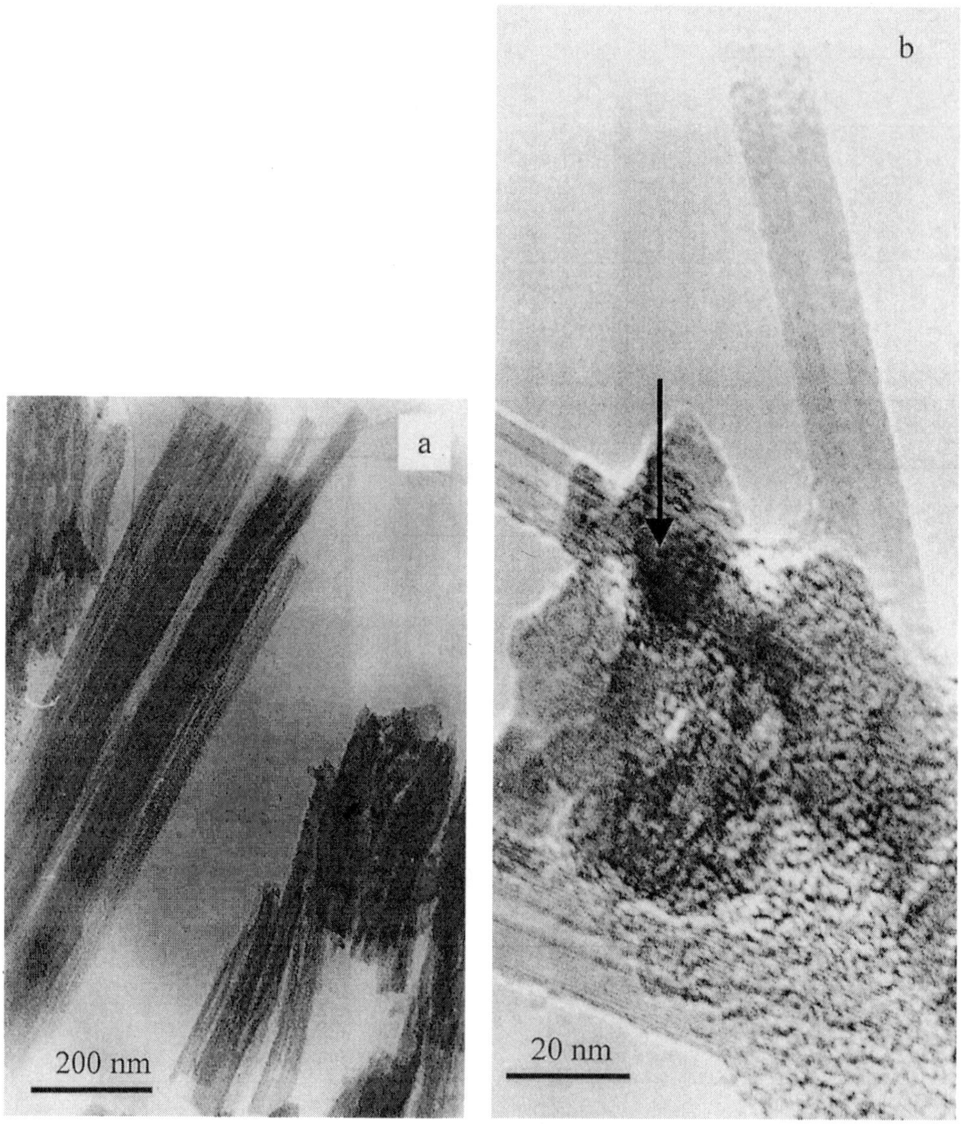

Figure 5. a) TEM image of nanofibers, and b) HRTEM image of well-crystallized carbon nanotubes with Fe slabs (arrowed).

4. Acknowledgement

The work was supported by the Committee for Scientific Research (KBN) through the Department of Chemistry, Warsaw University, under Grant No. 3 T09A 058 16, the Royal Society and the Leverhulme Trust.

References

3. Anisimova, I.A., Anisimov, V.P., Makarov, V.P. and Otorbaev, D.K. (1998), High Temp. Mat. Proc. 2, 143.
4. Bolt, H., Hemel, V., Koch, F., Nickel, H., Wiechers, C. and Avni, R. (1996), Surf. Coat. Technol. 88, 50.
5. Bower, C., Zhou, O., Zhu, W., Werder D.J. and Jin, S. (2000), Appl. Phys. Lett. 77, 2767.
6. Buuron, A., Koch, F. and Nöthe, M. and Bolt, H. (1999), Surf. Coat. Technol. 116-119, 755.
7. Choi, Y.C., Bae, D.J., Lee, Y.H., Lee, B.S., Han, I.T., Choi, W.B., Lee, N.S. and Kim, J.M. (2000), Synthetic Metals 108, 159.
8. Chung, S.J., Lim, S.H. and Jang, J. (2001), Thin Solid Films 383, 73.
9. Ebbesen, T. W. and Ajayan, P. M. (1992), Nature (London) 358, 220.
10. Ebbesen, T.W., Hiura, H., Fujita, J., Ochiai, Y., Matsui, S. and Tanigaki, K. (1993), Chem. Phys. Lett. 62, 1881.
11. Fauchais P., ed. (2001), Progress in Plasma Processing of Materials, Begell House Inc., New York.
12. Han, J.-H., Yang, W.-S., Yoo, J.-B. and Park, C.-Y. (2000) J.Appl. Phys. 88, 7363.
13. Ho, G.W., Wee, A.T.S., Lin, J. and Tjiu, W.C. (2001), Thin Solid Films 388, 73.
14. Holland, L. and Ojha, S.M.(1976), Thin Solid Films 38, L17.
15. Huang, S., Dai, L. and Mau, A.W.H. (1999), J. Phys. Chem. 103, 4223.
16. Huczko, A. (2001), Appl. Phys. A 73, 1.
17. Iijima, S. (1991), Nature (London) 354, 56.
18. Journet, C., and Bernier, P. (1998), Appl. Phys. A 67, 1.
19. Jung, C.O., Chi, K.K., Hwang, B.G., Moon, J. T., Lee, M.Y. and Lee, J.G. (1999), Thin Solid Films 341, 112.
20. Kogelschatz, U., Eliasson, B. and Egli, W. (1999), Pure Appl. Chem. 71, 1819.
21. Krätschmer, W., Lamb, L.D., Fostiropoulos, K. and Huffman, D.R. (1990), Nature (London) 347, 354.
22. Kroto, H.W., Heath, J.R., O'Brien, S.C., Curl, R.F. and Smalley, R.E. (1985), Nature (London) 318, 162.
23. Liang, G.-T. and Hong, F.C.-N. (2000), 11th Eur. Conf. Diamond, Diamond-Like Materials, Carbon Nanotubes, Nitrides and Silicon Carbide, 3-8 September 2000, Porto, Portugal, Abstract Book No. 5.6.23
24. Liu, J., Rinzler, A.G., Dai, H., Hafner, J.H., Bradley, R.K., Boul, P.J., Lu, A., Iverson, T., Shelimov, K., Huffman, C.B., Rodriguez-Macias, F., Shon, Y., Lee, T.R., Colbert, D.T. and Smalley, R.E. (1998), Science 101, 1253.
25. Marechal, C., Zeinert, A., Zellama, K., Lacaze, E., Zarrabian, M.and Turban, G. (1999) Solid State Comm. 109, 23.
26. Murakami, H., Hirakawa, M., Tanaka, C. and Yamakawa, H. (2000), Appl. Phys. Lett. 76, 1776.
27. Obraztsov, A.N., Pavlovsky, I., Volkov, A.P., Obraztsova, E.D. and Kuznetsov, V.L. (2000), J. Vac. Sci. Technol. B 18, 1059.
28. Okai, M., Muneyoshi, T., Yaguchi, T. and Sasaki, S. (2000) Appl. Phys. Lett. **77**, 3468.
29. Saito, Y., Hamaguchi, K., Uemura, S., Uchida, K., Tasaka, Y., Ikazaki, F., Yumura, M., Kasuya, A. and Nishina, Y. (1998), Appl. Phys. A 67, 95.
30. Shi, F.F. (1996), Surf. Coat. Technol. 82, 1.
31. Stark, R.H. and Schoenbach, K.H. (1999), J. Appl. Phys. 85, 2075.
32. Terrones, M., Grobert, N., Olivares, J., Zhang, J.P., Terrones, H., Kordatos, K., Hsu, W.K., Hare, J.P., Townsend, P.D., Prassides, K., Cheetham, A.K., Kroto, H.W. and D.R.M. Walton, (1997), Nature (London) 285, 52.

33. Terrones, M., Hsu, W.K., Schilder, A., Terrones, H., Grobert, N., Hare, J.P., Zhu, Y.Q., Schwoerer, M., Prassides, K., Kroto, H.W. and Walton, D.R.M. (1009) Appl. Phys. A 66, 307.
34. Tsai, S.H., Chiang, F.K., Tsai, T.G., Shieu, F.S. and Shih, H.C. (2000), Thin Solid Films 366, 11.
35. Wang, X., Hu, Z., Wu, Q., Chen, X. and Chen, Y. (2000), Thin Solid Films 390, 130.
36. Zhang, Q., Yoon, S.F. , Ahn, J., Rusli, B.G. and Yu, M.-B.(2000), J. Mater. Res. 15, 1749.

A NOVEL NETWORK STRUCTURE OF ORGANOMETALLIC CLUSTERS IN GAS PHASE

ATSUSHI NAKAJIMA
Department of Chemistry, Keio University,
3-14-1 Hiyoshi, Kohoku-ku, Yokohama 223-8522, Japan
KOJI KAYA
Institute for Molecular Science, Myodaiji, Okazaki, 444-8585, Japan

1. Introduction

Synthesis of organometallic compounds in gas phase presents a novel guideline to preparative chemists because they can be prepared without solvent and oxidation in air. Especially, the application of the laser vaporization to the gas-phase synthesis of organometallic compounds enables us to prepare the constituents in considerable density in a short time because there are no interfering effects of solvents, aggregation phenomena, and counter ions. This new approach should open up an entirely different aspect of organometallic chemistry and physics, which can be studied quite nicely in the gas phase, and indeed several groups independently have succeeded in the synthesis of novel organometallic complexes.[1-7] The technique used to produce and detect the gas-phase organometallic clusters involves a beam source with laser vaporization of metal rods and time-of-flight mass spectrometry whose distinct ability was well-demonstrated in the discovery of C_{60} [8] and metallo-carbohedrene,[9,10] a so-called met-car.

Furthermore, recently, formation of fullerene-based organometallic compounds suggests that new forms of materials can be synthesized because the fullerenes may prove to be highly versatile ligands due to their intriguing topography and aromaticity.[11-14] A great deal of effort has been spent in the past decade on modifying fullerenes by coordinating atoms both inside and outside the cage structures.[15-17] In particular, the finding of superconducting alkali metal fullerides[18-22] stimulated considerable interest, suggesting that new forms of materials and superstructures can be synthesized with important chemical and physical properties. Since C_{60} acts as not only an electron donor, but also an electron acceptor, these peculiar ligand properties of C_{60} make studies of its derivatives attractive. Interest in metal-C_{60} complexes has been driven by the potential for catalytic activity, and by the possibility of forming new supermolecular compounds with new physical and chemical properties.[23]

As well as organometallic compounds of transition metal elements, the f-block elements of lanthanide (Ln) and actinide (Ac) metals have been of considerable importance in many areas of modern technology.[24-27] After discovery of $Ac(COT)_2$ (COT=cyclo-octatetraene; C_8H_8),[28-30] studies on this criteria have extensively stimulated experimentalists and theoreticians, and $Ln(COT)_2$ were established to be

E. Buzaneva and P. Scharff (eds.), Frontiers of Multifunctional Nanosystems, 173–190.

trivalent complexes in bulk materials;[26,31] these findings are in agreement with theoretical calculations,[32-34] except for the case of divalent complexes of Ln = Eu and Yb. Lanthanide π-carbocylic complexes were synthesized as their salts of $K[Ln(COT)_2]$, where trivalency of Ln metals overcomes aromaticity of COT^{2-}, resulting in a highly ionic complex denoted as $Ln^{3+}(COT^{1.5-})_2$. This predominantly ionic character results from that the ligand of COT can act as stable dianion. These properties make it difficult to study neutral and multi-core complexes of $Ln_n(COT)_m$, and until now, very few synthetic studies on the larger clusters have only been reported except for a few examples.[35,36]

From the viewpoint of application of clusters to fabricating new materials, furthermore, nanostructure formation at surfaces would provide a promising route, using the controlled deposition of the size-selective clusters in the gas phase,. When cluster ions are size-selectively prepared, both the kinetic energy and the neutralization energy of the ion, i.e., the ionization energy of the neutral cluster, should be dissipated to ensure soft-landing of the clusters. Then, the synthesis of the nanostructure surfaces requires low kinetic energies to be released during the deposition to ensure a nondestructive deposition in which the clusters maintain their individual characteristics. Some of the clusters can be expected to work as a functional unit such as a catalyst, and organometallic clusters would be one of the promising candidates among functional molecular units.

We focuses on the formation and the characterization of the superstructures composed of multi-metal atoms and multi-organic ligands by use of laser vaporization, chemical probe, and photoionization methods, revealing the nature of the metal-ligand interface and their electronic structures, as well as the soft-landing of organometallic clusters with size selection. In these studies, 3d-transition metals and lanthanide metals are used as metal elements, and Bz, C_{60}, and COT were used as organic ligand molecules.

2. Experimental Setup

A couple of experimental methods were employed to characterize the electronic structures of organometallic clusters in gas phase. A chemical probe method using adsorption reaction was used to deduce geometric structures of the organometallic clusters, and measurement of ionization energy (E_i) by photoionization with a tunable laser provides electronic characters of the neutral organometallic clusters. Furthermore, photoelectron spectroscopy (PES) was used for size-selected Ln-COT anions.

2.1. *PRODUCTION, CHEMICAL PROBE, AND E_I MEASUREMENT*

Figure 1 schematically shows an experimental setup which consists of laser vaporization source and time-of-flight (TOF) mass spectrometer.[7,37,38] $M_n(Bz)_m$

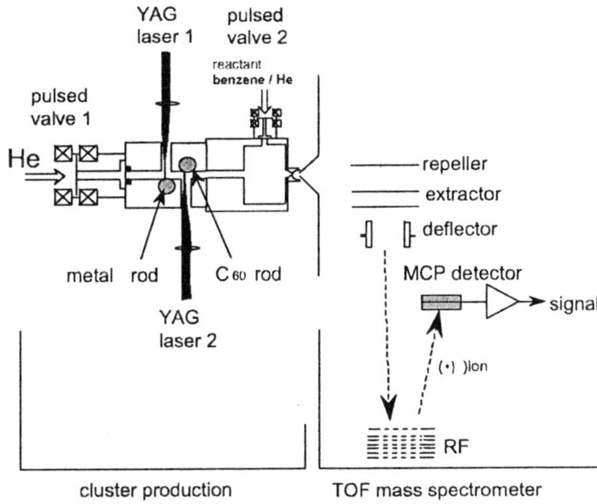

Figure 1.
A shcmatic experimental setup consisting of laser vaporization source and time-of-flight (TOF) mass spectrometer.

clusters (M = Sc - Ni) were synthesized by the combination of a laser-vaporization method and a flow-tube reactor.[39] Metal atoms were vaporized by the irradiation with the second harmonic of pulsed Nd^{3+}: YAG laser (532 nm) and the vaporized hot metal atoms were cooled to the room temperature by a pulsed He carrier gas (10 atm). Then, the metal atoms were sent into the flow-tube reactor (FTR) where benzene vapor seeded in a He gas was injected in synchronization with the flowing of the metal atoms. After skimming of the cluster beam, the $M_n(Bz)_m$ clusters were ionized by an ArF excimer laser (6.42 eV) or second harmonic of a dye laser pumped by an XeCl excimer laser. The photoions were mass-analyzed by a reflectron time-of-flight (TOF) mass spectrometer, whereas the cluster cations were accelerated with a pulsed electric potential to +3 kV without photo-ionization. The ions directly produced in the beam were mass-analyzed by a TOF mass spectrometer with a reflectron.

Similarly to the metal-benzene clusters, the binary clusters of both transition metal (M) - C_{60} and Ln - C_{60} were produced in the gas phase by laser vaporization; two lasers were used for metal and C_{60} rods, respectively, as shown in fig. 1. The second vaporization laser was fired with ~ 5 ms delay time after the first pulsed laser, which was adjusted to synchronize with the flow speed of the He carrier gas in order to mix them homogeneously. The fluence of the vaporization laser for the transition metal rod was 10-15 mJ/pulse while that for the C_{60} rod was 70-100 μJ/pulse to avoid "C_2 loss" fragmentation of C_{60}. The produced hot vapors of metal and C_{60} were quenched to room temperature by pulsed He carrier gas, and the M-C_{60} clusters were generated. The photoionized clusters were mass-analyzed by a time-of-flight (TOF) mass spectrometer with a reflectron. In order to detect heavy cluster ions efficiently, an ion detector known as "Even-Cup" was used,[40] in which cations accelerated to 20 kV hit a cup-like aluminum dynode and the ejected electrons extracted onto a grounded scintillator were converted into photons which were detected by a photomultiplier. The C_{60} rod was prepared by pressing C_{60} powder.

Lanthanide-cyclooctatetraene complexes, $Ln_n(COT)_m$ [Ln = Eu, Ho, and Tb], were similarly produced by the foregoing method. Instead of benzene in the formation of metal-benzene, COT vapor (~ 70 Torr; 70 °C) diluted with He carrier gas was injected into the FTR, and the $Ln_n(COT)_m$ clusters were generated.

In order to obtain information on the structure of $M_n(Bz)_m/M_n(C_{60})_m$, the clusters were further reacted with CO, O_2, C_2H_2, and NH_3 gas inside another FTR, which is added downstream of the source, and their adducts produced were also mass-analyzed. In the E_i measurement for $M_n(Bz)_m/M_n(C_{60})_m/Ln_n(COT)_m$, the second harmonic of the dye laser was used as the ionization laser. The photon energy was changed at 0.01–0.03 eV intervals in the range of 5.9 - 3.5 eV, while the abundance and composition of the clusters were monitored by the ionization of an ArF laser. The fluences of both the tunable ultra-violet (UV) laser and the ArF laser were monitored by a pyroelectric detector and were kept at ~200 μJ/cm^2 to avoid multiphoton processes. To obtain photoionization efficiency curves, the ion intensities of the mass spectra ionized by the tunable UV laser, were plotted as a function of photon energy with the normalization to both the laser fluence and the ion intensities of ArF ionization mass spectra. The E_is of the clusters were determined from the final decline of the photoionization efficiency curves.

2.2. *SOFT-LANDING APPARATUS OF SIZE-SELECTED CLUSTERS*

The apparatus used in the soft-landing experiment consists of a cluster ion source, ion optics with mass selection, and a deposition room for IR spectroscopy. The produced cluster ions were guided by a series of ion optics; electrostatic ion lens, a quadrupole (Q-pole) deflector, a quadrupole mass filter, and an octapole ion guide. The cluster ions were accelerated and focused to the Q-pole deflector through a couple of ion lenses, and were bent perpendicularly at the Q-pole deflector. This deflection can separate the cations from neutrals, anions, and electrons. Going through another set of ion lens, the cluster cations were mass-separated with a commercial Q-pole mass spectrometer (4-4000 amu). The size-selected cluster cations were further guided onto the deposition substrate by two sets of the octapole ion guides, which have larger transmission rate, compared to the Q-pole. The apparatus was partitioned into four differentially pumped units. The deposition substrate was a polycrystalline gold (Au), which was cooled down to 18 K by a liquid He cryostat. Before the deposition, the substrate was exposed to Ar gas to make an Ar layer of ~30 MLs on the Au substrate. The Ar layer works as a cushion for the soft-landing of the cluster ions, and also as a spectroscopically inert media for IR spectroscopy. The deposited cluster cations are successively neutralized by charge tunneling through the thin Ar layer.[41] The ion current can be monitored by the neutralizing charge and the number of deposited clusters was obtained by integrating the measured current on the substrate over the deposition time. The deposition energy was controlled by a difference of the bias voltages between the cluster source and the substrate, which was typically 20 eV. For the deposition energy below 50 eV, the fragmentation of the clusters can be excluded. For selected clusters, Fourier transform infrared (FTIR) studies have been performed to observe the vibrations of the clusters with a commercial FTIR spectrometer. The chamber windows for the IR radiation were KBr and the reflected IR radiation at a polished Au substrate was observed by a Mercury-Cadmium-Tellurium (MCT) detector, with which the IR absorption spectrum can be measured in the range of 700 – 6000 cm^{-1}.

3. Results and Discussion

3.1. *METAL-BENZENE CLUSTERS*

Figures 2(a) and 2(b) show typical mass spectra of the ArF laser ionization of vanadium (V) and benzene (Bz) clusters produced by the foregoing procedure.[7] The mass spectra of early transition metals of Sc,[42] Ti,[42] and V[43] are very similar one another; prominent peaks were observed at almost the same compositions, denoted as $M_n(Bz)_m$ [henceforth (n, n+1)], although production efficiency for larger complexes depends on the metal elements. These prominent peaks of (n, n+1) remained unchanged in the mass spectrum, even if (1) concentration of benzene vapor was changed to higher one and (2) CO reactant was exposed additionally. This result indicates that all the metal atoms of (n, n+1) are contained as an inner component, because an exterior metal atom should result in a site of CO adsorption. Furthermore, when the concentration of metal vapor

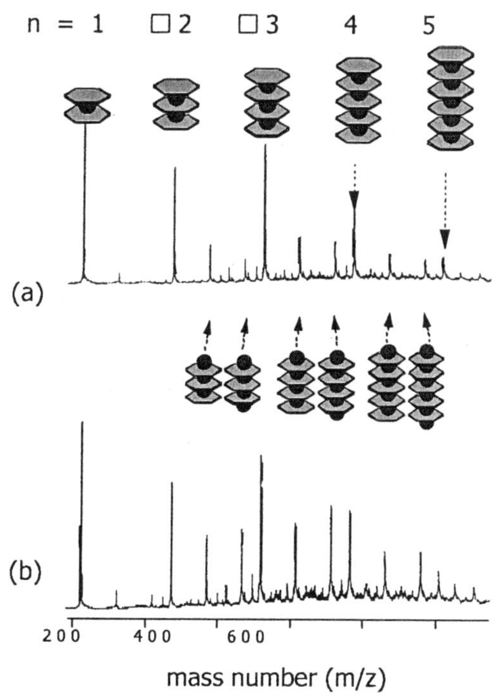

Figure 2. TOF mass spectra of V-benzene organometallic clusters: the spectrum (b) was measured under V rich-vapors compared to (a). For each peak, the plausible sutuctures of a multiple-decker sandwich structure are shown. The abundant (n, n+1) clusters have no exterior V atoms, while two more exterior V atoms are acceptable.

was changed to a higher value, new minor peaks denoted as (n, n) and (n+1, n) appeared in the spectrum, but no peaks for (n+2, n), as shown in fig. 2(b). These two kinds of additional clusters showed adsorption reaction toward CO, resulting in (n, n) + 3CO and (n+1, n) + 6CO for V-Bz. These results on mass spectra in the chemical probe experiment indicate that the (n, n+1) clusters are sandwich clusters in which metal atoms and benzene molecules are alternately piled up. Namely, the superfluous metal atom(s) over the (n, n+1) composition are exterior atoms and are placed at the terminal of clusters up to two atoms. These (n, n) and (n+1, n) clusters reasonably correspond to a sandwich cluster having one and two exterior metal atom(s), respectively. Because there are two sites for the exterior atoms, the (n+2, n) cluster is missing. This result of the chemical probe experiment leads us to conclude that the most explicable structure for $V_n(Bz)_{n+1}$ clusters is a multiple decker structure (fig. 2).[42,43] Indeed, Bowers and co-workers did an ion mobility experiment for the $V_n(Bz)_{n+1}$ clusters and reported that they really take the multiple sandwich structure.[44]

E_is of $V_n(Bz)_{n+1}$ clusters were measured by the photoionization method and the E_i values are tabulated in table 1.[42,43] Figure 3 shows the photoionization curves of n=1-4 plotted against the photon energy. The first onset of each curve corresponds to the E_i values which drastically decrease with cluster size n, although the clusters consist of metal atoms and benzenes having relatively high E_is; $E_i(V)$ = 6.74 eV, and $E_i(Bz)$ =

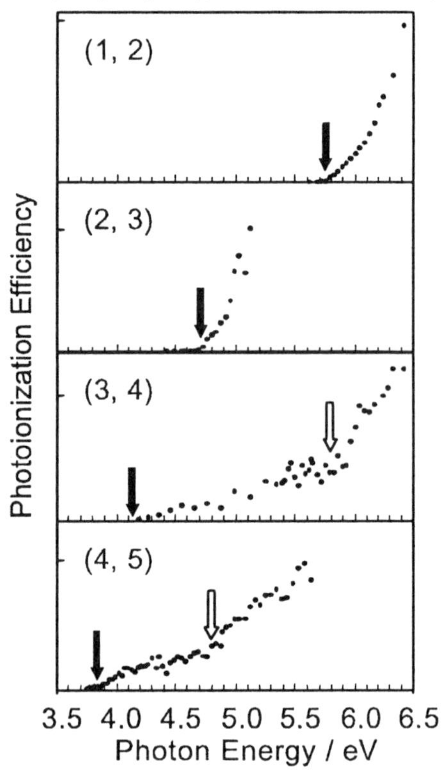

Figure 3.
Photoionization efficiency curves of $V_n(Bz)_{n+1}$ for n=1-4. The first onset in each curve correspond to ionization energy, which is shown by a solid arrow. For (3,4) and (4,5), the second onset is observed as the ionization channel from a deeper dδ orbitals, which is shown by an open arrow .

9.24 eV.[45] The trend of the E_i decrement is understood as a common feature for the multiple decker sandwich structure for Sc-, Ti-, and V-Bz. This phenomenon of E_i decrement has been theoretically elucidated by the DFT(B3LYP) method, showing that the ionization occurs from delocalized molecular orbital of metal-metal interaction interposed by π^* orbitals of benzene.[46] Along the molecular axis of the sandwich cluster, one dimensional delocalization of the metal $d\delta(e_{2g})$ orbitals occurs through the LUMO (e_{2g}) of bebzebe, giving a drastic decrement of E_i.

According to a simple Hückel method, ionization energies can be estimated as follows: the simple Hückel model treats only the interaction between the adjacent metal atom and benzene molecule. The orbital energies are given as the eigenvalues from the Hückel matrix. Figure 4 shows obtained quasi-band electronic structure of dδ orbitals. Since it can be assumed by the Koopmans theorem that the orbital energy of the top orbital corresponds to the lowest ionization energy, the shift of the orbital energy of the HOMO corresponds to the lowering of E_i. The experimentally obtained the lowest ionization energies were shown also in fig. 4. The ionization energies from unexplored orbitals are calculated from the experimentally obtained ionization energy values for (1, 2), (2, 3) and (3, 4). Surprisingly, the experimentally obtained the lowest E_i for (4, 5) is well reproduced by this model calculation.

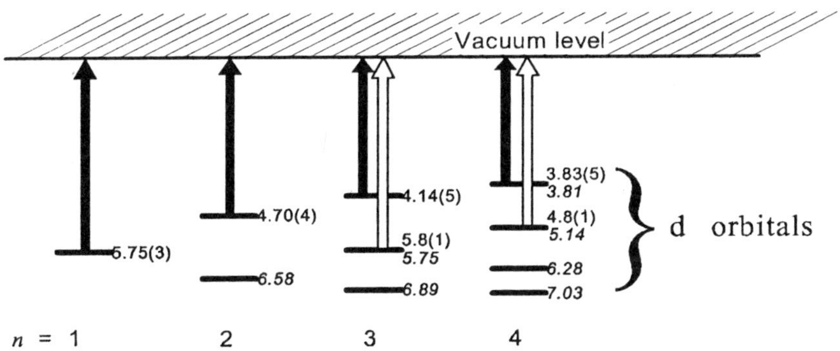

Figure 4. A quasi-band electronic structure of dδ orbitals by a Hückel model. The values in italics show predicted ones by experimental E_i of (1, 2), (2, 3) and (3, 4).

Furthermore, this model calculation predicts the next deeper HOMO orbital.[46] For (3, 4), the next HOMO is predicted to be located at the same level of the HOMO for (1, 2). Therefore the second lowest ionization energy for (3, 4) is expected to be the same as the lowest ionization energy for (1, 2). Indeed, this second lowest E_i is correspondingly found in the PIE curve for (3, 4), where the slope of the curve is changed to a steeper one. Position of the second onsets in PIE curves is good consistent with the theoretical prediction of the second lowest ionization from the next HOMO. Similarly, the change of the slope in the PIE curve for (4, 5) also corresponds to the second HOMO by the theoretical prediction. From these facts, the change in the slope of the curve can be reasonably thought as the second lowest ionization. This good agreement between experimental and calculated E_i values provides a distinct evidence that the vanadium-benzene clusters take a linear multiple-decker structure.

Late transition metal-benzene complexes of Fe, Co[47] and Ni were also produced. Figure 5 shows the mass spectra of $Co_n(Bz)_m$ clusters under relatively high concentration of benzene vapor. Even if the concentration of benzene was changed to higher one, the mass spectrum remained unchanged. For late transition metal complexes, $M_n(Bz)_m$ clusters exhibit the specific maximum m value (m_{max}) for each n. This behavior in the mass spectrum is completely different from that of the multiple decker sandwich clusters. When the magic-number clusters having m_{max} were exposed to NH_3 reactant gas, all of them were unreactive. The magic-number behavior of $Co_n(Bz)_m$ cannot be explained only by the sandwich structure, because the sandwich clusters having formulae of (3,3), (4,4), and so on, should have an exterior Co atom which is expected to be a reactive site toward NH_3. Taking into account of the two facts that every number of Co atoms (n) has a specific m_{max} and that they have no exterior Co atoms, the most plausible structure of $Co_n(Bz)_m$ is the structure of Co_n clusters covered with benzene molecules (m); a rice-ball structure with a grilled laver.

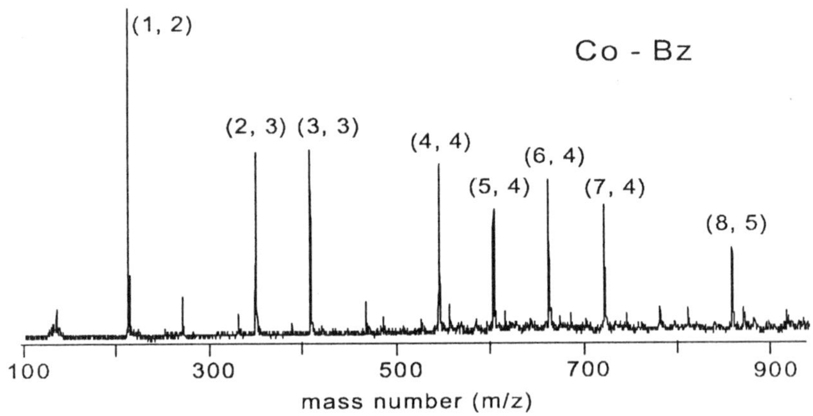

Figure 5. TOF mass spectrum of cobalt-benzene (Co-Bz) clusters.

3.2. *METAL-C_{60} CLUSTERS*

Figure 6 shows TOF mass spectra of $V_n(C_{60})_m^+$ cluster cations, which were directly extracted from the beam without photoionization.[48,49] Peaks of the cationic clusters are labeled according to the notation, (n, m) $^+$, denoting the number of metal atoms (n) and C_{60} (m). Under this condition, the contribution of V_n^+ cluster to $V_n(C_{60})_m^+$ formation was negligible because the abundance of the V_n^+ clusters (n 2) was less than 1/100 compared to that of the V^+ atom. The TOF mass spectrum indicates three common features: (i) main products are $(1, 2)^+$, $(2, 3)^+$, $(3, 4)^+$, $(4, 4)^+$, and $(5, 5)^+$, (ii) $(4, 4)^+$ is

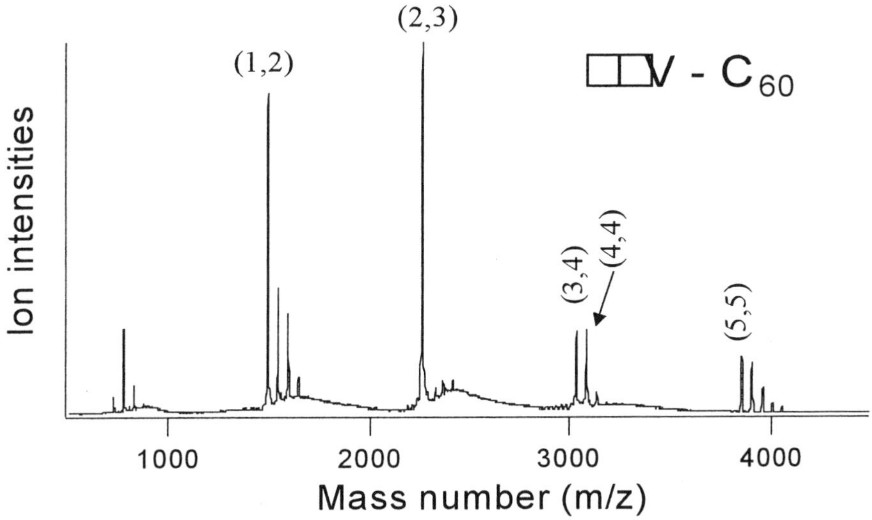

Figure 6. TOF mass spectrum of vanadium-C_{60} clusters.

more abundant than $(3, 4)^+$, while $(5, 4)^+$, is less abundant than $(4, 4)^+$, and (iii) $(4, 5)^+$ is scarcely produced, instead $(5, 5)^+$ is the first prominent peak in $(n, 5)^+$ series.

To deduce the structure of the V-C_{60} clusters by the chemical probe method, the reactivity of these abundant clusters toward O_2 gas was measured. The abundant clusters with (n, m) = $(1, 2)^+$, $(2, 3)^+$, $(3, 4)^+$, $(4, 4)^+$, and $(5, 5)^+$ were non-reactive toward O_2, whereas the other species drastically diminished with the injection of O_2, followed by O-atom adducts formation ($V_n(C_{60})_mO_k^+$). It is reasonably presumed that an exterior V atom in $V_n(C_{60})_mO_k^+$ is a reaction site to the oxidation, because C_{60}^+ itself does not show the oxidation reaction. Namely, these results indicate that the V-C_{60} clusters of $(1, 2)^+$, $(2, 3)^+$, $(3, 4)^+$, $(4, 4)^+$, and $(5, 5)^+$ have no exterior V atoms.

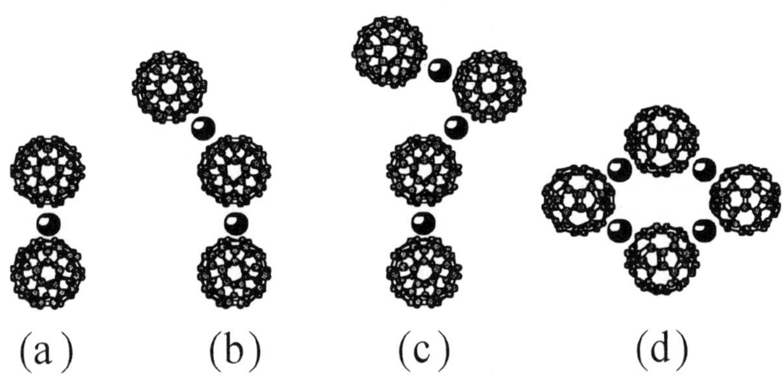

Figure 7. Plausible structures of V-C_{60} clusters; (a) (!,2), (b) (2,3), (c) (3,4), and (d) (4,4).

As well as the V-C_{60} clusters, the abundant species of the Sc-C_{60} and the Ti-C_{60} clusters also show no reactivity toward O_2. On the basis of no reactivities toward O_2 or CO gas, therefore, the presumed geometry of $(n, n+1)^+$ at n=1-3 is a multiple dumbbell structure (figs. 7(a)-(c)) and that of $(n, n)^+$ at n=4-5 is a ring structure (fig. 7(d)) where metal atom and C_{60} are alternatively piled up. There is no necessity for assuming linear chain structures of $(n, n+1)^+$ at n = 2 and 3, because C_{60} consists of many rings for bonding sites; 12 pentagonal rings and 20 hexagonal rings.

The ionization energies (E_is) of neutral $V_1(C_{60})_2$ and $V_1(C_{60})_1(Bz)_1$ clusters were determined as 5.93 ± 0.05 and 5.90 ± 0.05 eV by a photoionization spectroscopic method. The E_is of $V_1(C_{60})_1(Bz)_1$ was determined from the final decline of the PIE curves, although the others were determined from their appearance photon energy, because it was difficult to obtain the PIE curve for them being close to the limit of tunable range of the UV laser. The E_is for all of the $V_1(C_{60})_1$ and larger clusters having n 2 were higher than 5.9 2 eV, and lower than 6.42 eV.

The E_i values of the tabulated clusters are lower than those of a metal atom (V: 6.74 eV), C_{60} (7.61 eV), and C_6H_6 (9.24 eV).[45] Moreover, the E_i of the $V_1(C_6H_6)_2$ (5.75 eV) is similar to those of $V_1(C_{60})_1(Bz)_1$ and $V_1(C_{60})_2$. The similarity in E_is of $V_1(C_{60})_x(Bz)_y$ (x+y =2) evidently shows that the metal atom is sandwiched between the 6-membered rings of C_{60} rather than the 5-membered rings. In fact, E_i becomes much higher in a bis-cyclopentadienyl complex (i.e., $V_1(C_5H_5)_2$; E_i = 6.78 eV), where the V atom is sandwiched between the 5-membered rings. It generally can be seen that the interaction with η^6-C_{60} results in relatively low Ei of 5.5-5.7 eV. As shown in fig. 3,[42,43] the E_i value of the $M_n(Bz)_{n+1}$ sandwich cluster decreases greatly with the number n; the E_is of $V_n(Bz)_{n+1}$ 5.75, 4.70, 4.14, and 3.83 eV for n=1, 2, 3, and 4. However, the E_i value of the $V_n(C_{60})_{n+1}$ dumbbell cluster never decreases with the cluster size. The reason is probably attributed to the difference in the interaction between d electrons of the metal atom and π electron of the carbon ring; the d-π interaction is discontinuous in $M_n(C_{60})_m$, while it is continuous in $M_n(Bz)_m$ along a molecular axis.

3.3. *LANTHANIDE-COT CLUSTERS*

Figure 8 shows a typical example of the photoionization mass spectrum of $Ln_n(COT)_m$ for Ln = terbium (Tb) [COT = 1,3,5,7-cyclooctatetraene (C_8H_8); henceforth (n, m)] produced by the foregoing procedure. For other lanthanide elements of Nd, Eu, Yb, and Ho, prominent peaks in each spectrum showed almost the same compositions denoted as (n, n+1). Even when the concentration of COT vapor was changed to higher one, these main peaks remained unchanged in the mass spectra. Therefore, these (n, n+1) species are indeed stable clusters formed in the saturatedly high concentration of COT.

The most probable structure is a multiple-decker sandwich structure by analogy to the structure of the transition metal-benzene complexes,[43,44] in which Ln metal atoms and COT molecules are alternately pile up. In fact, it is natural to extend the single sandwich structure into multiple sandwich one, because triple decker structure of Ce-COT or Nd-COT was actually prepared in the condensed phase.[50,51] Since the

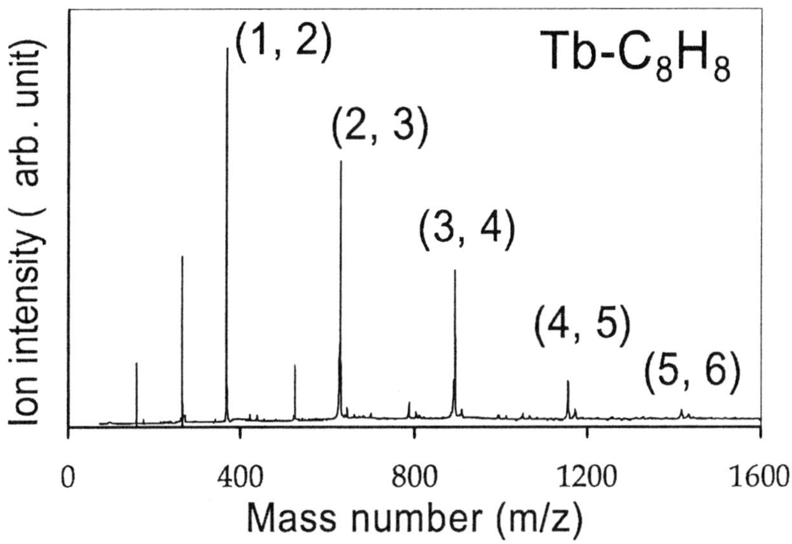

Figure 8. TOF mass spectrum of terbium-C_8H_8 (Tb-COT) clusters.

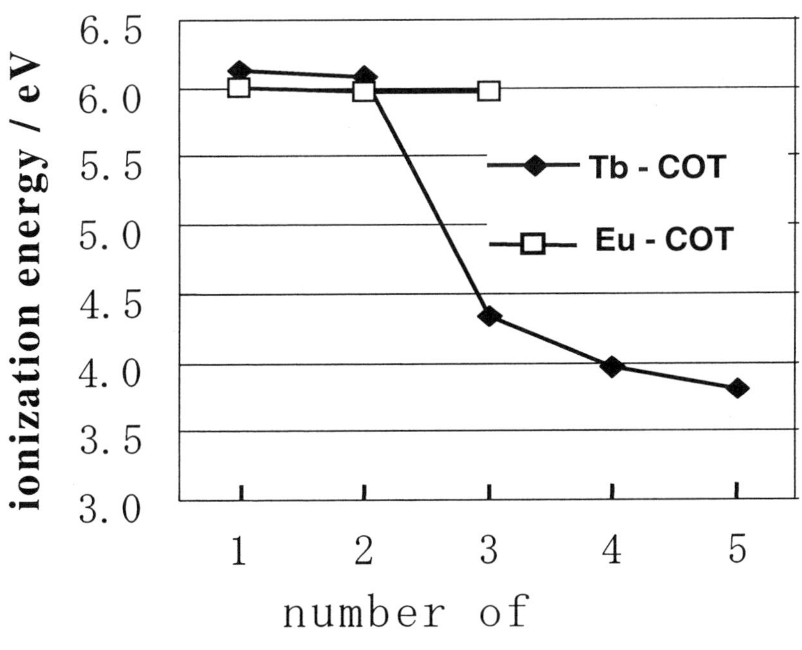

Figure 9. Charge distributions of Ln-COT: (a) Ln = +3 oxidation state and (b) Ln = +2 oxidation state.

advantage of the gas phase synthesis is no environmental factors such as organic solvents, it seems that these novel structures of Ln-COT can possibly come true as well as those of M-Bz and M-C_{60}.

As well known by 4n+2 rule of aromatics, a COT molecule can act as an electron acceptor by two electrons.[26] In fact, all of $Ce(COT)_2$ and $Nd(COT)_2$ in bulk materials have been prepared as alkali metal salts denoted as $M^+[Ln^{3+}(COT^{2-})]$.[26,31] Therefore, considering that Ln atoms favor the Ln^{3+} or Ln^{2+} state in the ligand field, the proposed structures and the distribution of valence electrons of multiple-decker $Ln_n(COT)_{n+1}$ clusters result in those shown in fig. 9.

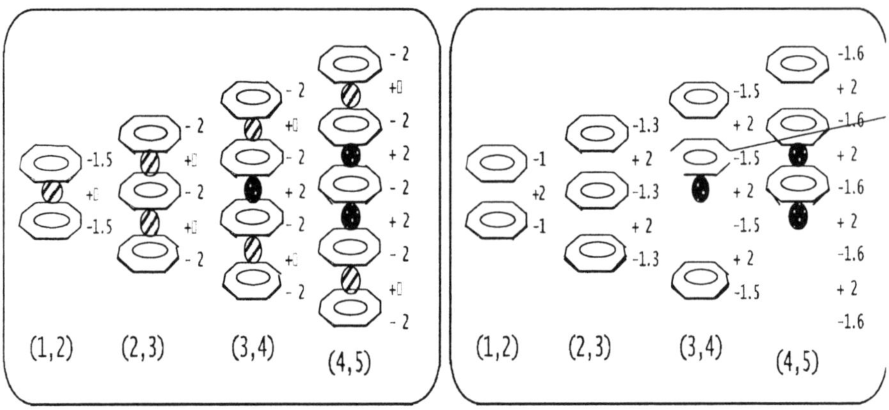

(a) +□ oxidation state

Figure 10. Size dependence of ionization energy for terbium (Tb)-COT and europium (Eu)-COT clusters.

To know the electronic properties the ionization energies (E_is) of the neutral $Ln_n(COT)_m$ were measured for Ln = Eu and Tb by using photoionization spectroscopy (figure 10). For Eu-complexes, they cannot be photoionized by 5.92 eV photons, but can be by 6.42 eV photons of the ArF laser. The values are constant within experimental uncertainties, and it can be attributed that the charge distributions in Eu-COT is kept similar among the sandwich structures as shown in fig. 9. As shown in fig. 10, however, E_is for $Tb_n(COT)_{n+1}$ exhibit size-dependent change to be 6.13, 6.09, 4.34, 3.97, and 3.87 eV for n=1-5, respectively, from the final decline of photoionization efficiency curves. The ionization energy is constant between (1, 2) and (2, 3), but decreases sharply in (3, 4), then decreased asymptotically in (4, 5) and (5, 6).

First, the sharp decrease of ionization energy from n = 2 to n = 3 is discussed. In both (1, 2) and (2, 3) clusters, all the Tb atoms take the stable +3 oxidation state in common. Since it takes a large amount of energy to oxidize a lanthanide from +3 oxidation state to +4 oxidation state, ionization in (1, 2) and (2, 3) clusters occurs from the orbital derived from the HOMO in COT. In (3, 4), one of the Tb atoms must be +2 oxidation state as shown in fig. 9, since each COT molecule accepts up to two electrons. Therefore, it is favorable to ionize +2 oxidation state into +3 by removing a 4f electron at (3, 4). Accordingly, it can be concluded that the sharp decrease of ionization from n = 2 to n = 3 is caused by the change in the ionization channel from ligand-derived orbital (COT) to metal-derived orbital (Tb).

Next, the asymptotic E_i decrease of ionization energy at larger n of n 3 is discussed. This asymptotic E_i decrease as the number of stacking has been observed in a similar experiment on $V_n(C_6H_6)_{n+1}$, as discussed above. In $V_n(C_6H_6)_{n+1}$, the reason for the asymptotic behavior has been explained by the formation of quasi-band electronic structure from one-dimensional delocalization of dδ electron.[46] The question then is whether this argument can be extended to the ionic-bonding $Tb_n(COT)_{n+1}$ clusters. The 4f [$f\phi(e_{3u})$] orbital of Tb^{2+} oxidation state can be interacted with neighboring COT LUMO having same symmetry [$L\phi(e_{3u})$]. In other words, only the COT LUMO can be interacted with the *f* orbital throughthe same symmetry of ϕ. This is very similar principle in V-Bz, where the interaction works between d orbital [$d\delta(e_{2g})$] of V and LUMO [$L\delta(e_{2g})$] of benzene. Therefore, for the clusters larger than (3, 4) with more than one +2 Tb oxidation state, the quasi-band electronic structure is formed by the one-dimensional delocalization of the 4f electrons of +2 oxidation state Tb through COT. Since f orbital is spacially so compact compared to d-orbital, there seems less interaction with the ligands in Ln-COT. This reasonably explains that the splitting of the quasi-fϕ band is smaller than that of the quasi-dδ band.

3.4. *SOFT-LANDING OF ORGANOMETALLIC CLUSTERS*

As discussed above, the pattern in the mass spectrum of V-Bz can be explained by the sandwich structure along with the results on the chemical reactivity toward CO molecules. However, this qualitative explanation needs some spectroscopic information to confirm the geometric structures, and for this purpose IR spectroscopy was performed on the size-selected organometallic clusters deposited on the Ar matrix.

With mass selection in the Q-mass spectrometer, the $V_n(Bz)_m^+$ clusters were size-selected and were deposited onto a Ar-preadsorbed gold substrate. The ion current for $V_n(Bz)_m^+$ were typically 10 nA for (1, 2) and 2 nA for (2, 3) at the deposition energy of 20 eV. It should be noted that the current of 10 nA equals to 7 x 10^{10} ions per a second (or 2.2 x 10^{14} ions per an hour), which yields 0.08 μg of the $V(Bz)_2$ product per an hour. Similarly, the laser ablation of a titanium (Ti) and a chromium (Cr) target provides the size-selected ions of $Ti(Bz)_2^+$ (7 nA) and $Cr(Bz)_2^+$ (5 nA), respectively. Figure 11 shows the IR absorption spectrum of $V(Bz)_2$ deposited onto the Au substrate with the thin Ar layer for 1 hour. Three distinct peaks were observed at 748, 958, and 990 cm^{-1}. A useful reference experiment for this cluster deposition is the thermal vapor condensation of chemically synthesized $V(Bz)_2$ into an Ar matrix.[52,53] The series of absorption peaks in our work correspond closely with those of the chemically synthesized $V(Bz)_2$ having a sandwich structure that has been revealed by X-ray crystalline analysis.[54] Thus, the IR spectrum shows that (1) the cluster ions were

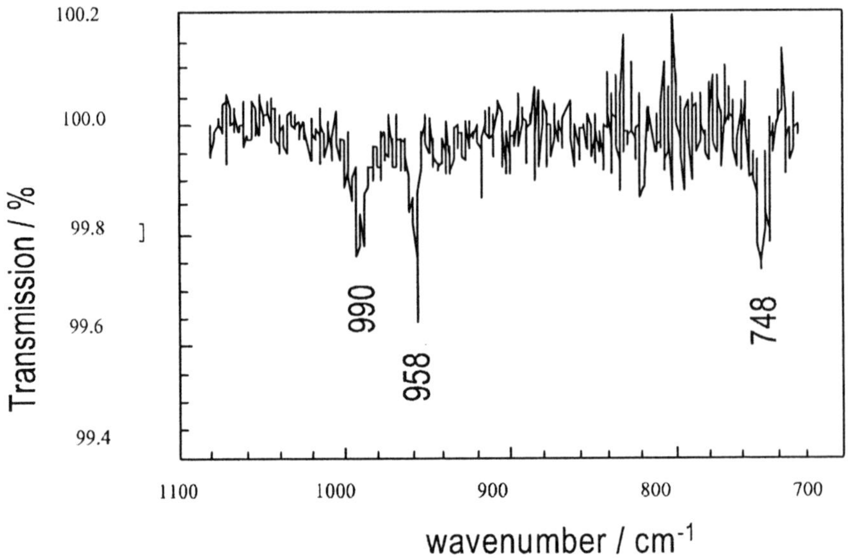

Figure 11. IR spectrum of soft-landed $V(Bz)_2$ clusters onto a Au substrate.

non-dissociatively deposited into the Ar matrix and that (2) the neutralized cluster has a sandwich structure.

When the Ar layer are thin below 50 MLs, the deposited cluster ions can be neutralized by the charge tunneling from the Au substrate.[55] For more than 100 MLs of Ar, however, the monitored current gradually decreased with the deposition time due to pile-up of the charge on the substrate. The current was immediately recovered when the Ar matrix is removed by heating the substrate. This technique for neutralization gives an experimentally important advantage; the ion current can be monitored directly by the tunneling current from the metal substrate beneath the matrix.

The metal substrate in this work plays an important role in IR spectroscopy with the reflection mode. As reported in a lot of studies,[56] the reflection mode provides higher sensitivity than the transmission mode, because the total path length for the IR light passing through the sample is longer in the former. For comparison, the IR absorption was measured in the transmission mode for $V(Bz)_2$ on a Ar-preadsorbed CsI which was prepared with the deposition of the corresponding ions under the electron shower. It turned out that the sensitivity of the reflection mode was roughly 3-5 times higher than that of the transmission mode.

As mentioned above, the soft-landing of the size-selected ions is accompanied by energy release in the two processes: (1) deposition, (2) neutralization of the cations. In this work, the IR spectrum of $V(Bz)_2$ clearly shows the cluster ions can be soft-landed in the processes of the deposition and the neutralization. Since both the impact energy for the deposition (20-50 eV) and the ionization energy (5. 75 eV) are higher than the binding energy of the sandwich cluster (2.42 – 2.55 eV for metal-benzene ring),[58] it is considered that the excess energies are dissipated by vaporization of the Ar matrix, resulting in the non-destructive deposition. Furthermore, although the charge combination generates the excess energy corresponding to the ionization energy, the cold Ar media seemingly prevents the hot cluster from dissociating itself.

This demonstration of the soft-landing of organometallic clusters coupled with IR spectroscopy can lead to getting information on structures of the clusters produced in the gas phase. Furthermore, when the matrix media of Ar are removed by heating-up of the substrate, the soft-landed clusters can be dispersed on the substrate. Therefore, this technique should open up an approach to make a nanostructured material with the controlled deposition of (multi-)functional units, such as a molecular magnet, an emissive cluster, and so on. For example, each V atom in $V_n(Bz)_{n+1}$ having a sandwich structure has an unpaired d electron. Although it depends on the amount of exchange interaction, the stacking of the $V_n(Bz)_{n+1}$ clusters might produce one-dimensional material showing high electron multiplicity with the selective deposition onto a substrate.

4. Conclusions

In the gas phase, various network structures of organometallic clusters of M-Bz/C_{60} and Ln-COT/C_{60} were produced by the laser vaporization method. From the pattern of the predominant mass peaks and the chemical reactivity toward the gas reactant their sandwich or dumbbell structure was revealed, where metal atoms are separated and connected with the organic ligands. The electronic structures were studied and discussed by photoionization or photoelectron spectroscopy. Although both early transition metal (M_E)-Bz and Ln-COT take on a similar multiple sandwich structure, the E_i of M_E-Bz is significantly lowered with the number of layers, while that of Ln-COT depends not on the number of layers, but on the charge distribution. This difference is ascribed to the bonding nature; M_E-Bz is formed through a covalent bond and d electrons are delocalized along the molecular axis, whereas Ln-COT is formed through an fairly ionic bond, and the electronic state is expressed by the stacking of a local sandwich unit. For larger Ln-COT clusters, asymptotic decrease of Eis suggests that f electrons in Ln^{2+} are delocalized. As well as benzene, C_{60} can also act as an organic ligand and M-C_{60} forms a network structure. The bonding in M_E-C_{60} arises from the interaction between d electrons of the metal atom and π electrons of the hexagonal carbon ring (η^6), giving superstructures of V-(η^6-C_{60}). The regular arrangement of metal atoms in organometallic clusters should introduce useful properties, because the properties of bare metal atoms are modified by deliberately adding organic ligands. It is important to develop a new methodology to form deposited cluster materials. One of the solutions is the soft-landing of the size-selected cluster ions. With the development of the soft-landing technique, the gas phase organometallic chemistry and physics can open up new aspects of chemical reaction, catalytic activity and electron spin chemistry, and this area is one of the most promising for cluster materials.

5. Acknowledgements

The authors are grateful to Prof. S. Yabushita, Dr. K. Hoshino, Dr. T. Kurikawa, Dr. T. Yasuike, Dr. K. Judai, Dr. Y. Negishi, Dr. S. Nagao, Mr. K. Miyajima, and Mr. M. Hashimoto for their great contributions. This work is supported by a program entitled "Research for the Future (RFTF)" of Japan Society for the Promotion of Science (98P01203) and by a Grant-in-Aid for Scientific Research on Priority Areas from the Ministry of Education, Science, Sports, and Culture.

References

1. Teh, C. S., Willey, K. F., Robbins, D. L., Pilgrim, J. S., and Duncan, M. A. *Chem. Phys, Lett.* 1992, ***196***, 233.
2. Higashide, H., Kaya, T., Kobayashi, M., Shinohara, H., and Sato, H. *Chem. Phys. Lett.* 1990, ***171***, 297.
3. Holland P. M., and Castleman Jr, A. W. *J. Chem. Phys.* 1982, ***76***, 4195.
4. Robels, E. S. J., Ellis, A. M., and Miller, T. A. *J. Chem. Phys.* 1992, ***96***, 8791.
5. Misaizu, F., Sanekata, M., Fuke, K., and Iwata, S. *J. Chem. Phys.* 1994, ***100***, 1161.
6. Mitchell, S. A., Blits, M. A., Siegbahn, P. E. M., Svensson, M. *J. Chem. Phys.* 1994, ***100***, 423.
7. Nakajima, A., and Kaya, K. *J. Phys.Chem. A*, 2000, ***104***, 176.
8. Kroto, H. W., Heath, J. R., O'Brien, S. C., Curl, R. F., and Smalley, R. E. *Nature*, 1985, ***318***, 162.
9. Guo, B. C., Wei, S., Purnell, J., Buzza, S., and Castleman, Jr, A. W. *Science* 1992, ***256***, 515.
10. B. C. Guo, K. P. Kerns, and A. W. Castleman, Jr., *Science* 1992, ***255***, 1411.
11. Fagan, P. J., Calabrese, and J. C., Malone, *Science* 1991, ***252***, 1160.
12. Hawkins, J. M. *Accounts Chem. Res.* 1992, ***25***, 150.
13. Balch, A. L., Catalano, V. J., and Lee, J. W. *Inorg. Chem.* 1991, ***30***, 3980.
14. Douthwaite, R. E., Green, M. L. H., Stephens, A. H. H., and Turner, J. F. C. *J. Chem. Soc., Chem. Commun.* 1993, 1522.
15. Chai, Y., Guo, T., Jin, C., Haufler, R. E., Chibanante, L. P., Fure, J., L. Wang, L., Alford, J. M., and Smalley, R. E. *J. Phys. Chem.* 1991, ***95***, 7564.
16. Caldwell, K. A., Giblin, D. E., Hsu, C. S., Cox. D., and Gross, M. L. *J. Am. Chem. Soc.* 1991, ***113***, 8519.
17. McElvany, S. W. *J. Phys. Chem.* 1993, ***96***, 4935.
18. Haddon, R. C., *Nature* 1991, ***350***, 320.
19. Hebard, A. F., Rosseinsky, M. J., Haddon, R. C., Murphy, D. W., Glarum, S. H., Palstra, T. T. M., Ramirez, A. P., and Kortan, A. R. *Nature* 1991, ***350***, 600.
20. Holczer, K., Klein, O., Huang, S. -M., Kaner, R. B., Fu, K. -J., Whetten, R. L., and Diederich, F. *Science* 1991, ***252***, 1154.

21. Benning, P. J., Martins, J. L., Weaver, J. H., Chibante, L. P. F., and Smalley, R. E. *Science* 1991, ***252***, 1417.

22. Benning, P. J., Poirier, D. M., Ohno, T. R., Chen, Y., Jost, M. B., Stepniak, F., Kroll, G. H., Weaver, J. H., Fure, J., and Smalley, R. E. *Phys. Rev. B* 1992, ***45***, 6899.

23. Hawkins, J. M., Meyer, A., Lewis, T. A., Loren, S., and Hollander, F. J. *Science* 1991, ***252***, 312.

24. Marks, T. J. *Progress in Inorganic Chemistry*, An Interscience Publication 1978, ***24***, 51.

25. Green, J. C. *Structure and Bonding*, Springer-Verlag, New York 1981, ***43***, 64.

26. Schumann, H., Meese-Marktscheffel, J. A., and Esser, L. *Chem. Rev.* 1995, ***95***, 865.

27. Marks, T. J., and Ernst, R. D. *Comprehensive Organometallic Chemistry*, Pergamon Press 1982, ***3***, 192.

28. Streitwieser, Jr., A., and Müller-Westerhoff, U. *J. Am. Chem. Soc.* 1968, ***90***, 7364.

29. Zalkin, A., and Raymond, K. N. *J. Am. Chem. Soc.* 1969, ***91***, 5667.

30. Streitwieser, Jr., A., and Yoshida, N. *J. Am. Chem. Soc.* 1969, ***91***, 7528.

31. Hodgson, K. O., and Raymond, K. N. *Inorg. Chem.* 1972, ***11***, 3030.

32. Dolg, M., Fulde, P., Küchle, W., Neumann, C., and Stoll, H. *J. Chem. Phys.* 1991, ***94***, 3011.

33. Dolg, M., Fulde, P., Stoll, H., Preuss, H., Chang, A., and Pitzer, R. M. *Chem. Phys.* 1995, ***195***, 71.

34. Liu, W., Dolg, M., and Fulde, P. *J. Chem. Phys.* 1997, ***107***, 3584.

35. Greco, A., Cesca, S., and Bertolini, G. *J. Organomet. Chem.* 1976, ***113***, 321.

36. (a) DeKock, C. W., Ely, S. R., Hopkins, T. E., and Brault, M. A. *Inorg. Chem.* 1978, ***17***, 625. (b) Ely, S. R., Hopkins, T. E., and DeKock, C. W. *J. Am. Chem. Soc.* 1976, ***98***, 1624.

37. Hettich, R. L., Jackson, T. C., Stanko, E. M., and Freiser, B. S. *J. Am. Chem. Soc.*, 1986, ***108***, 5086.

38. Hettich, R. L., and Freiser, B. S. *J. Am. Chem. Soc.*, 1987, ***109***, 3537.

39. Geusic, M. E., Morse, M. D., O'Brien, S. C., and Smalley, R. E. *Rev. Sci. Instrm.* 1985, ***56***, 2123.

40. Harberland, H. *"Clusters of Atoms and Molecules"*, (Springer-Verlag, Berlin, 1994), p. 230.

41. Bromann, K., Félix, C., Brune, H., Harbich, W., Monot, R., Buttet, J., and Kern, K. *Science*, 1996, ***274***, 956.

42. Kurikawa, T., Takeda, H., Nakajima, A., and Kaya, K. *Z. Phys. D* 1997, ***40***, 65.

43. Hoshino, K., Kurikawa, T., Takeda, H., Nakajima, A., and Kaya, K. *J. Phys. Chem.* 1995, ***99***, 3053 .

44. Weis, P., Kemper, P. R., and Bowers, M. T. *J. Phys. Chem. A*, 1997, ***101***, 8207.

45. (a) Moore, C. E. "Analysis of Optical Spectra", NSRDS-NBS 34, National Bureau of Standards, 1971.: (b) Weast, R. C. "Handbook of Chemistry and Physics," CRC Press 1980, ***61***, E-69.: (c) Robinson, J. W. "Handbook of Spectroscopy", CRC Press 1974, ***1***, 257.

46. Yasuike, T, and Yabushita, S, *J. Phys. Chem. A.* 1999, ***103***, 4533.

47. Kurikawa, T., Hirano, M., Takeda, H., Yagi, K., Hoshino, K., Nakajima, A., and Kaya, K. *J. Phys. Chem.* 1995, ***99***, 16248.

48. Nakajima, A., Nagao, S., Takeda, H., Kurikawa, T., and Kaya, K. *J. Chem. Phys.* 1997, ***107***, 6491.

49. Nagao, S., Kurikawa, T., Miyajima, K., Nakajima, A., and Kaya, K. *J. Phys. Chem. A* 1998, ***102***, 4495.

50. Kinsley, S. A., Streitwieser, Jr., A., and Zalkin, A. *Organometallics* 1985, ***4***, 52.

51. Kinsley, S. A., Streitwieser, Jr., A., and Zalkin, A. *Acta Cryst.* 1986, ***C42***, 1092.

52. Fritz, H. P., Lüttke, W., Stammerich, H., and Forneris, R. *Spectrochim. Acta*, 1961, ***17***, 1068.

53. Andrews, M. P., Mattar, S. M., and Ozin, G. A. *J. Phys. Chem.*, 1986, ***90***, 744.

54. Muetterties, E. L., Bleeke, J. R., and Wucherer, E. J. *Chem. Rev.* 1982, ***82***, 499.

55. Xu, C., and Goodman, D. W. *Chem. Phys. Lett.*. 1996, ***263***, 13.

56. Hoffmann, F. M. *Surf. Sci. Reports*, 1983, ***3***, 107.

57. Meyer, F., Khan, F. A., and Armentrout, P. B. *J. Am. Chem. Soc.* 1995, ***117***, 9740.

NANOTECHNOLOGY OF DNA/NANO – Si AND DNA/CARBON NANOTUBES/NANO – Si CHIPS

E. Buzaneva[1], A.Gorchynskyy[1], G. Popova[1], A. Karlash[1], Y.Shtogun[1], K.Yakovkin[1], D.Zherebetskiy[1] O.Matyshevska[2],
Yu. Prylutskyy[2], P. Scharff[3],
Kyiv National Taras Shevchenko University, [1]Radiophysical, Physical and [2]Biological faculties 64, Vladimirskaya str., 01033, Kyiv, Ukraine,
[3]Institute of Physics, TU Ilmenau, PF. 100565, D-98684 Ilmenau, Germany.

Abstract

DNA nanotechnology using DNA molecular assembly for the formation of the nanosystems from DNA and nanotubes in chip on the silicon surface, has been developed on the base of theoretical predictions of chemical activity changing of SWCNT's, nano-Si's under their localization into the biopolymer surrounding.
We report the direct investigation by AFM of DNA assembled on the surface porous Si with nano – Si system, UV-vis and IR spectroscopic study of partially unwrapped double helix DNA molecules with carbon nanotubes, which spontaneously self-assembled into blocks consisting of DNA molecule and carbon nanotube. The shift of absorption peak on 5,8 nm and the raised absorption in the presence of carbon nanotubes are testified about the change of hydrogen bonds between the bases of DNA molecules in the layer with carbon nanotubes. Both the decrease of the wavenumber of N-H···N, N-H···O groups on 12 cm^{-1} and the absence of C-CH_3 group vibrational mode at 1440 cm^{-1} in IR transmittance spectra from DNA/nanotubes layer are indicative of their self – assembly formation.
The electronic and photoelectronic properties of the formed nanosystems under the change of DNA/SWCNT molecular and DNA/nanocrystalline-Si (nano-Si) networks by DNA modification in the chips have been studied.

1. Introduction

There are known facts that DNA molecules, carbon nanotubes (CNT) as well as nano-Si can be used in nanotechnology as elements with novel properties [1 – 10]. Particularly, a single DNA molecules and CNT are able to conduct electrical charges [2 – 8].

Our idea about the use of DNA molecules bases on the result of investigations of the combination of intrinsic properties of DNA and nano-Si or DNA, nanotubes and nano-Si permits of achieve properties that cannot to be obtained using either components individually [9 - 22]. In [11] we have got evidences that transport properties of such systems are determined by the charge transfer through the barriers in the DNA double helix/DNA unwrapped double helix and in the DNA unwrapped double helix/SWCNT

E. Buzaneva and P. Scharff (eds.), Frontiers of Multifunctional Nanosystems, 191–212.

nanoheterostructures, in the DNA/Si, SWCNT/Si nanoheterostructures that was proved by tunneling spectroscopy investigations. In addition we considered the role of disorder effects on electronic states and conductance in SWCNT and nano-Si [23 – 27]. In this connection the contribution of interfacial interaction of nanotubes with the DNA molecules and nano-Si increased.

Nanotechnology of DNA/nano-Si, DNA/CNT/nano-Si blocks has a great significance in view of their using for the construction of "lab on a chip" electron system. The chips on the bases of nano-Si, DNA, could lead to the development of highly sensitive detectors, novel molecular photovoltaic and sensor systems. The obtained results could be used as base for further development of molecular computer engineering principles [2,5,8,11-23].

The use of DNA molecules for the formation of the multifunctional nanosystems from - DNA/SWCNT (Single Wall Carbon Nanotube) building blocks and nanocrystalline-Si (nano-Si) networks we have proved on the base of theoretical predictions of chemical activity changing of SWCNT's, nano-Si's under their localization into the biopolymer surrounding; the possibility of different DNA networks self-organization [2,9,10,14,15,17,19,22]; achievements of the nanotechnology of self-assembling of nanoblocks/molecules [29 – 35] and inorganic/organic nanoparticles [2, 3 – 36] as well as recognition template [37-40] and template imprinted nanostructural surfaces [41].

In this view experimental results supporting the proposed model of nanoblocks for engineering of DNA/SWCNT molecular, nano-Si networks and a self-organization of their electronic structure are present.

In our investigation we used the adsorbtion from DNA/NaOH buffer solution with and without SWCNT on the nanoporous Si substrate with nanocrystalline - Si. The irradiation by visible light is used to change the regular system of hydrogen bonds when the DNA molecule structure turns from double helix into chaotic ball structure. The adsorption on the substrate could be stimulated by this process and determinate molecular and molecular/nano-Si interfaces.

The electrical, optical properties and electronic structure of DNA/SWCNT molecular and nano-Si networks are formed as a result of changes of electron levels at DNA/DNA, DNA/SWCNT, DNA/nano-Si and SWCNT/nano-Si interfaces which were directly revealed by tunnel, photoluminescence (PL) and UV-vis spectroscopies.

The first direct experimental evidences for the organization of DNA/SWCNT molecular and nanocrystalline-Si networks, consisting of DNA double helix, SCWNT and nanocrystalline-Si building blocks, due to DNA molecules have been obtained. The nanotechnology of DNA, SWCNT molecular layers formation on the nanopor-Si surface with nanocrystalline-Si, DNA diffusion into pores and self-organization of DNA/nanocrystalline-Si interface have been developed. The electronic structure self-formation and charge transfer for the network systems from building blocks by tunnel and photoluminescence spectroscopies have been studied. We have obtained the evidences of the main role of the interface molecular interaction in a self-organization of electronic structure of these networks and the charge injection from nanocrystalline-Si explaining the features of diode I-V characteristics and photoluminescence spectra.

2. DNA as a construction material.

The properties that make DNA such an effective molecule for its biological role as genetic material also make it a superb molecule for nanoconstruction. One key to using DNA for this purpose is to produce stable complex motifs, such as branched molecules. Combining branched species by sticky ended interections, leads to N – connected stick figures whose edges consist of double helical DNA. Zero mode removal or reciprocal crossover, leads to complex fused motifs, such as rigid multi – crossover molecules and paranemic crossover molecules. Multi – crossover molecules have been used to produce 2D arrays and a nanomechanical device. Algorithmic assembly and the use of complex complementarities for joining units are goals in progress that are likely to produce new capabilities for DNA nanotechnology [4].

Watson and Crick's DNA structure and replication model placed the phenomenology of genetics on a firm molecular basis [1]. It is now well known that genetic information resides in the complementary of the bases: Adenine (A) froms two hydrogen bonds with thymine (T) and guanine (G) from three hydrogen bonds with cytosine (C) (Figure 1). In combination with backbone geometry, this specific pairing relationship leads to the double helical structure of the DNA polymer; the double helix is about 2 nm wide, and its repeat consists of about 10.5 nucleotides pairs, each separated by about 3.4 Å. The chemical polarities of the two strands are antiparallel, resulting in a series of twofold axes for the backbone structure, perpendicular to the helix axis, one axis roughly through the base pair planes, and another one halfway in between. It is important to recognize that the helix axis is linear in cellular DNA, not in the algebraic sense of being a straight line, but in the topological sense of being unbranched. This feature characterizes not only naturally occurring molecules, but also recombinant DNA species produced by molecular biotechnology.[4]

The molecular properties of the DNA molecule that make it such an effective molecule for storing genetic information also confer on it the characteristics of an outstanding nanoscale synthon tor a biomimetic nanotechnology. This fact has led to the development of both DNA nanotechnology [3,4] and the closely related field of DNA-based computation [5]. This article will describe the development of DNA nanotechnology by deriving the unusual motifs of DNA that have led to it. These motifs have been inspired by a variety of sources, and we have shown that they lead naturally to a generalization of Watson-Crick complementarity. The basic feature shared by both unusual DNA motifs and generalized complementarity is discontinuous complementarity. Conventional nucleic acid complementarity entails two continuous polynuclcotide backbones: The sequence on one strand is one-for-one complementary to the sequence on the other strand. It has long been known that the bases of nucleic acids can pair in many ways other than Watson-Crick complementarity, but our discussion here will be restricted to Watson-Crick interactions. Other investigators are using non-Watson-Crick pairing, such as G tetrads, in work on DNA nanotechnology [1-4].

Alivisatos et al. have demonstrated the use of DNA as a rigid template to direct the spatial disposition of nanoclusters [1,3]. Here DNA does not fulfill a biological role as a carrier of information, which direct the cluster to desired site via sequence specific hydrogen bonding interactions. Mirkinet al. Have shown that nanopaticles with several single strands of DNA attached can be used to reversibly create arrays of nanoparticles, when the appropriate complementary sequence is present, providing a quick testing

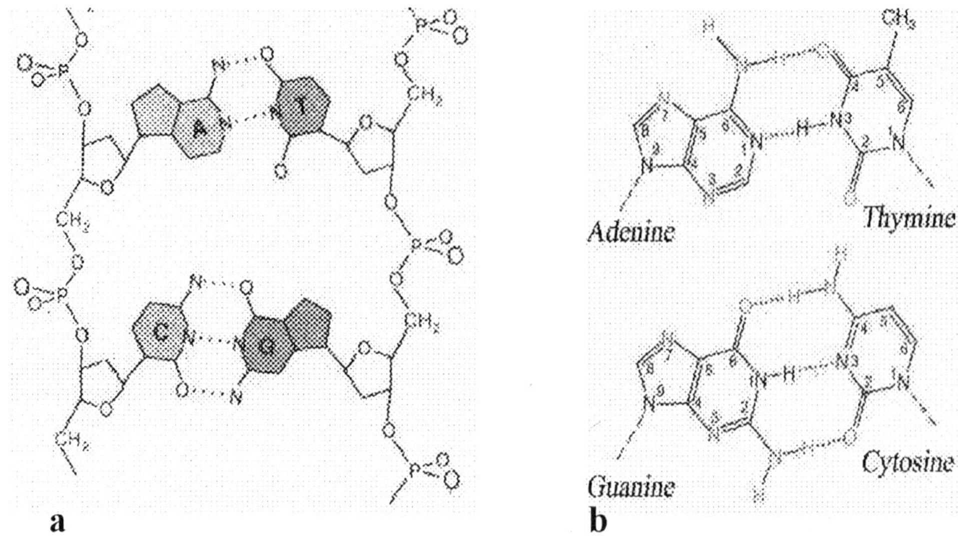

Figure1. **a**-Wrapped helix of DNA molecule. Sugar-phosphate bases $-CH_2$-O-P(OOH)-O- and "stairs" pairs of bases, the elements and bonds between its. The chains in double helix direct in opposed sides (the ends of chains mark 3' and 5'). Every pair of bases consist from puryn (A or G) and perimidine (T or C), which connect by hydrogen bonds (it has shown by points). A-adenine, G-guanine, T-tymine C-cytosine.
b- The image of two pair of the bases. The dotted line show glycoside bonds(N-C), which direct to sugar residuum, the points show hydrogen bonds inside pair s. Under the formation of hydrogen bonds with regular molecules of NH_2- group (in the site A- 6, G- 2, C- 4) these groups can play the role of hydrogen donor; the atoms of oxygen (T- 2, G-6, C-2) and nitrogen (G- 7, A- 3), play role of acceptors.

method for specific sequences (see Figure 2).

This example also highlight other important features of biomolecules as substrates for materials assembly. Some of the general structural motifs are predictable for these systems, but can be (re) engineered as desired. Thus DNA generally is a stiff rod (double helix) but H-bonding mismatches can create bulge structures, clover leaves, or other patterns. Proteins, with greater structural diversity, can exhibit a wide range of structural motifs. For example the simple helical structures with predictable structural patterns can further assemble into helical coiled coils with useful elastic properties e.g., collagen.

The DNA-cluster example also highlights the potential that biomolecules may provide in constructing novel organic- inorganic composites. Such examples abound in nature, in bones, shells, and related structural composites. Inorganic materials display a vast range of useful properties, but until now the ability to create complex spatial arrangements of inorganic matter has been possible principally through lithography, which operates on relatively large length scales of tens of nanometers. The patterning of inorganic materials using organic functional groups to direct the assembly is unimportant area for future work.

Finally, a hierarchy of (structural) length scales is natural in biology, and equally natural in biomolecule- derived materials. Amino acids assemble into polymers (proteins), which fold into structures of ca. nanometer size. These units may further

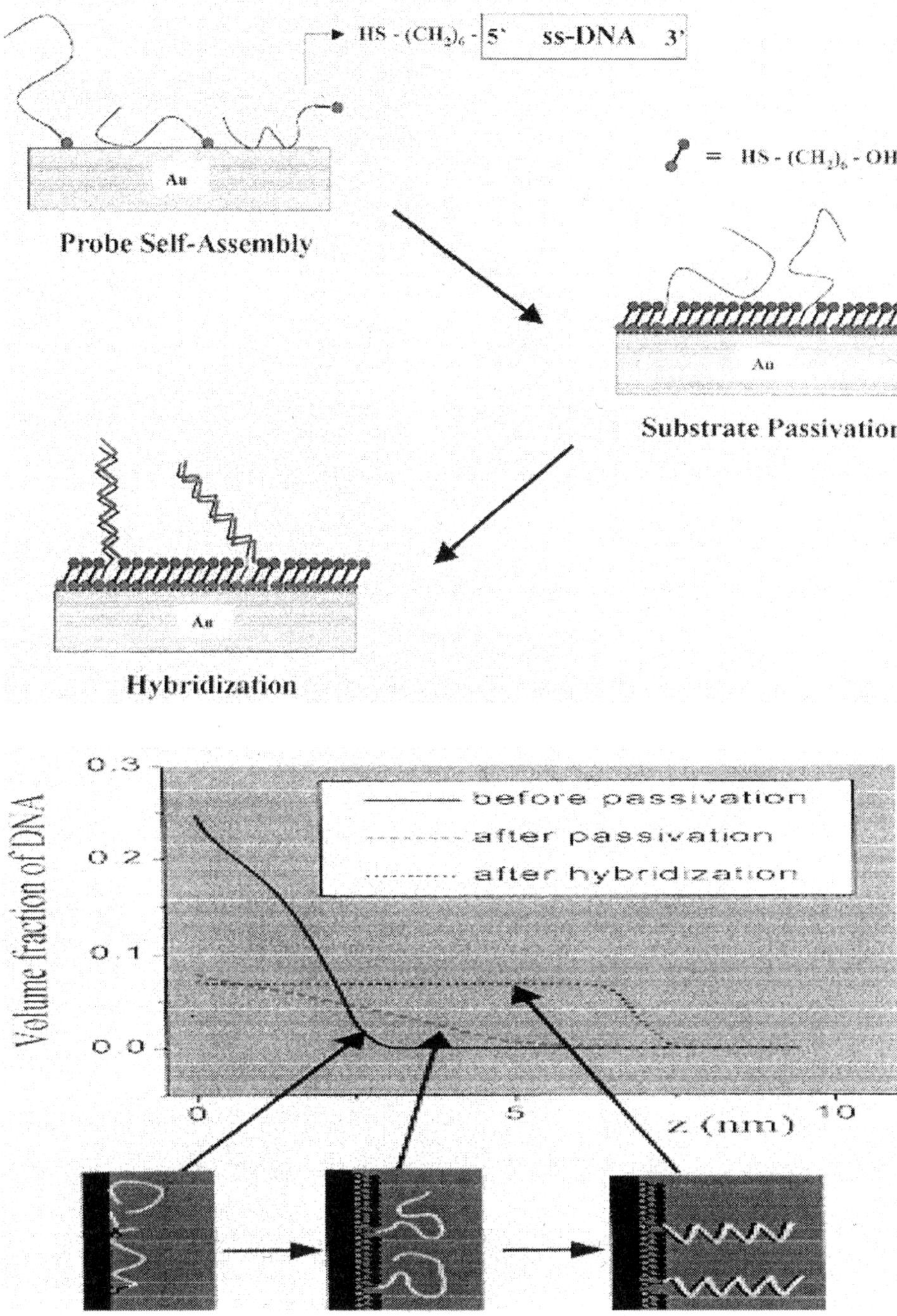

Figure 2 The assembly of DNA monolayers in DNA probe system.

assemble into repeating multimeric units, both static geometric structures as that of the icosahedral viruses and/or dynamic self- assembled structures such as that of actin, or microtubules, which can extend across the micrometer lengths of cells. Chemists have only begun to emulate such self- assemblies of biomolecular components. For example, Ghadiri has demonstrated the spontaneous formation of "tubules" from small cyclic peptide components that can assemble like "lock washers".[2,3] By extending molecul-based superstructures to macroscopic scales biology builds biomolecule- mediated cells interactions, such as the axons of nerves that control brain function or "smart materials" such as muscle, which reshape in response to environmental signals

3. DNA-Directed assembling: potential for nanofabrication

DNA chips and microarrays represent a technology which has immediate applications in genetic research and diagnostics. DNA array technology may also play a future role in enabling nanofabrication. DNA chips or arrays are devices in which different DNA sequences are arrayed in a microscopic format on a solid support (glass, silicon, plastic, etc.). DNA arrays can have anywhere from 100 to 100,000 different DNA sites (pixels) on the chip surface. Depending on the chip, the sites can range in size from 10 microns to over 100 microns (smaller sites are possible). Each DNA site can contain from 10^6 to 10^9 DNA sequences in a DNA hybridization assay (Figure 3).

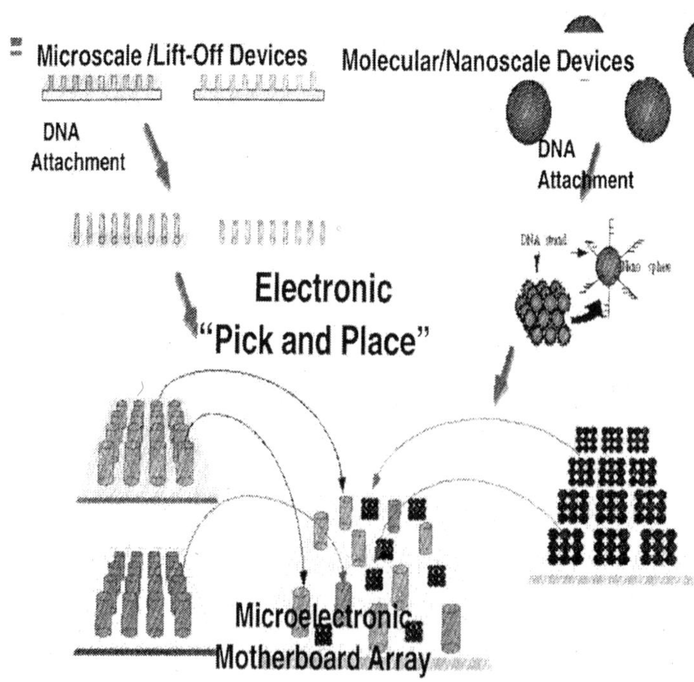

Figure 3. Directed nanofabrication on a chip (Nanogen, Inc.). [2]

Newer generations of electronically active DNA microarrays (under development of Nanogen) that produce controlled electric fields at each site may have potential applications for nanofabrication. These active microelectronic devices are able to transport charged molecules (DNA, RNA, proteins, enzymes), nanostructures, cells and micronscales structures to and from any test site on the devices surface. Then DNA hybridization reaction are carried out, these devices are actually using electric fields to direct the self-assembly of DNA molecules at specified sites on the chip surface. These active devices are serving as semiconductor hosts or motherboards for assembly of DNA molecules into more complex three-dimensional structures. The DNA molecules themselves have programmable and self-assembly properties and can be derivatized with a variety of molecular electronic and photonic moieties. DNA molecules can also be attached to large nanostructures, including metallic and organic particles, nanotubes, microstructures, and silicon surfaces. In principal, active microelectronic arrays and DNA-modified components may allow scientists and engineers to direct self-assembly of two- and three-dimensional molecular electronic circuits and devices within the defined perimeters of large silicon or semiconductor structures (Figure 2). Thus, electronically directed DNA self-assembly technology could encompass a broad area of potential applications from nearer term heterogeneous integration processes for photonic and microelectronic devices fabrication to the longer term nanofabrication of true molecular electronic circuits and devices.

4. Carbon Nanotubes: fundamental properties and self – organisations of DNA/Nanotube interface.

Carbon nanotubes are long, cylindrical molecules consisting of a circular array of sp^2 hybridized carbon atoms. They are capped at both ends and can be thought of as elongated fullerenes. Two types of nanotubes have been discovered: single-walled nanotubes[1] and multi-walled nanotubes[2](Bucky tubes). Single-walled nanotubes are single tubes that are approximately 1 nm in diameter and 1-100 microns in length. In multi-walled nanotubes, each tube is embedded within a larger tube (Figure 4).
Single-walled nanotubes are of interest as building blocks for nanoscale materials.3 More interesting is their potential as quantum wires in nanoscale electronic devices![1,5-8,11-21]
We have single-walled carbon nanotubes from CarboLex, Inc. of Lexington, Kentucky: a higher-purity SE-grade.
The SE-grade product consists of prepared bundles of single-walled carbon nanotubes, with 10-200 individual nanotubes per bundle. The average diameter of the nanotubes is 13A°, with almost all tubes falling within the diameter range of 12-15A°

Carbon nanotubes are giant molecular wires in which electrons can freely propagate, as in an ordinary metal. This strongly contrasts with conventional “conducting” polymers (like, for example, polythiophene) where the electrons localize. These molecules are actually insulators and become conductors only if heavily doped. Furthermore, conducting polymers have considerably shorter conjugation lengths. Each carbon atom of the graphene honeycomb lattice has four valence electrons of which three are strongly bound to neighbors giving graphene its in-plane rigidity. The fourth

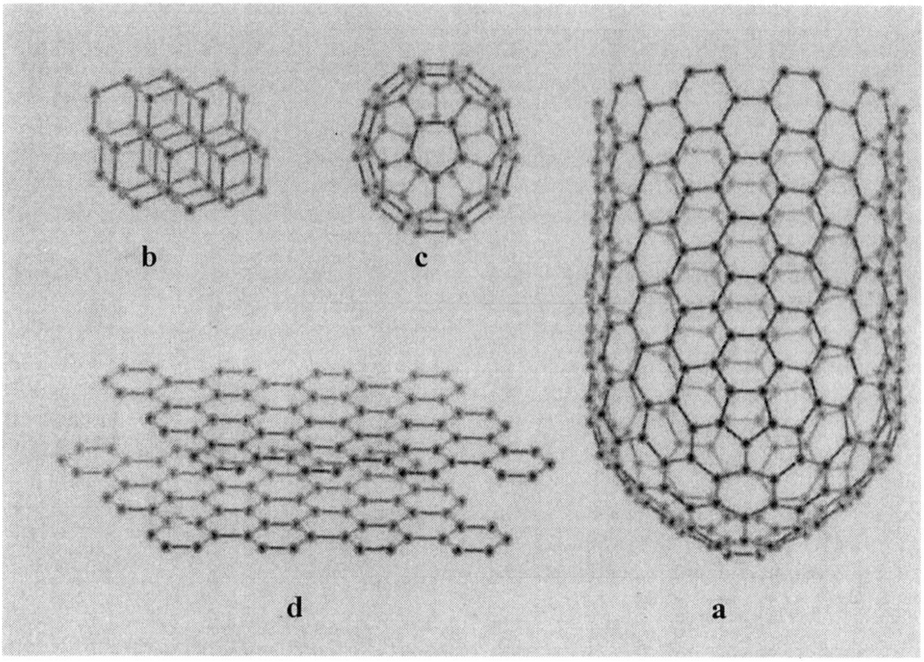

Figure 4. Fullerene nanotube (**a**) and systems of elements from carbon: **b-** diamond; **c-** C_{60} "buckmisterfullfullerene"; **d-** graphite) [1]

electron is delocalized and shared by all the atoms allowing for electrons to conduct. However, it turns out that a single sheet of graphite (graphene) is electronically a hybrid: it is a neither a semiconductor, not a metal. Graphene is a semimetal or a "zero-gap" semiconductor. This peculiarity renders the electronic states very sensitive to additional boundary conditions, as the periodic boundary condition along the circumference of a carbon nanotube.

4.1 VISUALIZATIONS OF DNA/NANOTUBE

Scanning tunneling microscopy (STM) and atomic force microscopy (AFM) are the main methods currently used for the visualization of DNA. In these cases, the DNA molecules are immobilized onto even-surfaced substrates such as gold and mica. For example, DNA oligonucleotides can be imaged in aqueous solution by adsorption onto a positively charged gold (lll) electrode substrate. This has provided images with helical structures and even indications of the major groove periodicity of the double helices. However, reproducible fixation of samples on an atomically even surface is still the primary challenge, and images with atomic resolution are difficult to achieve. The lack of reliable probe tips is another impediment to progress in atomic-level microscopy. Carbon nanotubes have the properties that satisfy perfectly the requirements as STM probe tips. Such tips arc likely to increase the resolution and reliability of SFM experiments.

Weak electron scattering and radiation damage have hindered the determination of the structures of biological molecules at atomic resolution by HRTEM. Therefore a total of seven heavy atoms were introduced into the double-stranded DNA oligomer. These heavy atoms can enhance the electron scattering and therefore enhance the contrast in the image. Good TEM images of amorphous materials with average thicknesses of 7-13 Å on the surfaces of carbon nanotubes have been obtained (Figure 5). Energy dispersive X-ray (EDX) analysis indicated that these materials contained Pt, P, and I, and therefore are likely to be the iodinated and plalinated DNA duplex. Interestingly the DNA molecules tended to cover the surface of the nanotubes evenly, suggesting a strong interaction with the carbon surface. The nature of such interactions is not yet clear. The electron conductivity of carbon nanolubes may play an important role in shielding the DNA from electron beam damage.

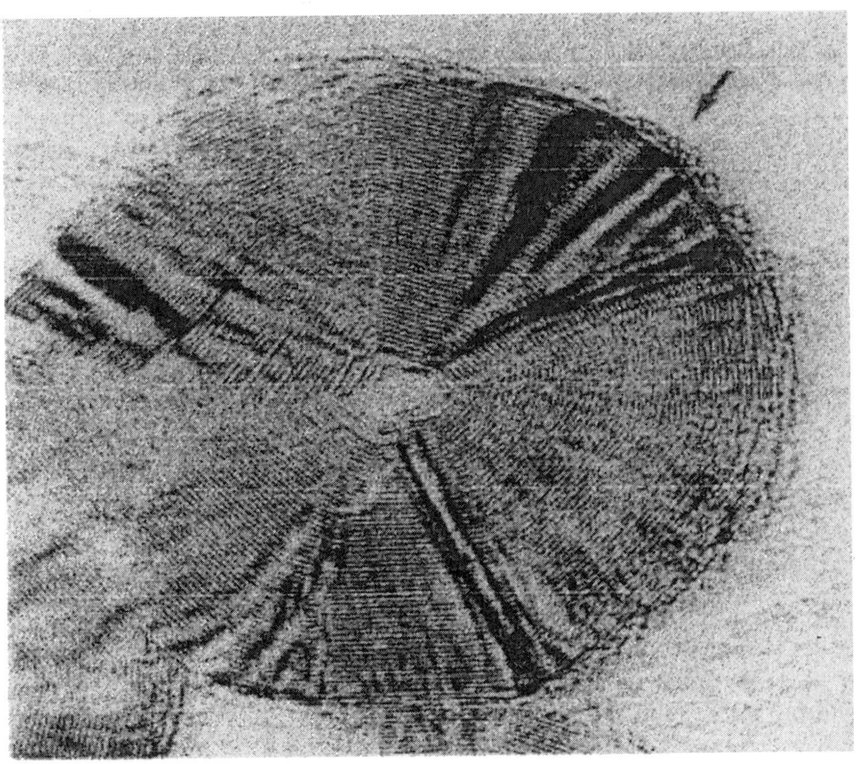

50 Å

Figure 5. TEM images of carbon nanotubes treated with iodinated and platinated DNA. The arrow indicates absorbed DNA material on the surface [10].

The rapid development of methods for the synthesis of DNA derivatives has raised the possibility of using **DNA** as a material for nanotechnology. Deposition of DNA molecules on robust carbon nanotubes and the reversibility of the Watson-Crick base pairing could promise further exciting advances in this field. Open carbon nanotubes filled with biomolccules could be delivered specifically to selected targets. Moreover, initial experiments with the enzyme P-lactamasel immobilized in or on carbon nanolubes have shown that a significant amount ol' the enzyme remains catalytically active. Carbon nanotubes could also find use in biomimetic systems, because tubular structures, such as tubulin, have important biological functions. It will also be interesting to study the recognition of biomolecules by singlc-walled carbon nanotubes, which can exhibit both metallic and semiconducting properties depending on the diameter and helicily of the arrangement of graphitic rings in the walls.

5. Nano-Si systems

To say that Si and other inorganic semiconducting materials have revolutionized electronics and optoelectronics is a severe understatement. It is better to say that inorganic semiconductors created electronics and optoelectronics. Science the report of the first transistor the speed and efficiency of semiconductor- based devices has been increasing at an astounding pace. The success of silicon and other inorganic semiconductors in electronic and optoelectronic applications is due to several factors. One important breakthrough was the ability to prepare silicon with impurities at part per billion (ppb) levels, allowing controlled doping to part per million (ppm) levels of selected dopants. Special techniques have been developed for the growth and patterning of semiconductors, metals, and insulators at submicrometer resolution, found in common devices. In order to design and produce high-speed devices it was very important to develop a detailed understanding of the electronic structure of semiconductors and how the use of different materials could control injection, migration, and recombination of charges. In addition it was very important to understand how interface or barrier layers may also effect the properties of a multilayer devices. A logical question is where will molecular materials fit into a world dominated by silicon and other semiconductors? In this section we will address the important issues of where molecular materials may have benefits over bulk inorganic semiconductors, and what requirements need to be met by molecular materials for them to be useful in electronic and optoelectronic applications.

A significant change in the properties of Si is seen when dimensions are shrunk to nanometer scales. Inorganic “nanocrystals”, consisting of a few hundred to as many as tens of thousands of atoms, display a vast range of size-dependent properties. Examples of size-dependent properties include the bandgap of a semiconductor. As nanometer-size crystals can now be made and observed routinely, this opens the prospect of creating materials with designed properties, not just by changing the chemical composition, as has been done in the past, but by controlling the size.[57-62]

The properties of nanocrystals change systematically with size, extrapolating slowly to the familiar properties of bulk solids. For most properties the size variation approximately follows a “scaling law”. The regular variation of properties arise because, once a crystal comprises a few tens of atoms, there is an identifiable core, or interior of the crystals, in which the local bonding geometry closely resembles that of the bulk solid. In larger sizes the core region is composed of more atoms, and the ratio of surface of interior atoms is smaller, leading to the systematic variations that are observed. As the scaling laws are hypothesized and refined it becomes possible to predict the properties of nanocrystals.

6. Experimental. DNA/Nanotubes and DNA/Nano-Si sysmems

The concentrated DNA from Amicon a GRACE company was used. SWCNTs from Dr. S. Fang, University of Kentucky were mixed with DNA/NaOH buffer solution. Single droplets of the upper layer of this mixture and DNA/NaOH buffer solution were deposited on the por-Si surface with Si-nanocrystallites (nano-Si) obtained by the method [11-13] as well as on KRS-5 (42% TlBr- 58% TlJ) surface and dried with and without 2 h irradiation by visible light. In the result the DNA, DNA/SWCNT and DNA/nano-Si/SWCNT layers are formed in the separated sections with common bounds on por-Si substrate as well as DNA/carbon nanotube molecular and nano-Si are formed in the single layer on KRS substrate.

I–V characteristics of Pt/Ir tip – DNA (or DNA/SWCNT, DNA/ nano-Si /SWCNT) layer – Pt/Ir tip nanostructures were registered with lamp lighting on c-Si in the por-Si/c-Si structures by Tunneling Spectrometer constructed at the University Laboratory. PL spectra of DNA/c-Si, DNA/por-Si/c-Si and DNA /SWCNT /por-Si/c-Si structures excited by the impulse of N–laser (wave length is 337 nm, impulse time and energy are 4 ns and 3,67 eV, respectively, intensity is up to 0,5 mJ/sm) in the spectral range of λ= 350-750 nm at T=293 K have been observed using Double Diffraction Monochromator. Optical reflectance spectroscopy measurements were used for the detection of Si nanocrystallite and electron transitions at DNA/nano-Si, DNA/ SWCNT interfaces. The normal-incidence reflectance spectra in the range of 400-800 nm are recorded by spectrometer with high-sensitive registration of the reflective signal [12].

7. Results and discussion

7.1 *THE EVIDENCES OF DNA SELF-ORGANIZATION OF NANOCLUSTERS ON NANO-SI SURFACE*

7.1.1. *AFM images.*

The images on Figures 6, 7 confirm that DNA molecules form the nanoclusters on nano-Si surface (see Figure 7 (d, e))

7.1.2 *UV-VIS and IR spectroscopy Optical reflection (absorption) by* DNA/ *nano*-Si *heterostructures*

In the wide absorption bands, observed in Fig.8 (b,c,d) the separated minima (maxima of absorbance) are presented.

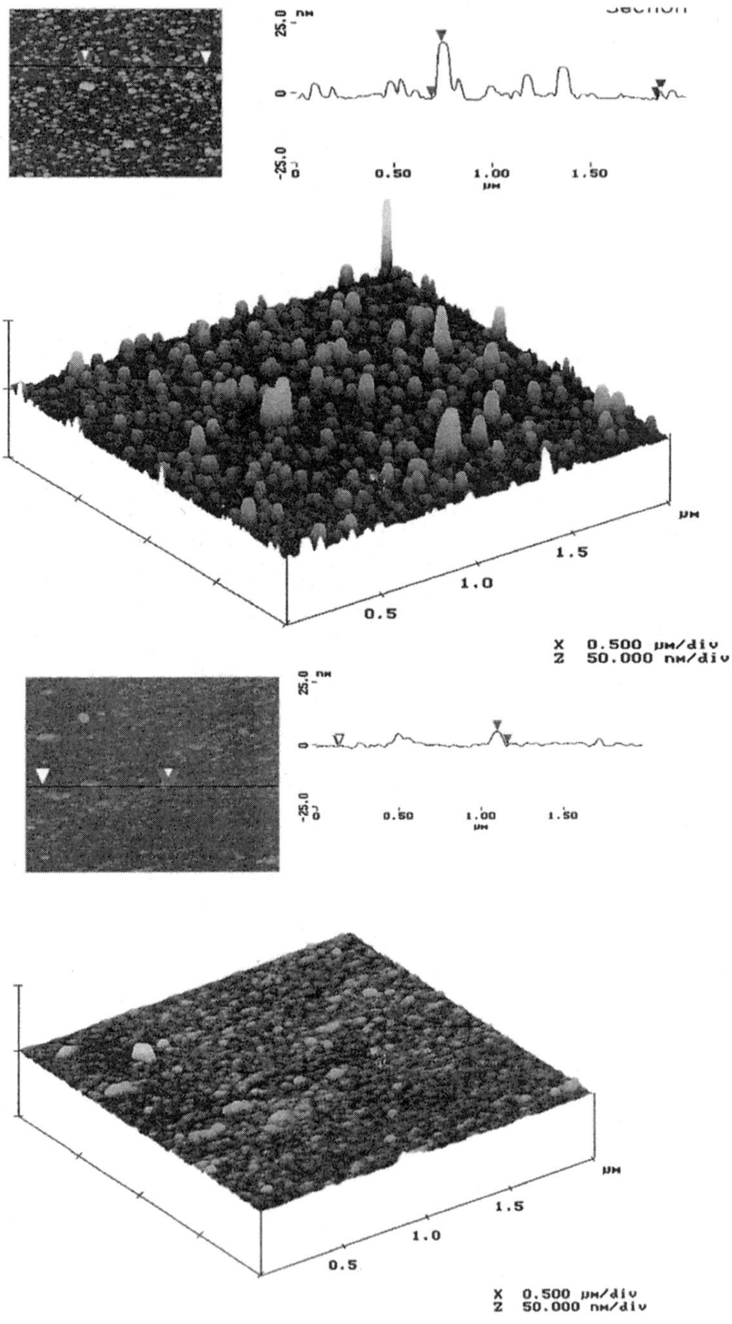

Figure 6. AFM phase images of nanostructured Si surface (**a**) and this surface with DNA layer (**d**).
AFM 3-D images of nanostructured Si surface (**c**) and this surface with DNA layer (**f**).
Profiles of relief of these surfaces (**b** and **e** correspondingly) along the line marked in images (**a** and **d**).

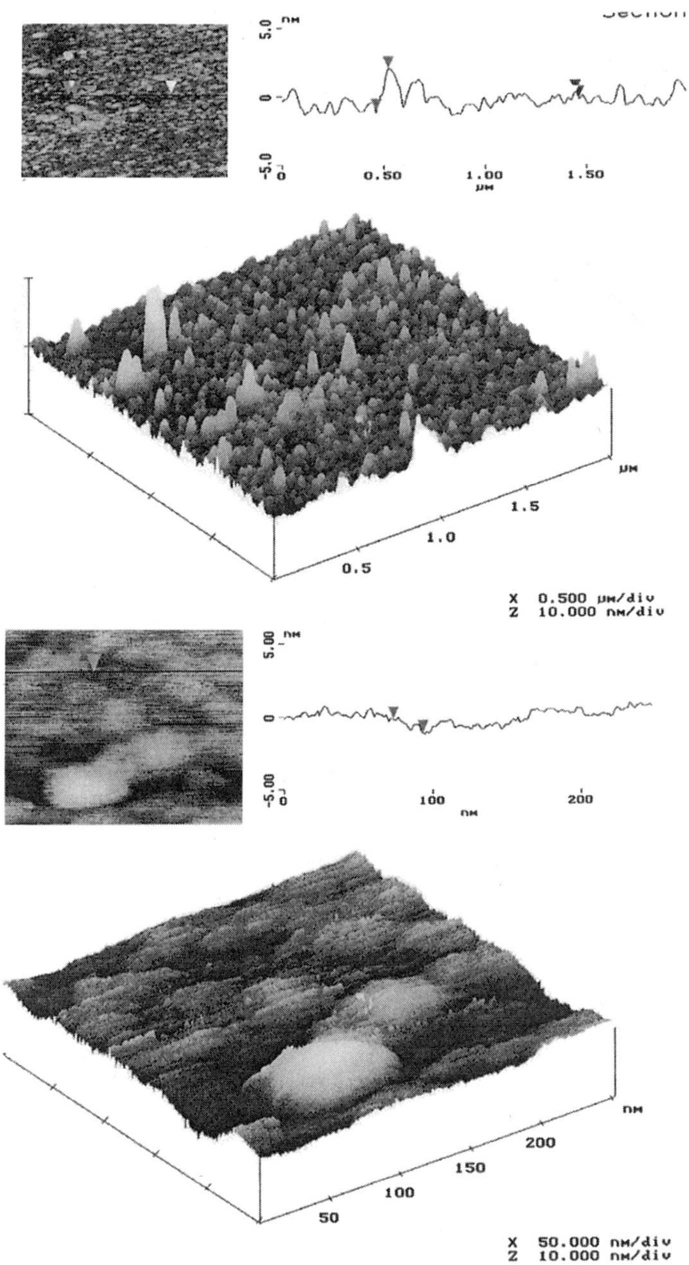

Figure 7. AFM phase images of nanostructured Si surface with DNA layer (**a**- scan area is 2X2 μm^2, **d**- scan area is 250X250 nm^2).
AFM 3-D images of nanostructured Si surface with DNA layer (**c**- scan area is 2X2 μm^2, **f**- scan area is 250X250 nm^2).
Profiles of relief of these surfaces (**b** and **e** correspondingly) along the line marked in images (**a** and **d**).

The sub-band with minimum at 627 nm (1.97 eV) which is intimate to 620nm (2.0eV) is associated with the optical absorption by the DNA/por-Si (with DNA) and DNA/c-Si interfaces. The decreasing of width and shape of the sub-band after SWCNT adding (curve d) testifies that it characterizes the light absorption at the DNA/Si interface.

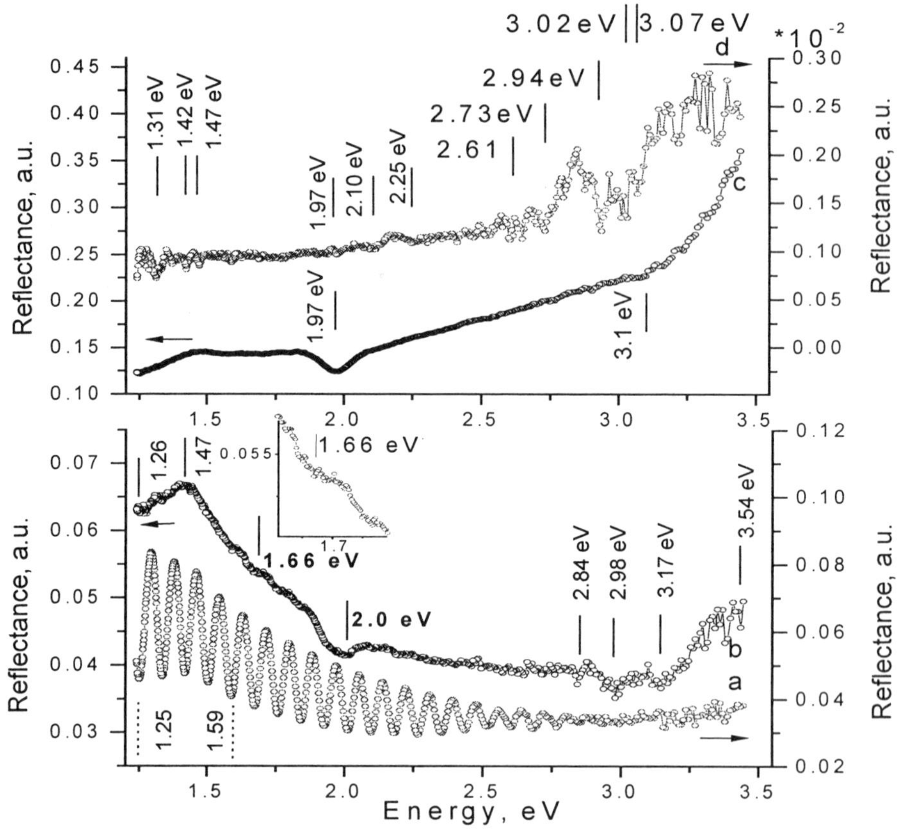

Figure 8. Reflectance spectra from por-Si/c-Si (a), DNA/por-Si (with DNA)/c-Si (b), DNA/c-Si (c), (DNA:SWCNT)/por-Si (with DNA)/c-Si structures.

The role of the Si porosity in the optical absorption at the DNA/por-Si interface is revealed by the existence of the 436nm (2.84eV) – 391nm (3.17eV) sub-band (curves b and d). Actually, this sub-band is not detected for the DNA/c-Si structure (curve c). Minima in optical reflectance spectrum (maxima of absorption) at 436nm (2.84eV), 416nm (2.98eV) and 391nm (3.17eV) testify about electron transitions in the self-organized level system at the DNA/por-Si interface. The role of SWCNT addition such a self-organization is detected by shifting of the minima at 436nm (2.84eV), 416nm

(2.98eV) and 398nm (3.17eV) on the values of 0.11, 0.04 and 0.1eV, respectively; the new minimum at the 410nm (3.02eV) is appeared.

Using these results we assert that system of electron levels is self-formed at the (DNA:SWCNT)/por-Si interface as a result of DNA-tube interaction and optical absorption at 410nm (3.02eV) is due to the SWCNT.

The separated minima at 496nm (1.31eV), 873nm (1.42eV), 844nm (1.47eV), 590nm (2.10eV), 551nm (2.25eV) and 475nm (2.61eV) (curve d) are concerned to the absorption caused by electron transitions in SWCNTs. Some types of these SWCNTs have been defined by the tunnel spectroscopy [11-16].

Minimum at 747nm (1.66eV) is associated with optical absorption in nano-Si crystallites of por-Si layer accordingly to [13].

8. The charge transport into nanoheterostructures of networks

I-V curves for Pt/Ir tip – (DNA/SWCNT)/(DNA/nano-Si/SWCNT) – Pt/Ir tip nanoheterostructure into the network in the range of bias (-1.5V) –(+1.5V) are nonlinear and asymmetrical (Fig.9). They are similar to I-V curves for Pt/Ir tip – (DNA/SWCNT) - Pt/Ir tip nanoheterostructure with DNA double helix [11-16].

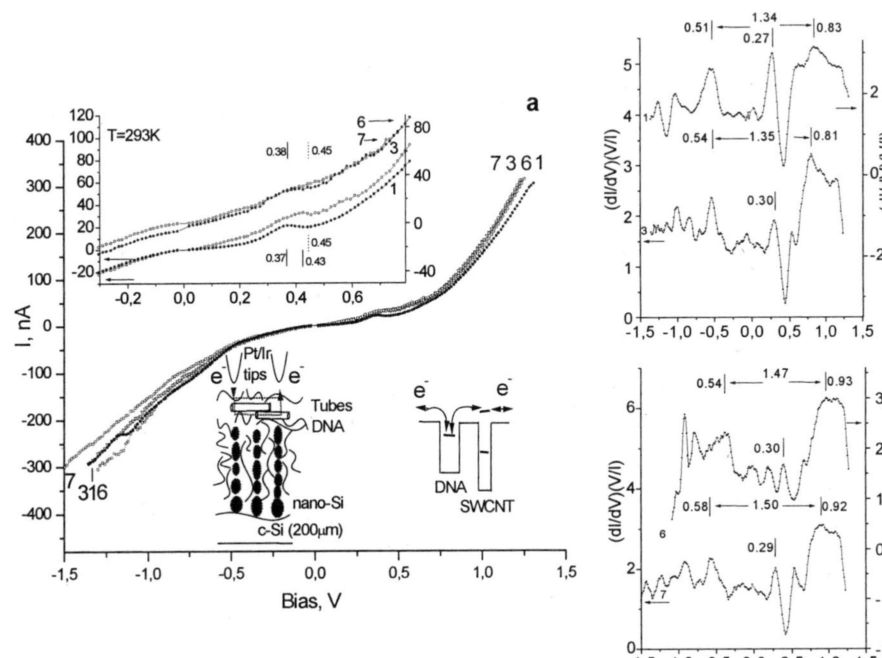

Figure 9. I-V curves (**a**) and normalized differential conductance vs applied bias (**b**) for Pt/Ir tip – (DNA/SWCNT)/(DNA/nano-Si/SWCNT) – Pt/Ir tip nanoheterostructure (the distance between tips is 1000 μm).

In Fig.2**a** I-V curves are measured without (1,3,7) and with (6) tungsten lamp lighting on c-Si in the por-Si/c-Si structures at four cycles of bias ("+","-"V). In Fig.2**b** the energy gap Eg=1.34, 1.35, 1.47, 1.5 eV and the singularities at 0.27, 0.3, 0.29 V are ascribed to SWCNT block and electronic level in it, respectively, that is fitted by the results of [13-18] for semiconducting SWCNT. In the inset the change of current in the range of the bias from –0.2V to 0.6V is shown.

The singularities i.e. the type of current decreasing at applied bias V= 0.37, 0.38, 0.43V (the first, third and seventh current measurements) are observed to be shifted their situation from V=0.37 to 0.43 V. After sixth current measurement with the illumination in the por-Si/c-Si structures (injection of carries into nano-Si) this singularity in the I-V characteristics is not detected. The shift of singularity after multi-bias application is conditioned by change of electronic structure of these nanoheterostructures in the network. It's confirmed by the analysis of (dI/dV)(V/I) vs V curves (which represent a good approximation of the shape of electronic state density and is proportional to it (Figure 9b)).

9. The photoluminescencs and photoresponse of DNA/nano-Si and DNA/nano-tubes systems

PL spectra ranged from 1.5 (827nm) to 3.5eV (354nm) for different type of nanoheterostructures into networks are shown in Figure 10. For the comparison the spectrum from DNA/c-Si structure is also given.

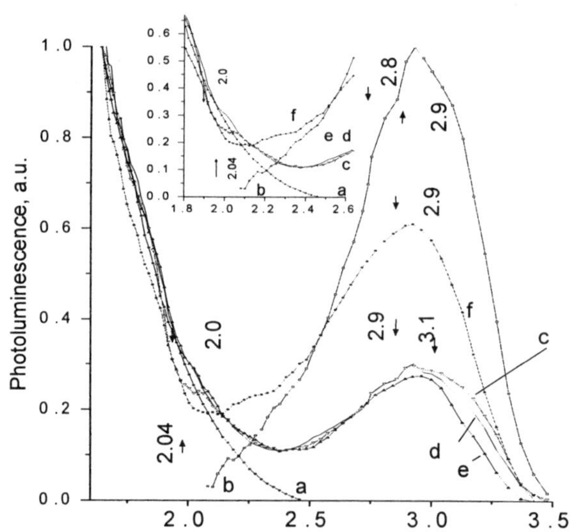

Figure 10. PL spectra of nanoheterostructures into networks: a – por-Si/c-Si, b – DNA/c-Si, c - DNA/por-Si (with incorporated DNA), d – DNA/SWCNT/nano-Si, e – DNA/por-Si with incorporated DNA (after 5 days), f - DNA/SWCNT/nano-Si (nano-Si size is 2 nm). In the inset the PL spectra ranged from 1.8 to 2.6 eV is shown.

As seen the intensive PL from nano-Si/c-Si structures into network of nano-Si is observed at 1.65eV (752nm) (band-band transition in nano-Si) and its intensity decreases fast to zero value at 2.5eV (496nm).

The intensive PL peak at 2.8-3.1 eV is present for all nano- heterostructures into networks with DNA. Additional PL maximum at 2.0 (613nm) is revealed for DNA/SWCNT/nano-Si nanoheterostructures into networks.

Using PL data and results of UV-vis spectroscopies as well as IR-spectroscopy [12] we built the scheme of electronic transitions into nanoheterostructures, namely, band-band transition with the recombination into nano-Si (Eg=1.65eV), into SWCNT (Eg=1.85eV), interlevel transitions into DNA molecular network - 2.04, 2.8, and 3.1eV and into the DNA/SWCNT/nano-Si network -2.9, and 2.8eV.

We also carried out the photoresponse measurements for studying the optoelectronic properties of the DNA/CNT and DNA layers on the Si substrate as well as DNA/CNT - DNA interface of these layers. This study can lead to a range of optoelectronic applications, for example, in solar cells or photodetectors. For the diode structures of Pt/Ir – DNA layer (with C_{60} or CNT) – Si substrate the dependence of the photoresponse versus wave length (500 – 1100 nm) at the incident excitation of the layer surface near Pt/Ir electrode is detected. Figure 11 shows the photoresponce versus wave length measurements for different layers deposited on the silicon substrate.

Maximum is observed at around 1040 nm and it corresponds to the photoemission from the DNA and DNA/CNT layers. Peak at around 550 nm is due to the photoemission from the Si with DNA layer (1) and with DNA/(DNA-CNT) layers (2) and DNA/CNT layer (3), on its surface, respectively. This peak is broadened and its intensity increases for DNA/CNT layer in comparison with the DNA layer. So one could successfully use the self-organized DNA/CNT layers deposited on the Si substrate in various optoelectronic applications.

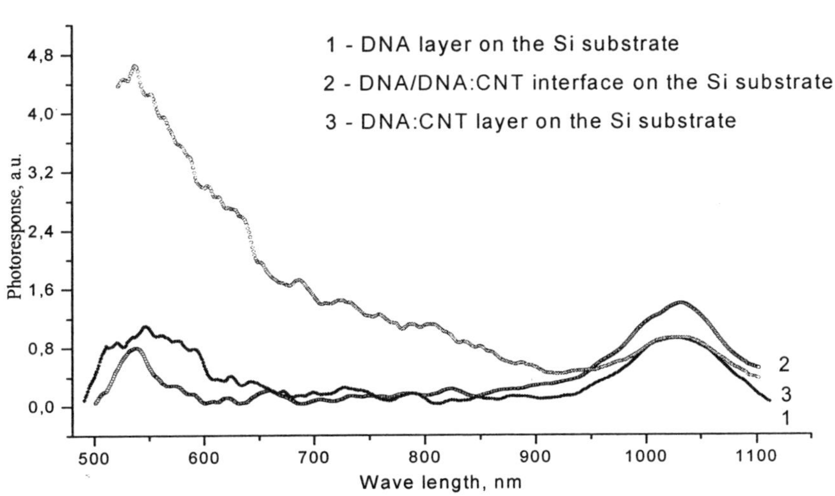

Figure.11 The dependence of the photoresponce versus wavelength for DNA (1), DNA/CNT (3) layers deposited on the Si substrate and the DNA/(DNA-CNT) layers (2).

10. Conclusions

The first direct experimental evidences for the organization of DNA/SWCNT molecular and nanocrystalline-Si networks, consisting of DNA double helix, SCWNT and nanocrystalline-Si building blocks, due to DNA molecules have been obtained: The nanotechnology of DNA, SWCNT molecular layers formation on the nanopor-Si surface with nanocrystalline-Si, DNA diffusion into pores and self-organization of DNA/nanocrystalline-Si interface have been developed. The electronic structure self-formation and charge transfer for the network systems from building blocks by tunnel and photoluminescence spectroscopies have been studied. We have obtained the evidences of the main role of the interface molecular interaction in a self-organization of electronic structure of these networks and the charge injection from nanocrystalline-Si explaining the features of diode I-V characteristics and photoluminescence spectra

We have realised our idea about construction of multifunctional nanosystems from DNA/carbon nanotube and DNA/nanocrystalline – Si blocks in chip by DNA nanotechnology using the theoretical predicted and experimental confirmed facts that single DNA, carbon nanotube molecules as well as nanocrystalline - Si are able to conduct electrical charges and have high photosensitivity.

We would like to present the resume about fundamental properties of DNA molecules as the block for the nanotechnolodgy:

- DNA has structure of double helix. The sides of the ladder have a periodic structure with alternating sugar and phosphate groups. The nucleotide base pairs form the rungs of the ladder, and the genetic information is encoded in the order of these base pairs.

- DNA contains 4 nucleotide bases – A, T, C, G – which are closely related to each – other in chemical structure. The bases are always paired as A – T and C – G along the rungs of the ladder by Hydrogen bonds. This base – pairing makes the two DNA strands complementary in character.

- The sugar and phosphate groups along the ladder are held together covalent bonds. Their bonding is completely to the bases attached to them, and takes place in the presence of DNA polymerase enzymes.

- During replication, the helices enzyme separates two strands of DNA by breaking the Hydrogen bonds, mach like opening a zipper. The unpaired bases along each of the separated strands find their partners from the surrounding molecules, producing two copies of the original DNA.

- The sides of the ladder have asymmetric ends. The replication process is directed, always proceeding from the 5′ end to the 3′ end (these numbers label the position of the carbon atoms in the sugar rings) of the intact strand. The DNA polymerase enzyme slides along the intact strand, adding one base at a time to the growing strand. During this process, base – pairing and sugar – phosphate bonding alternate.

• Hereditary DNA is accurately assembly, with an error rate of 10^{-7} per base pair, after the proof – reading exonuclease action. Proteins are assembled less accurately, with an error rate of 10^{-4} per amino acid.

This review paper charts the progress made over the last decade in realising a variety of optoelectronic devices using nanostructure porous silicon. Although weveguides, modulators and detectors are covered, emphasis is given to visible light – emitting diodes whose efficiency is now 1%. Although their performance is still inadequate for optical interconnect applications, two promising approaches for further major advances are concluded to be the fabrication of more monodisparse nanostructures and the exploitations of optical microcavities.

11. Acknowledgments

Support of this work by the EU ESPRIT Project, SBLED INCO – COPERNICUS 977037, Greece-Ukrainian Grant 2M/192 – 2001 and Ukrainian State Grant 01 BFO52 – 07 for is acknowledged.

References

1. Alivisatos, A. P., Barbara, P. F., Castleman, A. W., et all. (1998) From Molecules to Materials: Current Trends and Future Directions.*Adv. Mater.*, **10**, No. 16. 1297-1336
2. Nanotechnology Research Directions: IWGN Workshop Report. Vision for Nanotechnology R&D in the Next Decade. Edited by Roco M.C., Williams R.S., Alivisatos P., (2000), Kluwer Academic Publishers, Dordrecht/Boston/London, 316.
3. Alivisatos, A.P., K.P.Johnsson, X.G.Peng, T.E.Wilson, C.J. Loweth, M.P.Bruchez, P.G. Schultz. (1996) Organization of nanocrystal molecules using DNA. *Nature* **382**:609-611.
4. Nadrian C. Seeman, Chengde Mao, Furong Liu, et al., (2000). Nicks, Nodes, and New Motifs for DNA Nanotechnology. *Mathematics, Physics and Chemistry*, **6**, 177-208 .
5. Schonenberger C., Forro L. (2001), Physics of multiwalled carbon nanotubes In Book: Physics, Chemistry and Application of Nanostructures, Editors: V.E.Borisenko, S.V.Gaponenko, V.S.Gurin, World Scientific, Singapore, 86-93.
6. Dekker Cees, (May 1999) Carbon Nanotubes as Molecular Quantum Wires, *Physics Today*,. 22-28
7. Porath D., Bezryadin A., Simon de Vries & Dekker Cees, (2000) Direct Measurement of electrical transport through DNA molecules, *Nature* Vol. **403,** 10 February, 635-638.
8. Reed. M.A., C.Zhou, C.J.Muller, T.P.Burgin, and J.M.Tour. (1997). Conductance of a molecular junction. *Science* **278**: 252- 254
9. Yan Mao, Lambert N. Daniel, (1994). DNA Binding to Crystalline Silica Characterized by Fourier-Transform Infared Spectroscopy. Environ Health Perspect. Noel Whittaker and Umberto Saffiotti, *Environ Health Perspect,* **102**(Suppl), 165-171
10. By Zijian Guo, Salder Peter J., Tsang Chi Shik (1998) Immobilization and Visualization of DNA and Protein on Carbon Nanotubes. *Adv. Mater. **10**,* No.9
11. Buzaneva E.V., Karlash A.Yu., Putselyk S.O., Shtogun Ya.V., Gorchynskyy K.A., Prylutskyy Yu.I., Prylutska S.V., Matyshevska O.P., Scharff P., (2000) The Electronic Properties for DNA/CNT Chip on Si in *Electronic Properties of Novel Materials- Molecular Nanostructures, , Editors: Hans Kuzmany, Jorg Fink, Michael Mehring, Siegmar Roth, AIP Conference Proceedings* **544**, *Melville, New York: American Institute of Physics*, 468-472.
12. A.Karlash, S.Putselyk, Ya.Shtogun, K.Yakovkin, K.Gorchinsky, A.Benilov, S.Prylutska, O.Matyshevska, Yu.Prylutskyy, E.Buzaneva. Self-Organization of the DNA/(C_{60}) and DNA/Carbon

Nanotube Blocks. In Workbook Frontiers of Nano-Optoelectronic Systems: Molecular-Scale Engineering and Processes". NATO/EC, ARW/Spring School, Ukraine, Kyiv 22-26 May 2000, Nano T-7.

13. O.Matyshevska, A.Karlash, Ya.Shtogun, N.Procopchuk, A.Benilov, Yu.Kirgissov, A.Gorchinsky, T.Veblaya, N.Kharchenko, E.Buzaneva, Yu.Prilutskiy, O.Marchenko Self-Organizing DNA/Fullerene molecular films. In Book of abstracts, International Conference on Electronic Materials & European Materials Research Society Spring Meeting, Strasbourg, France, May 30 – June 2, 2000, Symposium E: Current Trends in Nanotechnologies, E – 31 (E – V/P07).
14. Ya.Shtogun, A.Karlash, S.Putselyk (2000) Patterned Films of DNA/Carbon Nanotubes. In Book of abstracts, International Conference on DNA Conformation, Modification and Recognition in Biomedicine, 2-5 July, Brno, Czech Republic,126.
15. A.Karlash, Ya.Shtogun, S.Prylutska, O.Matyshevska, E.Buzaneva (2000) Biology active DNA/Fullerenes/ Carbon Nanotubes Assemblies.in Book of Abstracts, International Conference on DNA Modification and Recognition in Biomedicine, 2-5 July , Brno, Czech Republic, 98.
16. K.Yakovkin, O.Nichiporuk, Ya.Shtogun, S.Prylutska, V.Scryshevskiy, S.Litvinenko, E.Buzaneva (2000) Photoactivity of DNA/Carbon Nanotubes layers in Pt/Ir – Silicon cell. In Book of abstracts, International Conference on DNA Conformation, Modification and Recognition in Biomedicine. 2-5 July, Brno, Czech Republic,138.
17. Shtogun Ya, Prylutska S., (2000) Infrared Identification of the Self–formation Route of a DNA/Protein/ Carbon nanotubes Molecular Layer. In Book of abstracts, International Young Scientists' Conference "Scientific Problems of Optics in the XXI century", October 5 – 6, Kiev, Ukraine, 56.
18. Buzaneva E., Gorchinskiy A., Popova G., Karlash A., Ciapa S., Shtogun Ya., Yakovkin K. (2001) Electrical and Optical Properties of DNA/Nano-Si, DNA/Fullerene/Nano-Si , DNA/ Carbon Nanotubes/ Nano-Si Chips. In Workbook. for Kiev NATO ARW/Autumn School, 9-12 September: Frontiers in Molecular-Scale Science and Technology of Fullerene, Nanotube, Nanosilicon and Biopolymer (DNA, Protein) Multifunctional Nanosystems. *Kiev University*, SB-L6
19. Karlash A., Shtogun Ya.,Veblay T.,Gorchinsky A., Matyshevska O.,Buzaneva E. (2000) Spectroscopy of IR-Transmittance of DNA Molecular layers. *Visnyk of Kyiv University, Series: Phys.-Math. Science*,№ **1**, 366-379
20. Karlash A., Shtogun Ya.,Veblay T.,Gorchinsky A., Matyshevska O.,Buzaneva E, Lositsky M., Prylutska S., Yashchuk V.,(2000) Self-Organized DNA Mole-cules/Carbon Nanotubes layer: UV, IR-Spectorscopy". Visnyk of Kyiv University, Series: Phys.-Math. Science, № **2**, 412-421
21. Buzaneva E., Karlash A., Putselyk S., Shtogun Ya., Gorchinskyy K., Prylutskyy Yu., Ogloblya O., Matyshevska O., Prylutska S., Eklund P., Scharff.P. (2000) Electronic properties of DNA/CNT chip on Si. // *Mol.Cryst.Liq.Cryst.*,
22. Anna Angelescu, Irina Kleps, Nina Kovtyukhova, Sergeiy Putselyk, Volodymyr Kudryavtsev, Alexsander Gorchinskiy, Galina Popova, Eugenia Buzaneva. (2001) Integration of Si biocomplex building blocks into porous Si for self-organization of multifunctional structures. *Materials Science and Engineering C,* 716.
23. Buzaneva E., Gorchinskiy A., Popova G., Veblaya T.,Zankovych, Boiko Yu., Zolotarenko P., Pogorelov V., Bukalo V., Benilov A., Lazaruk S., Beyliss S., Starovoitov A., Senkevich A. (2000) Photophysical properties of nano Si/SiO_x composites in Al/composite/mono Si structures for green light emitting and photodetector – Schottky diodes. *Materisls Science in Semiconductor Processing* **3** 529 –537
24. Bisi O., Stefano Osicini, Pavesi L. (2000) Porous silicon: a quantum sponge structure for silicon based optoelectronics. *Surface Science Reports,* **38**, 1-126
25. Prylutskyy Yu., Durov S, Ogloblya O, Buzaneva E., Scharff P., (2000) Molecular dynamics simulation of mechanical, vibrational and electronic properties of carbon nanotubes. *Comput.Mat.Sci.*, **V. 17**, N 2-4, 352 – 355
26. Prylutskyy Yu. Ogloblya O., Buzaneva E., Gorchinskiy A.,. Eklund P., Scharff P.. (2000) Optical properties of single-walled carbon nanotubes. *Func.Mat.*, Vol. **7**, No 4(1).652-654
27. Prylutskyy Yu. Ogloblya O., Buzaneva E., Eklund P., Scharff P.,Putselyk S. (2000) Effect of pressure on the electronic properties of carbon nanotubes. *In Book: AIP Conf. Proc.* ***544****, Electronic Properties of Novel Materials- Molecular Nanostructures XIV International Winterschool/ Euroconference, Kirchberg, Tirol, Austria, 4-11 March, Editors: Hans Kuzmany, Jorg Fink, Michael Mehring, Siegmar Roth. Melville, New York* 371-375.
28. Kovtyukhova N., Buzaneva E., Gorchinskiy A., Olliver P.J., Martin B., Waraksa C., Mallouk T.E., (2000) Self – Assembly of Nanoblocks and molecules in Optical Thin – Film Nanostructures. *Frontiers of Nano – Optoelectronics Systems: Molecular – Scale Engineering and Processes, Eds.,: Lorenza*

Pavesi and Eugenia Buzaneva, NATO ASI Series, Series E: Applied Science – Vol.**447** *Kluwer Academic Publish, Dordrech/Boston/London.* 160 – 174.

29. Simon P Newton, J. Fraser Stoddart and Wayne Hayes (1997), Self-assembled macromolecular and macrosupramolecular systems, Supramolecular *Science* **3**, 221-236.
30. Brune, H., Giovannini, M., Bromann, K. And Kern, K. (1998) Self – Organized Growth of Nanostructure Arrays on Strain – Relief Patterns, Nature **394**, 451 – 453.
31. Goodby J.W., (1999) Mesogenic molecular crystalline materials. Current Opinion. *Solid State and Materials Science*, Vol. **4,** 361-368.
32. Stupp, S.I., LeBonheur V., Walker K., L.S. Li, K.E. Huggins, M. Keser, and A.Amstutz. (1997). Supramolecular materials: self – organized nanostructures. *Science* **276**:384 – 389.
33. Kevin E. Healy. (1999) Molecular engineering of materials for bioreactivity. Current Opinion. *Solid State and Materials Science*, Vol. **4**, 381-387.
34. Brynda, E. and Houska, M. (1998), Preparation of organized protein multilayers. *Macromol. Rapid Commun.* **19**, 173–176.
35. Lin, X.M., Sorensen C.M.l,. Klabunde K.J,. (1999). Alkaline – earth oxide nanoparticles obtained by aerogel methods. Characterizations and ration for unexpectedly high surface chemical reactivites. *Chem. of Materials* **11**: 197 – 202.
36. Buddy D. Ratner, Huaiqiu Shi. (1999) Recognition templates for biomaterials with engineered bioreactivity. Current Opinion. *Solid State and Materials Science*, Vol. **4**,395-402.
37. Ratner D. (1997), The engineering of biomaterials exhibiting recognition and specificity. *J. Mol. Rec.* **9**, 617-625.
38. Shi, H, Ratner,BD (1999), Template recognition of protein imprited polymer surfaces. *J. Biomed.Mater. Res.* **50**, 1-11.
39. Shi, H, Ratner, BD (1999), Recognition templates for biomaterials with engineered bioreactivity, Current Options in Solid State & Materials *Science* **4**, 395-402.
40. Shi H., Tsai W-B, Ferrari S, Ratner BD (1999), Template imprinted nanostructural surfaces for protein recognition. *Nature* **398**, 593-597.
41. Feingold Mario. (2000) Single- molecule studies of DNA and DNA-protein interactions. *Physica E: Low-dimensional Systems and Nanostructures*, Vol. **9**, March,616-620.
42. Braun, E., Y.Eichen, U Sivan, G.Ben Yoseph. (1998). DNA-templed assembly and electrode attachment of a conducting silver wire. *Nature* **391**:775-778.
43. Mucic, R. C., J.J.Storhoff, C.A. Mirkin, and R.L. Letsinger. (1998). DNA- directed synthesis of binary nanoparticle network materials. *Journal of the American Chemistry Society* **120**:12674 – 12675.
44. Mirkin, C.A., R.L.Lestinger,R.C. Mucic, and J.J. Storhoff. (1996). A DNA-based method for ratinaly assembling nanoparticles into macroscopic materials. *Nature* **382**: 607 – 609.
45. Tanigawa, M. and Okada, T. (1998), Atomic force microscopy of supercoiled DNA structure on mica, *Analytica Chimica Acta* **365**, 19-25.
46. Fink H.-W. (2000) Electrical Conduction through DNA Molecules. *in Book Electronic Properties of Novel Materials- Molecular Nanostructures, Editors: Hans Kuzmany, Jorg Fink, Michael Mehring, Siegmar Roth, AIP Conference Proceedings* **544,** *Melville, New York: American Institute of Physics*, 457-461.
47. Chen. J., M.A.Hamon, H.Hu, Y.Chen, A.M. Rao, P.C. Eklund, and R.C. Haddon. (1998). Solution properties of single – walled carbon nanotubes. *Science* **282**:95-98.
48. Kikuo Hargaya, (2000) "Disorder Effects on Density of States and Electronic Conductance in Carbon Nanotubes." in *Electronic Properties of Novel Materials- Molecular Nanostructures, Editors: Hans Kuzmany, Jorg Fink, Michael Mehring, Siegmar Roth, AIP Conference Proceedings* **486**, *Melville, New York: American Institute of Physics*, .346-350.
49. Wong et all. (1998). Covalently functionalized nanotubes as nanometre – sized probes in chemistry and biology. *Nature*. **394**:52
50. Muris M., Dupont-Pavlovsky N., Bienfait M., Zeppenfeld P. (2001), Where the molecules absorbed on single-walled nanotubes?, *Surface Science*, Vol. **492**, Nos. 1-2, 67-74.
51. Xiangang Wan and Jinming Dong (1998), Optical properties of carbon nanotubes, *Physical Review B*, Vol. **58**, Nov. 11, 6756-6759.
52. Alvarez L., Guillard T.,Sauvajol J.L., Flamant G., Laplaze D.(2001) Growth mechanisms and diameter evolution of single wall carbon nanotubes. *Chemical Physics Letters* **342**, 7-14.
53. J. Hygard, D.H. Codben, M. Bockrath, P.L. McEuen, P.E. Lindelof, (1999) Electrical Transport Measurements on single-walled carbon nanotubes, *Appl.Phys.***A** **69**, 297-304

54. Wong S, Joselevich E, Woolley A, Cheung CL, Lieber C. (1998) Covalentnly functionalized nanotubes and nanomitre-sized probes in chemistry and biology. *Nature*;**394**:52-5.
55. Falvo M, Taylor Rhelser A, Chi V, Broks F, Washburn S, Superfine R. (1999) Nanometer-scale rolling and sliding of carbon nanotubes. *Nature*;**397**:236-8.
56. Walt A. De Heer. (1999) Recent developments in carbon nanotubes. Current Opinion. *Solid State and Materials Science*, Vol. **4**, 355-359.
57. Canham L.T. (1997) Biomedical applications of porous silicon. *Properies of Porous silicon. Edited by Leigh Canham.emis Datareviews Series* No. **18**. An INSPEC publication.London, United Kingdom.371-376.
58. Bayliss S. C., Buckberry L. D. And Mayne A. (2000), Interaction of biomaterials with porous silicon. *in Book Frontiers of Nano-Optoelectronic Systems, edited by Pavesi L. and Buzaneva E., NATO Science Series II. Mathematics, Physics and Chemistry* – Vol. **6**, *Kluwer Academic Publishers, Dordrech/Boston/London.* 199-208.
59. Bruchez, M., M.Moronne, P.Gin, S.Weiss, A.P.Alivisatos. (1998). Semiconductor nanocrystals as fluorescent biological labes. *Science* **1**:197-201.
60. Chan, W.C.W. and S.M. Nie. (1998). Quantum dot bioconjugates for ultrasensitive nonisotopic detection. *Science* **281**:2016-2018.
61. Mirkin, C.A. at all.(1997). Selective colorimetric detection of polynucliotides based on the distance – depended optical properties of gold nanoparticles. *Science* **277**:1078.

FUNDAMENTAL PROPERTIES AND APPLICATIONS OF FULLERENE AND CARBON NANOTUBE SYSTEMS

P. SCHARFF
Institute for Physics, Chemical Department
TU Ilmenau, Weimarer Str. 25 (Curie-Building)
D-98693 Ilmenau. Germany

Abstract

With the discovery of the fullerenes and carbon nanotubes, carbon science had a remarkable renaissance: Discrete spherical carbon molecules, being soluble in many conventional organic solvents, offered a broad variety of possible reactions to the chemists, whereas the nanotubes stand out because of their unique physical properties.
Beside these spectacular fields, traditional carbon materials, as for example diamond or graphite, have been modified in order to make available the basis for new applications.
In all cases the type of hybridisation of the C-atoms, connected with the atomic arrangements, as well as the crystal structures and textures, define the properties of the materials.

The differences in reactivity between the fullerenes and graphite first of all arise from the twelve pentagonal rings being necessary to build up a closed spherical shell, due to Euler's theorem. On the one hand, such pentagons introduce strain to the sphere, and on the other hand prevent a complete delocalisation of the π-electrons, enabling the quasi double bonds to undergo addition reactions. Therefore a whole new world of exohedral fullerene compounds with organic groups, halogens, hydrogen etc. was born. Beyond this, it was found that, for example, alkali metal atoms can be intercalated into the gaps of the fullerene fcc lattice. Such compounds exhibit superconducting properties at T_c up to about 40 K. Fullerenes and fullerene derivatives are currently used as dopants for polymers in photovoltaic cells. From the very beginning of fullerene chemistry a fascinating idea was the existence of endohedral complexes, in which atoms or molecules are enclosed within the fullerene cage. Meanwhile several preparational routes for endohedrals have been developed, however in most cases the yields are still low, which is a main problem with respect to possible applications. Recently polymerized fullerene phases have attracted considerable attention because of their exceptional hardness and their unusual electrical and magnetic properties.
Carbon nanotubes can be divided into two types, single-walled tubes (SWNTs) and multi-walled tubes (MWNTs). SWNTs can be imagined as rolled-up rectangular strips of graphene monolayers , whereas MWNTs consist of several such nested cylinders. The tube ends may be closed, for example by C_{60} half-spheres. Hence follows that

E. Buzaneva and P. Scharff (eds.), Frontiers of Multifunctional Nanosystems, 213–224.

pentagonal rings can at most be found at the tube ends. Therefore the chemical reactivity of the tube walls is more comparable to the chemical reactivity of graphite than that of the fullerenes, which means that preferably intercalation reactions instead of the formation of covalent bonds are observed. The electronic properties of SWNTs are determined by the tube's symmetry, that means by the orientation of the tube axis with respect to the hexagonal graphene lattice. Carbon nanotubes have been proposed for different applications, as for example cold-cathode flat-panel devices, microwave sources, materials for hydrogen storage, electrodes in rechargeable lithium-ion batteries, artificial muscles and others. However for large scale technical applications the discovery of a low cost production method is a challenge as before.

Industrial carbons went through a rapid development during the last years. Carbon disc-brakes, electrodes and membranes in fuel cells, high-temperature materials, structural materials, and materials for medical uses represent only a very limited selection of products resulting from well-aimed advancements. Many future demands for materials properties will be only realizable on the basis of carbon materials.

1. Introduction

The carbon allotropes graphite and diamond are known since long time. They were formed by nature under certain geological conditions, but both under thermodynamic control. The most stable carbon allotrope is the hexagonal form of graphite, whereas rhombohedral graphite is slightly (0.59±0.17 kJ mol^{-1} (Greenwood et al., 1988)) less stable. Diamond is preferably formed under high pressure, because the density of diamond (3.514 g cm^{-3}) is higher than the density of graphite (2.26 g cm^{-3}) (Hollemann et al., 1985). In diamond, the heat of formation amounts to 1.90 kJ mol^{-1} (Greenwood et al., 1988).

The so called novel forms of carbon, comprising fullerenes, carbon nanotubes, and also other nanocarbons, as for example carbynes, carbolite etc. are formed under kinetic control, that means by quenching a hot carbon plasma (3000-8000 K). The heat of formation per carbon atom in C_{60} has been determined experimentally by calorimetry to be 42.51 kJ mol^{-1} (Beckhaus et al., 1992). Hence follows that upon the formation of C_{60} only a local energy minimum is reached. With an increasing number of C-atoms in a closed shell structure, the heat of formation becomes lower and lower, indicating that the stress energy in the cage is reduced. Obviously the quenching conditions in the carbon plasma determine the type of nanostructures, which are formed in the preparation process. Therefore a fine tuning of the parameters is required to optimise the yield of a certain fullerene species or any other nanocarbon. In case of the preparation of carbon nanotubes, the use of catalysts (for example Ni or Co) proved to be favourable. The chemical reactivity of the novel forms of carbon can be traced back to the hybridisation state of the contained carbon atoms, the stress energy resulting from bond angles deviating from the ideal values, and the energy levels of HOMOs and LUMOs. In C_{60} the hybridisation state of the carbon atoms is $sp^{2.28}$, ranging between sp^3 in diamond and sp^2 in graphite. The double bonds in the C_{60} molecule are not completely delocalised – as in benzene or graphite. Therefore C_{60} behaves much more like an electron deficient olefine than an aromatic molecule. In contrast, carbon nanotubes, which contain mainly sp^2 hybridised carbon atoms, reveal graphite-like properties.

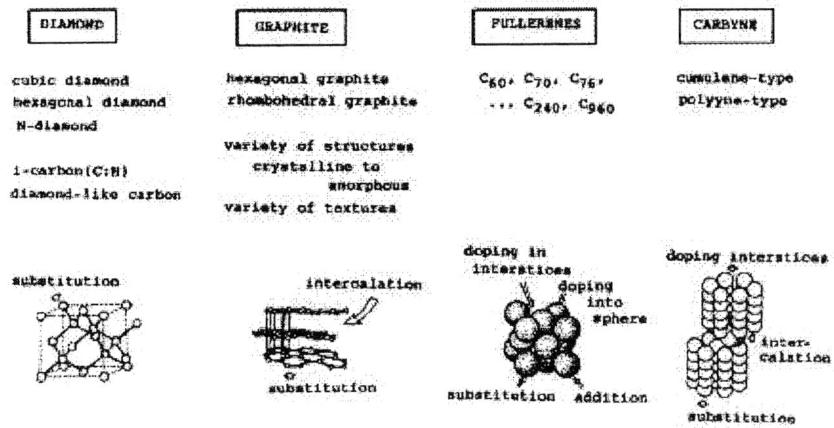

Figure 1. Typical reactions of different forms of carbon.

In figure1 (Inagaki, 2000) the typical reactions of several forms of carbon are displayed. As can be seen, a honeycomb carbon network, be it flat or rolled up, typically reacts via electron exchange between the П-bands and a substrate, whereby intercalation compounds are formed. In case of diamond with sp^3 hybridised carbon atoms the typical reaction is the substitution of a carbon atom by a hetero atom.

In fullerenes both these types of reaction, but also addition to the double bonds, and the formation of endohedral and exohedral complexes can be observed, enabling the fullerenes to act as the central unit for a whole new class of chemical compounds.

In materials science the novel forms of carbon have meanwhile attracted considerable attention for their unique physical and chemical properties. In many fields technological applications lie ahead.

2. The Fullerenes

The discovery of the fullerenes by Kroto, Smalley and Curl in 1985 was acknowledged by the Nobel prize award for Chemistry in 1996. However only in 1990 Krätschmer and Huffman succeeded in developping a preparational method suited for the production of bulk quantities of fullerene material. Since then the fullerenes have been widely studied. Fullerenes derivatives can be divided into three groups: endohedral compounds, exohedral compounds and heterofullerenes.

2.1. ENDOHEDRAL FULLERENES

In endohedral fullerene compounds at least one atom or ion is located inside the cage (figure 2). The C_{60} cage provides several flat potential troughs on it's inner surface, for which reason the included species does not keep a fixed position, but is moving from one potential trough to the other. The average position may be centered (in the case of atoms being included) or off-centered (in the case of ions being included). As a consequence of lower symmetry the higher fullerenes provide only few, comparatively deep potential troughs at their inner surfaces, which leads to a more pronounced localization of the included species.

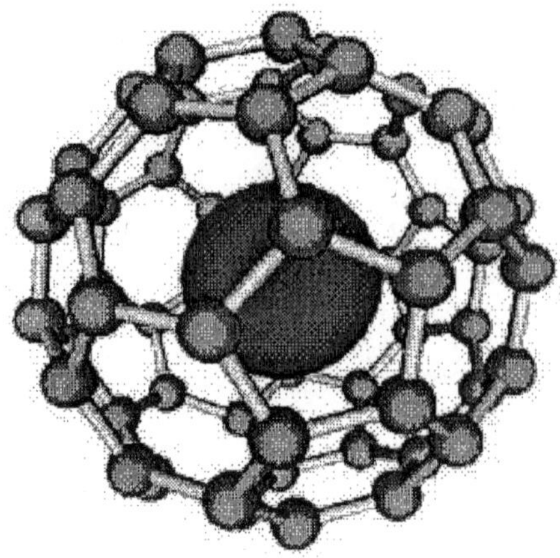

Figure 2. Endohedral Fullerene N@C_{60}

At least four principally different methods for the preparation of endohedrals have been described:

i) co-evaporisation of carbon and metals or metal oxides (laser, arc-evaporisation),

ii) heating of fullerene material with gases under pressure (window-mechanism),

iii) defined opening and closing of the fullerene cage by chemical reactions (not yet achieved), and

iiii) ion bombardment of fullerenes and fullerene films (comparatively high yield in case of alkali metal ions).

As the endohedral compounds are more polar than the parent fullerenes, separation can be achieved by chromatographic methods, especially HPLC. Very often repeated chromatographic separations are necessary in order to obtain the desired purification effect. For further purification sublimation can be used.

Beside endohedrals with noble gases or alkali metals preferably such with lanthanides have been prepared. Some time ago a compound N@C60 has been reported (Pietzak et al., 1997). This compound is very interesting, first of all because the contained nitrogen is in the atomic state. It is stable up to 260°C. The interaction between N and the carbon cage can be probed by ESR (hyperfine splitting by ^{13}C), and turned out to be extremely weak. In the meantime, it was found that larger fullerenes (for example C_{80}, C_{84}) can host metal clusters, being only stable inside the fullerene cage .

Endohedrals can be used as photoconductors or in non-linear optical units. Due to the possibility of tailoring the electronic properties of fullerenes by forming endohedral compounds, this class of compounds is propagated for the application in solar cells based on polymers. Endohedrals can also be used as contrasting agents in NMR-tomography. Quantum computers could be realized using the spin flop, for example in La@C_{82}, to represent the logical 1 and 0.

2.2 *EXOHEDRAL FULLERENES*

2.2.1 *Covalent bonding*

A large variety of functionalisations with organic addents has been performed, opening up the wide field of side group chemistry.

Covalently bonded inorganic derivatives have been obtained for example with electron acceptors as halogens, O_2, strong acids, sulphur, and dicyan.

Another group of covalently bonded fullerene derivatives is formed by 1-, 2-, and 3-D polymerised fullerenes. According to the T,P phase diagram shown in figure 3 (Blank et al., 1998a), choosing the appropriate conditions allows to obtain definite phases.

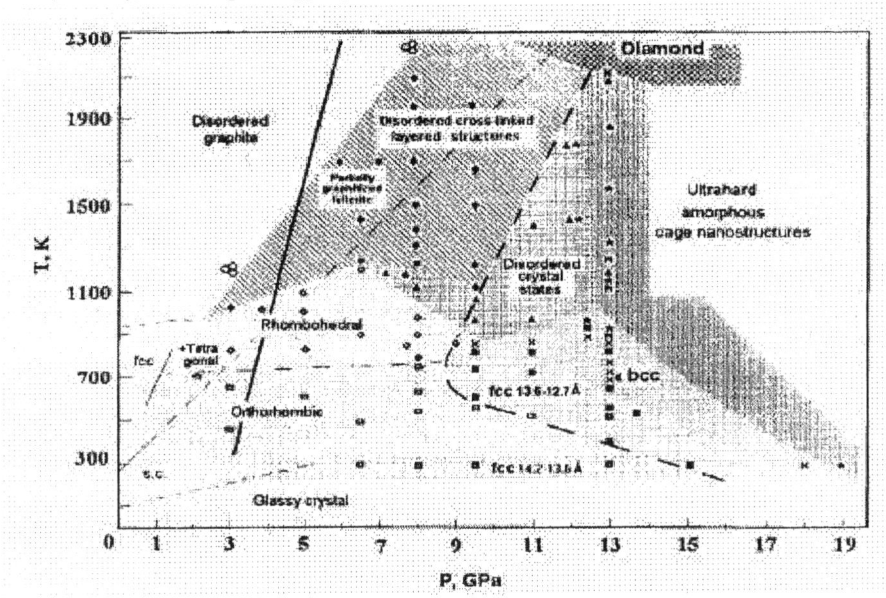

Figure 3. T,P phase diagram of C_{60} fullerene

By ESR measurements it was proved that with increasing pressure the electrical conductivity of 3-D polymeric fullerenes becomes more and more metallic, which is explained by an increasing overlap of molecular orbitals due to narrowing of the spheres.

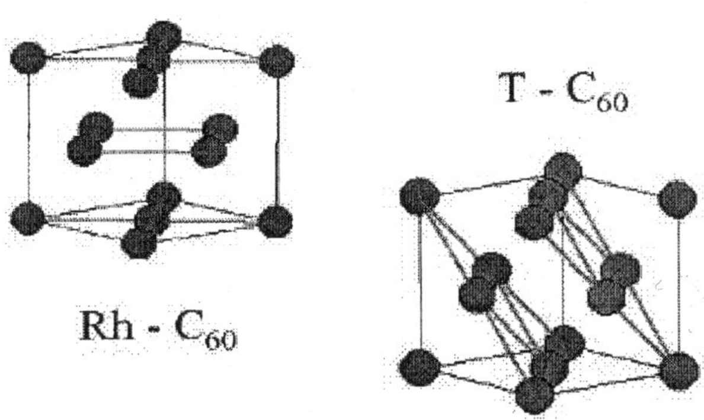

Figure 4. Rhombohedral and Tetragonal arrangement in 2-D C_{60} polymers

3-D polymers are known to form superhard phases, being harder than even diamond (Blank et al., 1998b). Therefore this class of substances is very attractive for being used as a new generation of superhard materials.

2-D polymeric phases can exist in a rhombohedral and in a tetragonal form, as shown in figure 4.

1-D and tetragonal phases are semiconductors, like pristine C_{60}, however the rhombohedral phase exhibits a phase transition to metallic behaviour at elevated preparation temperatures. The metallic conductivity can be observed only in the polymerisation plane, across the planes the resistivity was found to be 100.000 times higher. Recently, a very astonishing finding was made (Makarova et al., 2001): Whereas pristine C_{60} is diamagnetic – neglected the influence of oxygen impurities, which can add a paramagnetic contribution, especially in powder samples – and polymerised phases are all paramagnetic (figure 5), the 2-D rhombohedral phase exhibits a magnetically ordered state, showing features typical for ferromagnetic behaviour, as saturation magnetisation, and a hysteresis loop. The temperature dependence of the saturation and remanent magnetisation shows a Curie point near 500 K. This is the first observation of ferromagnetic behaviour in a pure carbon material.

2.2.2. *Intercalation*

Alkali metals, iodine and interhalogen compounds can be intercalated into the fullerene lattice (figure 6). The alkali metal compounds were found to be superconductors with transition temperatures up to 40 K (Cs_3C_{60}, b.c.c.).

Attempts are made to enhance the transition temperatures by co-insertion of voluminous molecules $((NH_3)_6Na_3C_{60}$, $(NH_3)_8Na_2C_{60}$, ...).

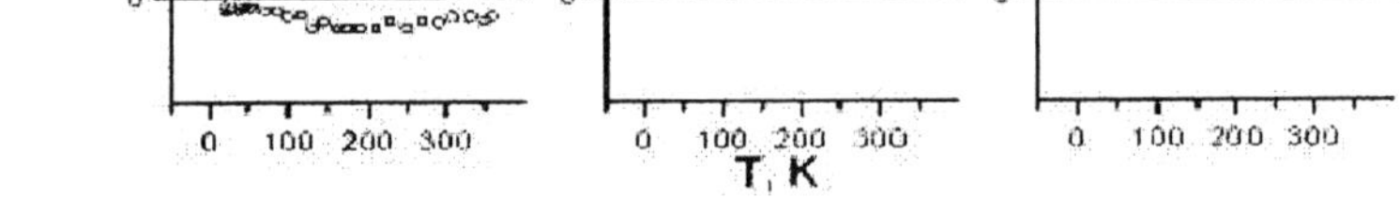

a - pristine C_{60} is diamagnetic with the superimposed Curie term due to paramagnetic contribution from intercalated oxygen.

b - one-dimensional polymer becomes paramagnetic.

c - paramagnetic properties are increased in two-dimensional phase

Figure 5. Magnetic behaviour of C_{60} and 1-D/2-D polymeric phases

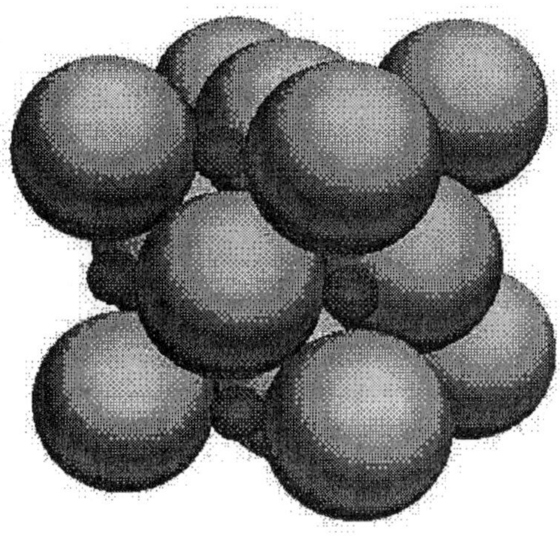

Figure 6. K_3C_{60}

Iodine does not interact strongly with the C_{60} spheres as can be concluded from Raman spectra, which are composed from the fundamental C_{60} vibrations and the characteristic I_2 mode at 200 cm^{-1}. In case of the insertion of IBr or ICl, evidence is found for a partly decomposition of the interha-logen compounds to give I_2 and Br_2 or Cl_2 respectively. Br_2 and Cl_2 add to the fullerene spheres whereas the I_2 molecules are present in the interstitial sites along with the intact interhalogen molecules (Scharff et al., 2001). In these compounds no superconductivity has been found so far.However, upon washing with ethanol, the product, resulting from the conversion of C_{60} with IBr, turned out to transform into C_{120}, a C_{60} dimer, in which the two C_{60} spheres are interconnected by a cyclobutane ring. The reaction mechanism is proposed as follows:

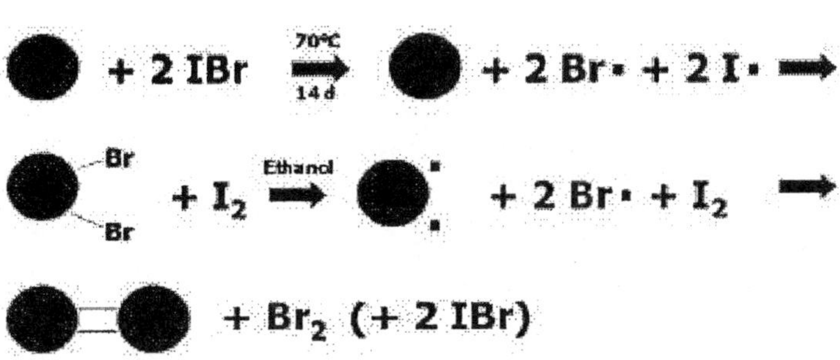

Figure 7. Reaction mechanism for dimerisation of C_{60} via conversion with IBr and ethanol

Recently it was found that hole doping of pure C_{60} in a FET arrangement leads to a superconducting phase, exhibiting a transition temperature of 52 K (Schön et al., 2000). By expanding the C_{60} lattice via the insertion of molecules such as chloroform or bromoform, the transition temperature can be raised up to 117 K.

2.2.3. *Heterofullerenes*

In heterofullerenes at least one carbon atom is substituted by a heteroatom such as nitrogen, sulphur, or boron. As a desired consequence, a doping effect should occur if the heteroatom does not have the same number of valence electrons as carbon. Recently $C_{59}N^+$ has been sythesized (Hummelen et al., 1995).
However in this compound a doping effect cannot be found, as it is only stable in the dimeric form $(C_{59}N)_2$.

2.2.4. *Applications of the parent fullerenes*

Parent fullerenes have been proposed for application as pharmaceuticals, for example as anti-Aids agent, or as radical catcher in cancer prophylaxis. As C_{60} can transform triplett O_2 to singulett O_2 under irradiation with light, it can be used for the treatment of tumors. For this purpose a water based fullerene solution is needed. Such solution can be obtained by ultrasonic treatment of a toluene/water/fullerene mixture, whereby water-soluble aggregates, consisting of 13 C_{60} molecules in form of primary particles, and water, are formed (Andrievsky et al., 1999).
The catalytical activity of C_{60} has been widely studied. For this application, the purity of the fullerene material may be low, so that a costly purification is not needed, what might be important for a large-scale technical process.

3. Nanotubes

The nanotubes were discovered in 1991 by Iijima as a by-product of the fullerene synthesis (Iijima, 1991). There are two types of nanotubes: single walled nanotubes and multi walled nanotubes (figure 8).

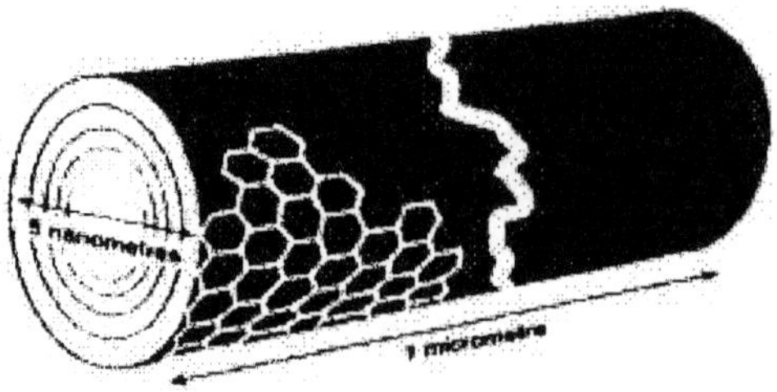

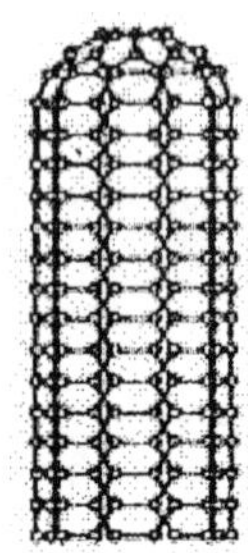

Figure 8. Carbon nanotubes

Multi walled tubes may consist of one rolled up graphene sheet or of a certain number of single walled tubes with growing diameters being arranged in concentric manner (like 'Russian doll'). Nanotubes are not soluble. For this reason, the separation and purification of nanotubes is still a problem, especially with respect to the production of big quantities. However nanotubes offer a variety of unique properties and thus are looked upon promising candidates in a lot of applications. The tensile strength of carbon nanotubes is higher than that of any other material. As a consequence it would be possible to construct a nanotubes based rope, which could be as long as would be necessary to connect a satellite in the orbit to the earth´s surface, a fact which stimulated people to have the idea of a "space-lift", illustrated on the cover of American Scientist July-August 1997 (figure 9).

Of course also other - more realistic - applications using the extremely high mechanical stability of carbon nanotubes, as for example amplifiers in carbon-carbon composites, are of great technological interest.

Another "hot topic", connected with carbon nanotubes, is the storage of Li and H_2. The question, whether carbon nanotubes offer advantages in Li-battery applications compared with the conventional graphitic materials, is still debated. For the introduction of H_2 either low temperature and high pressure, or electrolytic decomposition of water, with nanotubes used as the cathode, is needed. All storage applications depend on a previous opening of the tube caps, which can be achieved by selective oxidation (figure 10). In the case of electrochemical tube charging, the cap opening can be performed in situ.

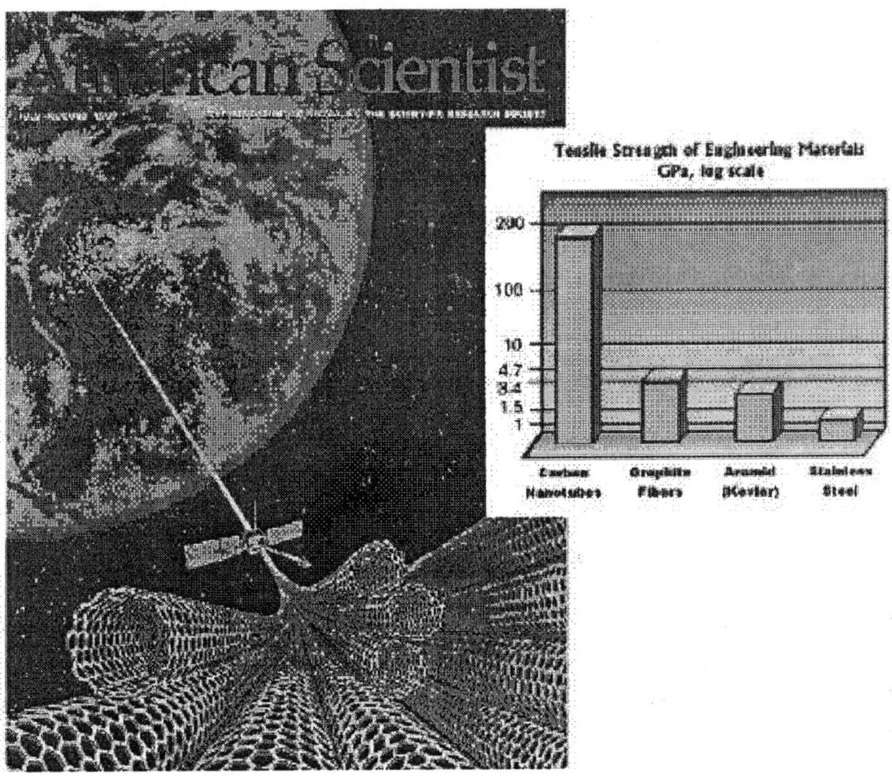

Figure 9. Extremely high tensile strength of carbon nanotubes

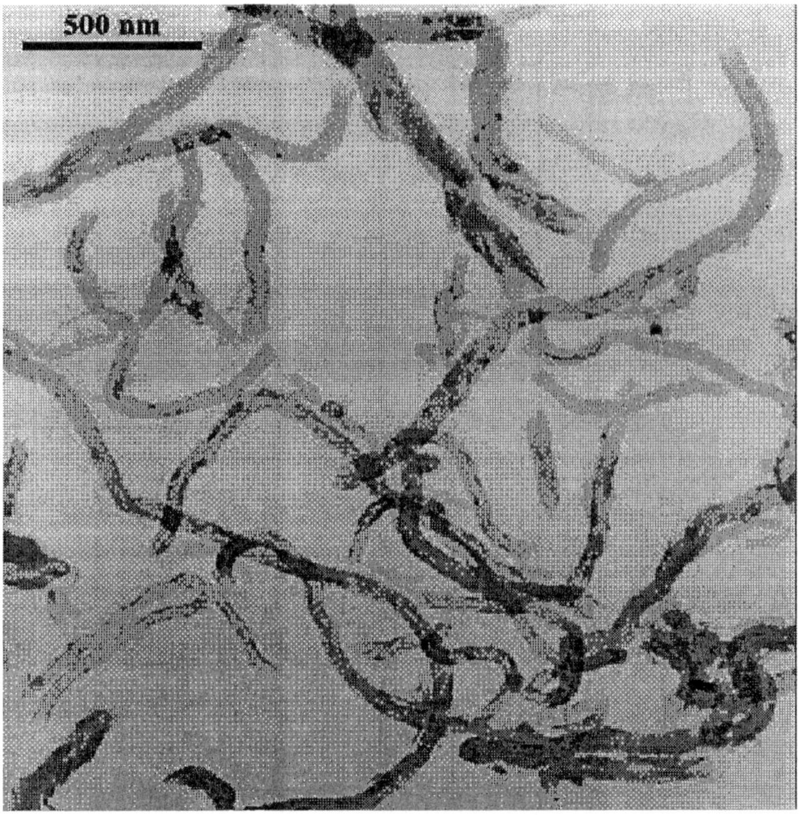

Figure 10. Opened multi walled carbon nanotubes

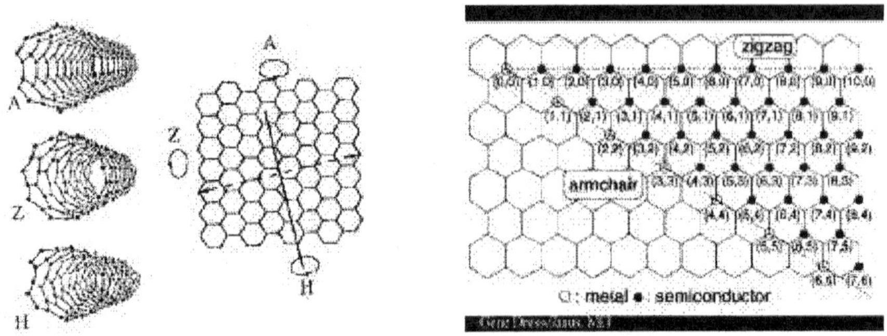

Figure 11. Electrical conductivity depending on chirality of nanotubes

A lot of proposed applications for carbon nanotubes are based on their special electronic properties. Depending on their chirality and diameter, carbon nanotubes can have metallic properties or can be semiconductors. Possible applications are the use as nanowires, for example on microchips (for further miniaturisation metals lack cohesion, application of nanotubes can overcome this problem), the use as field emitters for flat screens (prototype in DIN A4 format already introduced by SAMSUNG company, introduction to the market scheduled for 2003), as electronic modules (for example, bending of a tube by introduction of a heptagon-pentagon pair creates a Schottky barrier (figure 12)), artificial muscles (if negative potentials are applied to the tube, antbonding states are occupied, which is connected with prolonging of the tube and vice versa), and as materials for electrostatic shielding (NASA balloon).

Figure 12. Bending of a nanotube by insertion of a pentagon-heptagon pair

Nanotubes are obtained as cathode deposit in the arc vaporisation process of graphite (especially in the presence of transition metal catalysts), by laser evaporation of graphite, or by catalytic pyrolysis of hydrocarbons.

As another route to carbon nanotubes the template carbonization method was established (Kyotani et al., 1996). This method uses the inside of the channels in an aluminium plate, created by anodic oxidation of aluminium in sulfuric acid, for the pyrolytic deposition of carbon tubes. After the deposition reaction is complete, the anodic aluminium oxide template is removed by washing with HF solution, leaving monodisperse carbon nanotubes with uniform length, diameter and thickness. However, due to the low temperature carbonisation process, these tubes contain a comparatively large number of structural imperfections.

5. Carbolite

A new carbon phase has been obtained by quenching a high temperature carbon plasma on a copper substrate kept at room temperature (Tanuma et al., 1995). Due to the low specific weight of the material it has been named 'Carbolite'. Carbolite consists of quasi one dimensional carbon chains oriented along the c-axis of the hexagonal crystal lattice. The material can act as host lattice for the intercalation of alkali metals or halogens. The resulting intercalation compounds are expected to be superconductors. Carbolite proved to be a metastable phase under ambient conditions, undergoing an irreversible transformation to amorphous carbon when heated above 350 K.

6. Nanocarbons

Nanocarbons are originated besides fullerenes in the arc evaporization process of graphite. As further preparation routes laser ablation of fullerite material as well as molecular and ion beam techniques have been established. As typical examples for nanocarbons, nanotubes with helical winding of the graphene sheets, bucky onions, graphitic polyhedra, schwarzite, hollow diamonds and tetrahedrally bonded amorphous carbon are considered. Applications exist in the field of catalysis, as selective sorbents, and as optical coatings.

References

2 Andrievsky, G. V., Klochkov, V. K. Karyakina, E. L., Mchedlov-Petrossyan, N. O. (1999), *Chem. Phys. Lett.* **300**, 392

3 Beckhaus, H. D., Rüchardt, C., Kao, M., Diederich, F., Foote, C. S. (1992), *Angew. Chem.* **104**, 69

4 Blank, V. D., Buga, S. G., Serebryanaya, N. R., Dubitsky, G. A., Mavrin, B. N., Popov, M. Yu, Bagramov, R. H., Prokhorov, V. M., Sulyanov, S. N., Kulnitskiy, B. A., Tatyanin, Ye. V. (1998a), *Carbon* **36, 5-6**, 665

5 Blank, V. D., Popov, M., Pivovarov, G., Lvova, N., Gogolinsky, K., Reshetov, V. (1998b), *Diamond and Related Materials* **7**, 427

6 Greenwood, NN., Earnshaw, A. (1988), *Chemie der Elemente*, VCH, Weinheim Hollemann, A.F., Wiberg, E. (1985), *Lehrbuch der Anorganischen Chemie*, Walter de Gruyter, Berlin

7 Hummelen, J. C., Knight, B., Pavlovich, J., Gonzales, R., Wudl, F. (1995), *Science* **269**, 1554

8 Iijima, S. (1991), *Nature* **354**, 56

9 Inagaki, M. (2000), *New Carbons – Control of Structure and Functions*, Elsevier, New York

10 Kyotani, T., Tsai, L. F., Tomita, A. (1996) *Chemistry of Materials* **8**, 2109

11 Makarova, T. L., Sundquist, B., Höhne, R., Esquinazi, P., Kopelevich, Y., Scharff, P., Davydov, V. A., Kashevarova, L. S., Rakhmanina, A. V. (2001), *Nature* (in press)

12 Pietzak, B. Waiblinger, M., Almeida Murphy, T., Weidinger, A., Höhne, M., Dietl, E., Hirsch, A. (1997), in *Proc. ICAM'97, Symposium A, Strasbourg* Scharff, P., Carta-Abelmann, L. (2001), *Molec. Phys. Rep.* (in press)

13 Schön, J. H., Kloc, Ch., Batlogg, B. (2000), *Nature* **408**, 549

14 Tanuma, S., Palnichenko, A. V. (1995), *J. Mater. Res.* **10**, 1120

FUNDAMENTAL PROPERTIES OF DNA: SOME LESSONS FROM STUDIES ON THE MOLECULAR BASIS OF DRUG BINDING

MICHAEL WARING
University of Cambridge Department of Pharmacology, Tennis Court Road, Cambridge CB2 1QJ, England

Studies of drug binding to DNA have revealed much about the properties of the macromolecule, from effects on its supercoiling to kinetic factors involved in nucleotide sequence recognition and determinants of base pair discrimination. A spin-off has been the realisation that particular base pair substituents such as the 2-amino group of guanine strongly affect the disposition of ligand binding sites, the width of the helical minor groove, the flexibility of the polymer, its interaction with proteins, and the winding of DNA round the histone octamer in nucleosome core particles.

Almost 50 years ago when the structure of DNA was discovered it was known that various small molecules, especially aminoacridines, could interact with DNA which made them useful as histological stains or mutagens (and also for treating some microbial infections) but no-one knew how they worked, still less that they might one day become major 'leads' for drug design. The first scientific move in the right direction was the intercalation hypothesis advanced by Leonard Lerman. It was not universally acclaimed; indeed it was greeted with profound scepticism in certain quarters. Experiments were devised to verify or disprove it, most notably the circular DNA unwinding test (Figure 1), which confirmed drugs like ethidium as intercalators and others like netropsin or distamycin as something else -- minor groove binders as we now know. The test also unearthed certain DNA binders that unwind the double helix somewhat but do not intercalate: steroidal diamines and triphenylmethane dyes. They might reward further investigation.

A legacy of this early work is the detailed understanding of higher-order structure, especially circularity, of DNA which studies on drug interactions have helped to elucidate. Ethidium is still widely used to isolate plasmids.

Attention turned to the nature of drug binding sites, and to the possibility of sequence preferences. The first laborious experiments required determination of 'proper' binding isotherms, Scatchard plots, association constants, and the whole paraphernalia of physical chemistry. But it did become clear that various ligands could discriminate between potential binding sites and indeed could migrate between them to locate the best. Elegant kinetic studies suggested that this could occur without full dissociation of the drug from the nucleic acid.

E. Buzaneva and P. Scharff (eds.), Frontiers of Multifunctional Nanosystems, 225–228.

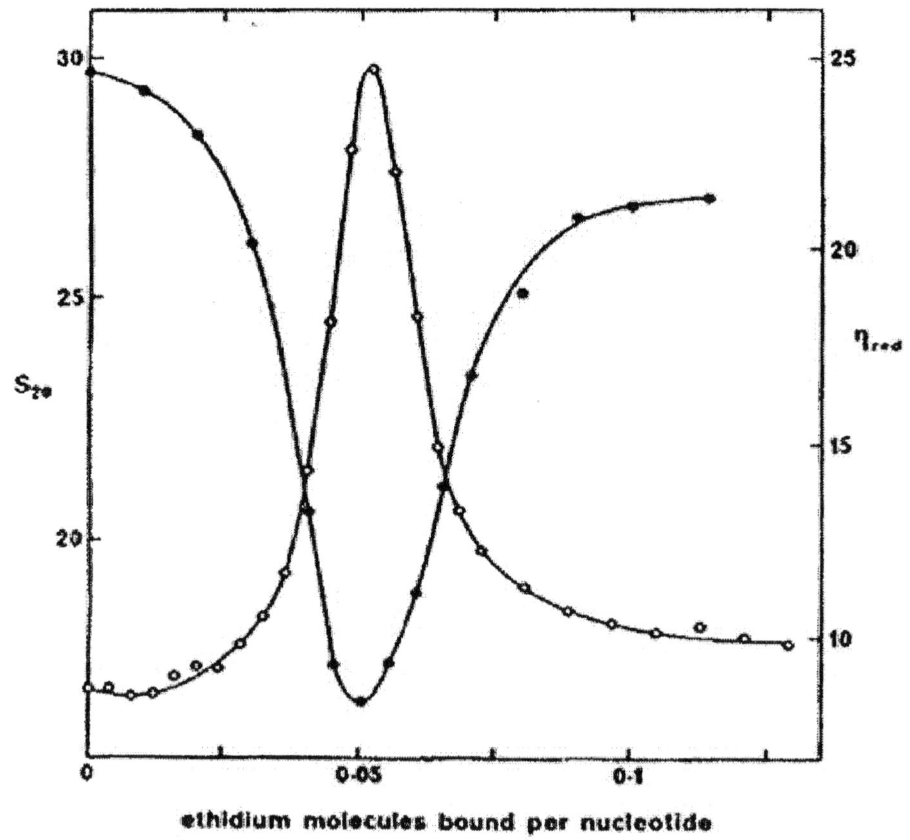

Figure 1. The circular DNA unwinding test illustrated by the effects of ethidium on the hydrodynamic behaviour of bacteriophage PM2 DNA. Intercalation of the drug causes local unwinding of the double helix, resulting in progressive removal and then reversal of the supercoiling of the circular duplex DNA molecules. Filled symbols show the effect on the sedimentation coefficient (left-hand ordinate); open symbols record the changes in reduced viscosity measured with a simple capillary viscometer (right-hand ordinate). The observed changes in supercoiling result from the accumulated unwinding of the double helix associated with binding of the drug. Details described in M.J. Waring and S.M. Henley, *Nucleic Acids Research* **2**, 567-586 (1975).

The next logical step was to seek molecules that could recognise longer, perhaps biologically meaningful base pair sequences. The prospects for success were enhanced with the discovery of bis-intercalation, originally by the antibiotic echinomycin, but synthetic bifunctional (and even polyfunctional) molecules soon followed.

A quantum leap occurred in the early 1980s with the invention of footprinting (Figure 2), which enabled exact identification of preferred binding sites and eventually quantitation to yield passable binding constants. Now a true thermodynamic comparison of ligand affinity and capacity to discriminate between different sites could be gained.

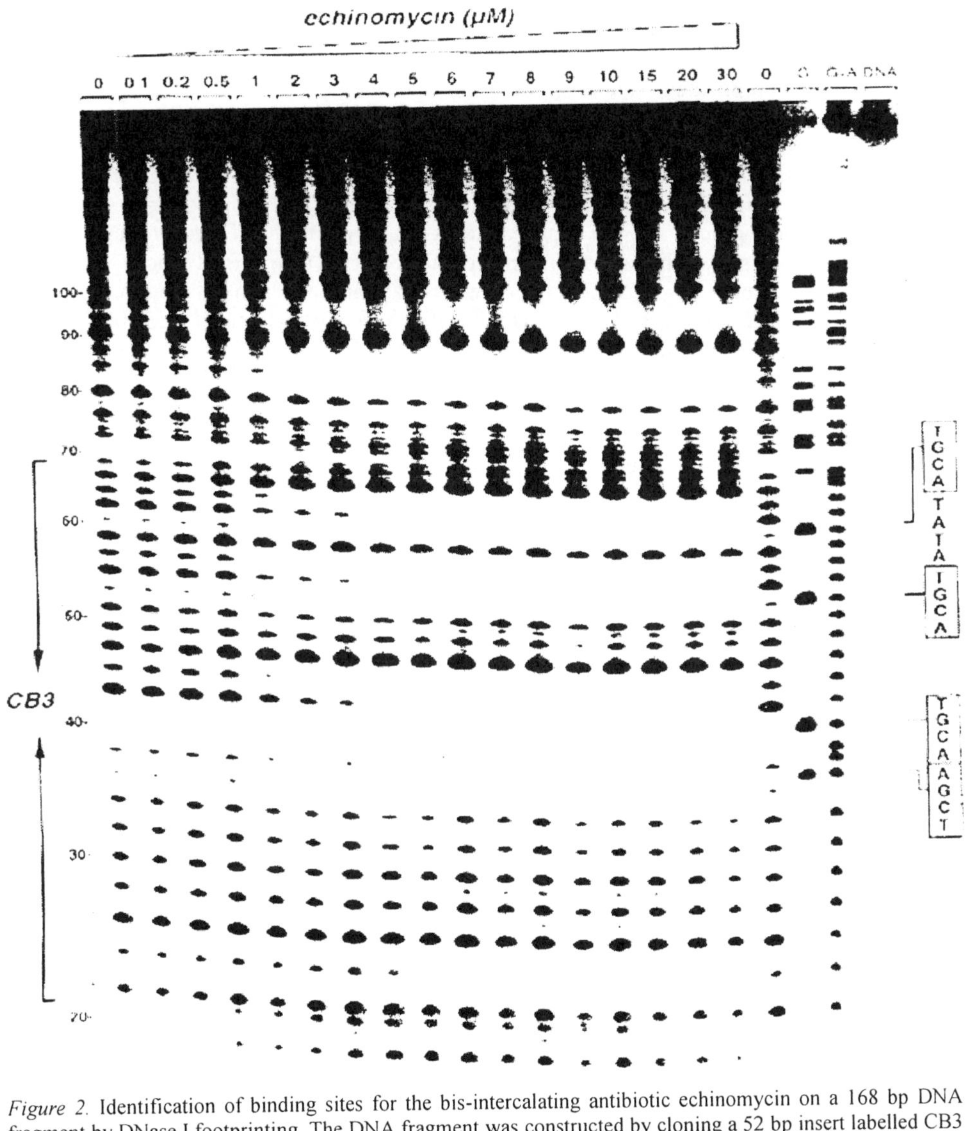

Figure 2. Identification of binding sites for the bis-intercalating antibiotic echinomycin on a 168 bp DNA fragment by DNase I footprinting. The DNA fragment was constructed by cloning a 52 bp insert labelled CB3 into pUC12; it contained four suspected strong echinomycin binding sites separated by chosen numbers of AT base pairs. Each site contained a central CpG dinucleotide step, and as the concentration of echinomycin was increased (see scale at top) footprints can be seen developing at each of the four sites around nucleotide positions 38, 42, 54 and 62. The latter two sites are well resolved, whereas the sites around positions 38 and 42 are fused into a single composite footprint extending from approximately position 35 to 46. Other binding sites are also visible in the vector around positions 85 and 95. For experimental details see C. Bailly, F. Hamy and M.J. Waring, *Biochemistry* **35**, 1150-1161 (1996).

Meanwhile, the application of X-ray diffraction and high-resolution NMR to drug-DNA complexes began to offer unparalleled insights into mechanisms of nucleotide sequence recognition, pointing to the 2-amino group of guanine as a critical substituent for discriminating GC from AT base pairs.

The prime route employed by pharmacologists to gain information about molecular recognition is through the study of structure-activity rela-tions. That is only possible where a diverse family of homologous variant molecules is available -- often not the case with complicated natural pro-ducts. Accordingly, in my laboratory we have adopted an unorthodox approach by deliberately making alterations to the *receptor* (DNA) and looking to see what are the consequences for ligand binding. Thereby we expect to identify functional groups on the base pairs that are critical for the recognition process. The approach works well for studying the role of the 2-amino group of guanine in the minor groove of the double helix. In-corporation of inosine and/or 2,6-diaminopurine shows that a 2-amino group on the purine ring acts as a positive effector to determine binding sites for GC-selective antibiotics such as actinomycin and echinomycin. By contrast, it serves as a negative effector that impedes interaction with minor groove binders like distamycin. Use of a similar approach to determine functional recognition sites in the major groove has been much less successful, though influences of the pyrimidine 5-methyl group and the purine N7 have been detected, for example on ditercalinium binding.

Fortuitously, the studies with substituted DNA species have thrown light on how base pair substituents affect DNA structure quite apart from drug binding.

Interfering with the purine 2-amino group seems to have a marked effect on the width of the minor groove. Shifting the purine 2-amino group and/or the pyrimidine 5-methyl group clearly affects the flexibility of the molecule, reflected in its response to helix-bending proteins and its winding around the histone octamer in nucleosome reconstitution experiments. Consistently we find that inosine substitution seems to produce a more bendable molecule whereas diaminopurine substitution makes the DNA behave more rigidly. As regards the rotational positioning of DNA on the nucleosome, strange effects are produced by shifting any of the exocyclic groups on the base pairs. An important lesson for drug design must be that 'analogue' readout of sequence information as opposed to 'digital' readout of base pair sequences is a real possibility. Moreover, the interaction of double-helical DNA with other large molecular species, be they proteins or carbon nanotubes, is likely to be modulated by its capacity to bend in a nucleotide sequence-dependent fashion.

References

1. Gale, E.F., Cundliffe, E., Reynolds, P.E., Richmond, M.H. and Waring, M.J. (1981) *The Molecular Basis of Antibiotic Action*, 2nd ed., pp 258-401, Wiley, London.
2. Calladine, C.R. and Drew, H.R. (1997) *Understanding DNA*, 2nd ed., Academic Press, London.
3. Neidle, S., ed. (1999) *Oxford Handbook of Nucleic Acid Structure*, Oxford University Press, London.
4. Chaires, J.B. and Waring, M.J., eds. (2001) *Methods in Enzymology* **340**: *Drug-Nucleic Acid Interactions*, Academic Press, New York.

DNA MODIFICATIONS BY NOVEL ANTITUMOR PLATINUM DRUGS

VIKTOR BRABEC
Institute of Biophysics, Academy of Sciences of the Czech Republic, CZ-61265 Brno, Czech Republic

Abstract

The development of metal-based antitumor drugs has been stimulated by the clinical success of *cis*-diamminedichloroplatinum(II) (cisplatin) and its analogues on one hand and by the clinical trials of other platinum complexes with activity against resistant tumors and reduced toxicity including orally available platinum drugs. Broadening the spectrum of antitumor drugs depends on understanding existing agents with a view toward developing new modes of attack. It is therefore of great interest to understand details of molecular and biochemical mechanisms underlying the biological efficacy of the platinum and other transition metal compounds. There is a large body of experimental evidence that the success of platinum complexes in killing tumor cells results from their ability to form on DNA various types of covalent adducts so that the research of DNA interactions of metal-based antitumor drugs has predominated. The results so far obtained support the view that platinum compounds which bind to DNA in a fundamentally different manner to that of "classical" cisplatin will have altered pharmacological properties. This concept has already led to the synthesis of several new unconventional platinum antitumor compounds that violate the original structure-activity relationships.

1. Introduction

More than thirty years have elapsed since accidental discovery of anticancer activity of cisplatin (*cis*-diamminedichloroplatinum(II)) (Rosenberg *et al.*, 1969). Cisplatin is the first purely inorganic antitumor drug introduced in the clinic. It is a very simple molecule consisting of only 11 atoms, six of them are hydrogen's (Fig. 1) and in spite of its simplicity belongs to the most potent drugs available for anticancer chemotherapy (Wong and Giandomenico, 1999). Cisplatin has been successfully applied in the treatment of a variety of human cancers, especially such as germ cell tumors, lung cancer, head and neck cancer, ovarian cancer and bladder cancer (Giaccone, 2000; O'Dwyer *et al.*, 1999). Despite its success, cisplatin has several disadvantages. Its applicability is still limited to a relatively narrow range of tumors - some tumors have natural resistance to cisplatin, while others develop resistance after the initial treatment. Cisplatin also has limited solubility in aqueous solution and is administered intravenously. There are also significant problems in terms

E. Buzaneva and P. Scharff (eds.), Frontiers of Multifunctional Nanosystems, 229–250.

of inducing severe side-effects (especially kidney damage and vomiting/nausea). The drawbacks coupled with cisplatin toxicity have been the impetus for the development of improved platinum drugs. Broadening the chemotherapeutic arsenal depends on understanding existing agents with a view toward developing new modes of attack. It is therefore of great interest to understand details of molecular and biochemical mechanisms underlying the biological efficacy of the platinum compounds.

The first step in the studies of molecular mechanisms of biological activity of chemotherapeutics is to determine their primary target in cells. The pharmacological target of platinum cytostatics may be defined as a site within the cell which is altered by the drug and whose modification leads to cell death. There is a large body of experimental evidence that DNA is the critical target for the cytostatic activity of platinum compounds (Jamieson and Lippard, 1999; Johnson *et al.*, 1989) and that the success of platinum complexes in killing tumor cells results from their ability to form on DNA various types of covalent adducts. There is also some evidence suggesting that other sites than DNA within the cell may be important in the mechanism of antitumor efficiency of platinum compounds. Nevertheless, it is generally accepted that the antitumor properties of platinum-containing drugs are attributable in large measure to their coordination to DNA so that the research of DNA interactions of platinum drugs has predominated.

2. DNA Interactions of "Classical" Antitumor Cisplatin and its Clinically Ineffective *trans* Isomer

2.1. *DNA ADDUCTS*

Cisplatin reacts with DNA in the cell nucleus, where the concentration of chlorides is markedly lower than in extracellular fluids. The drug loses its chloride ligands in media containing low concentrations of chloride to form positively charged monoaqua and diaqua species. It has been shown (Bancroft *et al.*, 1990) that only these aquated forms bind to DNA.

Cisplatin forms on DNA first monofunctional adducts preferentially at the guanine residues which subsequently close to intrastrand cross-links (CLs) between neighboring guanine residues or between adenine and guanine residues (~90 %) (1,2-GG or 1,2-AG intrastrand CLs) (Eastman, 1987; Fichtinger-Schepman *et al.*, 1985). Other minor adducts are intrastrand CLs between two purine nucleotides separated by one (1,3-GXG intrastrand CLs, X = A, C, T) or more nucleotides and few adducts remain monofunctional. Cisplatin also forms interstrand CLs (~6% in linear DNA) preferentially between guanine residues in the 5'-GC/5'-GC sequence (Brabec, 2000). Interestingly, the frequency of interstrand CLs is noticeably higher in negatively supercoiled DNA (Vrana *et al.*, 1996). In all adducts, cisplatin is coordinated to the N7 atom of purine residues.

Structure-activity studies often employ inactive compounds (Fig. 1) such as *trans*-diamminedichloroplatinum(II) (*trans* isomer of cisplatin, transplatin) and

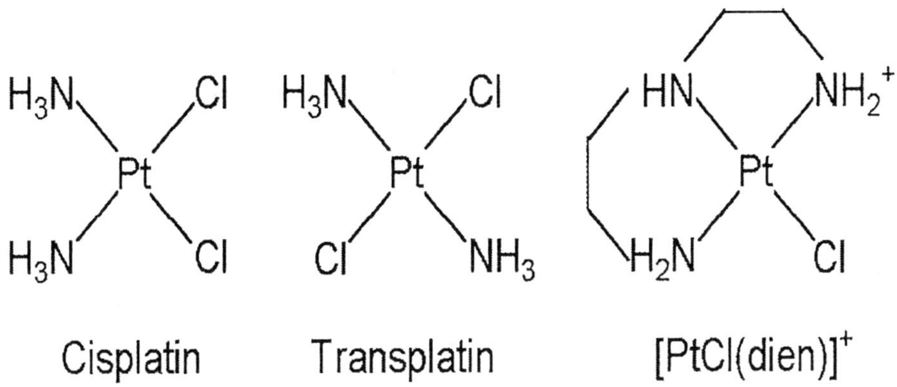

Figure. 1. Structures of "classical" platinum(II) compounds.

chlorodiethylenetriamineplatinum(II) chloride ([PtCl(dien)]Cl, the monodentate complex, used to afford DNA monofunctional adducts to simulate the first step of the bifunctional reaction of cisplatin or transplatin) (Fig. 1) (Brabec *et al.*, 1990). These compounds have been widely used to investigate the mechanism of action of cisplatin. In this approach, one searches for differences between active and inactive compounds that may be responsible for the pharmacological effect. Hence, the clinical inactivity of the *trans* isomer of cisplatin, is considered a paradigm for the classical structure–activity relationships of platinum drugs.

Transplatin-DNA adducts are mainly interstrand CLs (~12% in linear DNA) preferentially formed between guanine and complementary cytosine residues and a relatively large portion (~50%) of adducts remains monofunctional (Brabec and Leng, 1993; Eastman *et al.*, 1988). Intrastrand CLs between non-adjacent guanine residues or between guanine and either adenine or cytosine residues have been also deduced from the results of the analyses of isolated DNA incubated with transplatin at a relatively high extent (1 or more Pt atoms per 100 nucleotides). However, it has been suggested (Boudvillain *et al.*, 1995; Leng *et al.*, 2000) that in cells transplatin forms only a small amount of interstrand CLs because of the slow closure of the monofunctional adducts coupled to their trapping by intracellular sulfur nucleophiles.

The adducts formed between double-helical DNA and [PtCl(dien)]Cl comprise mainly monofunctional adducts at N7 atoms of guanine residues (Johnson *et al.*, 1982).

Because the percentage of the 1,2-intrastrand adducts formed by cisplatin is larger than statistically expected, this CL has generally been assumed to be the important adduct for anticancer activity. Further support for this conclusion came from transplatin's inability to form this type of intrastrand CLs because of geometric constraints and its low antitumor activity. Both isomers can form interstrand CLs, although of distinctly different nature (Brabec, 2000), and there is still support for interstrand CLs as the key lesion.

2.2. *CONFORMATIONAL ALTERATIONS INDUCED IN DNA BY THE ADDUCTS*

Much work has been also performed to reveal conformational distortions induced by the various DNA adducts of cisplatin, transplatin and [PtCl(dien)]Cl. Most of the structural information currently available pertains to the bidentate adducts. The survey of the conformational changes induced by the various adducts of cisplatin, transplatin and [PtCl(dien)]Cl described below reveal their predisposition to affect physical properties of DNA in distinctly different ways.

2.2.1. *Monofunctional Adducts*

Monofunctional adducts have been studied less thoroughly, although a knowledge of alterations induced in DNA by monofunctional binding of platinum complexes may also be important for better understanding how antitumor cisplatin forms the bifunctional genotoxic lesions. Before the more systematic study of monofunctional DNA adducts of [PtCl(dien)]Cl or cisplatin was started it was almost axiomatic that the monofunctional DNA adducts including those of [PtCl(dien)]Cl or cisplatin do not affect DNA conformation. It has been, however, shown that the monofunctional adducts of [PtCl(dien)]Cl and cisplatin at the guanine sites distort DNA and reduce its thermal stability in a sequence-dependent manner (Brabec *et al.*, 1994; Brabec *et al.*, 1992; Van Garderen *et al.*, 1989). The most pronounced effects are observed if the platinated guanine residue is flanked by single pyrimidines on both 3' and 5' sides. Also importantly, the conformational distortion was always more pronounced in the flanking base pairs containing the base on the 5' site of the monofunctional adduct. These distortions disturb stacking interactions in double-helical DNA and make bases around the adduct more accessible. In addition, the monofunctional adduct also unwinds the duplex (Keck and Lippard, 1992) - [PtCl(dien)]Cl-DNA structure exhibits an unwinding angle of 6°. On the other hand, no intrinsic bending is induced in DNA by monofunctional adducts of [PtCl(dien)]Cl (Brabec *et al.*, 1992).

2.2.2. *Cisplatin Cross-links*

Numerous studies have revealed marked conformational alterations induced in DNA by the major intrastrand CLs formed by cisplatin between adjacent purines [at d(GG) or 5'-d(AG)] (Gelasco and Lippard, 1999; Jamieson and Lippard, 1999). These adducts induce a roll of 26°-50° between the platinated purine residues, displacement of platinum atom from the planes of the purine rings, a directional, rigid bend of the helix axis towards the major groove, a local unwinding of 9-11°, severe perturbation of hydrogen-bonding within the 5'-coordinated GC base pair, widening and flattening of the minor groove opposite the cisplatin adduct, creation of a hydrophobic notch and global distortion extending over 4-5 base pairs (Fig. 2A). Additional helical parameters characteristic of A-form of DNA have been also reported although modification of B-DNA by cisplatin inhibits its transition to A-form (Balcarova and Brabec, 1986).

The minor 1,3-intrastrand CL formed by cisplatin bends helix axis towards major groove by ~30° and locally unwinds DNA (by ~19°) (Bellon and Lippard, 1990; Teuben *et al.*, 1999). In addition, DNA is locally denatured and flexible at the site of the adduct.

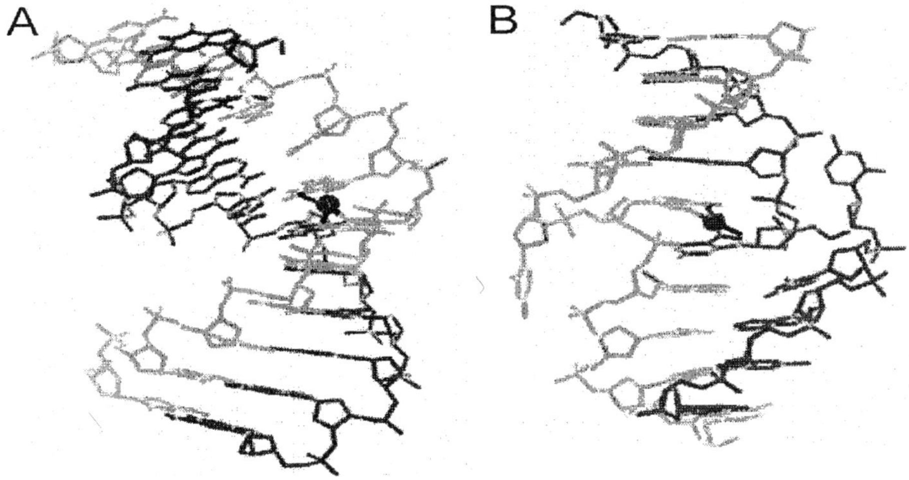

Figure. 2. (A) Crystal structure of a DNA duplex containing an 1,2-GG intrastrand CL of cisplatin. Image of the duplex d(CCTCTG*G*TCTCC).d(GGAGACCAGAGG), where the asterisks represent a cisplatin CL. The two strands are drawn in gray (the adducted strand) and black. (B) NMR structure of a DNA duplex containing an interstrand CL of cisplatin. Image of the duplex d(CATAG*CTATG). d(CATAG*CTATG), where the asterisks represent a cisplatin CL. The two strands are drawn in gray and black.

The important although less frequent interstrand CL, which is preferentially formed by cisplatin between opposite guanine residues in the 5'-GC/5'-GC sequence (Lemaire *et al.*, 1991), also induces several irregularities in DNA (Brabec, 2000; Coste *et al.*, 1999; Huang *et al.*, 1995) (Fig. 2B). The cross-linked guanine residues are not paired with hydrogen bonds to the complementary cytosines, which are located outside the duplex and not stacked with other aromatic rings. All other base residues are paired, but distortion extends over at least 4 base pairs at the site of the CL. In addition, the *cis*-diammineplatinum(II) bridge resides in the minor groove and the double helix is locally reversed to a left-handed, Z-DNA-like form. This adduct induces the helix unwinding by 76°-80° relative to B-DNA and also the bending of 20°-40° of the helix axis at the cross-linked site toward the minor groove. More detailed descriptions of structures of CLs of cisplatin may be found in other recent reviews (Brabec, 2000; Gelasco and Lippard, 1999; Jamieson and Lippard, 1999; Leng *et al.*, 2000).

2.2.3. *Transplatin Cross-links*

Clinically ineffective transplatin also forms in DNA various types of adducts which affect DNA conformation less severely than the CLs of cisplatin. Monofunctional lesions, which are the major lesions, affect DNA conformation very slightly similarly as the adducts of [PtCl(dien)]Cl (see Section 2.2.1.). The conformational alterations induced by the interstrand CLs are, however, much less severe than those induced by the CLs of cisplatin. The duplex is only slightly distorted on both sides of the CL but all bases are still paired and hydrogen bonded. The interstrand CL of transplatin unwinds the double helix by ~12° and induces a slight, flexible bending of ~20° of its axis towards minor groove (Brabec *et al.*, 1993; Malinge and Leng, 1999; Paquet *et al.*, 1999). The 1,3-intrastrand CL of transplatin, which is less likely lesion in cells (Boudvillain *et al.*, 1995), unwinds the double helix by 45°, bends the DNA by 26° and locally denatures imparting a flexibility to the DNA which acts like a hinge joint without producing a directed bend (Anin and Leng, 1990; Bellon and Lippard, 1990).

2.3. *EFFECTS OF THE ADDUCTS ON DNA REPLICATION AND TRANSCRIPTION*

DNA replication and transcription are essential processes in rapidly proliferating tumor cells so that their inhibition should result in cytostatic effects. As various adducts formed on DNA are capable of inhibiting DNA replication or transcription by DNA-dependent DNA- or RNA-polymerases the effects of platinum adducts on these processes have been thoroughly investigated.

2.3.1. *Replication*

Cisplatin inhibits DNA synthesis by prokaryotic and eukaryotic DNA polymerases both *in vitro* and *in vivo* (Comess *et al.*, 1992; Pinto and Lippard, 1985; Villani *et al.*, 1988). Inhibition of DNA synthesis occurs mainly at GG sites consistently with the fact that these sites are preferential binding sites of cisplatin. Transplatin produces on DNA the arrest sites less selectively and replication of platinated DNA is affected by the adducts of transplatin considerably less than by the lesions of cisplatin (Comess *et al.*, 1992; Pinto and Lippard, 1985). The monofunctional adducts of [PtCl(dien)]Cl, cisplatin or transplatin affect replication only negligibly. Importantly, the methodology based on the use of DNA polymerases and DNA templates globally modified by platinum compounds has been used in the "replication" mapping experiments to identify the arrest sites of DNA replication and consequently also preferential DNA binding sites of platinum compounds (Coluccia *et al.*, 1993; Murray *et al.*, 1992a; Murray *et al.*, 1992b; Murray *et al.*, 1997).

It has been demonstrated in the studies of replication of platinated DNA (Comess *et al.*, 1992) that the inhibition of DNA elongation is not complete, the polymerases are able to bypass the adducts. The frequency of replication bypass varies for the different polymerases. The capability of DNA polymerases to bypass the lesions of platinum compounds is consistent with the view that these compounds are mutagenic through such replication bypass. The mutagenic properties of drugs used in the clinic are important factors to consider since mutagenicity may lead to the development of secondary tumors. Interestingly, the 1,2-GG intrastrand CLs of cisplatin are considerably less mutagenic than 1,2-AG CLs in *E. coli* (Burnouf *et al.*, 1987) whereas controversial mutagenicity has been reported for the 1,3-GXG intrastrand CL (Burnouf *et al.*, 1987; Yarema *et al.*, 1995). Importantly, inhibition effects of cisplatin on DNA replication have been also observed in the *in vivo* studies (reviewed in ref. (Villani *et al.*, 1999)). Mutagenesis induced by cisplatin *in vitro* and *in vivo* is similar and shows a correlation with DNA binding sites of this drug. It would be reasonable to assume that inhibition of DNA replication by unrepaired cisplatin adducts is sufficient to explain cytostatic efficiency of cisplatin as it was concluded in several early studies. However, it becomes evident that inhibition of DNA replication cannot fully explain the antitumor efficiency of this drug (Sorenson and Eastman, 1988) and that this mechanism is more complex (Allday *et al.*, 1995; Chu, 1994). Nevertheless, there is no doubt that inhibition of DNA replication is an important part of the mechanism underlying antitumor effects of cisplatin.

2.3.2. *Transcription*

Another essential function of the cellular metabolism affected by lesions formed on DNA by platinum compounds is DNA transcription. Transcription by eukaryotic and prokaryotic RNA polymerases (RNA polymerase II and SP6 or T7 RNA polymerases, respectively) of DNA modified by cisplatin, transplatin and [PtCl(dien)]Cl *in vitro* has been investigated in detail (Brabec and Leng, 1993; Corda *et al.*, 1992; Corda *et al.*, 1991; Cullinane *et al.*, 1999; Lemaire *et al.*, 1991). The RNA polymerases react differently at the various platinum adducts. Bifunctional adducts of cisplatin strongly inhibit transcription of platinated DNA by RNA polymerases. In contrast, intrastrand CLs of transplatin did not provide an absolute block to this process so that RNA polymerase II bypasses transplatin adducts with a much higher efficiency than the adducts of cisplatin (Corda *et al.*, 1993). The RNA polymerases entirely bypass the monofunctional adducts of [PtCl(dien)]Cl (Brabec and Leng, 1993; Corda *et al.*, 1993; Lemaire *et al.*, 1991). It has been suggested (Corda *et al.*, 1993) that the platinum adducts which block the RNA polymerases not only constitute a physical barrier to the progress of the enzymes on the template but also specifically alter the properties of transcription complexes as a consequence of the specific conformational changes that they induce in template DNA. Similarly as in the case of DNA polymerases (see Section 2.3.1.), the phenomenon that RNA synthesis by DNA-dependent RNA polymerases is prematurely terminated by bidentate adducts of platinum compounds has been used in the "transcription" mapping experiments to identify the arrest sites of DNA transcription and consequently also preferential DNA binding sites of platinum compounds (Brabec and Leng, 1993; Lemaire *et al.*, 1991).

The effects of cisplatin and transplatin adducts on mammalian transcription *in vivo* have been also investigated (Jordan and Carmo-Fonseca, 1998; Mello *et al.*, 1995; Mymryk *et al.*, 1995). For instance, a significant decrease in transcription level has been observed when â-galactosidase reporter gene is transfected into human or hamster cells (Mello *et al.*, 1995). Hence, the data on inhibition of transcription by DNA adducts of platinum compounds support the view that this inhibition may also play a role in the mechanism of antitumor activity of platinum compounds.

2.4. *CELLULAR RESISTANCE TO CISPLATIN, REPAIR OF THE ADDUCTS*

2.4.1. *Mechanisms of Resistance*

The drawbacks coupled with cisplatin toxicity are also associated with resistance of tumor cells to this drug, i.e. that some tumors have natural resistance to cisplatin, while others develop resistance after the initial treatment. Resistance to cisplatin is multifactorial and in general it may consist of mechanisms either limiting the formation of DNA adducts and/or operating downstream of the interaction of cisplatin with DNA to promote cell survival. The formation of DNA adducts by cisplatin can be limited by

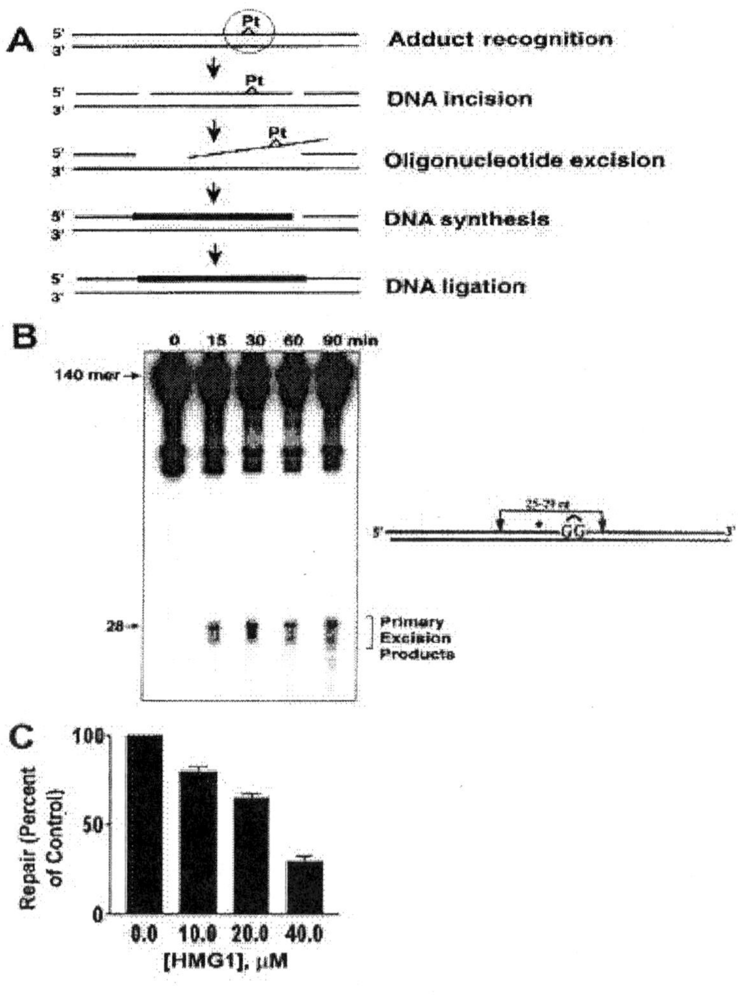

Figure. 3. (A) Scheme illustrating the primary biochemical steps of nucleotide excision repair. (B) Kinetics of removal of single 1,2-GG intrastrand CL of cisplatin centrally located in the linear duplex 140 base pairs long by HeLa excinuclease; after the excision reaction, DNA was resolved in 10% denaturing polyacrylamide gel to separate excision products from substrate DNA and visualized by autoradiography; the bands corresponding to 25-29 nucleotides located around the marker are excision products, as indicated on the scheme at the bottom, which marks the incision sites of the excinuclease. (C) Calf thymus HMGB1 inhibition of excision of the 1,2-GG intrastrand CL of cisplatin; the substrate (see Fig. 3A) was incubated with HMGB1 at the concentrations indicated for 10 min at 30 °C followed by incubation for additional 45 min in the HeLa cell-free extract. The level of repair without any HMGB1 was designated 100%.

reduced accumulation of the drug, enhanced drug efflux and cisplatin inactivation by coordination to glutathione and thiol-containing proteins including metallothioneins whose production may be increased as a consequence of cisplatin treatment. The second group of mechanisms includes enhanced repair of DNA adducts of cisplatin and increased tolerance of the resulting DNA damage.

2.4.2. *Nucleotide Excision Repair*

DNA repair pathways have a particularly intriguing role in the mechanism of biological activity of agents damaging DNA. Early studies (Beck and Brubaker, 1973; Beck *et al.*, 1985; Brouwer *et al.*, 1981; Janovska and Kleinwächter, 1986) indicate that the cells deficient in DNA repair are more sensitive to cisplatin than cells proficient in repair, the effect not observed with transplatin.

In human cells, cisplatin intrastrand adducts are removed from DNA mainly by the nucleotide excision repair (NER) (Fig. 3A), which is one of the many cellular defense mechanisms involved in elimination of the toxic effects of cisplatin. This type of repair includes removal of the damaged base by hydrolyzing phosphodiester bonds on both sides of the lesion. It has been found using human and rodent cell-free extracts or a reconstituted system containing highly purified NER factors that 1,2- and 1,3-intrastrand CLs of cisplatin are efficiently repaired (Zamble *et al.*, 1996) (Fig. 3B). Importantly, this repair of 1,2-, but not 1,3-intrastrand CL is blocked upon addition of an HMG-domain protein (HMG = high mobility group) (Fig. 3C) (see Section 2.5.1.).

An *in vitro* excision repair of a site-specific cisplatin interstrand CL has been also studied using mammalian cell-free extracts containing HMG-domain proteins at the levels not sufficient to block excision repair of the 1,2-intrastrand adducts (Zamble *et al.*, 1996). Repair of the interstrand CL formed by cisplatin has not been detected. Similarly in cell strains derived from patients with Fanconi's anemia NER of cisplatin-interstrand CLs is not observed, although NER can readily occur in these cells (Dijt *et al.*, 1988; Plooy *et al.*, 1985). Fanconi's anemia cells have been described as being extremely sensitive to cross-linking agents so that their noticeably high sensitivity to cisplatin is explained by incapability of these cells to repair cisplatin interstrand CLs (Fujiwara *et al.*, 1977). On the other hand, repair of these lesions has been detected with the aid of a repair synthesis assay, which measures the amount of new DNA synthesized after the damage removal in whole cell extracts (Čalsou *et al.*, 1992). In this way, however, the repair could also result from the mechanism different from that of nucleotide excision. The pathways for the repair of DNA interstrand CLs of cisplatin and other genotoxic agents in mammalian cells are poorly defined. DNA interstrand CLs pose a special challenge to repair enzymes because they involve both strands of DNA and therefore cannot be repaired using the information in the complementary strand for resynthesis.

2.4.3. *Mismatch Repair*

There is also an evidence that another cellular repair mechanisms, such as recombination or mismatch repair (MMR) can affect antitumor efficiency of cisplatin. Recent observations support the view that MMR mediates the cytotoxicity of cisplatin in tumor cells (Aebi *et al.*, 1997; Fink *et al.*, 1996) and that dysfunction of this type of DNA repair may result in the resistance of tumor cells to cisplatin or in the drug tolerance (Johnson *et al.*, 1998). The function of the MMR is to scan newly synthesized

DNA and remove mismatches that result from nucleotide incorporation errors made by the DNA polymerases.

To explain cisplatin tolerance, it is assumed that replication bypass of DNA adducts of cisplatin (see Section 2.3.1.) leads to mutations. During MMR, the strand to be corrected is nicked, a short fragment containing the mismatch is excised, and new DNA fragment is synthesized. The MMR system always replaces the incorrect sequence in the daughter strand, which would leave the cisplatin adduct unrepaired. This activity initiates a futile cycle. During DNA synthesis to replace the excised short fragment, the DNA polymerases would again incorporate mutations followed by attempts to remove them. The repeated nicks in DNA formed at each ineffective cycle of repair could trigger a cell death response. Thus, MMR recognition of cisplatin adducts on DNA may trigger a programmed cell death pathway rendering MMR-proficient cells more sensitive to DNA modification by cisplatin than MMR-deficient cells. In other words, loss of MMR would then increase the cell's ability to tolerate lesions formed on DNA by cisplatin. Also importantly, the MMR is likely to act mainly in the postreplicative phase, when the highest concentration of mispairs is expected in the newly synthesized DNA. This would be in contrast with NER, which probably acts independently of the cell cycle. Thus, rapidly proliferating tumor cells would be more prone to have futile repair attempts capable of triggering cell death. Although the level of resistance to cisplatin, which accompanies loss of MMR, is relatively small, loss of MMR has been shown to be sufficient to account for the failure of treatment with cisplatin in several model systems (Drummond *et al.*, 1996; Ferry *et al.*, 1999; Fink *et al.*, 1997).

2.5. *RECOGNITION OF THE ADDUCTS BY CELLULAR PROTEINS*

An important step in unraveling the mechanism of antitumor activity of cisplatin was also an identification of proteins that specifically bind to DNA modified by cisplatin. The search for these proteins has been initiated by the results suggesting that the DNA adducts formed by cisplatin and transplatin in cells might be differentially processed (Ciccarelli *et al.*, 1985).

To date several classes of these proteins have been identified and mechanisms have been proposed how they mediate antitumor effects of cisplatin. Extensive reviews addressing these questions have been recently published (Jamieson *et al.*, 1999; Zamble and Lippard, 1999; Zlatanova *et al.*, 1998).

2.5.1. *HMG-domain Proteins*

HMGB1 and HMGB2 proteins belong to architectural chromatin proteins that play some kind of structural role in the formation of functional higher order protein-DNA or protein-protein complexes or as signaling molecules in genetically regulated repair pathways. HMG-domain proteins bind selectively to the 1,2-GG or AG adducts of cisplatin, but not to its 1,3-intrastrand CLs and monofunctional adducts or to the adducts of transplatin.

The full-length HMGB1 protein contains two tandem HMG-box domains, A and B, and an acidic C-tail (Thomas and Travers, 2001). The domains A and B in HMGB1 protein (HMGB1a and HMGB1b) are linked by a short lysine-rich sequence (containing

A

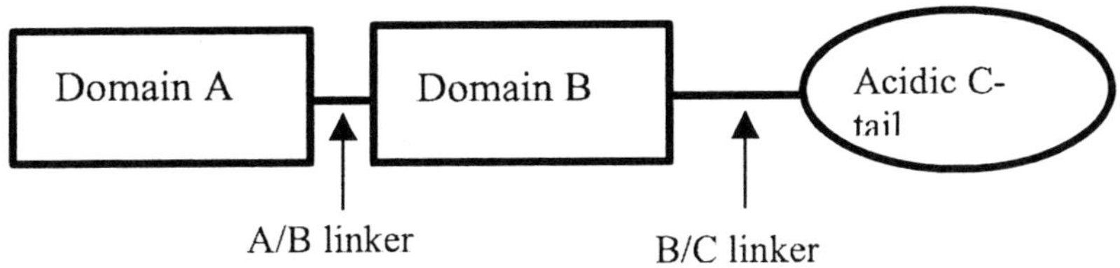

B

Figure. 4. (A) Schematic structure of HMGB1 protein. (B) The amino acid sequence of rat HMGB1 protein (GenBank accession no. Y00463).

7 amino acids, A/B linker region) (Fig. 4). Each "minimal" domain A or B alone specifically recognizes 1,2-GG intrastrand CL of cisplatin (Dunham and Lippard, 1997). The affinity of HMGB1a for DNA containing 1,2-GG intrastrand CL is generally higher than that of HMGB1b and it has been proposed that HMGB1a is the dominating domain in the full-length HMGB1 that binds to the site of the intrastrand CL, while HMGB1b facilitates binding providing additional protein-DNA interactions. Interestingly, the binding affinity of HMGB1a and HMGB1b with the duplex containing the site-specific 1,2-GG intrastrand CL is modulated by the nature of the bases that flank the platinum lesion by more than two orders of magnitude (Cohen *et al.*, 2000b; Dunham and Lippard, 1997). The impact of flanking-base sequences on HMG-domain protein recognition correlates with the modulatory effect of the flanking-bases on CL-induced alterations of duplex stability (Hofr *et al.*, 2001; Pilch *et al.*, 2000).

The HMG domain binds to the minor groove of the DNA double helix opposite the platinum 1,2-intrastrand CL located in the major groove (Ohndorf *et al.*, 1999). Distortions such as prebending, unwinding at the site of platination and preformation of a hydrophobic notch as a consequence of the 1,2-intrastrand CL formation are important for the recognition and affinity of the platinum-damaged DNA-binding proteins.

Moreover, the A/B linker region in the full-length HMGB1 protein attached to the N-terminus of the domain B markedly enhances binding of the B domain to DNA containing site-specific 1,2-GG intrastrand CL (Stros, 2001). It has been also proposed that binding of the A/B linker region within the major groove of DNA helped the two HMGB1 domains to anchor to the minor DNA groove facilitating their DNA binding.

These studies have been extended to DNA interstrand CLs produced by cisplatin or transplatin at the sites where these adducts are formed preferentially (Kasparkova and Brabec, 1995). Mammalian HMGB1 protein binds to the interstrand CL of cisplatin with a similar (slightly lower) affinity as to the 1,2-GG intrastrand CL. On the other hand, no binding of the HMGB1 protein to transplatin-interstrand CL has been noticed. Our most recent analysis has revealed that isolated HMGB1a has no affinity to the duplex interstrand cross-linked by cisplatin whereas the isolated HMG1b containing the A/B-linker region attached to its N-terminus binds with a noticeable affinity although lower than is that of the "minimal" HMGB1a to 1,2-GG intrastrand CL (unpublished results). 1,3-GXG intrastrand CL of cisplatin is not recognized by the HMGB1 in spite of the fact that bending and unwinding induced in DNA by this adduct are rather similar to those induced by 1,2-GG intrastrand adduct. This suggests that there are another factors which control the recognition of platinum adducts by the DNA-binding proteins.

At present the factors, which hinder the binding of these proteins to the platinum adducts capable of bending and unwinding of DNA are unknown. Some data on local conformation of DNA around the individual types of platinum adducts are consistent with the hypothesis that the recognition of DNA bending and/or unwinding induced by these lesions might be obscured, if the formation of the platinum-induced DNA lesion is accompanied by a local denaturation and/or by a lowered rigidity of the duplex around the adduct (Brabec, 2000; Kasparkova and Brabec, 1995).

The mechanism of full-length HMGB1 protein binding to cisplatin-modified DNA has not been completely resolved. The studies so far performed suggest that the binding of HMGB1-domain proteins to DNA modified by cisplatin can be modulated by conformational alterations induced by the CLs, changes in the stability and flexibility of the duplex, base-protein contacts, or some combination of these factors. Further studies are therefore warranted to reveal all details involved in the recognition of DNA adducts of platinum compounds by HMGB1 protein.

There is a variety of proteins other than HMGB1 or 2 that contain HMG boxes, a conserved 80-amino-acid domain found in a variety of eukaryotic DNA-binding proteins, including a number of transcription factors. Many. of these proteins (for instance HMG-D, human and *Drosophila* SSRP1, Cmb1, hUBF, Ixr1, mtTFA, LEF1, SRY, T160, tsHMG) have been shown to recognize and bind to DNA modified by cisplatin and several recent and excellent reviews describing these binding studies are available (Jamieson and Lippard, 1999; Zamble and Lippard, 1999; Zlatanova *et al.*, 1998).

2.5.2. *Proteins Without an HMG Domain*

The TATA-binding protein (TBP) is essential for transcription initiation in eukaryotes. TBP recognizes and binds to the minor groove of a consensus sequence, TATAAA, known as the TATA box (Orphanides *et al.*, 1996). The TBP binds selectively to and is sequestered by cisplatin-damaged DNA (Coin *et al.*, 1998; Vichi *et al.*, 1997). Interestingly, the structure of the TATA box in the human TBP-TATA complex is strikingly similar to the structure of 1,2-GG intrastrand CL. While the structure of the TATA box in the TBP-TATA complex is an induced fit, TBP binds to the 1,2-GG intrastrand CL to a pre-formed, bent DNA in a lock-and-key fashion. In this way, cisplatin turns the GG sequence into a potential site for TBP. In addition, it was also demonstrated recently (Cohen *et al.*, 2000a) that the 1,2-GG intrastrand CLs flanking the TATA box positions markedly enhances the binding affinity of TBP to the TATA box. Importantly, the binding affinity of TBP to 1,3-GXG intrastrand CLs of cisplatin and DNA modified globally by transplatin or [PtCl(dien)]Cl is noticeably weaker than that to DNA modified globally by cisplatin (Coin *et al.*, 1998) supporting the view that TBP binds selectively to specific three-dimensional structure.

Another protein lacking HMG-domain that has been shown to bind to cisplatin-modified DNA is Y-box binding protein 1 (Ise *et al.*, 1999). It has been suggested that this transcription factor may also function as a recognition protein for cisplatin-damaged DNA in DNA repair or in directing the cellular response to damage to DNA by cisplatin (Ise *et al.*, 1999; Ohga *et al.*, 1996).

Very abundant chromatin protein, the linker histone H1, also binds much more strongly to DNA modified by cisplatin than to transplatin-modified or nonmodified

DNA (Paneva *et al.*, 1998; Yaneva *et al.*, 1997). Interestingly, linker histones share with HMGB1/2 several important characteristics. Both bind to linker DNA in chromatin, specifically recognize four-way junction DNA and unwind DNA upon binding etc. (for review, see Bustin and Reeves (1996), Zlatanova and van Holde (1996) and Zlatanova and Van Holde (1998). The adduct responsible for the strong binding of histone H1 to cisplatin-modified DNA has not been determined, nevertheless the fact that this binding is stronger than that of the HMGB1/2 makes histone H1 a good candidate for participating in the mechanism of antitumor activity of cisplatin.

A relatively large number of proteins that specifically bind to cisplatin adducts belong to DNA repair proteins. The repair proteins that recognize cisplatin-modified DNA are mainly those involved in the first step in repair pathways, i.e. in damage recognition. Most of these proteins also recognize other types of DNA lesions than cisplatin adducts, such as for instance those created by exposure to other chemical agents or UV light. The repair proteins that probably have attracted the most attention in the studies of recognition of DNA modified by platinum drugs are those that are absent in patients suffering from the NER deficiency characteristic of the disease *Xeroderma pigmentosum* (XP). The minimal factors necessary for removal of damaged nucleotides also include XPA protein and replication protein A (RPA) which are among the major damage-recognition proteins involved in the early stage of NER. XPA (32-kDa protein) and RPA (composed of 70- and 34-kDa subunits) are able to bind damaged DNA independently, although RPA interaction stimulates XPA binding to damaged DNA. It has been suggested recently (Missura *et al.*, 2001) that XPA in conjuction with RPA constitute a regulatory factor that monitors DNA bending and unwinding to verify the damage-specific localization of NER complexes or control their three-dimensional assembly. These NER proteins have been tested for their capability to recognize and bind to cisplatin-modified DNA. Several studies have indicated that XPA binds to DNA globally modified by cisplatin being able to recognize selectively the rigid bend induced by the 1,2-GG intrastrand CL of cisplatin (Asahina *et al.*, 1994; Jones and Wood, 1993; Kuraoka *et al.*, 1996; Turchi *et al.*, 1999; Xu *et al.*, 1999). Low affinity of XPA to 1,3-GXG intrastrand CL (3-fold preference for platinated vs undamaged DNA) has been also reported (Hey *et al.*, 2001). Further analyses of the binding affinity of RPA have revealed differential recognition of site-specific DNA adducts of cisplatin by this protein (Clugston *et al.*, 1992; Patrick and Turchi, 1998; Patrick and Turchi, 1999; Turchi *et al.*, 1999). The adduct bound most efficiently is the 1,3-GXG intrastrand CL, with the 1,2-GG intrastrand CL also being bound preferentially compared to the undamaged control, but roughly 30% less compared to the 1,3-CL. The binding of a DNA substrate containing a single, site-specific interstrand CL is even less than that observed with the undamaged control DNA. The RPA affinity to individual cisplatin CLs has been correlated with thermal destabilization induced by these adducts (Patrick and Turchi, 1999; Turchi *et al.*, 1999) and it has been proposed that RPA may bind to cisplatin-modified DNA via the generation of transient single-stranded regions. Importantly, the rate of association of RPA with cisplatin-modified DNA is smaller than that of HMGB1 for the same substrate (Turchi *et al.*, 1999).

The proteins involved in the MMR pathway also display affinity for cisplatin-modified DNA. For instance, human MutS binds to DNA containing 1,2-GG intrastrand CL, but fails to bind to the 1,3-GXG intrastrand CL of transplatin (Duckett *et al.*, 1996). In addition, MSH2 (a component of MutS) also binds specifically to DNA containing cisplatin adducts and displays selectivity for the DNA adducts of therapeutically active platinum complexes (Mello *et al.*, 1996; Yamada *et al.*, 1997).

DNA photolyase involved in the repair of cyclobutane pyrimidine dimmers (Özer *et al.*, 1995) and Ku autoantigen, taking part in recombination and double-strand break repair are further examples of proteins lacking HMG-domain that bind to cisplatin-modified DNA. Interestingly, the latter protein binds to cisplatin-modified DNA but fails to stimulate DNA-activated protein kinase (Turchi and Henkels, 1996). Its activation requires translocation of its Ku subunits away from the DNA terminus and the adducts formed on DNA by cisplatin inhibit this translocation (Turchi *et al.*, 2000).

The repair protein that has been also thoroughly tested whether it recognizes specifically DNA adducts of cisplatin and transplatin is an enzyme with deoxyribonuclease activity, T4 endonuclease VII (Kasparkova and Brabec, 1995; Murchie and Lilley, 1993). This bacteriophage T4-encoded protein cleaves branched DNA structures, most notably four-way junctions, and is regarded as a repair enzyme. T4 endonuclease VII recognizes 1,2-intrastrand CLs of cisplatin, whereas 1,3-intrastrand CLs of both isomers are recognized much less efficiently (Murchie and Lilley, 1993). Later studies (Kasparkova and Brabec, 1995) have demonstrated that the DNA duplex containing a single interstrand CL of cisplatin is also precisely cleaved in both strands by this DNA-debranching enzyme with efficiency similar to that observed with DNA containing a single 1,2-GG intrastrand CL or a four-way junction. In contrast, the duplex containing the interstrand CL of clinically ineffective transplatin is cleaved considerably less efficiently, a property shared by HMG-domain proteins.

It has been hypothesized that sensitivity of tumor cells might be also associated with the processes involving p53 protein (Jordan and Carmo-Fonseca, 2000; Riva, 2000). The p53 gene encodes a nuclear phosphoprotein consisting of 393 amino acids that is biologically activated in response to genotoxic stresses including treatment with cisplatin. Also interestingly, on average, cells with mutant p53 are more resistant to the effect of cisplatin (O'Connor *et al.*, 1997). Hence, it seems reasonable to conclude that p53 can control the processing of DNA adducts of cisplatin depending on the cell type. DNA binding activity of p53 protein is crucial for its tumor suppressor function. DNA interactions of active wild-type human p53 protein with DNA modified by antitumor cisplatin and its clinically ineffective *trans* isomer (transplatin) has been investigated (Kasparkova *et al.*, 2001). DNA adducts of cisplatin reduce binding affinity of the consensus DNA sequence to p53 whereas transplatin adducts do not. This result has been interpreted to mean that the precise steric fit required for formation and stability of the tetrameric complex of p53 with the consensus sequence cannot be attained as a consequence of severe conformational perturbances induced in DNA by cisplatin adducts. The results also demonstrate an increase of the binding affinity of p53 to DNA lacking the consensus sequence and modified by cisplatin, but not by transplatin. In addition, major 1,2-GG intrastrand CLs of cisplatin are only responsible for this enhanced binding affinity of p53, which is, however, markedly lower than the affinity of p53 to unplatinated DNA containing the consensus sequence. The distinctive structural features of 1,2-intrastrand CLs of cisplatin has been suggested to play a unique role for this adduct in the binding of p53 to DNA lacking the consensus sequence.

The cisplatin-damaged DNA-binding proteins apparently occur in the nature for other purposes than for specific recognition of platinum adducts in DNA, since platinum compounds do not belong to natural components of our environment. The capability of cisplatin-damaged DNA to bind DNA-binding proteins, which may have a fundamental relevance to the antitumor activity of cisplatin and its simple antitumor analogues, is probably a coincidence when the formation of some platinum adducts in double-helical DNA adopts a structure that mimics the recognition signal for these proteins.

2.6. *HYPOTHESES FOR MECHANISM OF ANTITUMOR ACTIVITY OF CISPLATIN*

In more than 30 years after discovery of antitumor activity of cisplatin, basic research has provided extensive information how the drug exerts its antitumor effects and how some tumors are, or become, resistant to this drug. It is generally believed that the key intracellular target for cisplatin is DNA on which cisplatin forms various types of adducts. These adducts distort DNA conformation, but it still remains uncertain which of these adducts is the most important in terms of producing anticancer effects. As briefly summarized in the preceding sections of this article DNA adducts of cisplatin inhibit replication and transcription, but they are also bypassed by DNA or RNA polymerases. In addition, cisplatin adducts are removed from DNA mainly by NER. They are, however, also recognized by a number of proteins which could block cisplatin DNA adducts from damage recognition needed for repair. The other hypothesis based on the observation that a number of various proteins recognize cisplatin-modified DNA is that cisplatin-DNA adducts hijack proteins away from their normal binding sites, thereby disrupting fundamental cellular processes. Experimental support of thesehypothetical aspects of the mechanism underlying antitumor activity of cisplatin or resistance of some tumors to this drug has been recently thoroughly reviewed (Gelasco and Lippard, 1999; Guo and Sadler, 1999; Jamieson and Lippard, 1999; Johnson *et al.*, 1998; Jordan and Carmo-Fonseca, 2000; Kelland, 2000; Zamble and Lippard, 1999)

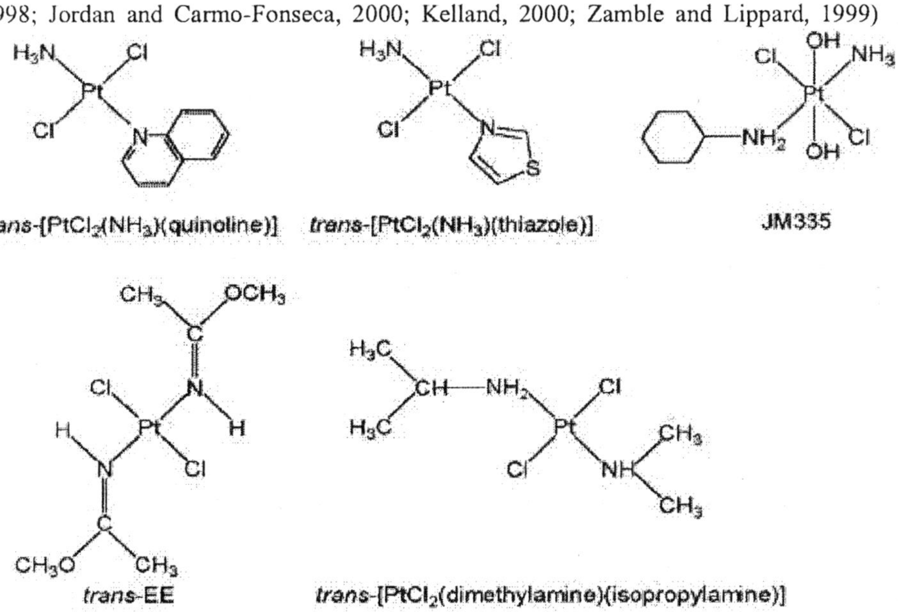

Figure. 5. Structures of antitumor transplatinum compounds.

Initially inhibition of DNA replication was considered to be a process very likely relevant to antitumor efficiency of cisplatin (Pinto and Lippard, 1985). However, later studies have indicated that cisplatin inhibits tumor cell growth at doses which are considerably lower than those needed to inhibit DNA synthesis (Sorenson and Eastman, 1988). Subsequent observations have revealed that cisplatin can trigger G2 cell-cycle arrest and programmed cell death (apoptosis) (Allday *et al.*, 1995; Chu, 1994) exposing a mechanism of cytotoxicity of this drug. However, since apoptosis is a very complex process, a number of possible pathways have to be still explored for a complete understanding of the process how cisplatin triggers apoptosis (Gonzalez *et al.*, 2001).

3. Activation of *trans* Geometry

Since discovery of antitumor activity of cisplatin, none of the structural analogues of cisplatin which have advanced to clinical trials are likely to display novel clinical properties in comparison to the parent drug. Therefore, the search continues for an improved platinum antitumor agent, motivated by the desire to design a less toxic compound that is non-cross-resistant with cisplatin or carboplatin and oxaliplatin. In this search the hypothesis that platinum drugs which bind to DNA in a fundamentally different manner to that of cisplatin will have altered pharmacological properties has been tested. This concept has already led to the synthesis of several new unconventional platinum antitumor compounds that violate the original structure-activity relationships. The clinical inactivity of transplatin is considered a paradigm for the classical structure–activity relationships of platinum drugs (Reedijk, 1996), but to this end several new analogues of transplatin which exhibit a different spectrum of cytostatic activity including activity in tumor cells resistant to cisplatin have been identified.

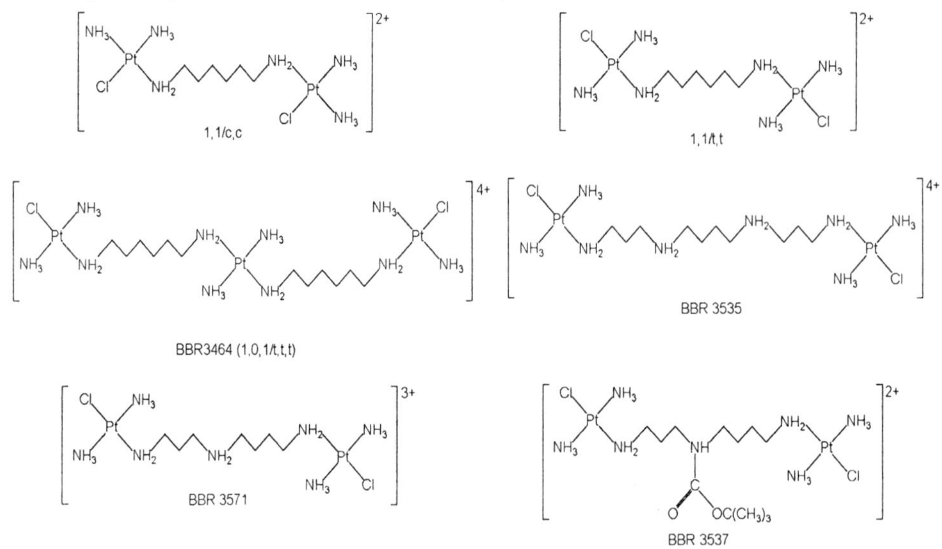

Figure. 6. Structures of antitumor dinuclear and trinuclear platinum(II) compounds. a) analogues containing planar amine ligand of general structure *trans*-$[PtCl_2(NH_3)(L)]$, where L = planar amine such as quinoline or thiazole; b) analogues containing iminoether groups of general formula *trans*-$[PtCl_2(E\text{-iminoether})_2]$ (*trans-EE*); c) analogues with asymmetric aliphatic ligands and *trans*-$[PtCl_2(NH_3)(L)]$, where L = cyclohexylamine (CHA).

The antitumor *trans*-platinum complexes whose DNA binding mode has been most extensively examined include three distinct series (Figure. 5):

4. Polynuclear Platinum Antitumor Drugs

Polynuclear platinum compounds represent a further class of antitumor drugs with chemical and biological properties different from cisplatin. The first drug of this class which already entered Phase II clinical trials is the trinuclear compound [{*trans*-$PtCl(NH_3)_2\}_2\mu$- *trans*-$Pt(NH_3)_2\{H_2N(CH_2)_6NH_2\}_2]^{4+}$ (Fig. 6). The compound, designated as BBR3464, is the lead representative of the entirely new structural class of DNA-modifying anticancer agents based on the poly(di,tri)nuclear platinum structural motif. Initial clinical trials have revealed activity in pancreatic, lung and melanoma cancers suggesting the complementary clinical anticancer activity of BBR3464 in comparison to cisplatin. The choice of BBR3464 as clinical candidate also comes from preclinical studies showing cytotoxicity at ten-fold lower concentration than cisplatin and collateral sensitivity in cisplatin-resistant cell lines. Importantly, BBR3464 also displays consistently high antitumor activity in human tumor xenografts characterized as mutant p53. This important feature suggests that the new agent may find utility in the over 60% of cancer cases where mutant p53 status has been indicated. DNA damage by chemotherapeutic agents is in many cases mediated through the p53 pathway. Consistently, cytotoxicity displayed in mutant cell lines would suggest an ability to bypass this pathway.

BBR3464 has been designed on the basis of systematic studies on various dinuclear compounds. History of development of this class of platinum drugs has been recently thoroughly reviewed (Farrell, 2000; Farrell *et al.*, 1999).

In summary, BBR3464 exhibits a unique profile of DNA binding, strengthening the original hypothesis that modification of DNA binding in manners distinct from that of cisplatin will also lead to a distinct and unique profile of antitumor activity.

5. Concluding Remarks

Platinum antitumor drugs represent important agents for the treatment of several different tumors. It appears that changing chemical structure of platinum and ruthenium compounds may substantially modulate their DNA binding mode, subsequent processing of DNA damage and consequently the mechanism of biological efficacy of these metal-based compounds. Thus, these structural alterations may also influence not only their spectrum of activity and the development of drug resistance, but also the toxicity profile of these drugs. A further understanding of how new platinum compounds modify DNA and how these modifications are further processed in cells should provide a rational basis for the design of new chemotherapeutic strategies and metal-based antitumor drugs rather than only searching for cisplatin analogues.

6. Acknowledgements

This work was supported by grant A5004101 from the Grant Agency of the Academy of Sciences of the Czech Republic.

References

1. Aebi, S., Fink, D., Gordon, R., Kim, H.K., Zheng, H., Fink, J.L. and Howell, S.B. (1997) Resistance to cytotoxic drugs in DNA mismatch repair-deficient cells, *Clin. Cancer Res.* **3**, 1763-1767.
2. Allday, M.J., Inman, G.J., Crawford, D.H. and Farrell, P.J. (1995) DNA damage in human B cells can induce apoptosis, proceeding from G1/S when p53 is transactivation competent and G2/M when it is transactivation defective, *EMBO J.* **14**, 4994-5005.
3. Anin, M.F. and Leng, M. (1990) Distortions induced in double-stranded oligonucleotides by the binding of *cis*-diamminedichloroplatinum(II) or *trans*-diamminedichloroplatinum(II) to the d(GTG) sequence, *Nucleic Acids Res.* **18**, 4395-4400.
4. Asahina, H., Kuraoka, I., Shirakawa, M., Morita, E., Miura, N., Miyamoto, I., Ohtsuka, E., Okada, Y. and Tanaka, K. (1994) The XPA protein is a zinc metalloprotein with an abilito to recognize various kinds of DNA-damage, *Mutat. Res.* **315**, 229-237.
5. Balcarova, Z. and Brabec, V. (1986) DNA modified by platinum derivatives cannot adopt the A-form, *Biochim. Biophys. Acta* **867**, 31-35.
6. Bancroft, D.P., Lepre, C.A. and Lippard, S.J. (1990) Pt-195 NMR kinetic and mechanistic studies of *cis*-diamminedichloroplatinum and trans-diamminedichloroplatinum(II) binding to DNA, *J. Am. Chem. Soc.* **112**, 6860-6871.
7. Beck, D.J. and Brubaker, R.R. (1973) Effect of *cis*-platinum(II)diamminodichloride on wild type and deoxyribonucleic acid repair-deficient mutants of *Escherichia coli*, *J. Bacteriol.* **116**, 1247-1252.
8. Beck, D.J., Popoff, S., Sancar, A. and Rupp, W.D. (1985) Reactions of the UVRABC excision nuclease with DNA damaged by diamminedichloroplatinum(II), *Nucleic Acids Res.* **13**, 7395-7412.
9. Bellon, S.F. and Lippard, S.J. (1990) Bending studies of DNA site-specifically modified by cisplatin, trans-diamminedichloroplatinum(II) and cis-Pt(NH_3)$_2$(*N3*-cytosine)Cl^+, *Biophys. Chem.* **35**, 179-188.
10. Boudvillain, M., Dalbies, R., Aussourd, C. and Leng, M. (1995) Intrastrand cross-links are not formed in the reaction between transplatin and native DNA: Relation with the clinical inefficiency of transplatin, *Nucleic Acids Res.* **23**, 2381-2388.
11. Brabec, V. (2000) Chemistry and structural biology of 1,2-interstrand adducts of cisplatin, in *Platinum-Based Drugs in Cancer Therapy* (L.R. Kelland and N.P. Farrell, eds.), pp. 37-61, Humana Press Inc, Totowa/NJ.
12. Brabec, V., Boudny, V. and Balcarova, Z. (1994) Monofunctional adducts of platinum(II) produce in DNA a sequence-dependent local denaturation, *Biochemistry* **32**, 1316-1322.
13. Brabec, V., Kleinwächter, V., Butour, J.L. and Johnson, N.P. (1990) Biophysical studies of the modification of DNA by antitumour platinum coordination complexes, *Biophys. Chem.* **35**, 129-141.
14. Brabec, V. and Leng, M. (1993) DNA interstrand cross-links of *trans*-diamminedichloroplatinum(II) are preferentially formed between guanine and complementary cytosine residues, *Proc. Natl. Acad. Sci. USA* **90**, 5345-5349.
15. Brabec, V., Reedijk, J. and Leng, M. (1992) Sequence-dependent distortions induced in DNA by monofunctional platinum(II) binding, *Biochemistry* **31**, 12397-12402.
16. Brabec, V., Sip, M. and Leng, M. (1993) DNA conformational distortion produced by site-specific interstrand cross-link of *trans*-diamminedichloroplatinum(II), *Biochemistry* **32**, 11676-11681.
17. Brouwer, J., Van de Putte, P., Fichtinger-Schepman, A.M.J. and Reedijk, J. (1981) Base-pair substitution hotspots in GAG and GCG nucleotide sequences in E.coli K-12 induced by *cis*-diamminedichloroplatinum(II), *Proc. Natl. Acad. Sci. USA* **78**, 7010-7014.
18. Burnouf, D., Daune, M. and Fuchs, R.P. (1987) Spectrum of cisplatin-induced mutations in *Escherichia coli*, *Proc. Natl. Acad. Sci. USA* **84**, 3758-3762.
19. Bustin, M. and Reeves, R. (1996) High-mobility-group chromosomal proteins: Architectural components that facilitate chromatin function, *Prog. Nucleic Acid Res. Mol. Biol.* **54**, 35-100.
20. Calsou, P., Frit, P. and Salles, B. (1992) Repair synthesis by human cell extracts in cisplatin-damaged DNA is prefentially determined by minor adducts, *Nucleic Acids Res.* **20**, 6363-6368.
21. Ciccarelli, R.B., Solomon, M.J., Varshavsky, A. and Lippard, S.J. (1985) In vivo effects of *cis*- and *trans*-diamminedichloroplatinum(II) on SV 40 chromosomes: Differential repair, DNA - protein cross-linking, and inhibition of replication, *Biochemistry* **24**, 7533-7540.
22. Clugston, C.K., McLaughlin, K., Kenny, M.K. and Brown, R. (1992) Binding of human single-stranded-DNA binding-protein to DNA damaged by the anticancer drug *cis*-diamminedichloroplatinum(II), *Cancer Res.* **52**, 6375-6379.
23. Cohen, S.M., Jamieson, E.R. and Lippard, S.J. (2000a) Enhanced binding of the TATA-binding protein to TATA boxes containing flanking cisplatin 1,2-cross-links, *Biochemistry* **39**, 8259-8265.
24. Cohen, S.M., Mikata, Y., He, Q. and Lippard, S.J. (2000b) HMG-Domain protein recognition of cisplatin 1,2-intrastrand d(GpG) cross-links in purine-rich sequence contexts, *Biochemistry* **39**, 11771-11776.

25. Coin, F., Frit, P., Viollet, B., Salles, B. and Egly, J.M. (1998) TATA binding protein discriminates between different lesions on DNA, resulting in a transcription decrease, *Mol. Cell. Biol.* **18**, 3907-3914.
26. Coluccia, M., Nassii, F., Loseto, F., Boccarelli, A., MarigiO, M.A., Giordano, D., Intini, F.P., Caputo, P. and Natile, G. (1993) A *trans*-platinum complex showing higher antitumor activity than the *cis* congeners, *J. Med. Chem.* **36**, 510-512.
27. Comess, K.M., Burstyn, J.N., Essigmann, J.M. and Lippard, S.J. (1992) Replication inhibition and translesion synthesis on templates containing site-specifically placed *cis*-diamminedichloroplatinum(II) DNA adducts, *Biochemistry* **31**, 3975-3990.
28. Corda, Y., Anin, M.F., Leng, M. and Job, D. (1992) RNA polymerases react differently at d(ApG) and d(GpG) adducts in DNA modified by *cis*-diamminedichloroplatinum(II), *Biochemistry* **31**, 1904-1908.
29. Corda, Y., Job, C., Anin, M.F., Leng, M. and Job, D. (1991) Transcription by eucaryotic and procaryotic RNA polymerases of DNA modified at a d(GG) or a d(AG) site by the antitumor drug *cis*-diamminedichloroplatinum(II), *Biochemistry* **30**, 222-230.
30. Corda, Y., Job, C., Anin, M.-F., Leng, M. and Job, D. (1993) Spectrum of DNA-platinum adduct recognition by prokaryotic and eukaryotic DNA-dependent RNA polymerases, *Biochemistry* **32**, 8582-8588.
31. Coste, F., Malinge, J.M., Serre, L., Shepard, W., Roth, M., Leng, M. and Zelwer, C. (1999) Crystal structure of a double-stranded DNA containing a cisplatin interstrand cross-link at 1.63 A resolution: hydration at the platinated site, *Nucleic Acids Res.* **27**, 1837-1846.
32. Cullinane, C., Mazur, S.J., Essigmann, J.M., Phillips, D.R. and Bohr, V.A. (1999) Inhibition of RNA polymerase II transcription in human cell extracts by cisplatin DNA damage, *Biochemistry* **38**, 6204-6212.
33. Dijt, F.J., Fichtinger-Schepman, A.M.J., Berends, F. and Reedijk, J. (1988) Formation and repair of cisplatin-induced adducts to DNA in cultured normal and repair-deficient human fibroblasts, *Cancer Res.* **48**, 6058-6062.
34. Drummond, J.T., Anthoney, A., Brown, R. and Modrich, P. (1996) Cisplatin and adriamycin resistance are associated with MutL alpha and mismatch repair deficiency in an ovarian tumor cell line, *J. Biol. Chem.* **271**, 19645-19648.
35. Duckett, D.R., Drummond, J.T., Murchie, A.I.H., Reardon, J.T., Sancar, A., Lilley, D.M. and Modrich, P. (1996) Human MutSa recognizes damaged DNA base pairs containing O^6-methylguanine, O^4-methylthymine, or the cisplatin-d(GpG) adduct, *Proc. Natl. Acad. Sci. USA* **93**, 6443-6447.
36. Dunham, S.U. and Lippard, S.J. (1997) DNA sequence context and protein composition modulate HMG-domain protein recognition of ciplatin-modified DNA, *Biochemistry* **36**, 11428-11436.
37. Eastman, A. (1987) The formation, isolation and characterization of DNA adducts produced by anticancer platinum complexes, *Pharmacol. Ther.* **34**, 155-166.
38. Eastman, A., Jennerwein, M.M. and Nagel, D.L. (1988) Characterization of bifunctional adducts produced in DNA by *trans*-diamminedichloroplatinum(II), *Chem.-Biol. Interactions* **67**, 71-80.
39. Farrell, N. (2000) Polynuclear charged platinum compounds as a new class of anticancer agents: Toward a new paradigm, in *Platinum-Based Drugs in Cancer Therapy* (L.R. Kelland and N.P. Farrell, eds.), pp. 321-338.
40. Farrell, N., Qu, Y., Bierbach, U., Valsecchi, M. and Menta, E. (1999) Structure-activity relationship within di- and trinuclear platinum phase I clinical agents, in *Cisplatin. Chemistry and biochemistry of a leading anticancer* drug (B. Lippert, B. ed.), pp. 479-496, VHCA, WILEY-VCH, Zurich, Weinheim.
41. Ferry, K.V., Fink, D., Johnson, S.W., Nebel, S., Hamilton, T.C. and Howell, S.B. (1999) Decreased cisplatin damage-dependent DNA synthesis in cellular extracts of mismatch repair deficient cells, *Biochem. Pharmacol.* **57**, 861-867.
42. Fichtinger-Schepman, A.M.J., Van der Veer, J.L., Den Hartog, J.H.J., Lohman, P.H.M. and Reedijk, J. (1985) Adducts of the antitumor drug cis-diamminedichloroplatinum(II) with DNA: Formation, identification, and quantitation, *Biochemistry* **24**, 707-713.
43. Fink, D., Nebel, S., Aebi, S., Nehme, A. and Howell, S.B. (1997) Loss of DNA mismatch repair due to knockout of MSH2 or PMS2 results in resistance to cisplatin and carboplatin, *Int. J. Oncol.* **11**, 539-542.
44. Fink, D., Nebel, S., Aebi, S., Zheng, H., Cenni, B., Nehme, A., Christen, R.D. and Howell, S.B. (1996) The role of DNA mismatch repair in platinum drug resistance, *Cancer Res.* **56**, 4881-4886.
45. Fujiwara, Y., Tatsumi, M. and Sasaki, M.S. (1977) Cross-link repair in human cells and its possible defect in Fanconi's anemia cells, *J. Mol. Biol.* **113**, 635-649.
46. Gelasco, A. and Lippard, S.J. (1999) Anticancer activity of cisplatin and related complexes, in *Metallopharmaceuticals I. DNA interactionsI.* (M.J. Clarke and P.J. Sadler, P.J. eds.), pp. 1-43, Springer, Berlin.
47. Giaccone, G. (2000) Clinical perspectives on platinum resistance, *Drugs* **59**, 9-17.

48. Gonzalez, V.M., Fuertes, M.A., Alonso, C. and Perez, J.M. (2001) Is cisplatin-induced cell death always produced by apoptosis?, *Mol. Pharmacol.* **59**, 657-663.
49. Guo, Z.J. and Sadler, P.J. (1999) Metals in medicine, *Angew. Chem. Int. Ed.* **38**, 1513-1531.
50. Hey, T., Lipps, G. and Krauss, G. (2001) Binding of XPA and RPA to damaged DNA investigated by fluorescence anisotrophy, *Biochemistry* **40**, 2901-2910.
51. Hofr, C., Farrell, N. and Brabec, V. (2001) Thermodynamic properties of duplex DNA containing a site-specific d(GpG) intrastrand crosslink formed by an antitumor dinuclear platinum complex, *Nucleic Acids Res.* **29**, 2034-2040.
52. Huang, H.F., Zhu, L.M., Reid, B.R., Drobny, G.P. and Hopkins, P.B. (1995) Solution structure of a cisplatin-induced DNA interstrand cross-link, *Science* **270**, 1842-1845.
53. Chu, G. (1994) Cellular responses to cisplatin. The roles of DNA-binding proteins and DNA repair, *J. Biol. Chem.* **269**, 787-790.
54. Ise, T., Nagatani, G., Imamura, T., Kato, K., Takano, H., Nomoto, M., Izumi, H., Ohmori, H., Okamoto, T., Ohga, T., Uchiumi, T., Kuwano, M. and Kohno, K. (1999) Transcription factor Y-box binding protein 1 binds preferentially to cisplatin-modified DNA and interacts with proliferating cell nuclear antigen, *Cancer Res.* **59**, 342-346.
55. Jamieson, E.R., Jacobson, M.P., Barnes, C.M., Chow, C.S. and Lippard, S.J. (1999) Structural and kinetic studies of a cisplatin-modified DNA icosamer binding to HMG1 domain B, *J. Biol. Chem.* **274**, 12346-12354.
56. Jamieson, E.R. and Lippard, S.J. (1999) Structure, recognition, and processing of cisplatin-DNA adducts, *Chem. Rev.* **99**, 2467-2498.
57. Janovska, E. and Kleinwächter, V. (1986) Inactivation effect of platinum(II) compounds on strains of Escherichia coli, *Studia Biophysica* **114**, 187-192.
58. Johnson, N.P., Butour, J.-L., Villani, G., Wimmer, F.L., Defais, M., Pierson, V. and Brabec, V. (1989) Metal antitumor compounds: The mechanism of action of platinum complexes, *Prog. Clin. Biochem. Med.* **10**, 1-24.
59. Johnson, N.P., Macquet, J.-P., Wiebers, J.L. and Monsarrat, B. (1982) Structures of adducts formed between [Pt(dien)Cl]Cl and DNA *in vitro*, *Nucleic Acids Res.* **10**, 5255-5271.
60. Johnson, S.W., Ferry, K.V. and Hamilton, T.C. (1998) Recent insights into platinum drug resistance in cancer, *Drug Resistance Updates* **1**, 243-254.
61. Jones, C.J. and Wood, R.D. (1993) Preferential binding of the Xeroderma Pigmentosum group A complementing protein to damaged DNA, *Biochemistry* **32**, 12096-12104.
62. Jordan, P. and Carmo-Fonseca, M. (1998) Cisplatin inhibits synthesis of ribosomal RNA in vivo, *Nucleic Acids Res.* **26**, 2831-2836.
63. Jordan, P. and Carmo-Fonseca, M. (2000) Molecular mechanisms involved in cisplatin cytotoxicity, *Cell. Mol. Life Sci.* **57**, 1229-1235.
64. Kasparkova, J. and Brabec, V. (1995) Recognition of DNA interstrand cross-links of *cis*-diamminedichloroplatinum(II) and its *trans* isomer by DNA-binding proteins, *Biochemistry* **34**, 12379-12387.
65. Kasparkova, J., Pospisilova, S. and Brabec, V. (2001) Different recognition of DNA modified by antitumor cisplatin and its clinically ineffective trans isomer by tumor suppressor protein p53, *J. Biol. Chem.* **276**, 16064-16069.
66. Keck, M.V. and Lippard, S.J. (1992) Unwinding of supercoiled DNA by platinum ethidium and related complexes, *J. Am. Chem. Soc.* **114**, 3386-3390.
67. Kelland, L.R. (2000) Preclinical perspectives on platinum resistance, *Drugs* **59**, 1-8.
68. Kuraoka, I., Morita, E.H., Saijo, M., Matsuda, T., Morikawa, K., Shirakawa, M. and Tanaka, K. (1996) Identification of a damaged-DNA binding domain of the XPA protein, *Mutat. Res.* **362**, 87-95.
69. Lemaire, M.A., Schwartz, A., Rahmouni, A.R. and Leng, M. (1991) Interstrand cross-links are preferentially formed at the d(GC) sites in the reaction between *cis*-diamminedichloroplatinum(II) and DNA, *Proc. Natl. Acad. Sci. USA* **88**, 1982-1985.
70. Leng, M., Schwartz, A. and Giraud-Panis, M.J. (2000) Transplatin-modified oligonucleotides as potential antitumor drugs, in *Platinum- Based Drugs in Cancer Therapy* L.R. Kelland N.P. Farrell eds.), pp. 63-85, Humana Press Inc, Totowa/NJ.
71. Malinge, J.-M. and Leng, M. (1999) Interstrand cross-links in cisplatin or transplatin-modified DNA, in *Cisplatin. Chemistry and Biochemistry of a Leading Anticancer Drug* (B. Lippert ed.), pp. 1589-180, Verlag Helvetica Chimica Acta, Wiley-VCH, Zürich, Weinheim.
72. Mello, J.A., Acharya, S., Fishel, R. and Essigmann, J.M. (1996) The mismatch-repair protein hMSH2 binds selectively to DNA adducts of the anticancer drug cisplatin, *Chem. & Biol.* **3**, 579-589.

73. Mello, J.A., Lippard, S.J. and Essigmann, J.M. (1995) DNA adducts of *cis*-diamminedichloroplatinum(II) and its *trans* isomer inhibit RNA polymerase II differentially *in vivo*, *Biochemistry* **34**, 14783-14791.

74. Missura, M., Buterin, T., Hindges, R., Hubscher, U., Kasparkova, J., Brabec, V. and Naegeli, H. (2001) Double-check probing of DNA bending and unwinding by XPA-RPA: an architectural function in DNA repair, *EMBO J.* **20**, 3554-3564.

75. Murchie, A.I.H. and Lilley, D.M.J. (1993) T4 endonuclease VII cleaves DNA containing a cisplatin adduct, *J. Mol. Biol.* **233**, 77-82.

76. Murray, V., Motyka, H., England, P.R., Wickham, G., Lee, H.H., Denny, W.A. and McFadyen, W.D. (1992a) An investigation of the sequence-specific interaction of *cis*-diamminedichloroplatinum(II) and four analogues, including two acridine-tethered complexes, with DNA inside human cells, *Biochemistry* **31**, 11812-11817.

77. Murray, V., Motyka, H., England, P.R., Wickham, G., Lee, H.H., Denny, W.A. and McFadyen, W.D. (1992b) The use of Taq DNA polymerase to determine the sequence specificity of DNA damage caused by *cis*-diamminedichloroplatinum(II), acridine-tethered platinum(II) diammine complexes or 2 analogues, *J. Biol. Chem.* **267**, 18805-18809.

78. Murray, V., Whittaker, J., Temple, M.D. and McFadyen, W.D. (1997) Interaction of 11 cisplatin analogues with DNA: characteristic pattern of damage with monofunctional analogues, *Biochim. Biophys. Acta* **1354**, 261-271.

79. Mymryk, J., Zaniewski, E. and Archer, T. (1995) Cisplatin inhibits chromatin remodeling, transcription factor binding, and transcription from the mouse mammary tumor virus promoter *in vivo*, *Proc. Natl. Acad. Sci. USA* **92**, 2076-2080.

80. O'Connor, P.M., Jackman, J., Bae, I., Myers, T.G., Fan, S., Mutoh, M., Scudiero, D.A., Monks, A., Sausville, E.A., Weinstein, J.N., Friend, S., Fornace, A.J. and Kohn, K.W. (1997) Characterization of the p53 tumor suppressor pathway in cell lines of the National Cancer Institute anticancer drug screen and correlations with the growth-inhibitory potency of 123 anticancer agents, *Cancer Res.* **57**, 4285-4300.

81. O'Dwyer, P.J., Stevenson, J.P. and Johnson, S.W. (1999) Clinical status of cisplatin, carboplatin, and other platinum-based antitumor drugs, in *Cisplatin. Chemistry and Biochemistry of a Leading Anticancer Drug* (B. Lippert ed.), pp. 31-72, VHCA, WILEY-VCH, Zürich, Weinheim.

82. Ohga, T., Koike, K., Ono, M., Makino, Y., Itagaki, Y., Tanimoto, M., Kuwano, M. and Kohno, K. (1996) Role of the human Y box-binding protein YB-1 in cellular sensitivity to the DNA-damaging agents cisplatin, mitomycin C, and ultraviolet light, *Cancer Res.* **56**, 4224-4228.

83. Ohndorf, U.M., Rould, M.A., He, Q., Pabo, C.O. and Lippard, S.J. (1999) Basis for recognition of cisplatin-modified DNA by high-mobility-group proteins, *Nature* **399**, 708-712.

84. Orphanides, G., Lagrange, T. and Reinberg, D. (1996) The general transcription factors of RNA polymerase II, *Genes Dev.* **10**, 2657-2683.

85. Özer, Z., Reardon, J.T., Hsu, D.S., Malhotra, K. and Sancar, A. (1995) The other function of DNA photolyase: Stimulation of excision repair of chemical damage to DNA, *Biochemistry* **34**, 15886-15889.

86. Paneva, E.G., Spassovska, N.C., Grancharov, K.C., Zlatanova, J.S. and Yaneva, J.N. (1998) Interaction of histone H1 with cis-platinum modified DNA, *Z. Naturforsch.* **53c**, 135-138.

87. Paquet, F., Boudvillain, M., Lancelot, G. and Leng, M. (1999) NMR solution structure of a DNA dodecamer containing a transplatin interstrand GN7-CN3 cross-link, *Nucleic Acids Res.* **27**, 4261-4268.

88. Patrick, S.M. and Turchi, J.J. (1998) Human replication protein a preferentially binds cisplatin-damaged duplex DNA in vitro, *Biochemistry* **37**, 8808-8815.

89. Patrick, S.M. and Turchi, J.J. (1999) Replication protein A (RPA) binding to duplex cisplatin-damaged DNA is mediated through the generation of single-stranded DNA, *J. Biol. Chem.* **274**, 14972-14978.

90. Pilch, D.S., Dunham, S.U., Jamieson, E.R., Lippard, S.J. and Breslauer, K.J. (2000) DNA sequence context modulates the impact of a cisplatin 1,2-d(GpG) intrastrand cross-link an the conformational and thermodynamic properties of duplex DNA, *J. Mol. Biol.* **296**, 803-812.

91. Pinto, A.L. and Lippard, S.J. (1985) Sequence-dependent termination of in vitro DNA synthesis by cis- and trans-diamminedichloroplatinum(II), *Proc. Natl. Acad. Sci. USA* **82**, 4616-4620.

92. Plooy, A.C.M., Van Dijk, M., Berends, F. and Lohman, P.H.M. (1985) Formation and repair of DNA interstrand cross-links in relation to cytotoxicity and unscheduled DNA synthesis induced in control and mutant human cells treated with cis-diamminedichloroplatinum(II), *Cancer Res.* **45**, 4178-4184.

93. Reedijk, J. (1996) Improved understanding in platinum antitumour chemistry, *Chem. Commun.*, 801-806.

94. Riva, C.M. (2000) Restoration of wild-type p53 activity enhances the sensitivity of pleural metastasis to cisplatin through an apoptotic mechanism, *Anticancer Res.* **20**, 4463-4471.

95. Rosenberg, B., Van Camp, L., Trosko, J.E. and Mansour, V.H. (1969) Platinum compounds: A new class of potent antitumor agents, *Nature* **222**, 385-386.
96. Sorenson, C.M. and Eastman, A. (1988) Mechanism of *cis*-diamminedichloroplatinum(II)-induced cytotoxicity - Role of G2 arrest and DNA double-strand breaks, *Cancer Res.* **48**, 4484-4488.
97. Stros, M. (2001) Two mutations of basic residues within the N-terminus of HMG-1 B domain with different effects on DNA supercoiling and binding to bent DNA, *Biochemistry* **40**, 4769-4779.
98. Teuben, J.M., Bauer, C., Wang, A.H.J. and Reedijk, J. (1999) Solution structure of a DNA duplex containing a cis-diammineplatinum(II) 1,3-d(GTG) intrastrand cross-link, a major adduct in cells treated with the anticancer drug carboplatin, *Biochemistry* **38**, 12305-12312.
99. Thomas, J.O. and Travers, A.A. (2001) HMG1 and 2, and related "architectural" DNA-binding proteins, *TIBS* **26**, 167-174.
100. Turchi, J.J. and Henkels, K. (1996) Human Ku autoantigen binds cisplatin-damaged DNA but fails to stimulate human DNA-activated protein kinase, *J. Biol. Chem.* **271**, 13861-13867.
101. Turchi, J.J., Henkels, K.M., Hermanson, I.L. and Patrick, S.M. (1999) Interactions of mammalian proteins with cisplatin-damaged DNA, *J. Inorg. Biochem.* **77**, 83-87.
102. Turchi, J.J., Henkels, K.M. and Zhou, Y. (2000) Cisplatin-DNA adducts inhibit translocation of the Ku subunits of DNA-PK, *Nucleic Acids Res.* **28**, 4634-4641.
103. Van Garderen, C.J., Van den Elst, H., Van Boom, J.H., Reedijk, J. and Van Houte, L.P.A. (1989) A double-stranded DNA fragment shows a significant decrease in double-helix stability after binding of monofunctional platinum amine compounds, *J. Am. Chem. Soc.* **111**, 4123-4125.
104. Vichi, P., Coin, F., Renaud, J.P., Vermeulen, W., Hoeijmakers, J.H.J., Moras, D. and Egly, J.M. (1997) Cisplatin- and UV-damaged DNA lure the basal transcription factor TFIID/TBP, *EMBO J.* **16**, 7444-7456.
105. Villani, G., Hubscher, U. and Butour, J.-L. (1988) Sites of termination of *in vitro* DNA synthesis on *cis*-diamminedichloroplatinum(II) treated single-stranded DNA: a comparison between E. coli DNA polymerase I and eucaryotic DNA polymerases alpha, *Nucleic Acids Res.* **16**, 4407-4418.
106. Villani, G., Le Gac, N.T. and Hoffmann, J.-S. (1999) Replication of platinated DNA and its mutagenic consequences, in *Cisplatin. Chemistry and biochemistry of a leading anticancer drug* (B. Lippert, ed.), pp. 135-157, VHCA, Wiley-CH, Zürich, Weinheim.
107. Vrana, O., Boudny, V. and Brabec, V. (1996) Superhelical torsion controls DNA interstrand cross-linking by antitumor *cis*-diamminedichloroplatinum(II), *Nucleic Acids Res.* **24**, 3918-3925.
108. Wong, E. and Giandomenico, C.M. (1999) Current status of platinum-based antitumor drugs, *Chem. Rev.* **99**, 2451-2466.
109. Xu, N.X., Pasa-Tolic, L., Smith, R.D., Ni, S.S. and Thrall, B.D. (1999) Electrospray ionization-mass spectrometry study of the interaction of cisplatin-adducted oligonucleotides with human XPA minimal binding domain protein, *Anal. Biochem.* **272**, 26-33.
110. Yamada, M., O'Regan, E., Brown, R. and Karran, P. (1997) Selective recognition of a cisplatin-DNA adduct by human mismatch repair proteins, *Nucleic Acids Res.* **25**, 491-495.
111. Yaneva, J., Leuba, S.H., van Holde, K. and Zlatanova, J. (1997) The major chromatin protein histone H1 binds preferentially to *cis*-platinum-damaged DNA, *Proc. Natl. Acad. Sci. USA* **94**, 13448-13451.
112. Yarema, K.J., Lippard, S.J. and Essigmann, J.M. (1995) Mutagenic and genotoxic effects of DNA adducts formed by the anticancer drug *cis*-diamminedichloroplatinum(II), *Nucleic Acids Res.* **23**, 4066-4072.
113. Zamble, D.B. and Lippard, S.J. (1999) The response of cellular proteins to cisplatin-damaged DNA, in *Cisplatin. Chemistry and biochemistry of a leading anticancer drug* (B. Lippert, ed.), pp. 73-110, VHCA, WILEY-VCH, Zürich, Weinheim.
114. Zamble, D.B., Mu, D., Reardon, J.T., Sancar, A. and Lippard, S.J. (1996) Repair of cisplatin-DNA adducts by the mammalian excision nuclease, *Biochemistry* **35**, 10004-10013.
115. Zlatanova, J. and van Holde, K. (1996) The linker histones and chromatin structure: New twists, *Prog. Nucleic Acid Res. Mol. Biol.* **52**, 217-259.
116. Zlatanova, J. and Van Holde, K. (1998) Binding to four-way junction DNA: A common property of architectural proteins?, *FASEB J.* **12**, 421-431.
117. Zlatanova, J., Yaneva, J. and Leuba, S.H. (1998) Proteins that specifically recognize cisplatin-damaged DNA: a clue to anticancer activity of cisplatin, *FASEB J.* **12**, 791-799.

STUDIES ON PROTEIN ELECTRON CARRIER COMPLEXES

ADRENODOXIN REDUCTASE – ADRENODOXIN COMPLEX IN STEROID BIOSYNTHESIS

SONA MARDANYAN AND YELIZAVETA SARGISOVA
H.Buniatyan Institute of Biochemistry of Armenian NAS, P.Sevak Str. 5/1 Yerevan 375014, Republic of Armenia

Electron-transfer reactions are characteristic features of a variety of fundamental biological processes that include energy metabolism, hormone biosynthesis and xenobiotic detoxification. For the proteins, involved in these processes, the active site is comprised of a metal center (most frequently Fe) as well as of organic cofactors (flavins, quinones). The great interest directed towards understanding of the biological electron-transfer processes in recent years reflects the importance of metabolic processes in which electron transfer is involved.

At an electron-transfer protein study, the relative positions of the electron donor and electron acceptor sites at the time of electron transfer have to be determined. To study electron-transfer reactions between the donor and acceptor centers located in different protein molecules, a solution containing the two proteins of interest is prepared at conditions optimal for their complex formation. The electron transfer from one to the other protein within the formed complex can be considered as unimolecular. The most important structural factors in determining the electronic coupling between the donor and acceptor centers in any case are the distance and the protein structure between them. The main problems of research in electron transfer are:

Whether the nature of the protein structure matters, or only the distance separating these sites?

Which structural elements of proteins may or may not facilitate the electronic coupling of the centers, and thereby influence the rate of electron transfer?

A special question arises on the role of aromatic groups in coupling of the donor and acceptor centers.

In several tissues of vertebrate, mono-oxygenase systems are located hydroxylating steroid molecules. They catalyse definite critical steps in the biosynthesis of steroid hormones in steroidogenic tissues (Lambeth et al., 1982; Hanukoglu, I. and Hanukoglu, Z., 1986), of bile acids in the liver (Wikvall, 1984; Waki et al., 1986) and vitamin D in the kidney (Hiwatashi et al., 1982). These systems consist of a ferredoxin family one-electron transport protein with a [2Fe-2S] cluster; a flavoprotein family NADPH-dependent membrane attached ferredoxin reductase with FAD molecule in the active centre, and a cytochrome P450 family membrane-integrated heme-containing steroid-

E. Buzaneva and P. Scharff (eds.), Frontiers of Multifunctional Nanosystems, 251–264.

hydroxylase. These proteins constitute an electron-transfer chain located in the matrix side of the inner mitochondrial membrane (Churchill et al., 1978; Farkash et al., 1986). The steroid hydroxylation system in adrenal cortex mitochondria consists, respectively, of flavoprotein NADPH-adrenodoxin reductase (AR), iron-sulfur protein adrenodoxin (Ad), and one of cytochromes P450: CYP11A1 (P450scc), CYP11B1 (P45011b) or CYP11B2 (P450aldo) (Bernhardt, 1996).
In the present paper some properties of a short electron transfer chain [AR x Ad] are presented. Particularly, the investigations on complex formation between two proteins are described. The probable role of tryptophan (Trp) residues of AR in intramolecular electron transfer and a possible mechanism of limitation of the rate of electron flow are discussed.

1. Adrenodoxin

Ad plays an essential role in the electron transport chain from AR to P450.

The acidic Ad residues Asp79, Asp76, Glu73 and Asp72 are important in the charge-driven molecular recognition mechanism. In human ferredoxin they contribute significantly to complex formation with P450 (Vickery et al., 1998). Asp76 and Asp 79 are also included in the main polar interaction with AR.

Aromatic amino acid, Tyr residue, is suggested to participate in electron transfer processes in a number of proteins (Carter et al., 1974, Adman et al., 1973, Debus et al., 1988). The unique Tyr82 of Ad was also proposed to be involved either in electron transfer or in interaction with AR (Taniguchi and Kimura, 1976). However, recently the results of site-directed mutagenesis show that Tyr82 of bovine Ad is not located in the immediate vicinity of the iron-sulfur cluster but between the cluster and the recognition sites for AR and P450. (Beckert et al., 1994). It seems not to be involved in intra- or intermolecular electron transfer. ^{1}HNMR studies also suggested that Tyr82 does not interact with the iron-sulfur center (Greenfield et al., 1989).

2. Adrenodoxin Reductase

The function of AR is to transfer electrons from NADPH to adrenodoxin, which transfers them one at a time to the P-450 (Lambeth et al., 1982; Omura et al., 1966; Hanukoglu. and Jefcoate, 1980). The same AR serves the different mitochondrial P-450 systems in steroidogenic tissues: it was demonstrated that in bovine and human steroidogenic tissues genomes it is encoded by a single gene (Hanukoglu and Gutfinger, 1989; Solish et al., 1988). AR and Ad form a short electron-transport chain, which transfers electrons from NADPH to cytochromes P-450 (Omura et al., 1966):

NADPH AR Ad P-450

In AR, Arg240 and Arg244 are the basic key residues for interaction with Ad, as derived from site directed mutation experiments (Vickery, 1998). Besides, the chemical modification of Lys243 inhibits the transfer of electrons from AR to cytochrome c via Ad (Hamamoto et al., 1988; Hamamoto.and Ichikawa, 1984). Hara. and Miyata, (1991) provided experimental evidence for 1:1 Ad-AR complex stabilized by a single amid bond between the ε-amino group of Lys66 in Ad and the γ-carboxyl group in Glu4 in AR which was introduced by chemical cross linking with 1-ethyl-3-(3-dimethylaminopropyl) carbodiimide (EDC). Using Edman degradation, mass spectrometry and X-ray crystallography, a cross link mediated by EDC between AR Lys27 and Ad Asp39 was detected, establishing a secondary polar interaction site between the molecules (Muller et al., 2001).

2.1. *ADRENODOXIN REDUCTASE TREATMENT WITH N-BROMOSUCCINIMIDE*

We studied a possible role of Trp in AR in the electron transfer between two proteins taking into account the probability of aromatic amino acids participation in such processes. Since it is impossible to make a small change in the Trp side chain by site directed mutagenesis, because no other amino acid closely resembles it (Okajima, 1990) a chemical modification technique was used.

Treatment of AR with N-bromosuccinimide (NBS) decreased the intensity of the characteristic band of Trp absorption at 280 nm (Fig.1), evidencing Trp oxidation to the oxyindol derivative (Spande and Witkop, 1967). From the absorption drop at 280 nm, the number of modified Trps was evaluated using the relation proposed in the mentioned work:

$$n = 1.31\ \varepsilon_{280}\ \Delta A_{280} / 5500\ A_{280} \qquad (1)$$

Here ε_{280} and A_{280}, respectively, are the molar extinction and initial intensity of protein solution at 280 nm. ΔA_{280} is the decrease of absorption at 280 nm in response to NBS treatment. 1.31 is an empirical coefficient that takes into account the oxyindol absorption at 280 nm. 5500 is the molar extinction of Trp at 280 nm.

AR was treated by NBS in two different conditions (Sargisova and Mardanyan, 1987). At first, the reaction mixture contained 6 M urea in acetic buffer, pH 4.5 to denature the protein molecule and make all Trp residues available for reagent. In this medium, the titration of AR by NBS brought to A_{280} decrease, evidencing nearly 10 Trps oxidation. In the conditions optimal for saving the AR activity (10 mM phosphate buffer, pH 7.4, without any denaturant), only inactivation because of NBS addition was observed. The maximal number of Trp residues modified in this medium was 3, evidencing the others immersed deep in the molecule.

2.1.1. Stability of the main characteristics of a protein molecule

Neither the absorption spectra nor the fluorescence of FAD were influenced by NBS treatment. Should the treatment remove the prosthetic group from the protein molecule, an increase of the fluorescence intensity would be registered. The stability of both the absorption (Fig. 1) and fluorescence spectra (Sargisova et al. 1990a) evidenced the stability of FAD attachment to the protein molecule.

The identity of Circular Dichroic spectra of native and NBS treated preparations (Sargisova and Mardanyan, 1987) allows to rule out the distortion of the secondary structure due to the chemical reaction.

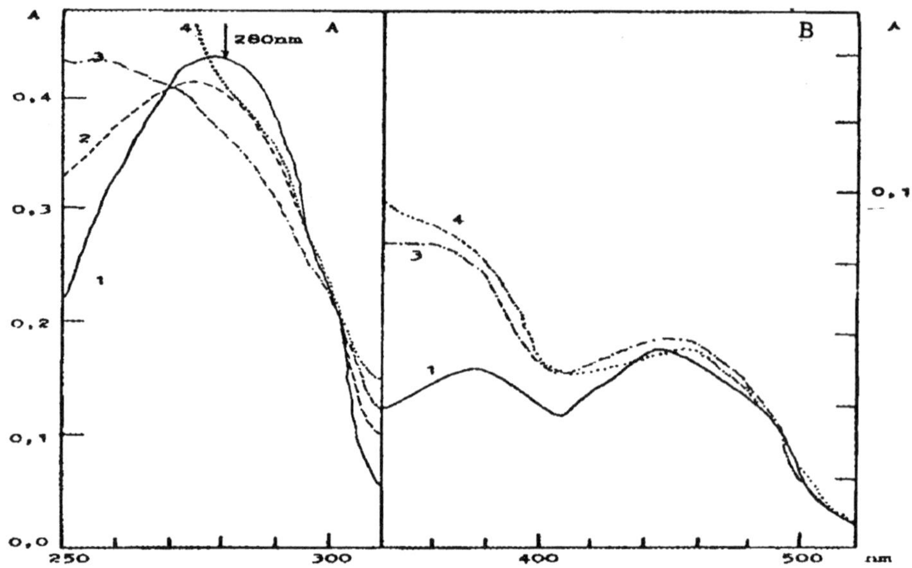

Figure 1. The absorption spectra of NADPH-adrenodoxin reductase solution in 50 mM potassium phosphate buffer, pH 7.4 (1); after addition of 0.12 (2) and 0.3 (3) mM NBS and after 3 hours of dialysis of the latter against buffer (4).

The possibility of NBS effect on several amino acids was excluded in the special experiments (Sargisova et al. 1990b). As probable targets of reagent, tyrosine, sulfhydryl and amino groups integrity were controlled. Using pyridoxal-5'-phosphate and o-phthalaldehyde, the stability of amino groups during NBS treatment was proved. Sulfhydryl groups specific reagent, DTNB, allowed to manifest that these groups are not responsible for the NBS effect on AR. The method for evaluation of Tyr number in a protein molecule based on pH-dependence of the residue's optical absorption (Edelhoch, 1967) demonstrated the stability of this amino acid over the used range of NBS concentration.

2.1.2. Suppression of electron transferring activity of AR.

AR titration with increasing concentration of NBS was accompanied by the protein electron transferring activity decrease. The activity decrease was in correlation with Trps oxidation (Fig. 2).

The analysis of the activity dependence on the number of modified Trps by the graphical method developed by Tsou (Clarke, 1987), allowed to estimate the number of Trp residues essential for AR activity. It was proved that in native AR one of three modifiable Trps is important for the protein electron transfer activity (Sargisova et al. 1990a).

The photo-oxidation under 290 nm light irradiation in the presence of thrichloroethanol specific for Trp (Toulme et al., 1984) also resulted in effective inactivation of AR (Sargisova et al., 1990b), one more time proving AR activity dependence on this amino acid.

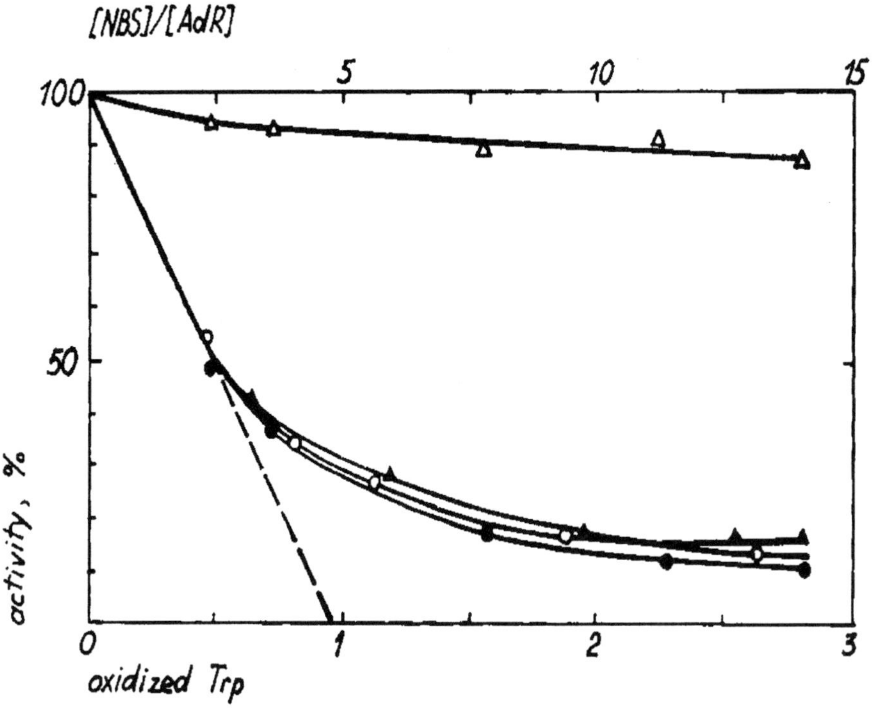

Figure 2. The rate of cytochrome c reduction in the reconstituted model electron transfer chain NADPH-AR-Ad-Cyt. c. Assay mixture in 2 ml of 50 mM potassium phosphate buffer, pH 7.4, contained 130 μM NADPH; 0.4 μM Ad; 30 μM cytochrome c. Reaction was started by addition of NBS-treated AR up to the final concentration 80 nM.
Preliminary titration of 3.5 μM AR with NBS was performed in 50 mM potassium phosphate buffer, pH 7.4 without any addition (o) and in the presence of equimolar Ad (Δ), bovine serum albumin (●) or cytochrome c (▲). The number of oxidized Trps in the sample without addition was estimated from the drop of absorption at 280 nm according to equation (1).

2.1.3. Ad protection of essential Trp in AR

The AR treatment in the above-mentioned two modification procedures was performed in the presence of Ad. Ad seems to protect AR against inactivation in both the NBS (Fig. 2) and the photo-oxidation reactions (Sargisova et al., 1990b). To exclude nonspecific interaction of NBS with any protein in the reaction mixture, the experiment was repeated in the presence of BSA and cytochrome c. The dependence of AR inactivation vs. NBS concentration in these cases coincided with that without any protein addition. Protection of AR activity against NBS-inactivation by Ad allowed us to suggest that the Trp essential for AR electron transferring activity is close to Ad binding site of the reductase.

This suggestion was proved by experiments on quenching by Ad of Trp fluorescence in AR native and modified to different extent. The analysis of the received data using Stern-Volmer relations proved the suggestion that the NBS treatment of AR, in the first place, affects the Trp residues more available to Ad (Sargisova et al., 1990a).

2.2. LUMINESCENCE PROPERTIES OF ADRENODOXIN REDUCTASE

The prosthetic group of AR, FAD, and Trp residues in protein molecule are the natural internal fluorophores, which can be used for investigation of their structural and functional relations at different states of the molecule. We studied the AR fluorimetric properties with the Perkin-Elmer MPF-44A spectrofluorimeter in rectangular cuvettes of side 1 cm. The concentration of a protein sample in 10 mM phosphate buffer, pH 7.4, was adjusted so as the optical density at an excitation wavelength be under 0.1. Fluorescence spectrum of AR at room temperature consists of two main bands (Fig. 3). Trp residues of the protein are responsible for one of them. At excitation in the range of Trp absorption, 280-295 nm, the emission maximum at 331 nm is observed (curve 1). The second emission band of AR is characteristic for flavin molecules. The emission maximum at 525 nm is observed at excitation by FAD absorption band at 450 nm (curve 3). In the fluorescence spectra at liquid nitrogen temperature, -196°C, besides sharpening and narrowing these emissions, the wide, structured phosphorescence emission of tryptophans with peaks at 415, 440 and 460 nm appears (Fig. 3, right).

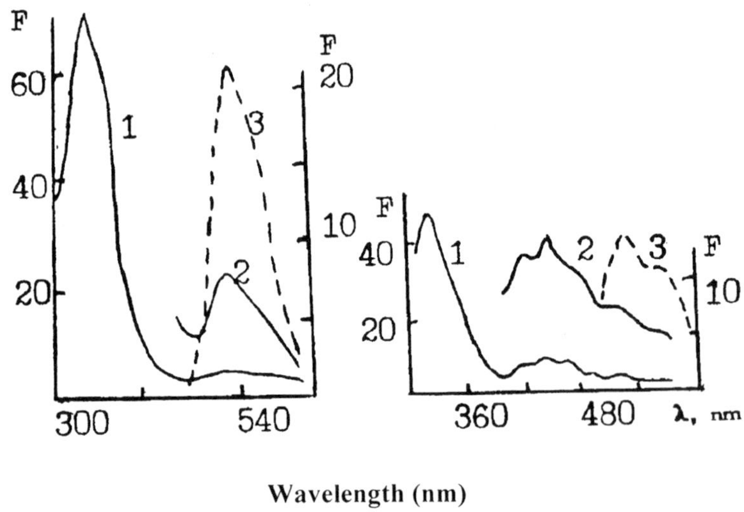

Figure 3. Luminescence spectra of AR at ambient temperature (left) and liquid nitrogen temperature, -196°C (right). Curves 1 and 2 are registered at excitation by 295 nm light; curves 3 are excited at 450 nm.

2.2.1 *Excitation spectra of visible emission of AR*

At excitation by 280-295 nm light in the visible region of emission spectra of both AR and free FAD, a low intensity band at 525 nm was observed (Mardanyan and Sargisova, 1995). In Fig. 4, A and B, respectively, the excitation spectrum of 525 nm emission band, and that in UV-region are shown for different samples. It is seen that the excitation spectra in visible region are the same except for the intensity decrease for FAD involved in protein molecule. But in UV-region, the free FAD spectrum (curve 1) essentially differs from that of FAD in AR molecule at any condition. The wide single band of free FAD splits into two expanded peaks at 260 and 290 nm in the protein molecule (curve 2). The intensity of the band at short wavelength increases if AR is in complex with Ad (curve 3). The spectrum is similar to that in AR at dissociation of [AR x Ad] complex at high ionic strength (in the presence of 0.4 M KCl), yet wider and more intensive (curve 4). The short wavelength excitation spectrum of free FAD emission at 525 nm was not influenced by addition of Ad and ionic strength.

These observations manifest the influence of protein environment on the electronic state of flavin molecule. Moreover, the electronic state of FAD in AR is influenced by [AR x Ad] complex formation and by ionic strength of the medium.

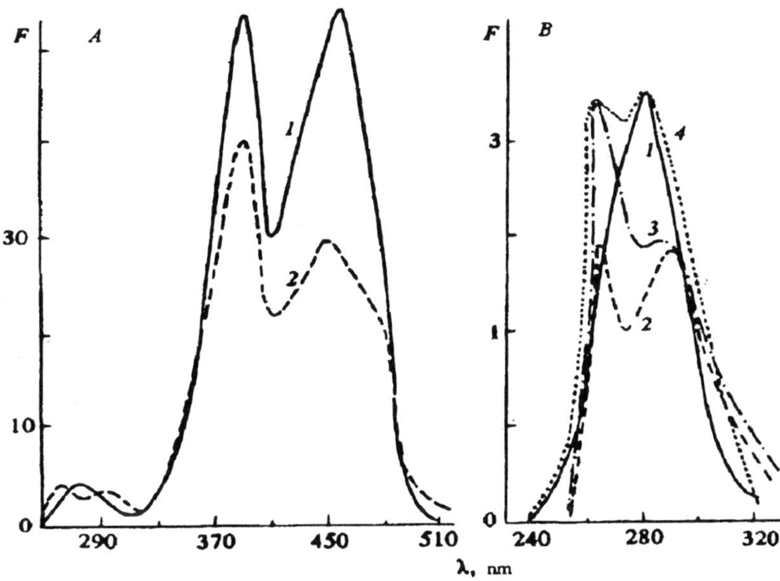

Figure 4. Excitation spectra of FAD fluorescence at emission maximum 525 nm
A – general view. B – the UV region of excitation.
1 – free FAD; 2 – FAD in AR; 3 – FAD in [AR x Ad] complex (equimolar amounts of proteins, 1.2 μM); 4 – the same as 3 in the presence of 0.4 M of KCl.

2.2.2 *Quenching of Trp and FAD fluorescence in AR*

Earlier it was shown that at reconstruction of holo-AR, the FAD incorporation into the apoprotein is accompanied by quenching of fluorescence both of FAD and Trps (Hiwatashi et al., 1976). The quenching of the emission of both fluorophores is also observed at [AR x Ad] complex formation (Hiwatashi et al., 1976; Sargisova et al., 1990a). This quenching allowed to suggest that Trps of AR are in the direct proximity of binding sites both for FAD and Ad (Hiwatashi et al., 1976) . Strictly speaking, the data presented in this work were insufficient for this conclusion. The incorporation of FAD into the apoprotein (as well as the complex formation) might also be accompanied by changes in the environs of both fluorophores leading to an increase in the non-radiation scatter of the excitation energy and, as a result, to a decrease of emission intensity. A direct indication of closeness of the Trps to the FAD binding site might be a demonstration of Förster energy transfer between them. However, the significant quenching of the fluorescence of both fluorophores at the holoprotein reconstruction and the complex formation with Ad (Sargisova et al., 1990a,b; Hiwatashi et al., 1976) made it impossible to follow the energy transfer based on the emission intensity change.

2.2.3. *Depolarization of FAD emission excited at 290 nm*

Then we studied the polarization of FAD emission at 525 nm in some preparations at excitation at 290 nm and 450 nm. In these experiments we used a polarization attachment of the MPF-44A instrument. The polarization was estimated as ratio $p = I_{II} - I \; / \; I_{II} + I$ where I is the emission intensity at perpendicular arrangement of excitation (0°) and emission (90°) polarizers, and I_{II} is the same at their parallel arrangement (usually 0°). The data in Table 1 testify that the polarization of emission in most of the cases is the same at excitation by the two wavelengths. The only exception is the complex [Ad x AR]. In this case, the 525 nm emission excited by 290 nm light is depolarized as compared with that excited by 450 nm. It means that, as a result of [AR x Ad] complex formation, an unidentified group near FAD absorbs 290 nm light and contributes to the emission of FAD.

TABLE 1. *Polarization of FAD emission at 525 nm upon excitation by light at 450 nm (p_1) and at 290 nm (p_2)*

Sample	p_1 (±10%)	p_2 (±10%)	p_2/p_1, %
1. Free FAD	0.3	0.29	96
2. Free FAD + Ad	0.3	0.28	93
3. FAD in AR	0.3	0.27	90
4. FAD in AR + Ad	0.25	0.14	56
5. FAD in AR + Ad + 0.4 M KCl	0.24	0.21	88

2.2.4 *Förster energy transfer between Trp and FAD*

It is worth to note that the spectral characteristics of FAD and Trp are favorable for Förster non-radiation energy transfer from Trp to FAD. Earlier from the integrals of Trp emission and FAD excitation spectra overlap in the 330-340 nm region, the critical distances of Förster transfer from tyrosine and tryptophan to flavins were calculated to be 20 Å (D'Anna and Tollin, 1971). However, the energy transfer requires not only a rather small distance between the fluorophores, but also a certain orientation of the acceptor and the donor dipoles (Weber, 1960a,b).

It is known that the transfer of the excitation energy from one fluorophore to another has to be accompanied by depolarization of the acceptor emission (Weber, 1960a). Hence, the observed depolarization of FAD emission in AR at excitation by 290 nm compared to that by 450 nm allowed us to suggest the Förster energy transfer in AR in complex with Ad. Presumably, in AR, the group(s) transferring emission energy to FAD might be Trps.

Comparing the AR emission at 525 nm at excitation by 450 and 290 nm (Armenyan et al., 1984; Table 1, line 3) we failed to note any depolarization of emission. Hence, we can conclude that the depolarization observed at [AR x Ad] complex formation may be a manifestation of rearrangement or reorientation of Trp towards FAD in the complex.

3. Adrenodoxin Reductase – Adrenodoxin Ñomplex

In the course of stopped flow experiments designed to measure the rates of anaerobic reduction of excess Ad by NADPH plus AR, a small but reproducible lag in reduction was observed, when Ad plus NADPH in one syringe was mixed with AR in the second syringe (Lambeth, 1980). A more reasonable interpretation suggested by author was that the lag represents some conformational change occurring subsequent to rapid association of the complex. This change might occur in either Ad or AR. Using apo- and cobalt-Ad instead of the native, they showed that a conformational change in AR caused the observed lag. The latter reflected a slow (1 sec^{-1}) conversion of flavoprotein from an inactive to an active state (AR AR*).

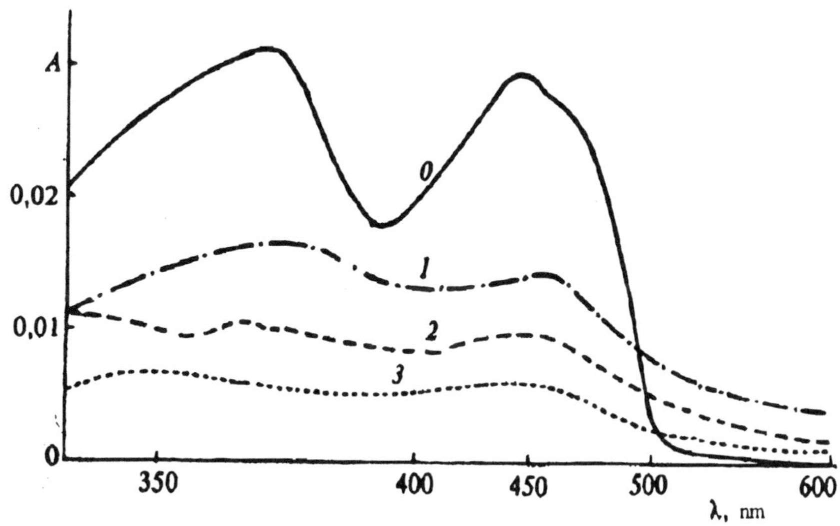

Figure 5. Absorption spectra of NADPH-adrenodoxin reductase (0) and of adrenodoxin complex with native (1) and two (2) and three (3) tryptophans modified reductase.

3.1. *DIFFERENTIAL SPECTRA OF [AR X AD] COMPLEX*

The complex formation between AR and Ad is accompanied by appearance of a characteristic optical spectrum registered in the absorption of the proteins mixture against their identical solutions in different parts of a tandem cuvette. In Fig. 5, curve 1 represents this absorption spectrum. It resembles the FAD spectrum (curve 0) widened and shifted to the longer wavelengths. A similar change in the spectrum of free flavin molecules was described when they interacted with aromatic amino acids, Tyr and Trp (Inoue et al., 1981). The same was observed for absorption of FAD in polypeptide molecule caused by Trp (McKenzie et al., 1969). We can conclude therefore that the absorption spectrum of the complex [AR x Ad] evidences FAD electronic structure perturbation by Trp (or, less probable, by Tyr).

In Fig. 5, curves 2 and 3 show the differential absorption spectra, registered in identical conditions, if native AR is replaced with that after Trps modification. The decrease of the intensity of these spectra testified weakening of FAD electronic state perturbation caused by Trp. This may be caused due to the following reasons:

1. The Trp residue participated in [AR x Ad] complex formation, and its oxidation hindered proteins interaction. A lot of investigations, however, have shown, that formation of complex [AR x Ad] is determined by electrostatic interaction of oppositely charged amino acids on the surfaces of two proteins, and neutral Trp could not be a participant of such interaction.
2. The Trp responsible for perturbation of FAD electronic structure at [AR x Ad] complex formation was .modified. This conclusion seems more likely taking into account the above described experimental results.

This circumstance allowed us to consider the probability of Trp role as a link of intra-complex electron transferring chain. In some other flavin-containing proteins functioning "switch down" of two- to one-electron flow, the presence of an aromatic acid close to the flavin molecule and its possible participation in the electron transfer process was also noted (Burnett et al., 1974, Falk et al., 1976, Ryan and Tollin, 1973, Strittmatter, 1961).

4. Trp Participation in Intramolecular Electron Transfer

The presence of an aromatic residue in close proximity of prosthetic group and its importance for the redox activity of several flavin-containing oxidoreductases had been shown earlier (Nisimoto and Shibata, 1982; Senda et al., 2000). Particularly interesting are the cases when the electrons have to be passed to the acceptor center at the same potential, as it is in the monooxygenase systems (Kamin et al., 1980). Let's note, that the quantum chemical calculations for charge transfer complexes of flavins and aromatic amino acids showed that the former act as electron acceptors and the latter as electron donors (Wantanabe et al., 1982). Thus, a preferential direction of electron transfer between two molecules is from Trp to FAD. It means that in the above stated cases the opposite direction of electron flow was suggested.

Our working hypothesis is that in these cases, the Trp molecule presence creates a barrier for electron flow, allowing biological regulation of a chemical process.

Recently, the 2-3-Å resolution crystal structure of a cross-linked 1:1 complex of Ad and AR was reported (Müller et al., 2001). In this complex, the closest approach of atoms belonging to the [2Fe-2S] cluster of Ad and the isoalloxazine ring of the FAD of AR was estimated 10 Å. The fractional packing density of protein groups between the redox centers was determined, 0.61. These parameters were used to evaluate the electron-tunneling rate between two centers, 10^8-10^9 s^{-1}. Whereas the flavin-to-iron transfer rate was determined experimentally as 3-4 s^{-1} (Lambeth et al., 1982). This discrepancy indicates, that some structural peculiarities were not taken into account in the evaluation of the electron-tunneling rate. It might be the presence of centers of groups in the space between electron donor and acceptor, which were not considered at calculations based on simple distance and density parameters. A Trp molecule may represent the mentioned group.

In AR, Trp376 is placed in Ad binding face of the molecule (Ziegler et al., 1999) confirming our suggestion about probability of its participation in electron transfer from FAD to [2Fe-2S] cluster. It explains also the protection by Ad of Trp against chemical modifications, described above.

The detailed study of the kinetic mechanism for cytochrome c reduction by the complex (Lambeth et al., 1982) brought to the conclusion that the [AR x Ad] complex cycles between a 1- and 3-electron reduced states. The FAD molecule in the course of the catalytic cycle keeps an electron remaining in a half-reduced, semiquinone state. Probably, this electron is necessary to overcome the barrier for electron transfer from FAD to Trp.

5. Conclusions

Based on the described information, we can conclude, that:

1. The differential absorption spectrum of the complex [AR x Ad] evidences the FAD electronic structure perturbation by Trp as a result of complex formation.
2. Depolarization of FAD emission at 525 nm excited by light of 290 nm wavelength in comparison with that excited by 450 nm wavelength in AR is a manifestation of mutual displacement or reorientation of Trp and FAD at [AR x Ad] complex formation.
3. The conformational change in AR noted by Lambeth et al., (1980) is a reflection of the above mentioned displacement or reorientation of Trp (or FAD). The active state of AR in the complex is the state with changed mutual dispositions of Trp and FAD molecules.
4. The presence of Trp on the pathway of the electron transfer from FAD to 2Fe-2S cluster creates a barrier in the electron pathway between the active centers, causing much slower electron rate than it was predicted by electron tunneling calculations (Müller et al., 2001).
5. To overcome this barrier, the FAD molecule in the course of the catalytic cycle must keep an electron, remaining in half-reduced, semiquinone state. Hence, this barrier determines the character of the functioning of the complex [AR x Ad] between the tri- and mono-electron states during the catalytic cycle (Lambeth et al., 1982).

The order of the FAD and Trp arrangement in [AR x Ad] complex and the suggested way of the intramolecular electron transfer from FAD to the site of AR interaction with Ad is proposed as a working hypothesis, and needs to be proved in further investigations.

References

1. Adman, E.T., Sieker, L.C. and Jensen, L.M. (1973) The Structure of a Bacterial Ferredoxin J.Biol.Chem., 248, 3987-3996.
2. Armenyan, A.G., Gasparyan, V.K., Mardanyan, S.S. and Nalbandyan, R.M. (1984) The Luminescent Properties of NADPH-adrenodoxin Reductase. Biokhimia (russian) 49, 1441-1448.
3. Beckert, V., Dettmer, R. and Bernhardt, R. (1994) Mutations of Tyrosine 82 in Bovine Adrenodoxin that Affect Binding to Cytochromes P45011A1 and P45011B1 but not Electron Transfer. J.Biol.Chem., 269, 2568-2573.
4. Bernhardt, R. (1996) Cytochrome P450: Structure, Function, and Generation of Reactive Oxygen Species. Rev. Physiol. Biochem. Pharmacol. 127, 137-221.
5. Burnett, R.M., Darling, G.D., Kendall, D.S, LeQuesne, E., Mayhew, S.G., Smith, W.W. and Ludwig, M.L. (1974) The Structure of the Oxidized form of Clostridial Flavodoxin at 1.9 Å Resolution. J.Biol.Chem. 249, 4383-4392.
6. Carter, C.W., Knaut, J., Freer, S.T., Alden, R.A. and Bartsch, R.G. (1974) Two-Angstrom Crystal Structure of Oxydized Chromatium High Potential Iron Protein. J.Biol.Chem. 249, 4212-4225.
7. Churchill, P.F., de Alvare, L.R.and Kimura, T. (1978) Topological Studies of the Steroid Hydroxylase Complexes in Bovine Adrenocortical Mitochondria. J.Biol. Chem. 253, 4924-4929.
8. Clarke, A. (1987) Essential Tryptophan Residues in the Functions of Cellulase from Schizophillum Commune. Biochim. Biophys. Acta 912, 424-431.
9. D'Anna, J.A. and Tollin, G. (1971) Protein Fluorescence and Solvent Perturbation Spectra, as Probe of Flavin-Protein Interactions in the Shethna Flavoprotein. Biochemistry 10, 57-64.
10. Debus, R.J., Barry, B.A., Sithole, I., Balcock, G.T. and McIntosh, L. (1988) Direct Mutagenesis Indicates that the Donor to P^{+}_{680} in Photosystem II is tyrosine-161 of the D_1 Polypeptide. Biochemistry, 27, 9071-9074.
11. Edelhoch, H., (1967) Spectroscopic Determination of Tryptophan and Tyrosine in Proteins. Biochemistry 6, 1948-1954.
12. Falk, M.C., Johnson, P.G and McCormick, D.B. (1976) Synthetic Flavinyl Peptides Related to the Active Site of Mitochondrial Monoamine Oxidase. I Chemical and Spectral Properties. Biochemistry 15, 639-645.
13. Farkash, Y., Timberg, R. and Orly, J. (1986) Preparation of Antiserum to Rat Cytochrome P450 Cholesterol Side-Chain Cleavage and Its Use for Ultrastructural Localization at the Immunoreactive Enzyme by Protein A-Gold Technique. Endocrinology 118, 1353-1385.
14. Greenfield , N.J., Wu, X. and Jordan, F. , (1989) Proton Magnetic Resonance Spectra of Adrenodoxin Features of the Aromatic Region. Biochim Biophys. Acta 995, 246-254.
15. Hamamoto, I. and Ichikawa, Y. (1984) Modification of a Lysine Residue of Adrenodoxin-Reductase, Essential for Complex Formation with Adrenodoxin. Biochem. Biophys. Acta, 786, 32-41.
16. Hamamoto, I., Kurokohchi, K., Tanaka, S. and Ichikawa, Y. (1988) Adrenoferredoxin-Binding Peptide of NADPH-Adrenoferredoxin Reductase. Biochem. Biophys. Acta, 953, 207-213.
17. Hanukoglu, I. and Jefcoate, C.R (1980) Mitochondrial Cytochrome P450. Mechanism of Electron Transport by Adrenodoxin. J.Biol. Chem. 225, 3057-3061.
18. Hanukoglu, I. and Hanukoglu, Z. (1986) Stoichiometry of Mitochondrial Cytochromes P450, Adrenodoxin and Adrenodoxin Reductase in Adrenal Cortex and Corpus Luteum – Implcations for Membrane Organization and Gene Regulation. Eur. J. Biochem. 157, 27-31.
19. Hanukoglu, I. and Gutfinger, T. (1989) cDNA sequence of Adrenodoxin Reductase Identification of NADPH-Binding Sites in Oxidoreductases. Eur. J. Biochem. 180, 479-484.
20. Hara, T. and Miyata, T. (1991) Identification of Cross-Linked Peptide of a Covalent Complex Between Adrenodoxin-Reductase and Adrenodoxin. J. Biochem. 110, 261-266.
21. Hiwatashi, A., Ichikawa, Y., Yamano, T. and Maruya, N. (1976) Properties of Crystalline Reduced Nicotinamide Adenine Dinucleotide Phosphate – Adrenodoxin Reductase from Bovine Adrenocortical Mitochondria. II. Essential Histidyl and Cysteinyl Residues at the NADPH-Adrenodoxin Reductase. Biochemistry, 15, 3091-3097.

22. Hiwatashi, A., Nishii, Y. and Ichikawa, Y. (1982) Purification of Cytochrome $P450_{D1\alpha}$(25-hydroxylation, D3-1α -hydroxylase) of bovine kidney mitochondria. Biochem. Biophys. Res. Commun. 105, 320-327.
23. Inoue, M., Shibata, M., Kondo, Y. and Ishida, T. (1981) Role of Tryptophanyl and Tyrosine Residues of Flavoproteins in binding with Flavin Coenzymes. X-ray Structural Studies using Model Complexes. Biochemistry, 20, 2936-2945.
24. Kamin, H., Lambeth, J.D. and Siegel, L.M. (1980) The Role of Flavins in Electron Transfer Between Two-Electron Donors and One-Electron Acceptors. In Flavins and Flavoproteins (Yagi, K and Yamano,T. eds.) pp. 341-348, Japan Scientific Societies Press, Tokyo.
25. Lambeth, J.D., Seybert, D.W. and Kamin, H, (1980) Adrenodoxin-Reductase-Adrenodoxin Complex. Rapid Formation and Breakdown of the Complex and a Slow Conformational Change in the Flavoprotein. J. Biol. Chem. 255, 4667-4672.
26. Lambeth, J.D., Seybert, D.W., Lancaster, J.R., Jr., Salerno, J.C. and Kamin, H. (1982) Steroidogenic Electron Transport in Adrenal Cortex Mitochondria. Mol. Cell. Biochem. 45, 13-31.
27. Mardanyan, S.S. and Sargisova, Y.G. (1995) Interaction of Flavin Adenine Dinucleotide and Tryptophan of NADPN-Adrenodoxin Reductase in Complex with Adrenodoxin. Biophysica, 40, 19-25.
28. McKenzie, R.E., Fory, W. and McCormick, D.B. (1969) Flavinyl Peptides. II. Intramolecular Interactions in Flavinyl Aromatic Amino Acid Peptides. Biochemistry 8, 1838-1844.
29. Muller, E.-C., Lapko, A., Otto, A., Müller, J.J., Ruckpaul, K. and Heineman, U. (2001) Covalently Crosslinked Complexes of Bovine Adrenodoxin with Adrenodoxin Reductase and Cytochrome P450scc. Mass Spectrometry and Edman Degradation of Complexes of the Steroidogenic Hydroxylase System. Eur. J. Biochem. 268, 1837-1843.
30. Müller, J.J., Lapko, A., Bourenkov, G., Ruckpaul, K. and Heineman U. (2001) Adrenodoxin Reductase-Adrenodoxin Complex Structure Suggests Electron Transfer Path in Steroid Biosynthesis. J.Biol.Chem. 276, 2786-2789.
31. Nisimoto, Y. and Shibata, Y. (1982) Studies on FAD- and FMN-Binding Domains in NADPH-Cytochrome P450 Reductase from Rabbit Live Microsomes. J. Biol. Chem. 257, 12532-12539.
32. Okajima, T., Kawata, Y. and Hamaguchi, K. (1990) Chemical Modification of Tryptophan Residues and Stability Changes in Proteins. Biochemistry 29, 9168-9175.
33. Omura,T., Sanders, E., Estabrook, R.W., Cooper, D.Y. and Rosenthal, O. (1966) Isolation from Adrenal Cortex of a Nonheme IronProtein and a Flavoprotein Functional as a Reduced Triphosphopyridine Nucleotide-Cytochrome P-450 Reductase. Archives Biochem. Biophys. 117, 660-673.
34. Ryan, J. and Tollin, G. (1973) Flavin-Protein Interactions in Flavoenzymes. Effect of Chemical Modification of Tryptophan Residues upon Flavin Mononucleotide Binding and Protein Fluorescence in Azotobacter flavodoxin. Biochemistry 12, 4550-4554.
35. Sargisova, Y.G. and Mardanyan, S.S. (1987) The role of Tryptophanyl Residues in Complex Formation of NADPH-Adrenodoxin Reductase with Adrenodoxin. Biokhimia (russ) 52, 1258-1262
36. Sargisova, Y.G., Mardanyan, S.S. and Haroutunian, A.V. (1990a) Significance of Trp in NADPH-Adrenodoxin Reductase Electron Transfer to Adrenodoxin. Biological J. of Armenia 43, 778-783.
37. Sargisova, Y.G., Mardanyan, S.S. and Haroutunian, A.V. (1990b) The role of Tryptophanyl Residues in Electron Transfer from NADPH-Adrenodoxin Reductase to Adrenodoxin. Biochemistry International 22, 977-982.
38. Senda, T., Yamada, T., Sakurai, N., Kubota, M., Nishizaki, T., Masai, E., Fukuda, M. and Mitsui, Y. (2000) Crystal Structure of NADPH-dependent Ferredoxin Reductase Component I Biphenyl Dioxygenase. J.Mol.Biol. 304, 397-410.
39. Solish, S.B., Leonard, P., Morely, Y., Kuhn, R.W., Mohaudos, T.K., Hanukoglu, I. and Miller, W.L. (1988) Human Adrenodoxin Reductase: Two mRNAs Encoded by a Single Gene on Chromosome 17cen-g25 Expressed in Steroidogenic Tissues. Proc.Nat.Acad.Sci. USA 85, 7104-7108.
40. Spande, T.P. and Witkop, B. (1967) Determination of the Tryptophan Content of Proteins with N-bromosuccinimide. Methods Enzymol. 11, 498-506.
41. Strittmatter, P. (1961) The Nature of the Flavin Binding in Microsomal Cytochrome b_5 Reductase. J.Biol.Chem. 236, 2329-2335.
42. Taniguchi, T. and Kimura, T. (1976) Studies of Nitrotyrosine-82 and Aminotyrosine-82 Derivatives of Adrenodoxin, Effect of Chemical Modification of the Complex Formation with Adrenodoxin Reductase. Biochemistry, 15, 2849-2853.

43. Toulme, J.J., Le Doan, T. and Helene, C. (1984) Role of Tryptophanyl Residues in the Binding of Gene 32 Protein from Phage T4 to Single-Stranded DNA. Photochemical Modification by Trichloroethanol. Biochemistry 23, 1195-1201.
44. Vickery, L.E. (1998) Molecular Recognition and Electron Transfer in Mitochondrial Steroid Hydroxylase Systems. Steroids 62, 124-127.
45. Waki, N., Hiwatashi, A. and Ichikawa, Y. (1986) Purification and Biochemical Characterization of Hepatic Ferredoxin (Hepatoferredoxin) from Bovine Liver Mitochondria. FEBS Lett. 195, 87-91.
46. Watanabe, Y., Nishimoto, K. and Kashiwagi, H. (1982) Ab Initio MO Study of the Stacking Complexes, Flavin-Tyrosine, Flavin-Tryptophan, and Flavin-NADH. In Flavins and Flavoproteins, p 537-540.
47. Weber, G. (1960a) Fluorescence Polarization Spectrum and Electronic-Energy transfer in Tyrosine, Tryptophan and Related Compounds. Biochem. J. 75, 335-345.
48. Weber, G. (1960b) Fluorescence Polarization Spectrum and Electronic-Energy Transfer in Proteins. Biochem. J. 75, 345-352.
49. Wikvall, K. (1984) Hydroxylations in Biosynthesis of Bile Oxides. J.Biol.Chem. 259, 3800-3804.
50. Ziegler, G.A., Vonrhein, C., Hanukoglu, I. and Schulz, G.E. (1999) The Structure of Adrenodoxin Reductase of Mitochondrial P450 Systems: Electron Transfer for Steroid Biosynthesis. J. Mol. Biol. 289, 981-990.

INTERACTION OF NUCLEIC ACIDS AND LIPIDS FROM TUMOUR CELLS WITH ANTICANCER DRUGS: AN SEIRA SPECTROSCOPY DATA.

G.I. DOVBESHKO, O.P. REPNYTSKA.
1-Institute of Physics of National Academy of Sciences of Ukraine, Prospect Nauki 46, Kyiv, 03039, Ukraine;
V.P. TRYNDIAK, I.M. TODOR, G.I. SOLYANIK, V.F. CHEKHUN.
2-Kavetsky Institute o f Experimental Pathology, Oncology and Radiobiology, Vasylkivska 45, Kyiv, 03022, Ukraine.

It is known that malignant tumour progression in the direction of the emergence of tumour drug resistance is one of the main factors resulting in the failure of cancer treatment. As tumour progression is based on the genetic instability, the study of the structure of nucleic acid from tumour cells is of great importance both for basic knowledge and biomedical application. We applied surface enhanced infrared spectroscopy (SEIRA) of nucleic acids on gold substrate and essentially increased the sensitivity of IR spectroscopy. We observed numerous changes in infrared spectra of DNA from sensitive and resistant cells that reflect drastic changes in molecular structure of DNA from tumour cells. The DNA from resistant cancer cells could be characterised as rigid structure, the structure of DNA from sensitive cancer strain seems to be flexible and after application of anticancer drugs drastically changes and approaches to the structure of helix forms. The molecular structure of lipids from resistant and sensitive cancers after application of anti-tumour drugs is modified also. Thus, we observed a disordering in the lipid chain packing from resistant cells after application of cisplatin and formation of phospholipid-Pt complex in some cases.

1. Introduction.

Despite the availability of approximately 60 clinically active anticancer agents eventual treatment failure is the outcome in the chemotherapy of the great majority of types of cancer. Among a number of factors resulting in the failure of cancer chemotherapy tumor drug resistance continues to be one of the major.

Multiple potential mechanisms of resistance have been identified at the cellular and molecular levels for the most active and widely used cytotoxic anticancer drugs cisplatin and doxorubicin (Kartalou et al., 2001).

Alterations in cellular pharmacology, including decreased drug accumulation, increased cellular thiol levels and increased repair of platinum-DNA damage, have been observed in numerous cisplatin-resistant model systems (Yoon et al.,2001; Haghighi et al., 1999). More recently, it has become apparent that an enhanced capacity to tolerate cisplatin-induced damage may also contribute to the emergence of tumor drug resistance against cisplatin action (Blommaert et al., 1998; Meijera et al., 2001). The doxorubicin resistance was shown to involve such as an impaired transport of drugs across the plasma membrane, an increased drug efflux, an increased intracellular detoxification of drugs, a reduced availability of intracellular drug targets, or a reduced

E. Buzaneva and P. Scharff (eds.), Frontiers of Multifunctional Nanosystems, 265–280.

ability of cells to induce apoptosis (Maestre et al., 2001).

In spite of the mentioned above multiplicity of the drug resistance mechanisms the evidence from a variety of sources implicates the genetic basis for most drug resistant phenotype (Richardson et al., 1997). By this is meant the genetic instability of cancer cells which includes mutations, deletions, transposition of genetic elements, gene amplifications, chromosomal rearrangements and translocations. So DNA is not only a macromolecular receptor for many important anticancer drugs (for instance, cisplatin forms covalent linkages with DNA while doxorubicin forms more noncovalent complexes with this macromolecule, but the changes in DNA structure is considered to play crucial role in the mechanisms of tumor drug resistance (Yang et al., 1999).

Earlier the clear changes in structure of tumour nucleic acids in comparison with those from normal tissues have been registered with FTIR spectroscopy by Mallins for DNA (Mallins et al., 1995) and by us for RNA (Dovbeshko et al., 1999; Dovbeshko et al., 2000). So, we already studied nucleic acid damage, and wanted to increase the sensitivity of FTIR spectroscopy. Thus the idea arose to apply surface enhanced infrared spectroscopy (SEIRA) in studies of nucleic acids and lipids structural peculiarities from sensitive and resistant cancer cells and their interaction with anticancer drugs (Dovbeshko et al., 2001). The SEIRA enhancement factor in the experiment could be reached up to 100 (Kosobukin, 1983).

2. Methods and materials.

FTIR methods, with their increased sensitivity (i.e. signal/noise ratio), are essential for structural studies of biological molecules. However, in modern genetic and biochemical applications where only very small amounts of materials are available, FTIR alone often cannot provide sufficient sensitivity. Nucleic acids belong to the materials whose isolation requires materials and time, so any technique for enhancing the FTIR signal is very important for analytical methods. A rough metal surface as substrate for studying absorbed molecules could lead to enhancement of the FTIR signal in reflectance or attenuated total reflectance modes. Consequently we have recorded enhanced FTIR signals from vibrations of nucleic acids deposited on noble metal surfaces in an experiment with reflectance attachment. The geometry of the experiment is presented on *Figure 1.*

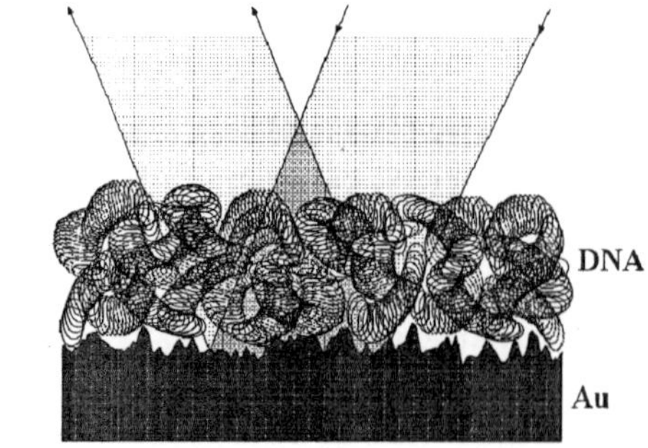

Figure 1. The geometry of SEIRA experiment

2.1 *FTIR SPECTRA EVALUATION.*

The FTIR-SEIRA spectra were collected in the 380-5300 cm^{-1} region with an IFS 48 Bruker instrument and the reflectance attachment (a light incidence angle close to 16.5°). After measuring the FTIR spectra we deconvoluted the bands using the Opus-2.2 program. This performs a Fourier-self-deconvolution of the spectrum. The aim of this is to enhance the apparent resolution of a spectrum or decrease the width of all lines contributing to the investigated spectral range. Positions of the bands have been estimated using the second derivative method and/or standard method. The bands have been assigned to the certain molecular groups in DNA and lipids according to references (Taillandier et al., 1985; Tajmir-Riahi et al., 1995; Schrader, 1995; Parker, 1983; Baret et al., 1978; Arrondo et al., 1998, Binder et al., 2000, Mantsch et al., 1991, Mendelson et al., 1986). For the estimation of halfwidth, integrated intensity and real frequencies of overlapping bands, the complex spectral bands in the regions 3800-2300 cm^{-1} (region of the OH-NH-CH stretching vibrations), 1800-1300 cm^{-1} (region of C=O, C=C, C=N stretching, C-H and N-H deformation vibrations), and 1300-1000 cm^{-1} (region of phosphate vibrations) were decomposed with the option "Curve fit" of Opus 2.2. Shapes of different spectral bands under decomposition have been approximated with sums of Lorenz and Gauss functions.

The enhancement factor of SEIRA for different vibrations was calculated as the ratio of integral areas of corresponding bands in SEIRA compared to the conventional geometry. The shapes of the main bands and their frequency positions for DNA and lipids were practically unchanged in SEIRA in comparison with conventional geometry.

2.2. *AFM IMAGING.*

In SEIRA experiment we used Au of 200-400Å thickness supported on a glass plate, with a roughness not greater than 50Å. The microphotograph of the Au surface used as substrate for SEIRA was obtained by atomic force microscopy (AFM) (*Figure 2*). We used the tapping mode under AFM imaging with a commercial Nanoscope IIIa (Digital Instrument, Santa Barbara, CA). Tapping force mode scans were performed using commercially available AFM tips (silicon nitride). The scanning frequency was approximately 1 Hz in all experiments.

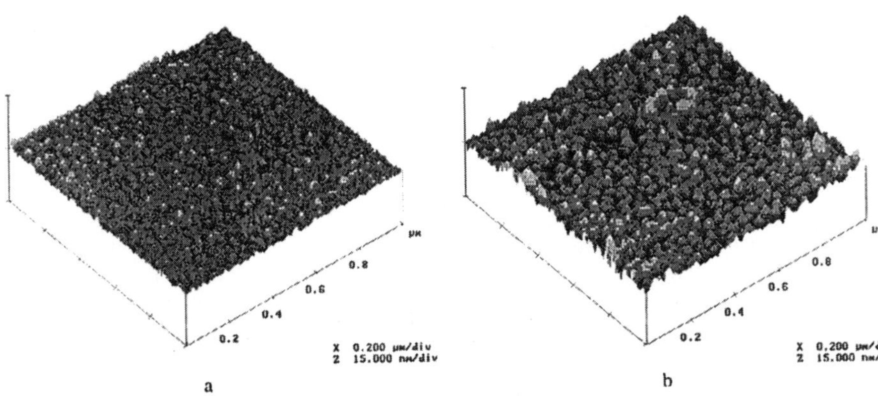

Figure 2. AFM image of the gold surface used in SEIRA experiment: a – thickness of Au film is 200Å, b - 300Å

2.3. *PREPARATION OF DNA AND PHOSPHOLIPIDS*

The samples were the total fraction of DNA isolated from frozen tissues of sensitive, doxorubicin-resistant and cisplatin-resistant strains of Carcinoma Guerina of Wistar line rats by conventional phenol method (Maniatis et al., 1984). The changes in the structure of DNA caused by application of the anticancer drug doxorubicin and cisplatin to these 2 types of cancer strains were also the subject of investigation. The DNA was isolated from cancer cells of rats exposed by 9.5 mg/kg doxorubicin and 6 mg/kg cisplatin during 1-4 hour. The samples of DNA from aqueous solution (1mg/ml) were deposited on the Au surface of 200-400Å thickness and were then liophilically dried. Protein and RNA impurity content of the DNA was less than 1%. Gel electrophoresis data do not show low-molecular fragments in the DNA under study.

The total fraction of phospholipids from plasma membranes of both strains of Carcinoma Guerina of Wistar line rats was isolated by the Folch method (Folch et al., 1957; Sips et al., 1978). Then they were deposited on Au surface from chloroform solution (10 mg/ml) and dried under room condition during 0.5-1 minute. The phospholipids isolated from plasma membrane of both strains after the application of anticancer drugs were the subject of investigation also. In experiment in vitro the plasma membrane of sensitive and resistant strains was incubated with 50 μM doxorubicine or cisplatine during 0.5 hour and then the total fraction of phospholipids was isolated.

All samples (DNA and phospholipids) were kept in closed boxes with P_2O_5 water adsorbent. The humidity of DNA under study was 60% and humidity the phospholipides was 7% in experiment in vivo and 15% in experiment in vitro. In addition the experiment on influence of cisplatin and doxorubicine on the phospholipids (with fraction of short peptides) of plasma membrane, isolated from sensitive strains, was done in vitro.

We studied 3 samples of each types (sensitive, cisplatin-resistant, doxorubicin-resistant and the same after application of anticancer drugs) from different rats. So, we have no enough statistical data, and presented here the individual spectra without any averaging.

3. SEIRA theoretical basis.

The effect of surface enhanced infrared absorption (SEIRA) has become well known during the last 20 years (Chang et al., 1982). Its essential effect is an increase (by a factor $\sim 10^2$) of the intensities of some IR bands of the molecules adsorbed on metal particles or surfaces. It can be accompanied by increase of intensities of some bands or even decrease of intensities of other bands (which can be characterized as «selection rules»). This effect was first reported by Hartstein et al. in 1980, who recognized its similarity to the well-known surface enhanced Raman spectra (SERS) effect. However, only recently it was named as SEIRA and saw practical application in the last decade (Masatoshi Osawa et al., 1991; Ken-ichi Ataka et al., 1996; Wanzenbuck et al., 1998; Kuhne et al., 1998; Roseler et al., 1998), mainly due to intensive studying of biological molecules. The interpretation of SEIRA, similar to SERS, is that it consists of at least 2 mechanisms: local enhancement of external electric field near the rough metal surface due to excitation of surface plasmons, and specific increase of the polarizability of the molecules at their adsorption on metal surface (Kosobukin, 1983).

The theoretical interpretation of the effect is connected with enhancement of external electric field due to excitation of local (surface) plasmon vibrations at the essential curvature of rough surface or on metal particles. In such systems, according to (Kosobukin, 1983), the electric field at a point $\boldsymbol{r}$ could be written as follows :

$$\boldsymbol{E}(\boldsymbol{r},t) = exp\ (-\ iùt)\,[\,\boldsymbol{E}_0(ù) + \boldsymbol{E}_1\,(\boldsymbol{r},ù)], \tag{1}$$

Here $ù$ is frequency, t is time. The additional term $\boldsymbol{E}_1\,(\boldsymbol{r},ù)$ is connected with excitation of local (surface) plasmon vibrations. The enhancement of the external field can be presented as:

$$E_\alpha(\boldsymbol{r},\omega) = g_{\alpha\hat{a}}\,(\boldsymbol{r},\omega)E_{0\hat{a}}(\omega)\,, \tag{2}$$

where $g_{\alpha\hat{a}}\,(\boldsymbol{r},\omega))$ is coefficient of the enhancement.

For the transition dipole moment matrix elements connecting states i (with energy ε_i) and f of the adsorbed molecule, we have:

$$\langle f\,|\,d_\alpha(\boldsymbol{r})\,|\,i\rangle = h_{\alpha\beta}{}^{(fi)}\,(\boldsymbol{r})\ \langle f\,|\,d_{0,\beta}\,|\,i\rangle\ , \tag{3}$$

where $\langle f\,|\,\boldsymbol{d}_0\,|\,i\rangle$ is the transition dipole moment for an isolated molecule.

In a classical approximation, the matrix elements of the Hamiltonian for the interaction of the adsorbed molecule with field are:

$$\langle f\,|\,H\,(r,\omega)\ i\rangle = -\langle f\,|\,d_\alpha(r)\,|\,i\rangle E_\alpha(r,\omega) = -h_{\alpha\beta}{}^{(fi)}\,(r)g_{\alpha\gamma}(r,\omega)\ \langle f\,|\,d_{0,\beta}\,|\,i\rangle\,E_{0,\gamma}\,(\omega)\ , \tag{4}$$

Having performed substitution from Eq. (4) into the expression

$$\omega_{fi}(r,\omega) = (2\eth/\text{ž}\ \ |\langle f\,|\,H(r,\omega)\,|\,i\rangle|^2\ \delta(h\omega - \varepsilon_f + \varepsilon_i)\ , \tag{5}$$

for the rate of the $i\rightarrow f$ transition, multiplied by the energy $h\omega$ of absorbed quantum, and divided by the density of the incident energy flow $cE_0{}^2/8\pi$ (where c is the velocity of light), we obtain the expression for the effective cross-section of light absorbance by an adsorbed molecule:

$$\sigma_\alpha(r,\omega) = \frac{16\pi^2\omega_{fi}}{cE_0^2}\,|\langle f\,|\,H(r,\omega)\,|\,i\rangle\,|^2\delta\,[\,h\,(\omega\text{-}\omega_{fi})] \sim \sigma_\alpha{}^{(0)}\,|\,g(r,\omega)\,|^2\,|\,h(r)\,|^2\ , \tag{6}$$

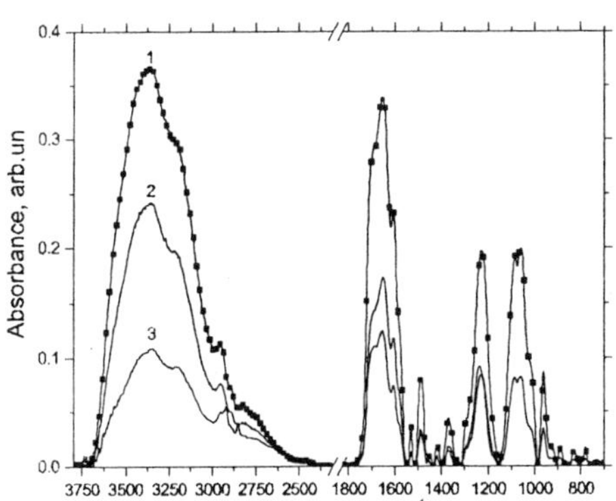

Figure 3. SEIRA spectra of calf thymus Na-DNA (Serva): 1 – on gold substrate with 200Å thickness, 2 – 300Å thickness, 3 – on CaF_2 substrate.

where $\sigma_\alpha{}^{(0)}$ is the cross-section for the case when metal is absent *(d=d_0 E=E_0)*. The right - hand part of the expression (6) is an order-of-magnitude expression for the cross-section, which shows the dependence of the cross-section on the (tensor-averaged) coefficients of amplification for the electric field g and dipole moment h of the transition near the metal surface. The influence of the last factor is probably not essential in some experiments where the transition moments for the absorbed molecules remain unchanged.

The estimation shows that the enhancement effect is increased in the case when the characteristic size of roughness of the surface is close to 50-100Å. In our experiment, we used a gold substrate with just this roughness (*Figure 2*) and obtained enhancement of SEIRA by factors of 3-5. The most enhancement was obtained for gold substrate of 200Å thickness, which has less roughness of gold surface than the same for 300Å thickness Au film. We supposed that maximum in enhancement is due to the fact that nucleic acid has more crossing places with gold in the case of the 200Å film in comparison with those on the 300Å Au film. We assume that this is not the maximal enhancement that could be obtained in such system. In the most cases amplification of the electric field g could be expressed by the ratio $\varepsilon'/\varepsilon''$ of the metal surface, where ε' and ε'' are the real and imaginary part of dielectric permeability of the metal surface (Kosobukin, 1983). For our Au film ε' equals to 566 and ε'' equals to -35.7 at 2800 cm^{-1}.

So, in this case the factor of theoretical enhancement is estimated as 16, in our experiment this value is 5. As it is seen from *Figure 3*, there is no deformation of shape of the bands for DNA on metal substrate, the position of the majority of the bands on the metal substrate are close to those on a CaF_2 substrate (Dovbeshko et al., 2001). So, we can conclude that SEIRA can be used for nucleic acid studies. The SEIRA method was already applied for studies of biological membranes (Kuhne et al., 1998).

4. Result and discussion.

Earlier D. Mallins recorded some features for DNA isolated from breast cancer and tissues exposed by chemicals (Malins et al., 1995; Malins et al., 1997). Some years ago we studied RNA isolated from rats (normal and with tumours) and found a number of specific features for RNA from tumour tissue also (Dovbeshko et al., 1999; Dovbeshko et al., 2000). These features could be connected with structural damage of nucleic acids (Dovbeshko et al., 2000). In (Dovbeshko et al., 1999; Dovbeshko et al., 2000) we have registered the FTIR absorbance spectra of tumour brain nucleic acids on CaF_2 and Si substrates, so it wasn't possible to study Phosphate I and sugar vibrations in the 600-1000 cm^{-1} region due to restricted transmittance of these substrates. Mallins (Malins et al., 1995; Mallins et al., 1997) also paid no special attention to this region. The application of SEIRA to nucleic acids on an Au/SiO_2 substrate resulted in good quality spectra in the 1100-600 cm^{-1} range with enhancement factor of 3-5 and revealed additional structural features of tumour nucleic acids (Dovbeshko et. al., 2001). Based on our previous data we used SEIRA spectroscopy for studies of DNA isolated from sensitive and resistant strains before and after application of anticancer drugs. SEIRA spectra have been recorded in four main regions: sugar moieties (600-900 cm^{-1}), sugar-phosphate backbone and bases (900-1500 cm^{-1}), bases (1500-1800 cm^{-1}) and the NH-OH-CH stretching vibrations of all molecular groups (2600-3600 cm^{-1}).

4.1. *VIBRATIONAL AND STRUCTURAL PECULIARITIES OF NUCLEIC ACIDS, ISOLATED FROM TUMOUR CELLS.*

In the spectral region corresponding to the sugar moieties of tumour brain RNA, we have identified both north-type (*$C_{3'}$ endo*) and south-type (*$C_{2'}$ endo*) sugar conformations (Dovbeshko, 2001). As usual in total fraction of RNA isolated from a normal brain, we have not observed the $C_{2'}$ *endo* sugar conformation. In other tumour RNA we observed suppression of the sugar vibrations in the 600-900 cm^{-1} region. According to (Saenger, 1989), in the A form of the helix RNA bases and sugars can adopt a $C_{3'}$ *endo/anti* conformation; in the Z-form they adopt $C_{3'}$ *endo/syn* conformation, seen in purine (guanosine and adenosine); and the $C_{2'}$ *endo/anti* conformation in pyrimidine (uridine and cytidine). However, recently it was observed that in RNA hairpins the secondary structure of ribonucleosides could be flexible and adopt both $C_{3'}$ *endo/syn* and $C_{2'}$ *endo/anti* conformations for all bases in the Z-form helix (Gaigeot et al., 2000).

In the spectra of DNA and RNA, we have studied two wide bands near 1090 (Phosphate I, symmetric vibration) and 1230 cm^{-1} (Phosphate II, asymmetric vibration) attributed to vibrations of sugar-phosphate backbone. Each of these features is in fact several overlapping bands. Decomposition allows one to estimate the positions and intensities of these band components. In the FTIR spectra of tumour RNA on Au/SiO_2 we found band at 1147-1155 cm^{-1} with increased intensity in the case of the more malignant tumour. In the absorption spectra of reference RNA, this band is practically absent. This band was recorded in the spectra of the DNA, isolated from tumour cells by Mallins (Mallins et. al., 1995). However the shoulder at 1150 cm^{-1} was only perceptible in his experiment using deconvolution techniques. We have achieved a much-improved observation of this band due to using a gold substrate, as well as due to the peculiarities of RNA structure (Dovbeshko et. al., 2001).

The spectral parameters of the Phosphate I band and Phosphate II band was drastically changed in RNA from tumour cells in comparison with those for reference RNA. The redistribution of intensities in the main peaks of (near 1243 cm^{-1} and 1223 cm^{-1} for reference RNA) and high frequency shift of the peaks in the case of the most malignant tumour has also been observed in tumour RNA. More vibrational states were obtained under the decomposition of these phosphate bands in the case of RNA and DNA. In (Dovbeshko et.al., 2000) we should apply the method of deconvolution for this purpose, as we recorded spectra of nucleic acids from tumour cells on Si and CaF_2 substrates. In the case of application SEIRA (Dovbeshko et al., 2001), the gold substrate clearly leads to increased spectral resolution, and hence the possibility of observing new components.

In the region of OH -NH-CH stretching vibrations in tumour DNA and RNA we observed a disappearance and/or decreasing of weak OH-bonds and increasing of NH bonds. In some cases we observed the appearance of drastic increasing of intensity of CH-bands. In the region of bases we observed the changes in intensity and frequency of C=O...H-N as well as N-H...N bonds of the bases.

4.2. *GENERAL SPECTRAL FEATURES OF DNA FROM TUMOUR CELLS.*

Generally, tumour DNA looks like tumour RNA with less prominent spectral features. The main features are the following:

i) redistribution of the components of the Phosphate bands and a shift in their peak positions;

ii) appearance of un-usual sugar and base conformations;

iii) redistribution of the H-bond nets;

iv) appearance of additional vibrational states.

Along with the afore-mentioned, in DNA isolated from tumour cells we have observed a decrease (by a factor of 1.5-2) in intensity of the band at 964 cm^{-1} which could be assigned to vibrations of C-C and C-O in deoxyribose (Baret et al., 1978; Parker, 1983)

4.3. *PECULIARITIES OF STRUCTURE OF DNA PERCIPITATED ON GOLD SUBSTRATE.*

The DNA right-handed helix blocks adopts the $C_{3'}$ *endo/anti* conformation in the A form and $C_{2'}$ *endo/anti* in the B form (Saenger, 1989). Their spectral band positions are 890 cm^{-1}, 878 cm^{-1}, 860 cm^{-1} and 805 cm^{-1} for the $C_{3'}$ *endo/anti* (A-form helix) conformation as well as 890 cm^{-1} and 835 cm^{-1} for the $C_{2'}$ *endo/anti* (B-form helix) conformation (Taboury, 1985). In the left-handed helix DNA (Z-form) the positions of the IR bands are 925-929 cm^{-1}, 868 cm^{-1}, 835-840 cm^{-1} and 802-805 cm^{-1} in the $C_{3'}$ *endo/syn* and $C_{2'}$ *endo/anti* sugar-bases geometries (Taboury et al., 1985; Schrader, 1995).

It is known that DNA conformations in the film and their IR spectra (especially the symmetrical phosphate band) strongly depend on the humidity of the film (Taillandier et al., 1985; Boal et al., 2000; Blagoi et al., 1994; Semenov et al., 1994). Consequently, we repeatedly tested the humidity of the sample (20 times per hour), and repeated the measurements the following days. In all our experiments it was 60% humidity, and we recorded reproducible spectra. However, for Na-DNA from a calf thymus (Serva) on the gold substrate we have observed the main bands in the sugar region at the following frequencies: strong 929 cm^{-1} (Z-form), strong 889 cm^{-1} (both A and B-forms), 860 cm^{-1} (A-form) and strong 830 cm^{-1} (B-form or Z-form). If we estimate according to E. Tailander et al., 1985 the contribution of the A and B forms, we find a big content the B-form. This seems impossible in our experimental conditions of 60% humidity, where a disordered (a coil) structure could be expected with a much greater contribution from the A-helix form (the majority of the spectral markers for Na-DNA on the gold substrate is close to the markers of A-form of DNA according to Schrader, 1995, namely, 1090 (in parents are presented data from Schrader, 1995), 1238 (1240), 1273 (1275), 1370 (1375), 1419 (1418), 1700 (1705).

We must assume that any substrate can strongly influence the sugar conformation of DNA, even in the case when other structural DNA components does not feel this influence (see *Figure 1*, insert). So the influence of substrate on the nucleic acid spectra should be estimated separately in every case. In our case, we deduce that the gold substrate induces conformational change of the sugars if they come closely to gold peaks (*Figure 2 a, b*). The band observed at 830 cm^{-1} is presumably due to reformation of intermolecular H-bonding (NH, OH and CH groups) near to the gold substrate (Boal et al., 2000), as the result of the bending DNA at gold peaks. It is also possible that water molecules take active part in this process.

A similar process was discussed in Gaigeot, et al. 2000, in connection with non-coincidence of the data obtained with inelastic scattering of nucleic acid blocks in solid state and calculations with density functional theory for the spectral region under 900 cm^{-1}. They proposed that intermolecular H-bonds are formed in the film involving N_1-H and N_3-H groups of the bases, giving strong band at the 830 cm^{-1}.

In the case of RNA from normal tissue we had not recorded the band at 830 cm^{-1}, the molecule of RNA shows its rigid properties and the gold substrate does not influence on the IR spectra in the sugar range. Only in the case when tumour process introduced drastic changes in structure of RNA, we have registered the appearance of this band (due to structural peculiarities of RNA, however IR spectra manifest themselves due to gold substrate). We tried to use this spectral region for DNA on gold substrate as analytical tool for detection of interaction of DNA with anticancer-drugs applied in vivo.

4.4 *DNA FROM DOXORUBICINE AND CISPLATIN SENSITIVE AND RESISTANT CELLS.*

The doxorubicin *(Figure 4a)* and cisplatin *(Figure 4b)* cancer cell resistance is accompanied by the special features of the infrared spectra of DNA in the region of phosphate and sugar vibrations. For DNA from resistant strain the tendency of the increasing of the Phosphate asymmetrical intensity in comparison with those for DNA from sensitive strain is observed *(Figure 5)*. The component analysis of complex band of Phosphate symmetric (ratio of integral intensity band at 1052 cm^{-1} to integral intensity at 1071 cm^{-1}) showed that the deviation of DNA structure from helix canonical form (A and B form) are less in DNA-resistance strain (for both cisplatin and doxorubicin) in comparison with DNA-sensitive strain *(Figure 6)*. The SEIRA spectra studies of sugar moieties confirm the diversity of conformation forms of DNA structures of sensitive and resistance strains for both anticancer drugs under study *(see 4.5.2)*.

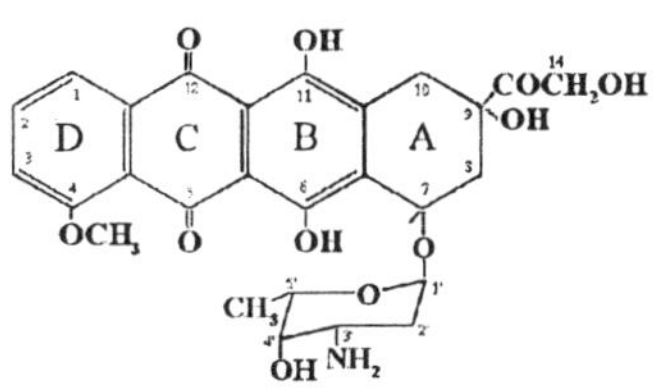

a: doxorubicin

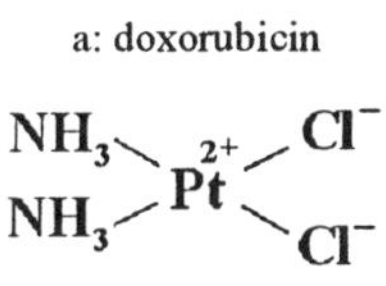

b: cisplatin

Figure 4. The chemical structure of anticancer drugs.

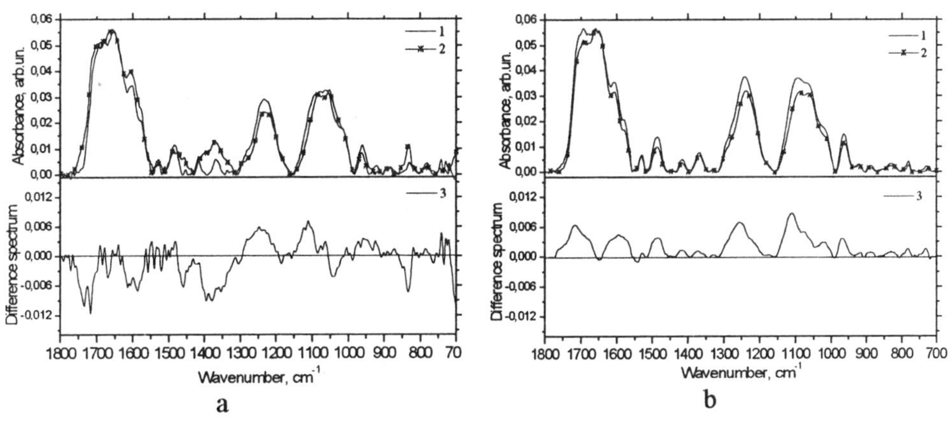

a b

Figure 5. FTIR spectra of DNA from resistant and sensitive strains:
a) 1 – doxorubicin-resistant, 2 – doxorubicin-sensitive, 3 – the difference spectrum
b) 1 – cisplatin-resistant, 2 – cisplatin-sensitive, 3 – the difference spectrum

4.5. DOXORUBICINE-INDUCED DNA STRUCTURE OF SENSITIVE AND DOXORUBICINE- RESISTANT STRAIN.

4.5.1. Phosphate region

Drastic anticancer drug - induced structural difference between DNA from resistant and sensitive strains have been observed. The doxorubicin application to sensitive cancer cells resulted in more changes in Phosphate asymmetrical intensity of DNA in comparison with those for resistant cells *(Figure 7)*. The integral intensity increasing of Phosphate asymmetric band is probably connected with intercalation of doxorubicin in DNA double strand shown in H-bonding formation of C14-OH of doxorubicine with PO_2^- of the DNA. Earlier the authors (Pohle et al., 1986; Beljebbar et al., 1995) studied the interaction of DNA with doxorubicine in vitro with the IR spectroscopy as well in RS spectroscopy and got in IR the similar results in intensity increasing. Their observed the low-frequency shift of the band. We can not conclude about the low-frequency shift, as the difference spectra show not only one enhanced component in this region.

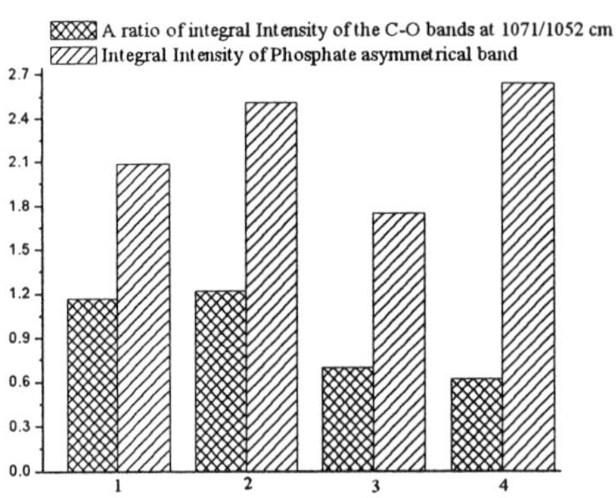

Figure 6. The spectroscopic parameters for DNA from doxorubicin-resistant tumour cells (1,2) and doxorubicin-sensitive (3,4) tumour cells before (1,3) and after (2,4) application of doxorubicin.

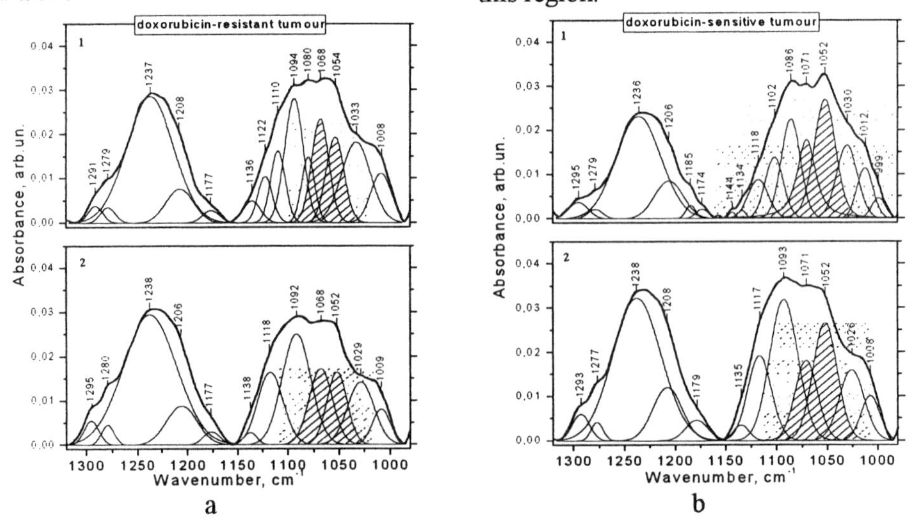

Figure 7. SEIRA spectra of DNA from doxorubicin-resistans (a) and doxorubicin-sensitive (b) tumour cells: 1 - before and 2 - after application of doxorubicin

In the Phosphate asymmetrical band the doxorubicin induced the structural changes in DNA from resistant strain are less. The dynamics of the Phosphate symmetrical band of DNA under the application of doxorubicin to sensitive and resistant strain show the similar features as the Phosphate asymmetrical. *Figure 6* confirms the above-mentioned influence of doxorubicin on the structure of the sensitive and resistant cancer cells in the Phosphate I band for two different C-O vibrations of deoxyribose at 1052 cm^{-1} (structure of DNA in A and B-forms of helix or ordering helix structure) and 1071 cm^{-1} (disordering structure or non-helix of DNA) (Semenov et al., 1994; Litvinov, 1991). It is usually difficult to see these two bands at 1052 cm^{-1} and 1071 cm^{-1} simultaneously using a conventional geometry. Only the SEIRA spectra clearly distinguish them. Calculating the integral intensity ratio of these two vibrations, we can characterise the level of disorder (non-helix form) in the DNA structure.

Component analysis of the Phosphate symmetrical band showed that the diversity of conformation states of the DNA from doxorubicin-resistant sells did not change after the action of doxorubicin. Otherwise the doxorubicin action upon the doxorubicin-sensitive cells showed that the conformation states of DNA had a tendency to shift to canonical helix forms (or more ordering form).

The peculiarities of the spectra of sugar region correlate with the above mention data.

4.5.2. *Sugar region*

After the application of doxorubicin to the sensitive cancer strain, the structure of DNA in the sugar region looks close to the structure of helix DNA (*Figure 8*). The intensity of the band at 830 cm^{-1} decreased by factor 2, numerous states in the 900-700 cm^{-1} region are now missing, and the intensity of the band at 963 cm^{-1} increased; the overall effect being to shift the spectrum in the direction of helix DNA form.

The spectrum of resistant tumour DNA in the sugar range looks close to the spectrum of helix DNA form, and their changes after application of doxorubicin are negligible small.

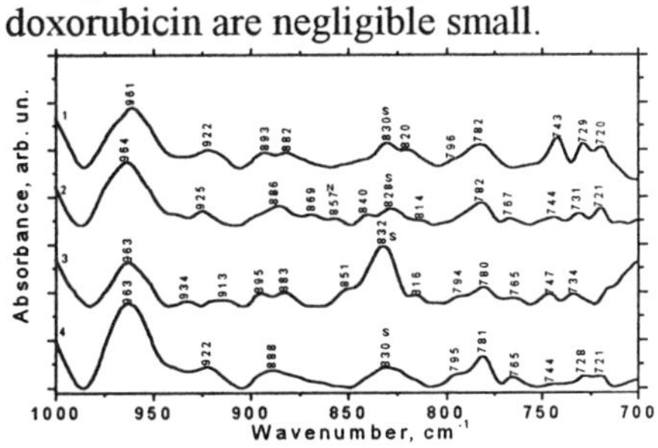

Figure 8. SEIRA spectra of DNA from doxorubicin-resistant tumour cells (1,2) and doxorubicin-sensitive (3,4) tumour cells before (1,3) and after (2,4) application of doxorubicin

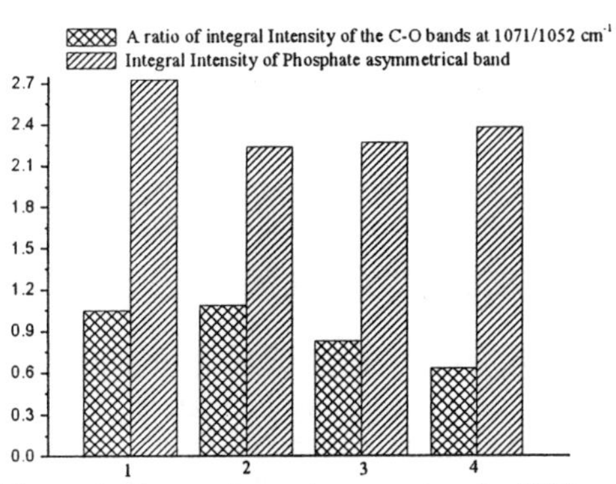

Figure 9. The spectroscopic parameters for DNA from cisplatin-resistant tumour cells (1,2) and cisplatin-sencitive (3,4) tumour cells before (1,3) and after (2,4) application of cispaltin.

4.6. *CISPLATIN-INDUCED DNA STRUCTU-RAL CHANGES IN SENSITIVE AND RESISTANT STRAINS.*

The application of cisplatin to cisplatin-sensitive cancer cells causes to both less and more changes in the DNA spectra in the phosphate range depending on the initial state of DNA from sensitive strain.

The application of cisplatin to cisplatine-resistant strain influences more the phosphate bands of DNA from resistant cells in comparison with the influence the doxorubicin on the doxorubicin-resistant strain. Namely, in the case of the application of cisplatine to cisplatine-resistant strain we observed the changing (decrease or increase depending on the initial state) of intensity of the Phosphate asymmetrical band *(Figure 9)* that could be connected with covalent interaction of cisplatine with the components of DNA. It is known that cisplatin interact with DNA on the N7 position of guanine, however not only metal coordination, the H-bonding (N-H to phosphate groups) may play a key role here also (Reedijk, 2000). In cisplatin-resistant strain after application of cisplatin we observed the decreasing of the integral intensity of the phosphate symmetrical band from 1.16 at 1096 cm^{-1} to 0.82 at 1094 cm^{-1} with the redistribution of the intensities of the C-O components in the direction of the increasing of the diversity of conformation forms (destabilisation of the structure). In the DNA from sensitive strain after application of cisplatin a little stabilisation of structure was observed.

Thus, the DNA from resistant cancer cells could be characterised as rigid structure and practically does not change after application of anticancer drugs. The structure of sensitive cancer DNA seems to be flexible, and after application of anticancer drugs it drastically changes and become close to the structure of helix DNA form (stable structure).

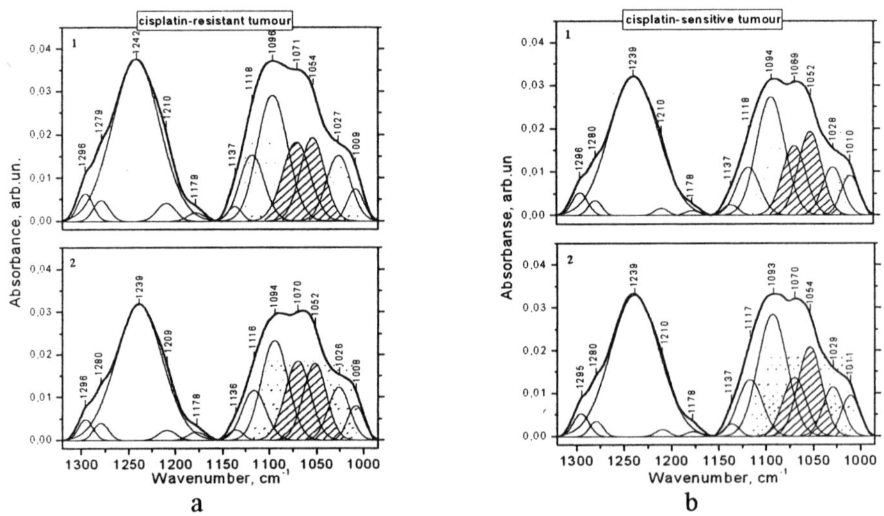

Figure 10. SEIRA spectra of DNA from cisplatin-resistans (a) and cisplatin-sensitive (b) tumour cells: 1 - before and 2 - after application of cisplatin.

4.7. *PHOSPOLIPIDS FROM SENSITIVE AND RESISTANT STRAINS.*

The comparison the IR spectra of phospholipids from resistant and sensitive to doxorubicin strain showed a manifestation of the liquid crystal phase (seems, orthorhombic) for resistant strain *(Figure 11)*. This supposition was made on the basis of the splitting of deformation (1473 cm^{-1} and 1463 cm^{-1} and libration (719 cm^{-1} and 730 cm^{-1}) of CH_2 vibrations (Mantsch et al., 1991). In the IR spectra of phospholipids from sensitive strain we have not observed the splitting of the CH_2 bands. In the region

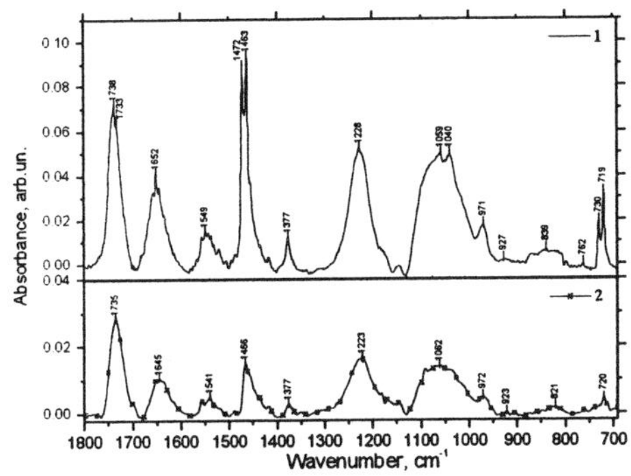

Figure 11. The SEIRA spectra of phoshpolipides from doxorubicin-resistant (1) and doxorubicin sensitive (2) tumour cells.

of stretching CH vibrations, the frequencies of phospholipids from resistant strain are shifted in the high frequency range (1-1.5 cm^{-1}) in comparison with those for sensitive strain, that could be the marker of increasing the contribution of gouche of lipids (less stable form in comparison with trans-form of lipids). In the region of C=O stretching vibration of the phospolipids from resistant cells we observed 2 components- 1738 cm^{-1} (gel-like phase) and 1733 (liquid crystalline phase). In the phospolipds from sensitive cells we observed C=O vibration at 1735 cm^{-1}. On the basis of spectroscopic data (presence of 2 phases after isolation, content of water estimated with IR spectra) we supposed also, that the phase transition for phospolipids from resistant cells has a lower temperature that those from sensitive cells. Earlier Ramu, 1984 noted the higher degree of structural order of plasma membrane lipids in doxorubicine-resistant P388 murine leukaemia cells than those for sensitive strain.

It is known that the phospholipids from resistant cells are more heterogeneous in comparison with those from sensitive cells. Therefore, the above mention features of the structure of total fraction of phospholipids from resistant and sensitive cells seems to be connected with the deviation in the composition of phospholipids.

4.8. *CISPLATIN-INDUCED STRUCTURE OF PHOSPHOLIPIDS FROM CISPLATIN-RESISTANT STRAIN.*

The IR spectra of phospholipids from cisplatin-resistant cells did no show the splitting in the CH_2 components, which we observed in the case of lipids from doxorubicin-resistant cells.

The application of cisplatin to cisplatin-resistant cells causes to the formation of lipid-platin complex. It could be proved by the appearance of the number of new bands in the phosphate range *(Figure 12)*. Some other vibrations in this range are narrowing, shifting, splitting and changing in intensity similar to Binder, 2000.

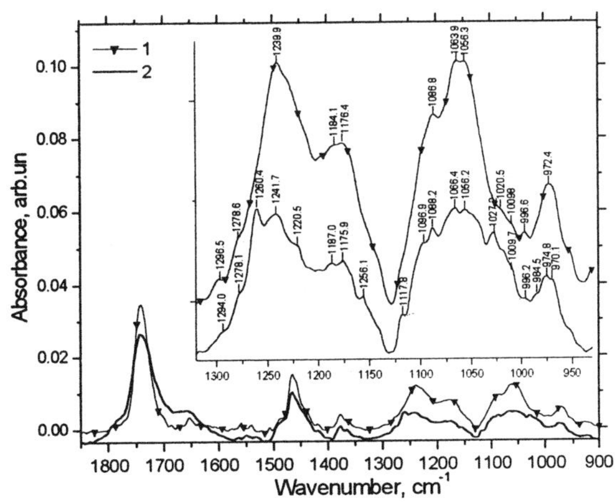

Figure 12. The SEIRA spectra of phoshpolipides from cisplatin-resistant tumour cells before (1) and after (2) application of cisplatin.

4.9. *CISPLATIN-INDUCED STRUCTURAL CHANGES IN PHOSPHOLIPIDS FROM SENSITIVE CELLS (AN EXPERIMENT IN VITRO).*

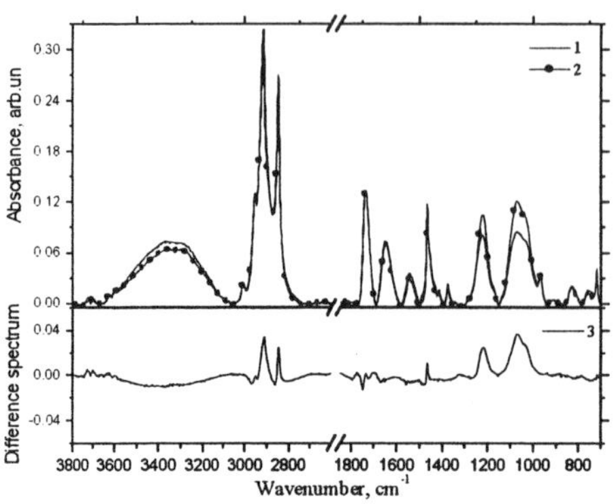

Figure 13. The SEIRA spectra of pospholipides from plasmatic membranes before (1) and after (2) incubation with cisplatin in vitro, 3- the difference spectrum.

The SEIRA spectra of the total fraction of phospholipids (with short peptides) isolated from sensitive cells plasma membrane incubated with 50 μM of cisplatin during 0.5 hour at 37 C° is presented in *Figure 13.* The most prominent differences after application of cisplatin could be note in the region of phosphate band: increasing intensity and 5 cm^{-1} shift to low-frequency range. The low-frequency cisplatin-induced shift in CH stretching vibration (about 1 cm^{-1}) is probably to indicate on some ordering of phospholipid structure. The experiment in vivo showed more prominent platinum - induced differences in structure of phospholipids than the experiment in vitro.

5. Conclusion.

SEIRA method could be applied for studies of anticancer drug-resistant induced structure of DNA and lipids as well as a test of anticancer drug action upon the tumour cells in cancer chemotherapy. This potentially is quantitative evidence for structural damage in nucleic acids isolated from different cancer cells, using the technique of vibrational spectroscopy.

These study has a preliminary character, however after the obtaining more statistical data, the SEIRA could be used as a test for detection of individual sensitivity of cancer cells to the influence of anticancer drugs.

6. Acknowledgments.

We are greatly indebted to Dr. Litvin P.M. for AFM images, Dr . Chegel V.I. and Prof. Shirshov Yu.M. for fruitful discussion and assistance.

References:

1. Arrondo, Jose Luis R. and Goni, Felix M., (1998), Infrared studies of protein-induced perturbation of lipids in lipoproteins and membranes, *Chemistry and Physics of Lipids,* **96**, 53-68.
2. Ataka Ken-Ichi, Yotsuyanagi Takao and Osawa Masatoshi, (1996), Potential-Dependent Reorientation of Water Molecules at an Electrode/Electrolyte Interface Studied by Surface-Enhanced Infrared Absorption Spectroscopy, *J. Phys. Chem.* **100**, 10664-10672.
3. Baret, F., Carbone, G. P. and Penon, P., (1978), Infrared dichroism of polynucleotides of repeated sequence. I. Poly(dA)·Poly(dT) and Poly[d(A-T)]·Poly[d(A-T)], *Biopolimers,* **17**, 2319-2339.

4. Beljebbar, A., Sockalingum, G.D., Angibous, J.F. and Manfait, M., (1995), Comparative FT SERS, resonance Raman and SERRS studies of doxorubicin and its complex with DNA, *Spectrochimica Acta Part (A)*, **51**, 2083-2090.
5. Binder, H., Arnold, K., Ulrich, A.S. and Zschörnig, O., (2000), The effect of Zn^{2+} on the secondary structure of a histidine-rich fusogenic peptide and its interaction with lipid membranes, *Biochimica et Biophysica Acta (BBA)/Biomembranes*, **1468, (1-2)**, 345-358.
6. Blagoi, Yu.P., Kornilova, C.V., Leont'ev V.C., et al., (1994), Structural and physicochemical characteristics of DNA of animal tissues exposed to prolonged chronic irradiation in the Chernobyl zone, *Biophysics*, **39 (4)**, 637-644.
7. Blommaert, F.A., Floot, B.G., van Dijk-Knijnenburg, H.C., Berends F, Baan, R.A., Schornagel J.H., den Engelse, L. and Fichtinger-Schepman, A.M., (1998), The formation and repair of cisplatin-DNA adducts in wild-type and cisplatin-resistant L1210 cells: comparison of immunocytochemical determination with detection in isolated DNA, *Chem Biol Interact.*, **108 (3)**, 209-225.
8. Boal, A.K. and Rotello, V.M., (2000), Intra-and Intermonolayer Hydrogen bonding in Amide-Functionalizes Alkanethiol Self-Assebled Monolayers on gold substrate," *Langmiur*, **16**, 9527-9532.
9. Chang, R.K. and Furtak T.E., (1982), *Surface Enhanced Raman Scattering*, Plenum Press, New York – London.
10. Dovbeshko G.I., Chegel V.I., Gridina N.Y., Repnytska O.P., Shirshov Y.M., Tryndiak V.P., Todor I.M., (2001), Surface enhanced infrared absorption of nucleic acids on gold substrate, *Semiconductor Physics, Quantum Electronics and Optoelectronics*, **4 (3)**, 202-206.
11. Dovbeshko G.I., Gridina N.Ya., Kruglova E.B., Paschuk O.P., (2000), FTIR spectroscopy studies of nucleic acids damage, *Talanta*, **53**, 233-246.
12. Dovbeshko, G.I. and Gridina, N.Ya., (1999), Infrared spectra of tumour nucleic acids, *Biophysical Bulletin*, **450**, 85-87, (Ukr.).
13. Folch, J., Lees, M. and Sloane-Stanley, G.H., (1957), A simple method for the isolation and purification of total lipids from animal tissues, *J. Biol. Chem.*, **226**, 497-509.
14. Gaigeot, P., Leulliot, N., Ghomi, M., Jobic, H., Coulombeau, C. and Bouloussa, O., (2000), Analysis of the structural and vibrational properties of RNA building blocks by means of neutron inelastic scattering and density functional theory calculations, *Chemical Physics*, **261**, 217-237.
15. Gaigneaux, A., Ruysschaert, J.M. and Goormaghtigh E., (2001), FTIR spectroscopy as a tool for screening differences between sensitive and daunorubicin-resistant leukemic cells, *in Proc. of 9th European conference on the spectroscopy of biological molecules, pp-251*, Charles University in Prague.
16. Haghighi, A., Lebedeva, S. and Gjerset, R.A., (1999), Preferential platination of an activated cellular promoter by cis-diamminedichloroplatinum, *Biochemistry*, **38 (38)**, 12432-12438.
17. Harstain, A., Kirtley, J.R. and Tsang, J.C., (1980), Infrared Absorption of the Molecular Monolayers with Thin Metal Overlayers, *Phys. Rev. Lett.*, **45**, 201-209.
18. Kartalou, M. and Essigmann, J.M., (2001), Mechanisms of resistance to cisplatin, *Mutat Res*, **478 (1-2)**, 23-43.
19. Kosobukin V.A., (1983), Effect of enhancement of external electric field near metal surface and its manifestation in spectroscopy, *Surface. Physics, chemistry, mechanics*, **12**, 5-20.
20. Kuhne, Ch., Steiner, G., Fischer, W.B. and Salzer, R., (1998), Surface Enhanced FTIR Spectroscopy on membranes, *Fresenius J. Anal. Chem.*, **360**, 750-754.
21. Litvinov, G.S., (1991), Infrared vibration spectra of plant viruses and its molecular compounds, *Biopolimers and cells*, **7 (6)**, 32-47.
22. Maestre, N., Tritton, T.R., Laurent, G. and Jaffrezou, J.P., Cell surface-directed interaction of anthracyclines leads to cytotoxicity and nuclear factor kappaB activation but not apoptosis signaling, (2001), *Cancer Res*, **61 (6)**, 2558-2561.
23. Malins, D.C., Polissar, N.L. and Gunselman, S.J., (1997), Infrared spectra demonstrate that exposure to environmental chemicals leads to a new forms of DNA, *Proc. Natl. Acad. Sci. USA*, **94**, 3611-3614.
24. Mallins, D.C., Polissar, N.L., Nishikida K., Holmes, E.H., Gardner, H.S. and Gunselman, S.J., (1995), The etiology and prediction of breast cancer: Fourier transform-infrared spectroscopy reveals progressive alterations in breast DNA leading to a cancer-like phenotype in a high proportion of normal women. *Cancer*, **75**, 503- 517.
25. Maniatis, T., Fritsch, E.E., and Sambrook, J., (1984), *Molecular cloning*, Moskva, "Mir".
26. Mantsch, H.H., McElhaney, R.N., (1991), Phospholipid phase transition in model and biological membranes as studied by infrared spectroscopy, *Chem. Phys. Lipids*, **57**, 213-226.
27. Masatoshi Osawa, and Masahiko Ikeda, (1991), Surface-Enhanced Infrared Absorption of p-Nitrobenzoic Acid Deposited on Silver Island Films: Contributions of Electromagnetic and Chemical Mechanisms, *J. Phys. Chem.*, **95**, 9914-9919.

28. Meijera, C., van Luyn, M.J., Nienhuis, E.F., Blom, N., Mulder, N.H. and de Vries, E.G., (2001), Ultrastructural morphology and localisation of cisplatin-induced platinum-DNA adducts in a cisplatin-sensitive and -resistant human small cell lung cancer cell line using electron microscopy, *Biochem. Pharmacol*, **61(5)**, 573-578.
29. Mendelsohn, R., Mantsch, H.H., (1986), Fourier transform infrared studies of lipid-protein interaction, In: Watts, A., DePont, J.J.H.H.M., (eds.), Progress in Protein-Lipid Interaction, **2**, Elsevier, Amsterdam, pp.103-146.
30. Parker, F.S., (1983), *Biochemical application of Raman and resonance Raman spectroscopies*, New-York, London, Acad. Press.
31. Pohle, W. and Flemming, J., (1986), Interaction of the C_{14}-OH group of adriamycin with DNA phosphate as spectroscopically evidenced, *Journal of Biomol. Struct. & Dyn*, ISSN 0739-1102, **4 (2)**, 243-250.
32. Ramu, A., Glaubiger, D. and Weintraub, H., (1984), Differences in lipid composition of doxorubicin-sensitive and –resistant P388 cells, *Cancer treatment reports*, **68 (4)**, 637-640.
33. Reedijk, J., (2000), How can cisplatin reach cell nuclei for DNA binding?, bock of abstract International conference on DNA conformation, modification and recognition in biomedicine, p.44.
34. Richardson, M.E. and Siemann, D.W., (1997), Tumor cell heterogeneity: impact on mechanisms of therapeutic drug resistance, *Int. J. Radiat. Oncol. Biol. Phys.*, **39 (4)**, 789-95.
35. Roseler, A. and Korte, E.H., (1998), Surface enhanced infared absorption observed with attenuated total reflection (ATR-SEIRA): modeling the optical responce, *Fresenius J. Anal. Chem.*, **362**, 51-57.
36. Saenger, W., (1989), *Principles of nucleic acid structure*, Moscow, Mir.
37. Schrader, B., (1995), *Infrared and Raman Spectroscopy*, Weinheim-New York-Basel-Cambridge-Tokio.
38. Semenov, M.A., Bol'buh, T.V., Krasnitskaya, A.A. and Maleev, V.Ya., (1994), Hydration and structural state of liver DNA from rats exposed within the Chernobyl NPP zone, *Radiation Biology. Radioekology*, **34 (3)**, 328-335.
39. Sips H.J., and K. van Dam, (1978), Sodium-dependent alanine transport in plasma-membrane vesicles from rat liver, *Biochem. J.* 174, 1083-1086.
40. Taboury, J.A., Liquier, J., Taillandier, E., (1985), Characterization of DNA structure by infrared spectroscopy: double helical forms of poly(dG-dC)poly(dG-dC),poly(dD8G-dC)poly(dD8-dC),and poly(dG-dm5C)poly(dG-dm5C), *Can.J.Chem.*, **63**,1909-1904
41. Taillandier, E., Liquier, J. and Taboury, J.A., Infrared spectral studies of DNA conformation, (1985), *Advances in Infrared and Raman Spectroscopy*, (Clarc, R. J. H., Hester, R. E. and Wiley Heyden, eds.), pp. 65-114,
42. Tajmir-Riahi, H.A., Neault, J.F. and Naoui, M., (1995), Does DNA acid fixation produce left-handed Z structure?, *FEBS Letters*, **370**, 105-108.
43. Wanzenbuck D., Mizaikoff, B., Weissenbacher, N. and Kellner, R., (1998), Surface Enhanced Infrared Absorption Spectroscopy (SEIRA) using external reflection on low-cost substrates, *Fresenius J. Anal. Chem.*, **362**, 15-20.
44. Yang, X.L. and Wang, A.H., (1999), Structural studies of atom-specific anticancer drugs acting on DNA, *Pharmacol. Ther.*, **83 (3)**, 181-215.
45. Yoon,, S..S., Ahn, K.S., Kim, S.H., Shim, Y.M. and Kim, J., (2001), In vitro establishment of cis-diammine-dichloroplatinum(II) resistant lung cancer cell line and modulation of apoptotic gene expression as a mechanism of resistant phenotype, *Lung Cancer*, **33 (2-3)**, 221-8.

AGGREGATION OF FULLERENES IN PYRIDINE/WATER SOLUTIONS

V.L. AKSENOV, M.V. AVDEEV
Frank Laboratory of Neutron Physics,
Joint Institute for Nuclear Research
141980 Dubna, Moscow Region, Russia

A.A. TIMCHENKO, I.N. SERDYUK
Institute of Protein Research RAS
142292 Puschino, Moscow Region, Russia

R.P. May
Institute Laue-Langevin
F-38042 Grenoble, France

1. Introduction

The behavior of fullerenes in solutions is of current interest (Bezmelnitsyn et al., 1998). This is partially explained by the fact that fullerene solutions show a number of unusual properties, in particular, the solvatochromism—a sharp change of the absorption spectrum with a small change in the solvent composition (Bezmelnitsyn et al., 1998; Mrzel et al., 2000; Aksenov et al., 2001b). Thus, fullerenes (C_{60}) in pyridine/water solutions exhibit a strong solvatochromatic effect when changing pyridine/water ratio (Mrzel et al., 2000). This effect is assumed (Mrzel et al., 2000) to be connected with the formation of specific aggregates. The results of our investigations by means of different techniques, which are reviewed in the present paper, give insight into the structure of these aggregates and their behavior in the solutions.

2. Experimental

2.1. *SAMPLE PREPARATION*

C_{60} (Hoecht) was dissolved in pyridine (Merck) with the concentration of 0.5 mg/ml (maximal solubility is 0.89 mg/ml). Then, this solution was added to water up to the concentration of 2 and 10% resulting in the total C_{60} concentration of 0.01 and 0.05 mg/ml, respectively. The change in color from mauve for the C_{60}/pyridine solution to lemon-yellow for the final solution was observed in agreement with (Mrzel et al., 2000). All measurements were made at 20°C.

E. Buzaneva and P. Scharff (eds.), Frontiers of Multifunctional Nanosystems, 281–286.

2.2. *UV-VIS SPECTROSCOPY*

The UV-Vis spectra were recorded on the HITACHI U-2000 UV/Vis spectrophotometer using 10 mm path length quartz cell.

2.3. *ELECTRON MICROSCOPY (EM)*

The EM images were obtained by the JEM-100C electron microscope at 80 kV. Samples were prepared by negative staining with uranyl acetate.

2.4. *DYNAMIC LIGHT SCATTERING (DLS)*

The DLS experiments were carried out using He-Ne laser (wavelength 628 nm) and time-of-arrival photon digital correlator developed at the Institute of Protein Research RAS (Puschino, Moscow Region).

2.5. *SEDIMENTATION ANALYSIS*

Velocity sedimantation curves were obtained using the Beckman-E ultracentrifuge (Moscow State University) at 14000 rev/min ($\lambda = 340$ nm).

2.6. *SMALL-ANGLE X-RAY SCATTERING (SAXS)*

The SAXS experiments were performed on the BL-15A small-angle camera of the Photon Factory (Tsukuba, Japan) in a q-range $0.08 \div 0.8$ nm^{-1}.

2.7. *SMALL-ANGLE NEUTRON SCATTERING (SANS)*

The SANS experiments were carried out on the YuMO small-angle diffractometer at the IBR-2 pulsed reactor at the Frank Laboratory of Neutron Physics of the Joint Institute for Nuclear Research (Dubna, Russia) and the D22 small-angle diffractometer at the Institute Laue-Langevin (Grenoble, France). On the YuMO, scattering curves from different combinations of protonated/deutarated components in the solutions were obtained in a q-range $0.08 \div 0.8$ nm^{-1}. On the D22, measurements at q-values down to 0.03 nm^{-1} were made for the fully protonated solution.

3. Results and discussions

Here, we present the results obtained by the techniques described above. Some details of the treatment of experimental data can be found in our previous works (Aksenov et al., 2001a; Aksenov et al., 2001b).

UV-Vis spectra of the studied solutions are characterized by the absorption maximum at the wavelength of about 340 nm. (Mrzel et al., 2000; Aksenov et al., 2001b). This peak was assumed (Mrzel et al., 2000) to be connected with the formation of specific aggregates in the solutions. The EM image presented in Figure 1 confirms the presence of approximately spherical polydisperse aggregates with a radius of $10 \div 50$ nm in the studied solutions. Additionally, bigger complexes comprising several aggregates are observed. The size distribution functions of the aggregates obtained by DLS (Aksenov et al., 2001b) are shown in Figure 2. From the left peaks one can conclude that the mean

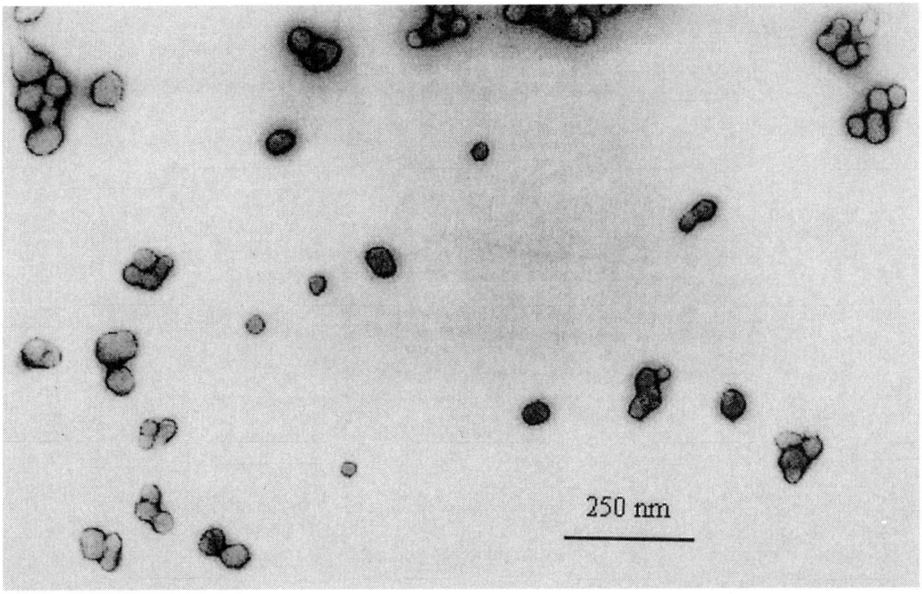

Figure1. Electron microscopy image for C_{60} in 2% pyridine aqueous solution. The C_{60} concentration is 0.01 mg/ml.

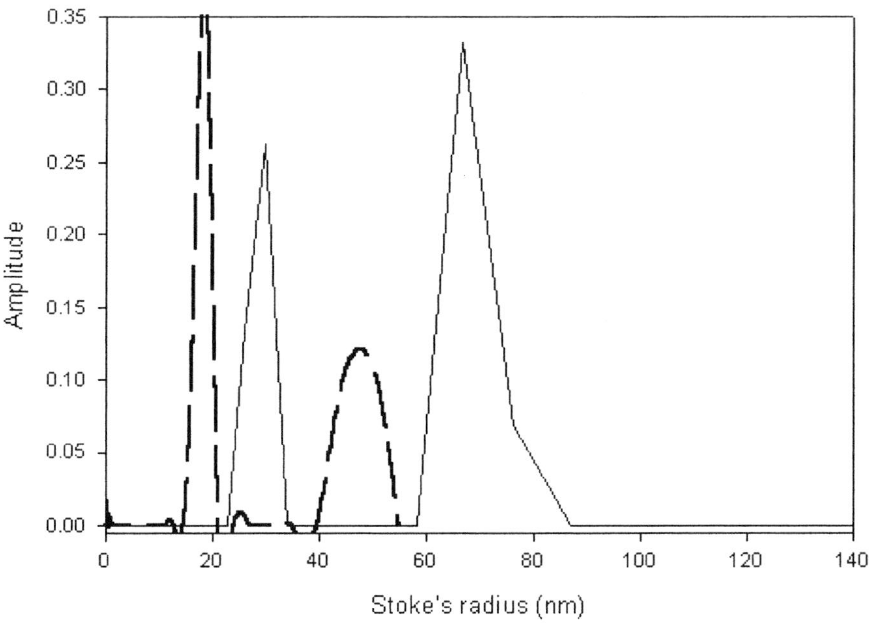

Figure 2. Size distribution function from DLS for C_{60} in 2% (dashed line) and 10% (solid line) pyridine. The C_{60} concentration is 0.01 and 0.05 mg/ml, respectively.

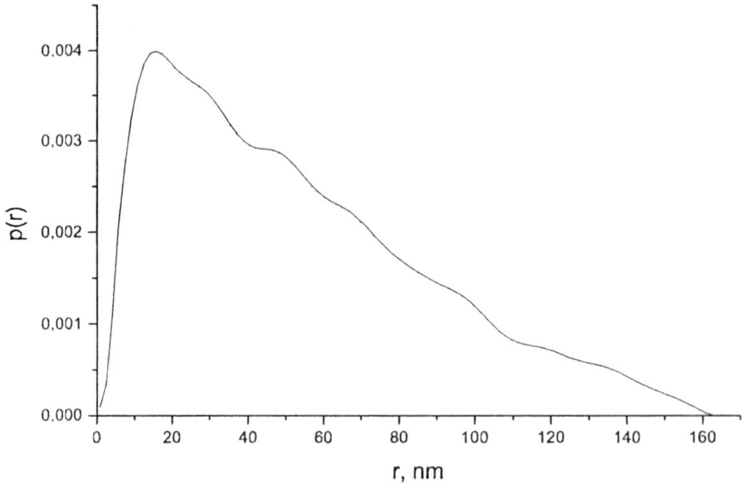

Figure 3. The p(r) pair distance distribution function for C_{60} in 10% pyridine aqueous solution. The C_{60} concentration is 0.05 mg/ml.

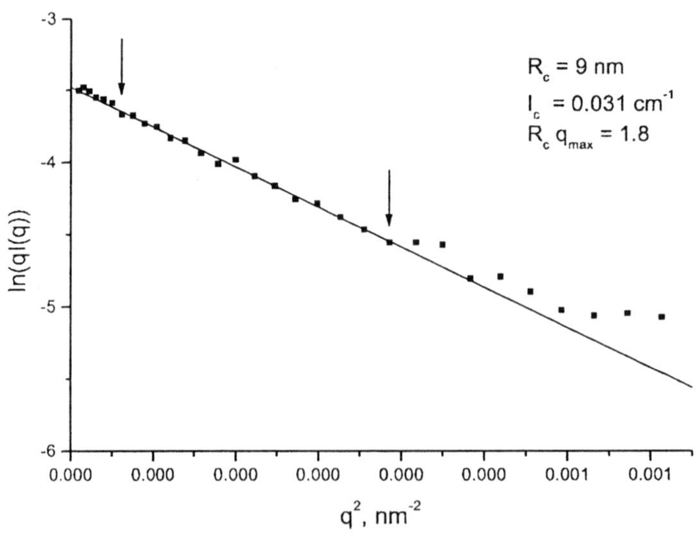

Figure 4. The $\ln(qI(q))\sim q^2$ graph of the scattering intensity I(q) for C_{60} in 10% pyridine aqueous solutions. The C_{60} concentration is 0.05 mg/ml. The arrows show where the linear dependence was fitted to the experimental data.

hydrodynamic radius of the aggregates is about 18 and 30 nm, and the polydispersity is about 25 and 30%, for the studied C_{60} concentrations, respectively. The right peaks in Figure 2 correspond to the aggregate complexes in the EM image in Fig.1. These complexes are also polydisperse and their radius exceeds 40 nm. It can be clearly seen that the size and polydispersity of both aggregates and their complexes increase with the C_{60} concentration. SAXS data (Aksenov et al., 2001b) give 23 and 26 nm for the characteristic radius of the aggregates for the two concentrations, respectively. From the sedimentation analysis (Aksenov et al., 2001b) the values of 97 and 220 Swedberg for the sedimentation constant, *s*, were obtained in the case of the C_{60} concentrations of 0.01 and 0.05 mg/ml, respectively. Using these *s*-values and the hydrodynamic radius from DLS, as well as the radius from SAXS one obtains the aggregate density within 2÷3 g/cm^3, which is closer to the density of fullerenes (≈ 1.7 g/cm^3) than that of pyridine (0.98 g/cm^3). This points to a small content of pyridine inside the aggregates. Taking into account the results of chemical tests (Mrzel et al., 2000), which show the inertness of C_{60} in the solutions, one can conclude that the aggregates consist of a massive fullerene core covered by a pyridine shell. The results of SANS experiments with the contrast variation (Aksenov et al., 2001a) are in qualitative agreement with the later model. Thus, the relative position of the scattering curves from different combinations of deuterated/protonated solution components in absolute scale corresponds to this model. Along with this, the quantitative analysis using SANS data faces a problem. The secondary aggregation in the solutions complicates scattering curves, so that the signal from the inner structure of the aggregates becomes insignificant in comparison with that from the aggregate complexes. In Figure 3 the p(r) pair distance distribution function obtained by the indirect Fourier transform (IFT) (Skov Pedersen, 1997) of the scattering curve from the fully protonated solution with the total C_{60} concentration of 0.05 mg/ml, is presented. Peculiarities of the p(r) graph allows one to conclude that polydisperse rod-like complexes with the length of more than 100 nm and the effective radius within 10÷20 nm are formed in the solution. The graph $\ln(qI(q))\sim q^2$ in Figure 4 shows a linear dependence in a q-interval 0.05÷0.2 nm^{-1} with R_c = 9 nm and I_c = 0.031 cm^{-1}. The obtained radius of gyration R_c corresponds to the mean radius of the "rods" in the solution of about 13 nm. This value must be close to the mean radius of the aggregates forming the complexes. However, it is twice less than the radius value from the electron microscopy, DLS and SAXS experiments. This difference may be connected with a possible large systematic error in the effective aggregate radius for all techniques involved.

4. Conclusions

From the results of our complex investigations we can conclude the following.

1) Approximately spherical polydisperse aggregates of the radius 13÷35 nm are formed when dissolving C_{60} in pyridine/water solutions.
2) Both the size and polydispersity increase with the C_{60} concentration.
3) The aggregates consist of a massive fullerene core covered by a thin pyridine shell.
4) An assosiation of the aggregates into bigger rod-like complexes of the length of about 100 nm takes place.

References

1. Aksenov, V.L., Avdeev, M.V., Belushkin, A.V., Mihailovic, D., Mrzel, A., Rosta, L., Serdyuk, I.N. and Timchenko, A.A. (2001a), "Structure of fullerene aggregates in pyridine/water solutions by small-angle neutron scattering", *JINR Communications* E14-2001-62, Dubna, Russia.
2. V Aksenov, V.L., Avdeev, M.V., Mihailovic, D., Mrzel, A., Vasiliev, V.D., Timchenko, A.A. and Serdyuk, I.N. (2001b), "Study of fullerene aggregates in pyridine/water solutions", in Proc. of Interanational School and Euroconference on Electronic Properties of Novel Materials-molecular Nanostructures,Tirol, Austria, March 5-12, 2001, in press.
3. Bezmelnitsyn, V., Eletskii, A. and Okun', M. (1998), "Fullerenes in solutions", *Uspekhi Fizicheskikh Nauk* **168**, 1195-1220
4. Mrzel, A., Mertelj, A., Omerzu, A., Eopie, M. and Mihailovic, D. (1999), "Investigation of encapsulation and solvatochromism of fullerenes in binary solvent mixtures", *J. Phys. Chem.* **103**, 11256-11260
5. Skov Pedersen, J. (1997), "Analysis of small-angle scattering data from colloids and polymer solutions: modeling and least-squares fitting", *Adv. Colloid and Interface Sci.* **70**, 171-210

INFRARED SPECTRUM OF FULLERENE C_{60} AGGREGATES IN WATER SOLUTION

A.A. GOLUB[1], Yu.I. PRYLUTSKYY[2], S.S. DUROV[3],
E.V. BUZANEVA[4],
Department of [1]Chemistry, [2]Biophysics, [3]Physics and [4]Radiophysics, Kyiv National Shevchenko University, Volodymyrska Str., 64, 01033 Kyiv, Ukraine
T. BRAUN
Dept. of Inorganic and Analytical Chemistry, L. Eotvos University Budapest, 1443 Budapest, Hungary
A. GRAJA
Institute of Molecular Physics, 60-179 Poznan, Poland
P. SCHARFF[7]
Institute of Physics, TU Ilmenau, D-98684 Ilmenau, Germany

Abstract

The IR transmission spectrum of fullerene C_{60} aggregates (cluster of I_h symmetry and microcrystal of T_h symmetry) in water solution was measured for the first time. The results obtained were compared with available theoretical calculations.

1. Introduction

In our previous paper [1] we have investigated in detail the structure of fullerene C_{60} in aqueous solution. In particular we have demonstrated that the fullerene C_{60} in water solution could aggregate rapidly at room temperature. The kinetic study showed that the structure of C_{60} aggregates (the average size 3-36 nm in a hydrated state) could be described as a fractal with a dimension of 1.8-2.0. This fact is in a good agreement with the fractal concept [2]. The spherical hydrated cluster (I_h symmetry group, diameter 3.4 nm) formed of 13 C_{60} molecules was shown to be the smallest stable form among all the possible aggregates $(C_{60})_N$ ($13 \leq N \leq 240$). This theoretical result is in excellent agreement with the experimental data [3-4]. Finally, the optical absorption spectra obtained confirm the STM data [4] on the existence in water of the crystalline phase formed of

E. Buzaneva and P. Scharff (eds.), Frontiers of Multifunctional Nanosystems, 287–290.

sphere-shaped C_{60} aggregates of 1-4 nm size. In this papar we have presented the additional experimental data such as the IR spectral ones which confirm the structure of fullerene C_{60} in aqueous solution (with highest concentration up to 2.4 g/l) described above.

2. Results and discussion

Earlier we found [5] that the values of vibrational Raman frequencies of the individual C_{60} molecules in aqueous solution increase in comparison with results for the single fullerene C_{60} [6] by the magnitude of (2-6)%. This increase we associate with strengthening the intramolecular bonds under forming hydrated fullerene C_{60} aggregates in aqueous solution. In other words *the effect of water* comes to external hydrostatic compression of the individual molecules C_{60} in the aggregate leading to the negligible decrease (up to 5%) of its geometrical sizes and increase of bond energy.

The vibrational spectrum of the smallest stable spherical-like $(C_{60})_{13}$ cluster in aqueous solution was calculated from the first principles [7] using the symmetry approach [8]. It was found that the 33 normal vibrations are degenerate and distributed among only 9 frequencies. Note that only one fundamental intramolecular vibrations is IR active: the threefold-degenerate (F) mode. The results obtained are represented in Table 1. As we can see the calculated vibrational frequencies for the $(C_{60})_{13}$ cluster in aqueous solution are located considerably lower than the fundamental modes for a single molecule C_{60} [6]: 273 - 1575 cm^{-1}.

Table 1. *Frequencies (in cm^{-1}) of intramolecular modes*

SYMMETRY	$(C_{60})_{13}$ cluster in water solution	
	theory	IR experiment
H	23.7	
G	32.0	
F	34.8	
H	38.1	
G	43.9	
F	75.0	79.3
H	76.2	
A	81.1	
F	158.0	

It is known that C_{60} molecules freely rotate in the solid at room temperature [9]. In our opinion, *the effect of water* on the forming crystal structure is connected with the fixation of hydrated C_{60} molecules orientation in the unit cell of the lattice (i.e., the transition from disordered structure to orientationally ordered crystalline phase [10] takes place). Thus, we shall be consider the following structure of C_{60} crystal in aqueous solution: the number of hydrated C_{60} molecules (I_h symmetry group, diameter ~1.2 nm) in the unit cell is Z=4, the space group is T_h and the lattice constant is a≅1.85 nm.

Taking into consideration *the effect of water,* we assume [7, 11] that intermolecular interactions are described only by a Lennard-Jones (12-6) atom-atom potential and electrostatic potential of charges located on single and double bonds of the C_{60} molecules. The values of fundamental intermolecular frequencies of C_{60} crystal in aqueous solution in the center of the Brillouin zone were calculated based on the analytical formulas [11] obtained by us for solid C_{60}. The results obtained are represented in Table 2. It is seen that the calculated vibrational frequencies for C_{60} crystal in aqueous solution are located lower than the calculated fundamental modes for solid C_{60} [11]. Note that only two F_u modes are IR active.

The IR transmission spectrum of C_{60} aqueous solution (see Fig. 1) was recorded with a Bomem DA3 Fourier spectrometer with resolution 2 cm^{-1} in the range of (20 – 100) cm^{-1} using a grid polarization interferometer. At room temperature three clear bands are observed at 36.6, 41.9 and 79.3 cm^{-1}. The lowest two modes correspond to the lattice absorption in the C_{60} aqueous solution and the last high-frequency mode is the intramolecular one according to the theoretical predictions [7]. Hence, in all probability the forming of crystalline structure and smallest spherical $(C_{60})_{13}$ clusters in C_{60} aqueous solution takes place.

Table 2. *Frequencies (in cm^{-1}) of intermolecular modes in the center of the Brillouin zone*

SYMMETRY	solid C_{60}	crystal C_{60} in aqueous solution	
	theory [9]	theory	IR experiment
A_u	35.0	26.9	
E_u	46.7	35.9	
F_u	42.9	33.0	36.6
F_u	51.0	39.2	41.9
A_g	18.6	14.3	
E_g	18.5	14.2	
F_g	17.1	13.2	
F_g	17.5	13.5	
F_g	22.0	16.9	

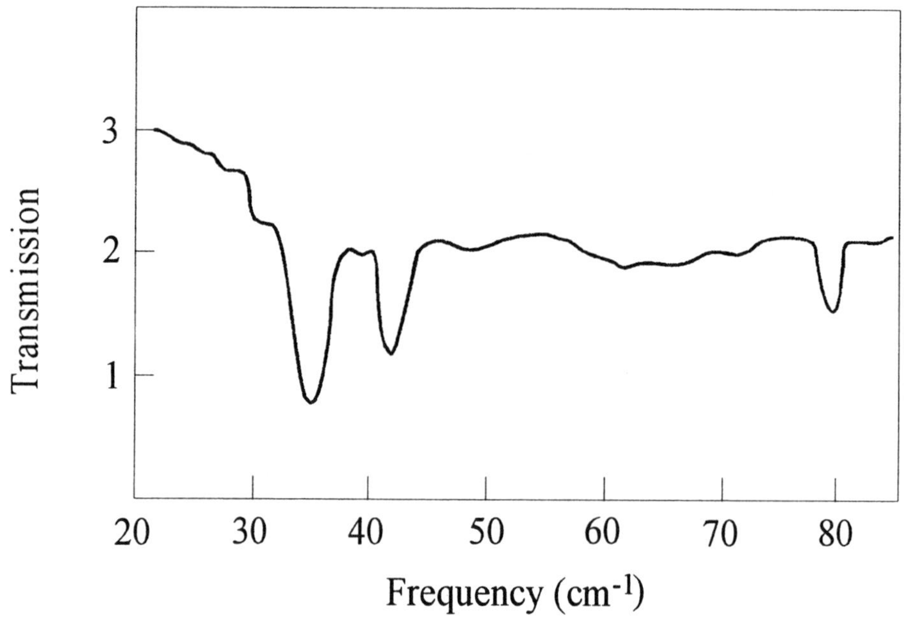

Figure 1. IR transmission spectrum of C_{60} water solution.

As one can see from Table 1 and Table 2 the predicted earlier theoretical values differ from the experimental data by up to 11%.

3. Acknowledgement

This research work was carried out within the framework of collaboration between the Ukrainian and Hungarian Academies of Sciences (Grant N 24).

References

1. Bulavin, L., Adamenko, I., Prylutskyy, Yu. et al. (2000), *Phys.Chem.Chem.Phys.*, 2, 1627-1629.
2. Smirnov, B.M. (1991), *Physics of Fractal Aggregates*, Nauka, Moscow.
3. Mchedlov-Petrosyan, N.O., Klochkov, V.K. and Andrievsky, G.V. (1997), *J.Chem.Soc., Faraday Trans.*, 93, 4343-4345.
4. Andrievsky, G.V., Klochkov, V.K., Karyakina, E.L. and Mchedlov-Petrosyan, N.O. (1999), *Chem.Phys.Lett.*, 300, 392-396.
5. Prilutski, Yu.I., Durov, S.S., Yashchuk, V.N. et al. (1999), *Europ.Phys.J. D*, 9, 341-344.
6. Bethune, D.S., Meijer, G., Tang, W.C. et al. (1991), *Chem.Phys.Lett.*, 179, 181-186.
7. Prilutski, Yu.I. and Durov, S.S. (1999), *Fiz.Tverd.Tela*, 41, 885-888.
8. Poulet, H. and Mathie, J.-P. (1973), *Spectres de vibration et symetrie des cristaux*, Gordon and Breach, Paris.
9. Kratschmer, W., Lamb, L.D., Fostiropoulos, K. and Huffman, D.R. (1990), *Nature*, 347, 354-359.
10. David, W.I.F., Ibberson, R.M., Matthewman, J.C. et al. (1991), *Nature*, 353, 147-152.
11. Prilutski, Yu.I. and Shapovalov, G.G. (1997), *Phys.St.Sol.(b)*, 201, 361-368.
12. Hirschfelder, J.O., Curtiss, Ch.F. and Bird, R.B. (1954), *Molecular theory of gases and liquids*, John Wiley and Sons, Inc., New York.

ELECTRICAL AND MAGNETIC PROPERTIES OF UNDOPED FULLERENE POLYMERS

TATIANA MAKAROVA [1,2], *BERTIL SUNDQVIST* [2]
[1] *Ioffe PTI, 194021, St.Petersburg, Russia*
[2] *Umeå University, 90187, Umeå, Sweden*

Carbon has a wide range of transport properties which are strongly dependent on the graphitic (sp^2) to tetrahedral (sp^3) carbon bonding ratio. Fullerenes comprise the third class of pure carbon, after diamond and graphite. The conductivity of van der Waals fullerene solids is governed by a thermal activation mechanism. Polymerization modifies the band structure with a decrease of activation energy. At certain conditions a transition to a metallic behavior is observed experimentally. Polymerization of fullerenes also results in a transition from diamagnetic to paramagnetic behavior.

Keywords: *fullerene, polymerization, high pressure, transport properties*

1. Introduction

The unusual spherical all-carbon molecules now known as fullerenes were discovered by Kroto and co-workers [1] in 1985. This discovery led to the 1996 Nobel Prize in chemistry. Five years later Krätschmer and co-authors [2] found the way to synthesize macroscopic amounts of fullerenes, which gave rise to a period of intensive fullerene research. Among the most exciting discoveries of this period are superconductivity in alkali-doped fullerenes [3] and its dependence on the structure [4]; ferromagnetism at 17 K in TDAE-C_{60} [5], which is a high critical temperature for organic ferromagnets and an absolute record for non-polymeric organic materials; and the creation of the fullerene-based superhard forms of carbon [6].

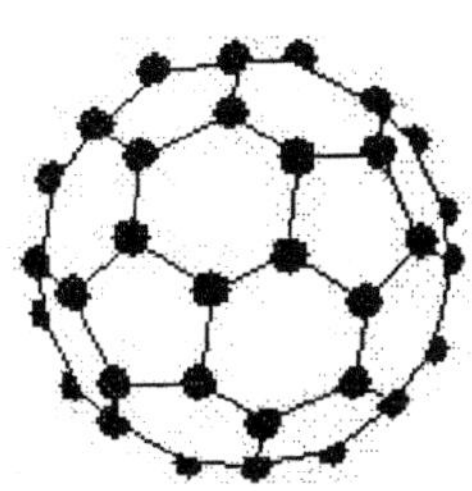

Figure 1. Fullerene C_{60}

Shortly after their discovery, fullerenes were called “the XXI century materials”. Since that time there has been a vital interest in practical applications of fullerenes. As carbon is the basic element of living substances, the discovery of his new

E. Buzaneva and P. Scharff (eds.), Frontiers of Multifunctional Nanosystems, 291–312.

form is of special significance. Fullerenes are closed-shell (exotic shape), hollow (can be filled) all-carbon (hydrogen free) molecules, which ought to result in a material-related breakthrough. Answering the question about its widespread market applications, we must say "no" or "not yet".

Why this sorrowful answer?

Fullerenes are considered as promising materials in many industrial fields. There have been ideas of constructing single-molecule electronic devices, using fullerenes in integrated circuits. Fullerenes can be used as superconductors or lubricants, they find use in pharmaceutics, catalysis, and diamond synthesis. They have already been applied for deleting atherogene lipoproteids from blood and for treatment of virus diseases.

However, the relatively high cost of the product makes it commercially non-competitive and therefore requires that the material should have remarkable physical properties, which differentiate it from other known materials. Good lubricants, catalysts, or superconductors do already exist.

Strikingly, the new wave of interest in fullerenes and other carbon-based materials arose exactly at the beginning of the XXI century. The temperature of the superconducting transition, which reaches 40 K in alkali-metal doped fullerenes, increased to 52 K for hole-doped C_{60} in the field-effect transistor geometry [7]. The authors of Ref.[7] promised to break this record by expanding the lattice parameter of the fullerene matrix, while still compelling the crystal to exhibit the hole conductivity – and they have already succeeded in creating a C_{60} - Br_3CH compound with a transition temperature of 117 K [8]. A substantial contribution has been made to the understanding of magnetism in TDAE-C_{60} [9], and experimental evidence of a magnetically ordered state in a two-dimensional C_{60} polymer has recently been published [10].

Coincidently, several contributions appeared reporting uncommon properties of graphite: a superconductor-insulator-type transition [11], and observation of ferromagnetic- and superconducting-like magnetization hysteresis loops below and above room temperature [12]. Magnetism in activated carbon fibers [13] and spin-glass-like disordered magnetism in nanographite [14] have been experimentally studied. It has been predicted theoretically that the electron-electron interactions in a graphene layer can lead to p-wave superconductivity [15] and magnetic ordering [16, 17]. Superconductivity has also been found in carbon nanotubes [18] at temperatures below 20 K [19].

Traditional fundamental solid-state physics has been almost exclusively studying the inorganic elements and simple compounds possessing semiconductor or metallic properties: Si, Ge, GaAs, and metals. The most exciting studies in solid-state physics are now being made in the field of organic materials, which are chemically very complicated. All the basic physical phenomena have already been observed on chemically complex organic molecular systems, among them superconductivity and magnetism. The Nobel Prize in Chemistry 2000 was awarded for the discovery and development of conducting polymers.

In this paper we consider polymers, but very simple ones, namely the polymers which consist of one element only, and this element is carbon. The abovementioned experimental evidence and theoretical predictions make us scrutinize the electrical and magnetic properties of undoped fullerene polymers.

2. Electrical properties of pristine fullerenes

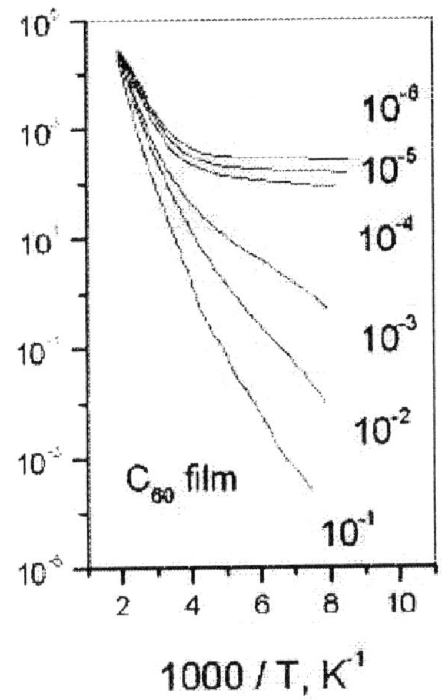

Figure 2. Temperature dependence of conductivity for the C_{60} film. Numbers near the curves stand for the oxygen pressure (Torr) in the measuring cell.

Electrical properties of van der Waals fullerene crystals and films are described in a recent review [20]. Briefly, both fullerene films and crystals are n-type semiconductors, but the measured values of activation energies are essentially lower than half the energy gap, reaching this value only at high temperatures. The room-temperature conductivity values for oxygen-exposed fullerenes are four orders of magnitude less than that for oxygen-free material. Electrical properties of vacuum-sublimed amorphous fullerene films are shown in Fig. 2. This Figure illustrates the main features of the electrical behavior of fullerenes, and this behavior keeps within the framework of the known theories for amorphous semiconductors. The predominant conductivity mechanism is determined by the temperature: at low temperatures variable-range-hopping (VRH) is observed, at higher temperatures activated hopping via band tail states is added, and only at very high temperatures approaching the sublimation temperature is band conduction via delocalized states observed. It is also seen from Fig. 2 that oxygen does not affect the band conductivity, but creates trap levels for carriers and neutralizes the defects forming localized electronic states.

3. Magnetic properties of pristine fullerenes

Magnetic properties of fullerenes are determined by the following factors: (1) they consist of carbon and (2) they contain pentagons and hexagons. The diamagnetism of the C^{4+} ion is $-1.2 \cdot 10^{-8}$ emu/g [21]. Hexagonal rings contribute to a diamagnetic term, while pentagonal rings contribute to a paramagnetic term. As the paramagnetic term is nearly equal to the diamagnetic one, the fullerenes show a small diamagnetic susceptibility: the diamagnetic susceptibility of neutral C_{60} is $-3.4 \cdot 10^{-7}$ emu/g [22]. The paramagnetic upturn at low temperatures is always observed in the AC susceptibility of pristine C_{60} [23, 24] except in the case of specially prepared C_{60} single crystals, never exposed to oxygen, which quickly penetrates into the van der Waals bonded C_{60}.

In low fields the magnetic susceptibility of C_{60} is found to be strongly nonlinear and field-dependent, which is interpreted in terms of the field dependence of the spectral energy separations of the van Vleck paramagnetic term of the susceptibility [26].

Unlike undoped C_{60}, which has a weak temperature independent diamagnetic susceptibility, a greater variety of magnetic behaviors occur in doped fullerenes. Fullerenes show a paramagnetic behavior if a magnetic doping ion with an unfilled *d* or *f* shell is introduced, forming endohedral or exohedral complexes. The AC susceptibility of the one-dimensional polymeric phase of RbC_{60} is positive [26] with a large Curie contribution. After the correction by the Curie term and the Pauli term, the remaining temperature-dependent term is positive and increases with the temperature up to 50 K. This term is believed to be intrinsic for the RbC_{60} polymeric phase. The effect is interpreted in terms of a spin-density wave.

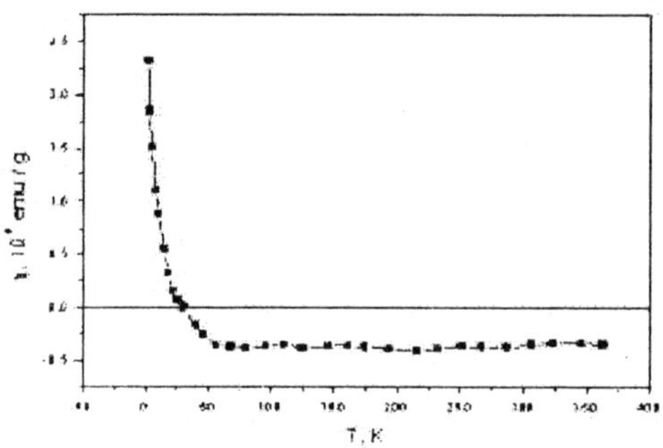

Figure 3. AC susceptibility of the C_{60} powder.

There have been reports on the existence of magnetic ordering in ammoniated fullerides $(ND_3)K_3C_{60}$ [27], hydrogenated fullerene, $C_{60}H_{36}$ [28], and palladium fullerides, $C_{60}Pd_n$ [29]. A ferromagnetic correlation develops at low temperatures in $Eu_xSr_{6-x}C_{60}$ (x = 1,3,6) and seems to be caused by the π - f interaction through C_{60} rather than the direct interaction between Eu atoms [30]. A coexistence of the ferromagnetic and diamagnetic (superconducting) phases was observed in fullerides Ce_xC_{60} with a clear superconducting transition at 10.5 K [31].

It is interesting to compare the magnetic properties of fullerenes with other forms of carbon. The diamagnetic susceptibility of graphite is strongly anisotropic, $\chi_{/\!/} = -5 \cdot 10^{-7}$ emu/g, $\chi_{\perp} = -5 \cdot 10^{-5}$ emu/g [32, 33]. Magnetic properties are different for the hexagonal and rhombohedral forms [34]. Ferromagnetic phase is theoretically predicted in graphite in the case of strong topological disorder: introduction of two pentagons and two heptagons instead of four hexagons in graphite modifies the band structure, and a structure with a flat band is exactly shown to be ferromagnetic for electron doping up to half-filling of this band [35]. Weak ferromagnetism is predicted in layered graphite at low doping [36]. Nanographite systems show spin-glass like behavior [37]. A ferromagnetic phase of pure carbon, which lies exactly in the middle between graphite and diamond, was predicted theoretically. This hypothetical material should contain equal numbers of sp^2 and sp^3 hybridized carbon atoms and consist of quasi-graphite layers arranged in a rhombic lattice [38]. The amorphous-like carbon compound prepared by direct pyrolysis of 1,2-diaminopropane is shown to be a stable strong magnet, the apparent saturation magnetization M_s of which is 10.5 emu/g (0.022 μ_B per carbon atom), being comparable to ca. one-third of magnetite Fe_3O_4 with 30.7 emu/g [39].

The magnentic properties of graphite change when graphite intercalation compounds (GIC) are formed. It remains diamagnetic for acceptor GICs and tends to be paramagnetic for donor GICs. Magnetic properties result from (1) a diamagnetic contribution from the core electrons; (2) a Pauli contribution from the free carriers and (3) an orbital contribution from the valence electrons [40]. The transition from diamagnetic to paramagnetic behavior is observed in fluorinated graphite: paramagnetism is caused by local magnetic moments of dangling bonds [41]. These bonds result from the reaction of a fluorine atom with a carbon atom, which converts a graphite sp^2 bond into a sp^3 bond at the expense of a π-bond. In partially fluorinated regions displaying more paramagnetic centers, strong exchange coupling between localized moments (possibly via a C - C bond) is observed. Electron paramagnetic resonance in $(C_2 F)_n$ shows a single narrow Lorentzian line (g_{iso} = 2.0058 ± 0.0005, ΔHpp = 0.968 ± 0.005 mT). Such a signal usually appears in solids containing many paramagnetic centers coupled by strong exchange interaction.

For the carbon nanotubes χ is divided into 3 parts: Pauli paramagnetism from the conduction electrons (10^{-8} from EPR), diamagnetism of the C^{4+} ion and the orbital diamagnetism of the conduction electrons due to a change in the band energy [42]. The third term is ca. 100 times larger then the others. The magnetic response is diamagnetic and anisotropic with the component along the nanotubes less diamagnetic than that of the perpendicular one [43]. The observed anisotropy is consistent with theoretical predictions, but contradicts earlier experimental findings [44, 45]: $\chi_{/\!/}$ = - $10.75 \cdot 10^{-6}$ emu/g for the magnetic field parallel to the buckybundle axis, which is approximately

Table 1. *Magnetic properties of carbon allotropes.*

	Susceptibility 10^{-6} emu / g
C_{60}	- 0.35
C_{70}	- 0.6
Diamond	- 5
Nanotubes	- 10
Graphite	- $0.5_{\parallel}$ - $50_{\perp}$

1.1 times the perpendicular value and 30 times larger than that of C_{60} [46].

4. Polymerization of fullerenes under pressure

Undoped fullerenes can be polymerized by different methods: exposure to light with a photon energy at least 2 eV, irradiation of electron or ion beam, or applying an electric discharge. Polymerization of fullerenes under pressure is described in two recent reviews [47, 48]. High pressure and high temperature treatment of fullerenes has the advantage over other polymerization methods of being capable to produce different types of polymerization: one-, two- or three-dimensional polymers. These phases occur due to the ability of the C_{60} molecules to form bonds by 2+2 cycloaddition, when double bonds on neighboring molecules break to form a sp^3-bonded fullerene dimer. The polymerization type is governed by the pressure and temperature conditions, which can

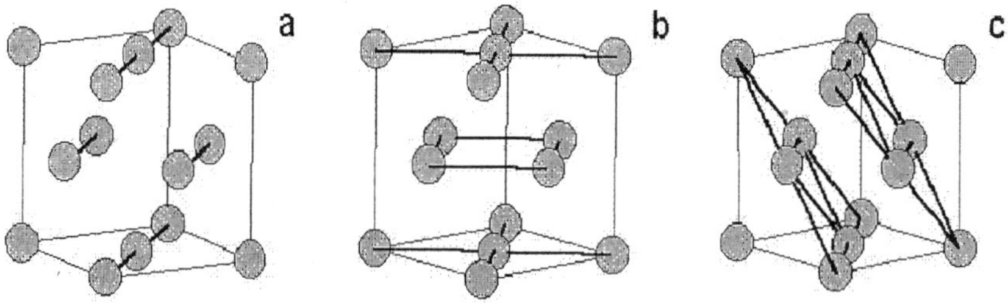

Figure 4. Formation of orthorhombic (a), tetragonal (b) and rhombohedral (c) C_{60} phase during polymerization.

be estimated from the pressure – temperature phase diagram. An orthorhombic phase (Fig. 4, a) forms at the pressures 1 – 9 GPa and low temperatures (below 650 K). Higher temperatures lead to an increase in the number of interfullerene bonds, and two types of two-dimensional polymers can be formed. The tetragonal phase with the interfullerene bonds along the <110> direction (Fig. 4, b) requires pressures about 2 GPa, whilst the optimal pressure for the rhombohedral phase (Fig. 4, c) is 6 GPa. This phase is more dense than the tetragonal, and polymerization occurs in the close-packed <111> plane. At temperatures above 1000 – 1100 K the fullerene cages break down, and a new phase appears described usually as a disordered cross-linked layered structure, disordered crystalline carbon or partially graphitized fullerene (Fig. 5).

5. Raman studies of the rhombohedral C_{60} phase

There exists a region at the pressure-temperature diagram of sample preparation, where physical properties of the material undergo break-neck changes. We speak about the transitional zone from fullerite to the peculiar crystalline form of carbon,

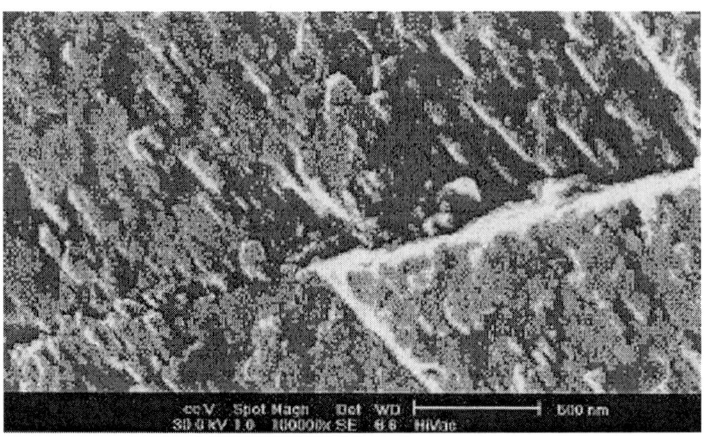

Figure 5. Carbon phase formed about the temperature stability limit for the fullerene molecules.

which was mentioned in the end of the previous section. This form differs from the known carbon solids like amorphous carbon, diamond-like carbon, disordered graphite, and deserves special attention .

In this section we discuss the structural changes in fullerite due to an increase of preparation temperature as studied by the Raman spectroscopy. The pressure concerned is 6 GPa, which is favorable for the formation of the rhombohedral phase. We present experimental results obtained on samples made by Davydov, Kashevarova and Rakhmanina [49, 50]. The products of high-pressure treatment are usually inhomogeneous in depth, and we show spectra taken from the very center of the sample.

The Raman spectra of the polymeric phases are well studied both experimentally [51, 52] and theoretically [53], and here we use these results to describe the transition from rhombohedral C_{60} to the carbon phase containing the broken fullerene cages.

The Raman spectra were obtained at room temperature using a Renishaw Ramascope with a 20× objective. The excitation wavelengths are 541, 633, and 782 nm.

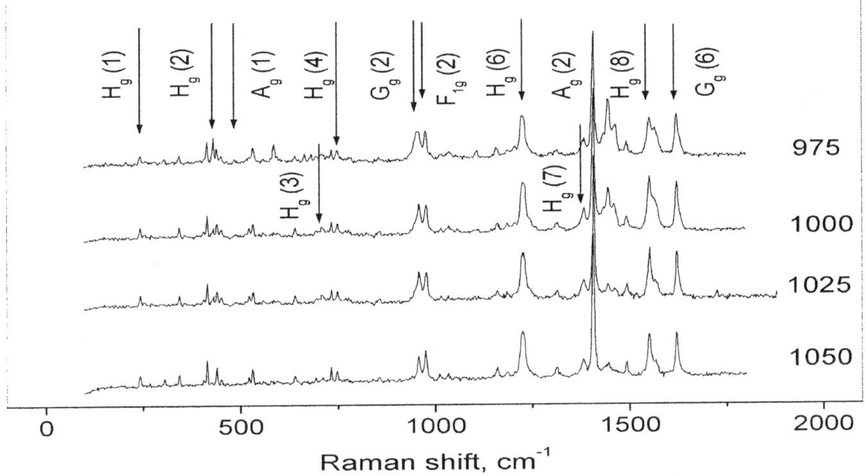

Figure 6. Raman spectra for the rhombohedral C_{60} phase prepared at different temperatures.

The spectra were recorded at low excitation intensities, as the 514-nm beam is able to depolymerize or repolymerize the sample, while the 633- or 782-nm beams burn it. It is well known that the Raman spectra of the carbon structures containing both sp^2- and sp^3-bonded carbon are sensitive to the excitation wavelength: an increase in photon energy blueshifts some peaks and notably changes their relative intensities [54].

The Raman spectrum of the isolated C_{60} molecule contains 10 peaks: the breathing $A_g(1)$ mode, the pentagonal $A_g(2)$ mode and 8 squashing H_g modes. The polymerization-induced cage distortion leads to more complex spectra.

Raman spectra (Fig .6) manifest the formation of an almost single-Rh-phase material with the pentagonal pinch mode $A_g(2)$, which is quite sensitive to the type and degree of polymerization, peaking at 1407 cm^{-1}. All other modes can be attributed to the calculated, modes for rhombohedral C_{60} as shown in Table 2. The spectra do not contain any indications of C_{60} cage destruction.

Table 2. *The vibrational modes of the rhombohedral C_{60} phase for different excitation wavelengths. w – weak; vw – very weak; s – strong.*

	Calculated [52]	541 nm	633 nm	782 nm
$H_g(1)$ pumpkin	243	241		
	273	267		241
	306	304	304	306
$H_u(1)$	344	343	344	344
$H_g(2)$	415	414 s	415 vw	415
	440	440	440	440
	453	453	453	
$A_g(1)$ breathing	470			
	487		492 vw	
$F_{1u}(1)$	523	521		523
$F_{2g}(1)$	533	531		532
$F_{1g}(1)$	559	559 vw	559	559
	595	594	595	
$H_g(3)$	695	694	694	695
	709			713
$G_u(2)$	733 dimer	732		733
$H_g(4)$	749	748	748 vw	748
	768			
$H_u(4)$	858	857	857	858
$G_g(2)$	956	958 s	958 vw	956
$F_{1g}(2)$	977	976 s	976 vw	976
$F_{2u}(4)$	1012	1012		1013
$H_g(5)$	1077		1075 vw	1077
$G_g(3)$	1158	1106		
	1186	1187 w		
$F_{2g}(3)$	1208			
$H_g(6)$	1222	1224 s	1221 vw	1222
	1229			
$G_g(4)$	1292		1292 vw	
	1315	1312		
$G_g(5)$	1333			
	1354			
$H_g(7)$	1381	1381	1379	
$A_g(2)$ pentagonal	1406	1406 s	1406	1409
$F_{1g}(3)$	1493	1493	1493	
$F_{2g}(4)$	1538			
$H_g(8)$	1550	1551 s	1550	
	1567			
$G_g(6)$	1620	1620 s	1619	1622
	1626			

A comparison of the spectra obtained for different excitation wavelengths show that all three excitation lines (541, 633, 782) are equally sensitive to the vibrational modes of rhombohedrally polymerized fullerenes. There exists only a slight difference: the breathing Ag (1) mode is seen only at the wavelength of 633 nm, but this line is unable to detect a very small amount of dimers, which are responsible for the feature at 733 cm^{-1} (a G_u (2) mode, which becomes Raman-active only in the case of dimers).

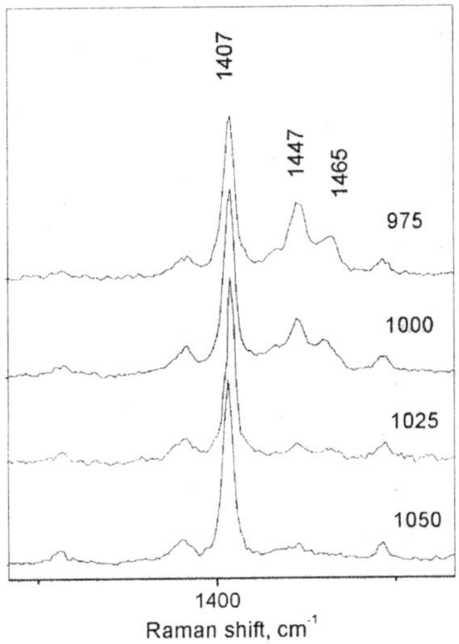

Figure 7. Raman spectra in the vicinity of the A_g (2) mode.

The influence of the sample preparation temperature on the polymerization type and degree is better seen in the near-pentagonal-pinch frequency region (Fig. 7). The behavior of the A_g(2) peak is used to detect the polymerization. For the pristine C_{60} its frequency is about 1469 cm^{-1}, for different types of one-dimensional polymers (chains, zigzag chains) it is downshifted by 5 – 15 cm^{-1}. In the case of tetragonal and rhombohedral polymerization the peak positions are 1447 and 1407 cm^{-1}, respectively .

It is clearly seen that the polymer obtained at 975 K is not a single-phase material. Rhombohedral polymerization dominates, but the contribution of the tetragonal phase is noticeable. An increase in the polymerization temperature leads to the formation of almost pure rhombohedral phase.

Analyzing the Raman spectra for the phases obtained at higher temperatures, close to the stability limit of the fullerene cages, we have found a curious phenomenon: unlike the previous case, the spectra are completely different for different wavelengths.

An example of this difference is shown in Fig. 8. Measurements at the

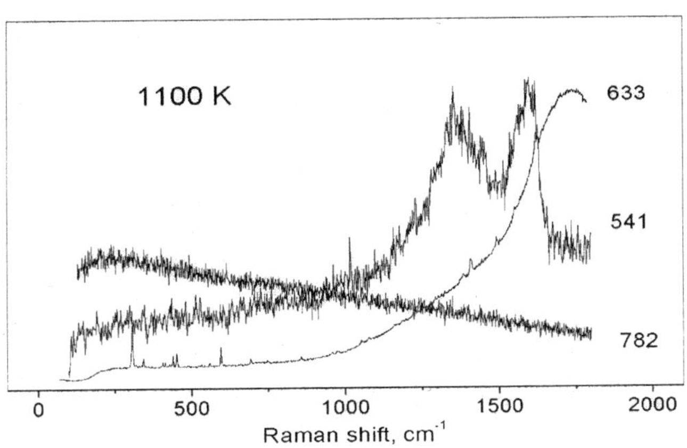

Figure 8. Raman spectra taken at 541, 633, and 782 nm excitation wavelength for the sample prepared at 1100 K.

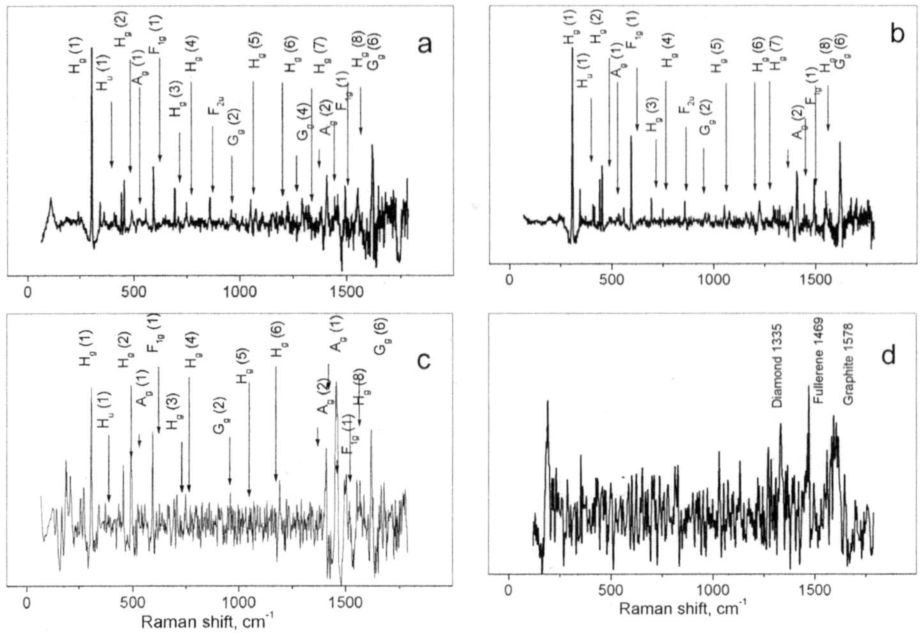

Figure 9. Raman spectra for the rhombohedral C_{60} phase prepared at different temperatures: a - 1075 K; b - 1100 K; c - 1125 K; d - 1150 K. The excitation wavelength is 633 nm.

"traditional" green line (541 nm) of the Ar ion laser show weak fullerene modes on a of strong M-shaped background. This two-humped shape is typical for mixture graphitic (1540 cm^{-1}) and disordered (1350 cm^{-1}) carbon. The infrared line of the diode laser (782 nm) proved to be insensible to these structures. In contrast, the spectrum taken using the red line of the He-Ne laser (633 nm) contained very strong fullerene modes without any indications of disordered or graphitic structures.

Fig. 9 shows the spectra taken using the 633 nm excitation wavelength after subtracting the luminescent background.

It is obvious from Fig. 9 that the carbon phase under investigation still exhibits the features of polymerized fullerene right up to the preparation temperature of 1150 K. In these phases the $H_g(1)$ "pumpkin" mode becomes unusually strong, and the other squashing modes (especially $H_g(2)$, $H_g(3)$, $H_g(4)$, $H_g(5)$) increase in the intensity. The breathing $A_g(1)$ mode, which was absent for the samples prepared at $T < 1050$ K, appears, and the intensity of this mode increases with the preparation temperature. The same is true for the silent $F_{1g}(1)$ mode. On the contrary, the intensity of the pentagonal pinch $A_g(2)$ mode decreases, and it acquires a satellite at 1456, revealing one-dimensional polymerization. The spectrum of the sample prepared at 1150 K contains only features of graphite, diamond-like carbon and unpolymerized fullerene. The abovementioned features are better seen in the Table 3.

From the Raman measurements using different excitation wavelengths we can conclude that the treatment of fullerene C_{60} at a pressure of 6 GPa and at temperatures higher than 1050 K leads to the formation of a material containing both rhombohedrally

Table 3. *The vibrational modes of the rhombohedral C_{60} phase prepared at different temperatures. The excitation wavelength is 633 nm.*

Mode	Calc.	1050	1075	1100	1125	1150
$H_g(1)$ pumpkin	306 273 243	304	306 s	306 s	306 s 271	
$H_u(1)$	344	344	343	345	341	
$H_g(2)$	415 440 453	415 vw 440 453	440 453	440 453	453	
$A_g(1)$	470 487		489	487 w	489s	
$F_{1u}(1)$	523					
$F_{2g}(1)$	533					
$F_{1g}(1)$	559 595	559 595	559 595 s	559 595 s	559 vw 595	
$H_g(3)$	695 709	694	694 s 709 w	694	695 vw 709 vw	
$H_g(4)$	749 768	748 vw	748	749	749 vw	
$H_u(4)$	858	857	858	858		
$G_g(2)$	956	958 vw	960			
$F_{1g}(2)$	977	976 vw	980 w			
$F_{2u}(4)$	1012		1013 w	1012 vw		
$H_g(5)$	1077	1075 vw	1077 w	1077 vw		
$G_g(3)$	1158 1186			1158 vw	1180 vw	
$F_{2g}(3)$	1208					
$H_g(6)$	1222 1229	1221 vw	1223 w	1224		
$G_g(4)$	1292 1315	1292 vw	1290 w	1292 vw		
$G_g(5)$	1333 1354					
$H_g(7)$	1381	1379	1381 w	1382 vw		
$A_g(2)$	1406	1406	1406	1406	1407 1456	1469
$F_{1g}(3)$	1493	1493	1492	1492	1493 w	
$F_{2g}(4)$	1538					
$H_g(8)$	1550 1567	1550	1552	1550 1568 w		
$G_g(6)$	1620 1626	1619	1620	1621	1617	

polymerized C_{60} and a graphite-like phase. The latter results from broken fullerene cages and can be detected from measurements using high photon excitation energies. For C_{60}, preresonant excitation conditions occur in the visible and near-infrared spectral ranges, and measurements at 633 nm are best suited for not only detecting the presence of fullerenes, but also to determine the type of polymerization.

6. High-pressure treatment: problem of sample homogeneity.

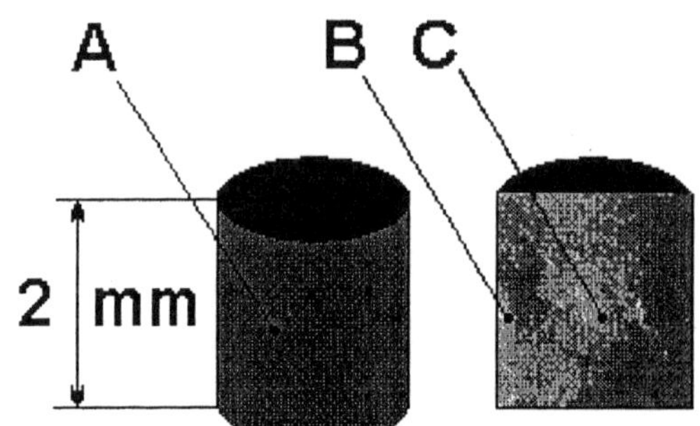

Figure 10. A typical sample of high-pressure polymerized fullerene.

In this section we draw attention to the fact that no one-to-one correspondence exists between the spectra presented in Figs. 6 – 9 and the conditions of sample preparation. In other words, the values of the sample preparation temperature, which are given in the figure captions, can be considered only as nominal ones.

There are two reasons for this statement. First, under the high pressure conditions there are difficulties in measuring the absolute values of temperature. The accuracy within one series of samples is ± 10 K, whereas samples from different series, being produced at nominally identical conditions, can differ by ± 25 K in the preparation temperature.

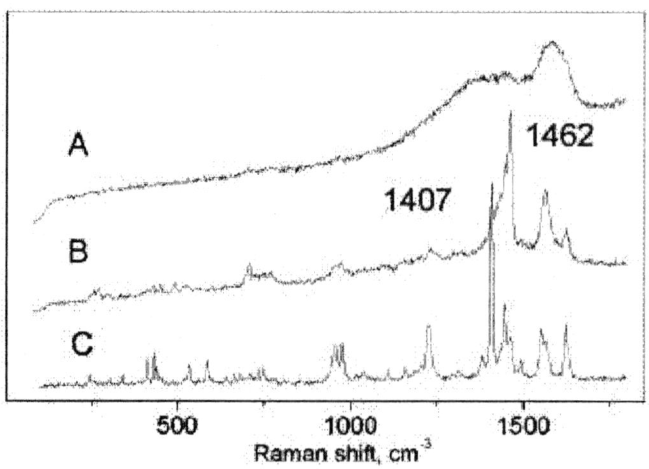

Figure 11. Raman spectra taken at the points A, B C for the sample prepared at p = 6 GPa and T = 1000 K. λ = 541 nm.

Another reason is more important. The products of high-pressure treatment are inevitably inhomogeneous due to the non-uniform distribution of pressure and temperature. The samples under investigation are small cylinders of approximately 2 mm diameter and height (Fig. 10). All the spectra shown in the previous section were taken in the central part of the samples (C). The Raman spectra taken at the cylinder generatrix (A) are noticeably different from those taken in the inner part of cleaved cylinders (B, C). Fig. 11 compares the spectra taken at the points A, B, and C for the sample prepared at p = 6 GPa and T = 1000 K. Obviously, the rhombohedral polymerization is achieved only in the very center of the sample. At the periphery a mixture of polymerized states is observed, and the outer part or the sample is mainly graphitic. This fact essentially complicates the investigation of the electrical and magnetic properties of the individual polymerized phases.

7. Electrical properties of pressure-polymerized fullerenes.

Electrical properties of fullerenes polymerized at the pressures 2 – 6 GPa and temperatures 500 – 1200 K have been described earlier [50, 55, 56]. Here we give a more detailed description of the properties of the rhombohedral C_{60} phase prepared at different temperatures. All the samples have layered structures, which is clearly seen in a microscope, and exhibit large anisotropy easily detected with a multimeter.

In very anisotropic materials the usual 4-probe method becomes inapplicable, but it is possible to establish the correct value of the anisotropy of the resistivity tensor using the six-contact Montgomery method [57]. Taking into account inhomogeneity, samples for the electrical measurements were specially prepared by cleaving the cylinders into the small pieces with characteristic dimensions of 100 – 500 μm.

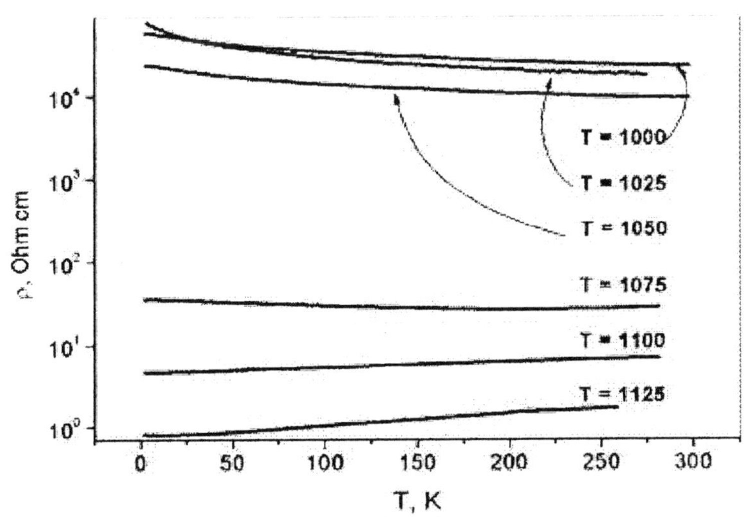

Figure 12. Temperature dependencies of conductivity for the samples prepared at 6 GPa and different temperatures.

The samples synthesized at p = 6 GPa and temperatures in the range 1000 – 1075 K revealed a gradual transition from the semiconductor to semi-metallic behavior (Fig. 12). Samples prepared at T = 1000 K showed the activated behavior with activation energies less than 0.1 eV. Samples prepared at higher temperatures exhibit great anisotropy in resistivity. The main features of the behaviour of these samples are: (i) in the z-direction the resistivity is of the order of several kOhm cm and decreases with the temperature; (ii) in the x-y direction resistivity is less than 1 Ohm cm and shows a minimum at a certain temperature.

Fig. 13 shows the experimental results for the in-plane resistivity. For each sample, the resistivity first decreases with decreasing temperature, then exhibits a minimum at a given temperature T_{min}, below which the resistivity starts to increase. For the preparation temperature of 1000 K (Fig 13, a) the dependence is linear in the $\rho = T^{1/2}$ coordinates which is usually attributable to 3D weak localization. For the samples prepared at 1050 K (Fig. 13, a) the resistivity shows a logarithmic increase with decreasing temperature which can be explained as 2D weak localization, and at the temperature of 1025 K an intermediate case is observed (Fig. 13 b).

There exists a correlation between the positions of the resistivity minima and the absolute values of resistivity. The common trend observed is a decrease in the temperature of the minimum point with a decrease of resistivity.

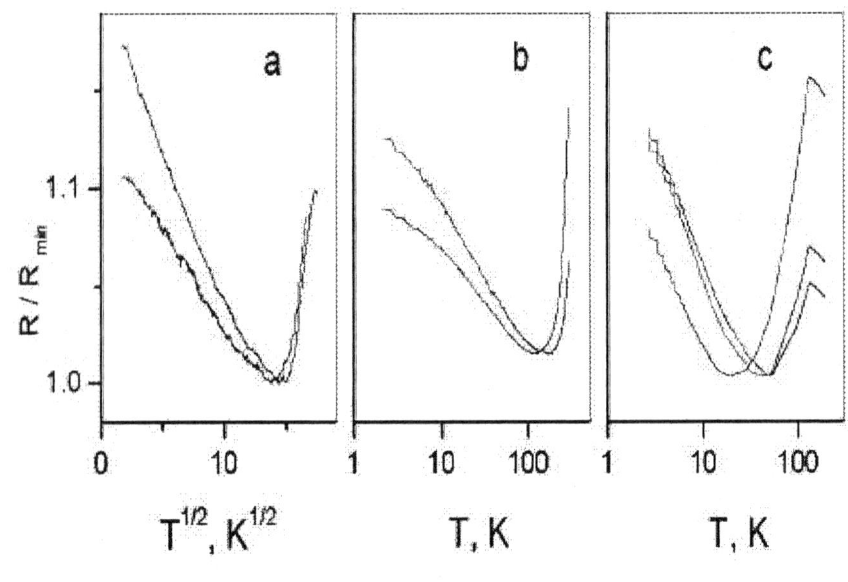

Figure 13. Temperature dependencies of in-plane conductivity for the samples prepared at 1000 K (a), 1025 K (b), 1050 K (c).

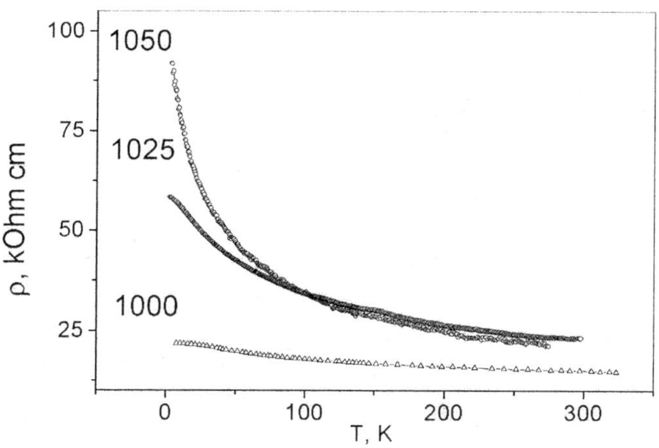

Figure 14. Temperature dependencies of the c-axis conductivity for the samples prepared at 1000 K (a), 1025 K (b), 1050 K (c).

The transport properties in the direction normal to the planes of polymerization are presented in Fig. 14. The resistivity exhibits a negative temperature coefficient $d\rho/dt < 0$ over the whole temperature range studied. This behavior can be attributed to an increase in a carrier density. However, the law is far from being of the Arrhenius type.

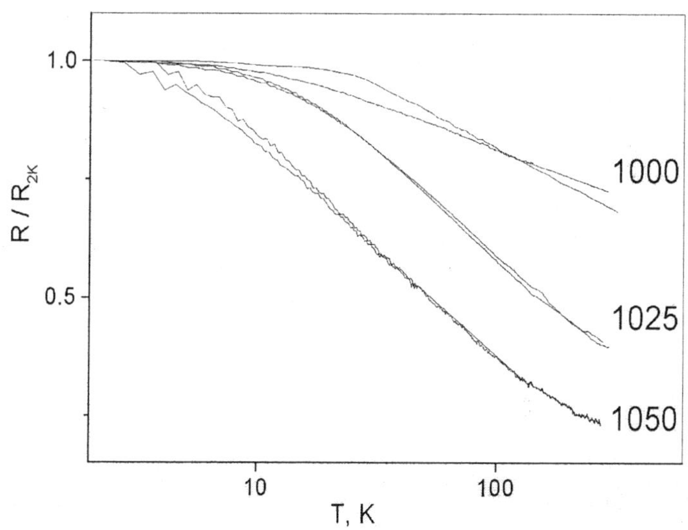

Figure 15. Reduced resistivity across the polymerization planes.

Experimental curves have proved to be linear in the coordinates $\rho = a - b \ln T$ (Fig. 15). This behavior, however, does not lead to an infinite resistivity as $T \rightarrow 0$. The resistivity curves show saturation below 10 K. Absolute values of conductivity are by several orders of magnitude higher than for C_{60} single crystals. This gives grounds for believing that conduction occurs via conductivity paths allowing the charge carriers to be transferred from one polymerized plane to another; the conduction paths are thus attributed to structural defects. The peculiarities in the transport properties in the z and x-y directions result in a high anisotropy ratio which even exceeds known values for graphite (R_z/R_{xy} = $10^4 - 10^5$).

Closing this section we want to draw attention to Fig.13 once more. One can easily see steps on the low-temperature portions on the curves in Fig. 13, b and Fig. 13, c. These steps have nothing to do with the device sensitivity. For the samples shown in Fig. 13 the absolute values of resistivity near 2 K are about 0.02 Ohm, which is far from the sensitivity limit. Exactly the same values of conductivity are observed at temperatures about 100 K, but there is no allusion to the stepwise function. The very position of the experimental points (Fig. 16) demonstrates that we deal with a sample-conditioned rather than a device-conditioned effect. At present we cannot give the explanation for the stepwise temperature dependencies of conductivity, but we emphasize that this behavior was observed only on the sample with pronounced magnetic properties.

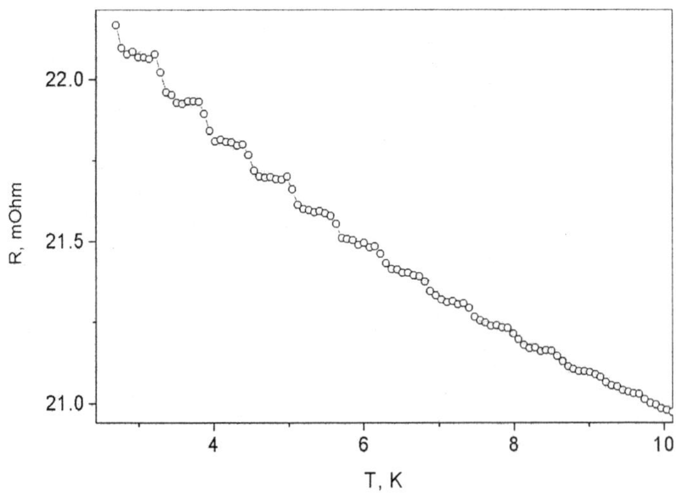

Figure 16. Low-temperature part of the in-plane conductivity curves.

8. Magnetic properties of polymerized fullerenes.

Magnetic properties of polymerized fullerenes are poorly studied. One of the rare examples is the study of static susceptibility in polymeric RbC_{60} [26] which has shown that the susceptibility can be decomposed into three kinds of terms:

$$\chi = \chi_{TD} + \chi_{Curie} + \chi_{TI}$$

where χ_{TD} and χ_{TI} are temperature dependent and temperature independent terms, and χ_{Curie} stands for the Curie contribution. χ_{TI} originates from the Pauli paramagnetic term, and χ_{TD} is regarded as an intrinsic property of the RbC_{60} polymer. This term increases with temperature up to T = 60 K and is interpreted in terms of a spin-density wave.

Some features of the AC susceptibility of neutral polymerized fullerenes were described in Ref [58]. As shown in Fig. 3, the AC susceptibility of pristine C_{60} is negative, except for the low-temperature region where the Curie contribution dominates. In contrast to pristine C_{60}, the susceptibility of photopolymerized C_{60} is positive (Fig. 17, a). It shows a similar upturn at the same temperature, i.e., the low-temperature behavior is preserved in the photopolymerized phase, and the Curie-term introduces the same contribution. Thus the paramagnetic impurities affect the magnetic behavior of pristine and photopolymerized fullerenes equally.

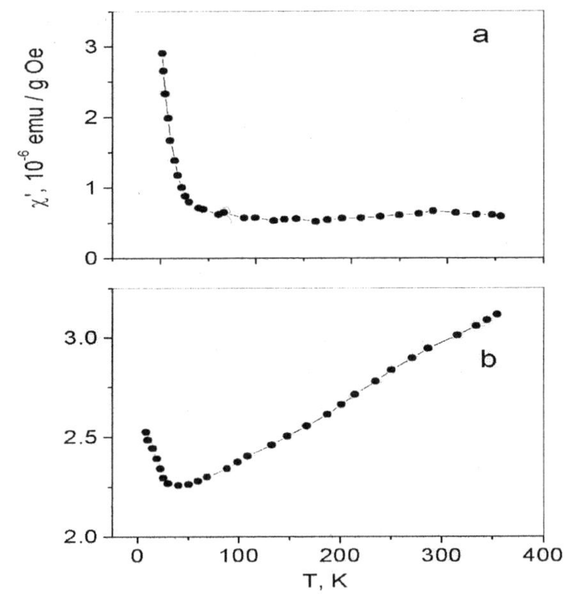

Figure 17. AC susceptibility of photopolymerized C_{60} powder (a) and pressure-polymerized tetragonal phase of C_{60} (b).

When allowance is made for the superimposed Curie term, the susceptibility of polymerized fullerenes increases slightly with temperature. The absolute value of the susceptibility is twice larger for the photopolymers than for pristine C_{60}. This value becomes even greater for the pressure-polymerized phases. Fig. 17, b shows the dependence obtained for the tetragonal C_{60} phase. Similar dependencies were found for the rhombohedral phase prepared at temperatures less than 1000 K.

The increase in polymerization temperature for the rhombohedral C_{60} phase leads to substantial changes in the magnetic properties.

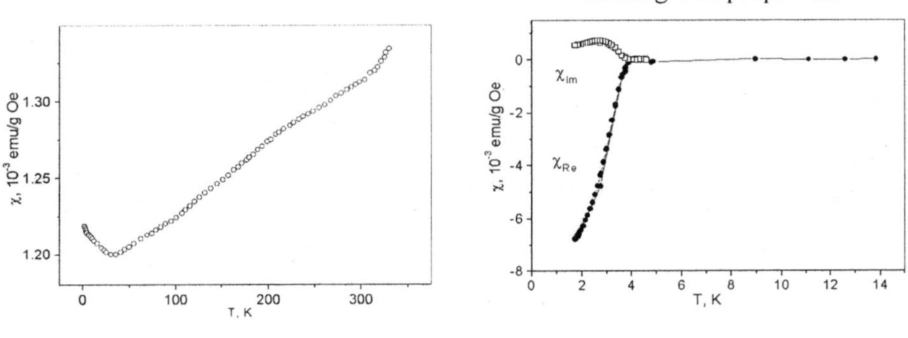

a

b

Figure 18. Temperature dependencies of AC susceptibility for the C_{60} polymers prepared at p = 6 GPa and T = 1025 K (a) and 1075 K (b).

The temperature dependence of the AC susceptibility for the rhombohedral C_{60} shown in Fig. 18, a is quite similar to that for the tetragonal phase (Fig. 17, b). The only difference is the 50–fold increase in the magnitude. SQUID measurements on this phase have revealed the features of ferromagnetic behavior with a Curie temperature about 500 K [10].

Quite the opposite situation is observed on the samples prepared from the same starting material, but at 50 K higher temperature. At room temperature and below the behavior is similar to that shown in Fig. 17, a: a weak and nearly temperature independent positive diamagnetic susceptibility. Below 2.5 K a strong enhancement of diamagnetism is observed, which is difficult to explain in any other way than by the presence of a superconducting phase in the sample.

Finally, we emphasize once more that the preparation conditions indicated for each sample in the sections 7 and 8 should be regarded as nominal (see section 6). Correlation between the sample properties and structural characterization is still an open question.

9. Conclusions. Role of experimental observation.

Studies of electrical and magnetic properties of pressure-polymerized fullerenes have revealed many unusual phenomena connected with the electronic phase transitions in this material. A transition from semiconducting to metallic behavior, a superconducting transition, and the formation of a ferromagnetic phase in this pure carbon material have all been observed experimentally. In this contribution we restricted ourselves to descriptions of the experimental evidence. Possible explanation of some effects can be found in the papers [10, 56, 59], but we believe that understanding the properties of this unusual form of pure carbon requires the simultaneous consideration of all transport properties and cooperative phenomena.

The experimental results obtained on the undoped fullerene polymers provide evidence for the fact that fullerenes can supplement the family of organic conductors, superconductors, and organic magnets.

Organic conductors are molecular-based materials with low resistance, and they are subdivided into two major classes: charge-transfer compounds and conducting polymers. Conducting polymers consisting of long carbon-based chains composed of simple repeating units (monomers) have attracted the attention since the early 1980s. They are different from other polymeric materials (which become conducting by the incorporation of carbon black or metal particles) in having an inherent electrical conductivity. Films of conducting polymers on electrodes can be switched electrochemically between the insulating and conductive state, which is followed by the changes in the electrical, optical, and chemical properties. Alan J. Heeger, Alan G. MacDiarmid, Hideki Shirakawa have been awarded The Nobel Prize in Chemistry 2000 "for the discovery and development of conductive polymers". Actually, they created a new breed of plastic electronic circuit, which works even when bent. Among the first organic devices already created [60] are important types of lasers and transistors and a prototype computer display. The organic molecules polyacenes have also exhibited superconductivity and fractional quantum Hall effect [61] seen previously only in inorganic semiconductors.

Organic superconductors are part of the organic conductor family, which includes molecular salts, polymers and pure carbon systems. The first organic superconductor, a so-called Bechgaard salt $(TMTSF)_2PF_6$ was synthesized in 1979 [62]. Although organic superconductors usually have low T_c values, they offer unusual properties like giant magnetoresistance, rapid oscillations, and quantum Hall effect. $(TMTSF)_2PF_6$ was shown to keep superconducting properties in magnetic fields up to 6 Tesla – usually such fields completely kill superconductivity in a material. During the last 20 years of the previous century, over 100 superconducting organic salts have been discovered. In the beginning of year 2000, i.e. before the appearance of works on hole-doped fullerenes, the record holder was Cs_3C_{60} with $T_c = 40$ K, which was 17 K higher than for all the known non-copper oxide superconductors. (39 K superconductivity in MgB_2 was discovered in May 2001 [63]). Bertram Batlogg, Hendrik Schön and Christian Kloc from the Bell labs inserted molecules of chloroform and bromoform into fullerene matrix to create a crystal with an increased lattice parameter [8]. These materials showed a superconducting transition at 80 and 117 K correspondingly, which approaches the record values for the copper oxides (138 K). Fullerene superconductors have the advantage over the conventional high-T_c materials: their behavior, to all appearance, is described by the classical theories for low-temperature superconductors.

Considered to be a scientific impossibility, organic magnets (containing little or no metallic material) were discovered by chemist Joel Miller and physicist Arthur Epstein [64]. This material has an ordering temperature of about 5 K, whilst the first organic ferromagnet consisting only of light elements (C, H, N, O), shows long-range ferromagnetic order below 0.65 K [65]. Now metalloorganic materials which order at temperatures 350 – 400 K are known [66]. Molecule-based magnets include materials with spins only on *p*-orbitals; on both metal ions (*d* or *f* orbital spins) and *p*-orbitals; as well as systems with spins on metal ions with an exchange path provided by organic passive fragments. There are three origins of magnetism: orthogonality of orbitals, configuration interaction and dipole-dipole interaction [67].

Organic magnets offer the advantages of low weight, flexibility, transparency, semiconductor or insulating electrical properties and non-metallurgical processing technology. Photoresponsiveness, i.e. the ability to change the magnetic properties when exposed to light, makes them candidates for high-density optical data storage systems. Magnetism is a field of study which has recently undergone a renaissance due to the discovery of giant and colossal magnetoresistance, design of microstructures containing hard and soft magnets, and also due to the discovery of molecular-based magnets with novel combinations of properties.

What comes next?

It is not so rarely, that the scientific discoveries are made by accident.

Richard Smalley and Harold Kroto were simulating the atmospheric conditions in red giant stars looking for the formation of linear cyanopolyynes. The significant point is that they refused to ignore large peaks of 720 and 840 in the mass spectra. Thus the completely new field of fullerene science was opened.

Another important landmark in the history of fullerenes is Wolfgang Krätschmer and Donald Huffman's breakthrough discovery of how to synthesize fullerenes in macroscopic quantities. The initial aim of their experiments was producing soot for comparison with the interstellar dust. They observed four “historical” bands in the infrared spectra. The recognition of the significance of these bands resulted in the enormous surge of activity in fullerene studies in the 1990s.

The metallic form of polyacetylene was discovered by accident when a Japanese student mistakenly used the "wrong" amount of catalyst while polymerizing acetylene. He converted the standard single-bond carbon chains to polymer backbones with alternating single and double bonds, a change that provided a pathway for free-electron charge carriers. Alan G. MacDiarmid and Hideki Shirakawa's hunch that something extraordinary had happened in the student's reaction flask led to a whole new area of materials science, with enormous commercial potential

Searching for evidence of superconductivity in undoped pressure-polymerized fullerenes, we have instead discovered typical features of ferromagnetic behaviour: saturation magnetisation, a hysteresis loop, and attachment to a magnet at room temperature [10].

Who is the next to make a mistake?

10. Acknowledgements.

The authors take pleasure in thanking V. A. Davydov, L. S. Kashevarova, and A. V. Rakhmanina for preparing the samples. The continued support of the Swedish Research Council and the Russian Federation for Basic Research (project "Novel electronic phase transitions in fullerene-based materials") is gratefully acknowledged.

References

1. Kroto, H.W.; Heath, J.R.; O'Brien, S.C.; Curl, R.F.; Smalley, R.E. C_{60}: Buckminsterfullerene, *Nature*, **318**, 162- 163 (1985).
2. Krätschmer, W.; Lamb, L.D.; Fostiropoulos, K.; Huffman, D.R. "Solid C_{60}: A New Form of Carbon", *Nature*, **347**, 354 - 358 (1990)
3. Hebard, A.F.; Rosseinsky, M.J.; Haddon, R.C.; Murphy, D.W.; Glarum, S.H.; Palstra, T.T.M.; Ramirez, A.P.; Kortan, A.R. Superconductivity at 18 K in potassium-doped C_{60}, *Nature*, **350**, 600 – 601 (1991).
4. Fleming, R.M.; Ramirez, A.P.; Rosseinsky, M.J.; Murphy, D.W.; Haddon, R.C.; Zahurak, S.M.; Makhija, A.V. Relation of Structure and Superconducting Transition Temperatures in AC_{60}, *Nature* **352**, 787 – 788 (1991).
5. Allemand P.-M., Khemani K.C., Koch A., Wudl F., Holczer K., Donovan S., Grüner G., and Thompson J.D., Organic molecular soft ferromagnetism in a fullerene C_{60}, *Science*, **253**, 301- 303(1991).
6. Blank, V.D., Buga, S.G., Serebryanaya, N.R., Denisov, V.N., Dubitsky, G.A., Ivlev, A.N., Mavrin, B.N., Popov, M.Y., Ultrahard and superhard carbon phases produced from C_{60} by heating at high pressure – structural and Raman studies. *Phys. Lett. A* **205**: (2-3) 208-216 (1995).
7. Schön J. H., Kloc C., Batlogg B. Superconductivity at 52 K in hole-doped C_{60}. *Nature* **408**: 549-552 (2000).
8. Schön J. H., Kloc Ch., Batlogg B., High-temperature superconductivity in lattice-expanded C_{60}, *Science* **293**, 2432 – 2434; published online 30 August 2001 (10.1126 / science. 1064773).
9. Narymbetov B, Omerzu A, Kabanov VV, Tokumoto M, Kobayashi H, Mihailovic D, Origin of ferromagnetic exchange interactions in a fullerene–organic compound. *Nature* **407**, 883 - 885 (2000)
10. Makarova T.L., Sundqvist B., Esquinazi P., Höhne R., Kopelevich Y., Scharff P., Davydov V. À., Kashevarova L. S., and Rakhmanina A. V.. Magnetic carbon. *Nature* (2001), in print.
11. Kempa H., Kopelevich Y., Mrowka F., Setzer A., Torres J.H.S., Höhne R, Esquinazi P. Magnetic-field-driven superconductor-insulator-type transition in graphite. *Sol. State Commun.* **115**, 539-542 (2000).
12. Kopelevich, Y., Esquinazi, P., Torres, J.H.S., & Moehlecke, S. Ferromagnetic- and superconducting-like behavior of graphite. *J.Low Temp.Phys.* **119**, 691-702 (2000).
13. Shibayama Y., Sato H., Enoki T., and Endo M. Disordered magnetism at the metal-insulator threshold in nano-graphite-based carbon materials. *Phys. Rev. Lett.* **84**, 1744-1747 (2000)
14. Enoki, T; Kawatsu, N; Shibayama, Y; Sato, H; Kobori, R; Maruyama, S; Kaneko, K. Magnetism of nano-graphite and its assembly, *Polyhedron* **20** 1311-1315 (2001).
15. González, J., Guinea, F., and Vozmediano, M. A. H. Electron-electron interactions in graphene sheets, *Phys. Rev. B* **63**, 134421-29 (2001).
16. Harigaya, K. The mechanism of magnetism in stacked nanographite: theoretical study. *J. Phys. Cond. Mat.* **13**, 1295-1302 (2001).

17. Harigaya K., New type of antiferromagnetic state in stacked nanographite, *Chem. Phys. Lett.* **340**, 123-128 (2001).
18. Kasumov A.Y., Deblock R., Kociak M., Reulet B., Bouchiat H, Khodos I.I., Gorbatov Y.B., Volkov V.T., Journet C., Burghard M., Supercurrents through single-walled carbon nanotubes. *Science* **284**, 1508-1511 (1999).
19. Tang Z.K., Zhang L.Y., Wang N., Zhang X.X., Wen G.H., Li GD, Wang J.N., Chan C.T., Sheng P., Superconductivity in 4 angstrom single-walled carbon nanotubes. *Science* **292**, 2462-2465 (2001).
20. Makarova T.L. Electrical and optical properties of pristine and polymerized fullerenes. *Semiconductors* **35**, 243-278 (2001).
21. R.R.Gupta, Diamagnetic susceptibility, in *Landolt-Börnstein New Series* II/16, edited by K.-H.Hellwege (Springer, Berlin, 1986), Vol.16, p.7.
22. Luo, W.L., Wang, H., Ruoff, R.C., Cioslowski, J., & Phelps, S. Susceptibility discontinuity in single-crystal C_{60}. *Phys. Rev. Lett.* **73**, 186 – 188 (1994)
23. Haddon, R.C., Scheemeyer, L.F., Waszczak, J.V., Glarum, S.H., Tycko, R., Dabbah, G., Kortan, A.R., Muller, A.J., Mujsce, A.M., Rosseinsky, M.J., Zahurak, S.M., Makhija, A.V., Thiel, F.A., Raghavachari, K., Cockayne, E., Elser V., Experimental and theoretical determination of the magnetic susceptibility of C_{60} and C_{70}. *Nature* **350**, 46-47 (1991)
24. Ruoff, R.S., Beach, D., Cuomo, J., McGuire, T., Whetten, R.L., Diederich, F., Confirmation of a vanishingly small ring-current magnetic susceptibility of icosahedral C_{60}. *J.Phys.Chem.* **95**, 3457-3459 (1991)
25. Ling M.F., Finlayson T.R., Raston C.L., Nonlinear field-dependent susceptibilities of C_{60} and carbon nanotubes, *Australian J. Phys.* **52**, 913-918 (1999).
26. Saito T., Akita Y., Tanaka K., Magnetic susceptibility of the one-dimensional polymeric phase of RbC_{60}, *Phys. Rev. B* **61**, 16091-16096 (2000).
27. Prassides K., Margadonna S., Arcon D., Lappas A., Shimoda H., Iwasa Y. Magnetic ordering in the ammoniated fulleride $(ND_3)\,K_3C_{60}$. *J. Am. Chem. Soc.* **121**, 11227-11228 (1999).
28. Lobach A.S., Shul'ga Y.M., Roshchupkina O.S., Rebrov A.I., Perov A.A., Morozov Y.G., Spector V.N., Ovchinnikov A.A. $C_{60}H_{18}$, $C_{60}H_{36}$ and $C_{70}H_{36}$ fullerene hydrides: Study by methods of IR, NMR, XPS, EELS and magnetochemistry. *Fullerene Sci. Technol.* **6** 375-391 (1998).
29. Shul'ga Y.M., Lobach A.S., Ivleva I.N., Morozov Y.G., Spector V.N., Ovchinnikov A.A., Magnetic properties of $C_{60}Pd_n$ fullerides, *Mol. Cryst. Liq. Cryst. C* **10**: 201-206 (1998)
30. Ishii K., Ootoshi H., Nishi Y,. Fujiwara A, Suematsu H., Kubozono Y., Crystal Structure and Magnetic Properties of $Eu_xSr_{6-x}C_{60}$, *AIP Conf. Proc.* **544** 29-32 (2000).
31. Maruyama Y., Motohashi S., Suzuki K., Takagi S. and Ogata H., Structures and magnetic properties of high-temperature reaction products of cerium metal and C_{60} solid. *Sol. State Commun.* **115**, 457-462 (2000).
32. McClure J.W., Diamagnetism in graphite, *Phys.Rev.* **104**, 666 - 671 (1956);
33. McClure J.W., Theory of diamagnetism in graphite, *Phys.Rev.* 119, 606 - 613 (1960).
34. Chehab S., Guerin K., Amiell J., and Flandrois S. Magnetic properties of mixed graphite containing both hexagonal and rhombohedral forms. *Eur. Phys. J. B* **13**, 235-243 (2000)
35. Kusakabe K., Wakabayashi K., Igami M., Nakada K., Fujita M,. Magnetism of nanometer-scale graphite with edge or topological defects, *Mol. Cryst. Liq. Cryst. A* **305**: 445-454 (1997).
36. D. V. Khveshchenko. Ghost excitonic insulator transition in layered graphite. *arXiv: cond- mat/* 0101306 v2 22 (2001).
37. Y. Shibayama, H. Sato, T. Enoki, and M. Endo, Disordered Magnetism at the Metal-Insulator Threshold in Nano-Graphite-Based Carbon Materials, *Phys. Rev. Lett.* **84**, 1744 - 1747 (2000).
38. Ovchinnikov, A.A., and Shamovsky, I.L., The structure of the ferromagnetic phase of carbon. *Theochem – J. Molec.Struct.* **83**, 133-140 (1991).
39. Murata, K., Ushijima, H., Ueda, H., Kawaguchi, K., A stable carbon-based organic magnet. *J. Chem. Soc. Chem. Commun.* **7**, 567-569 (1992).
40. M.S. Dresselhaus and G. Dresselhaus, "Intercalation Compounds of Graphite," *Adv. Phys.* **30**, 139- 326 (1981).
41. Panich M., Shames A.I, and Nakajima T., On paramagnetism in fluorinated graphite: EPR and solid state NMR study. *J.Phys.Chem.Solids* **62**, 959-964 (2001).
42. Bandow S., Magnetic properties of nested carbon nanostructures studied by electron spin resonance and magnetic susceptibility measurements *J. Appl. Phys.* **80**, 1020-1027 (1996).
43. Tsui F., Jin L., Zhou O., Anisotropic magnetic susceptibility of multiwalled carbon nanotubes, *Appl. Phys. Lett.* **76** 1452-1454 (2000).

44. Byszewski P., Baran M., Magnetic susceptibility of carbon nanotubes, *Europhys. Lett.* **31** 363-366 (1995).
45. Chauvet O, Forro L., Basca W., Ugarte D., Doudin B., Decher W.A., Magnetic anisotropies of aligned carbon nanotubes. *Phys. Rev. B* **52** R6963-R6966 (1995).
46. Wang X.K., Chang R.P.H., Patashinski A., Ketterson J.B., Magnetic susceptibility of buckytubes, *J. Mater. Research* **9,** 1578-1582 (1994).
47. Sundqvist B. Fullerenes under high pressures. *Adv.Phys.* **48**, 1-134 (1999).
48. Blank V. D., Buga S. G., Dubitsky G. A., Serebryanaya N. R., and Popov M. Yu. and Sundqvist B, High-pressure polymerized phases of C_{60}, *Carbon* **36**, 319-343 (1998)
49. Davydov V.A., Kashevarova L.S., Rakhmanina A.V., Senyavin V.M., Agafonov V., Ceolin R., Allouchi H., Dzyabchenko A.V., and Szwarc H., Tetragonal polymerized phase of C_{60}, *Phys. Rev. B* **58**, 14786-14790, (1998).
50. Makarova T.L., Wagberg T., Sundqvist B., Zhu X.-M., Nyeanchi E.B., Gaevski M.E., Olsson E., Agafonov V., Davydov V.A., Kashevarova L.S., and Rakhmanina A.V., Conductivity of two-dimensional C_{60} polymers, *Mol. Cryst. Liq. Cryst. C: Mol. Mat.* **13**, 157-162, (2000).
51. Rao A. M., Eklund P. C., Hodeau J–L., Marques L., Nunez–Regueiro M. Infrared and Raman studies of pressure-polymerized C_{60}, *Phys. Rev. B* **55**, 4766–4773 (1997).
52. Davydov, V.A., Kashevarova L. S., Rakhmanina A. V., Senyavin V. M., Céolin R. Szwarc H., Allouchi H. and Agafonov V. Spectroscopic study of pressure-polymerized phases of C_{60}. *Phys. Rev. B* **61**, 11936-45 (2000).
53. Adams, G.B., and Page, J.B., Theoretical studies of Raman spectra for planar polymerized C_{60}, *phys. stat. sol. (b)* **226**, 1996 – 2106 (2001)
54. Wagner, J., Ramsteiner, M., Wild, C.H., Koidl, P., Resonant Raman scattering of amorphous carbon and polycrystalline diamond films, *Phys. Rev. B* **40**, 1817 – 1824 (1989)
55. Makarova, T. L., Scharff P., Sundqvist B., Narymbetov B., Tokumoto M. Kobayashi H., Davydov V. A., Rakhmanina A. V., Kashevarova L. S. Anisotropic metallic properties of highly-oriented C_{60} polymer. *Synth. Metals* **121**, 1099-1100 (2001).
56. Makarova, T. L., Scharff P., Sundqvist B., Gaevski M.E., Davydov V. A., Rakhmanina A. V., and Kashevarova L. S.. Electrical properties of two-dimensional fullerene matrices. *Carbon* **39,** 2203 - 2209 (2001).
57. Corraze B., Ribault M. Extensive characterization of anisotropic conductors in the Montgomery geometry. *J. de Physique III* **4**, 741-750 (1994).
58. Makarova T.L., Liu B.-B., Sundqvist B. AC Susceptibility of Some Neutral C_{60} Polymers. *AIP Conference proceedings* (2001), in print.
59. Okotrub A.V., Bulusheva L.G., Belavin V.V., Davydov V.A., Makarova T.L., Tomanek D. Electronic structure and properties of the rhombohedral C_{60} polymer. *J. Chem. Phys.* 115, 5637-5641(2001).
60. Schön J.H., Kloc C., Batlogg B, Ambipolar organic devices for complementary logic, *Synth. Metals* **122**, 195-197 (2001).
61. Schön J.H., Kloc C., Batlogg B. Fractional quantum Hall effect in organic molecular semiconductors. *Science* **288**, 2338-2340 (2000).
62. Jerome D., Mazaud A., Ribauld M., Bechgaard K., *J. Physique Lett.* **41**, L95-98 (1980)
63. Eom, C. B., Lee M. K., Choi, J.H., Belenky L.J., Song X., Cooley L.D., Naus M.T., Patnaik S., Jiang J., Rikel M., Polyanskii A., Gurevich A., Cai X.Y., Bu D., Babcock S.E., Hellstrom E.E., Larbalesttier D.C., Rogado N., Regan K.A., Hayward M.A., He T., Slusky K., Inumaru K., Haas M.K., and Cava R.J., High critical current density and enhanced irreversibility field in superconducting MgB2 thin films, *Nature* **411**, 558-560 (2001).
64. Miller J.S., Calabrese J.C., Epstein A.J., Bigelow R.W., Zhang J.H., Reiff W.M. Ferromagnetic properties of one-dimensional decamethylferrocenium tetracyanoethylenide (1:1): $[Fe(C_5Me_5)_2]^+$ [TCNE]. *J. Chem. Soc. Chem. Commun.* **13** 1026-1028 (1986).
65. Takahashi M., Turek P., Nakazawa Y., Tamura M., Nozawa K., Shiomi D., Ishikawa M., Kinoshita M., Discovery of a quasi-1D organic ferromagnet, P-NPNN, *Phys. Rev. Lett.* **67,** 746-748 (1991).
66. Manriquez J.M., Yee G.T., Mclean R.S., Epstein A.J., Miller J.S. A room-temperature molecular organic based magnet. *Science* **252**, 1415-1417 (1991).
67. Miller J.C., Epstein A.J. Molecular-based magnets – an overview. *MRS Bulletin*, **25** 21 – 30 (2000).

DIRECT TRANSITION IN THE POROUS NANOSILICON MEASURED BY ELECTROREFLECTANCE

R. YU. HOLINEY, L.A. MATVEEVA, E.F. VENGER
Institute of Semiconductor Physics
pr. Nauky 45, Kyiv 03028, Ukraine

Abstract

Electronic structure of the crystalline nanoparticles of porous silicon is under discussion till now. It is known that the electroreflectance is the best method to determined direct transitions in the semiconductor materials. In present paper, porous silicon that was prepared by electrochemical and stain etching were investigated by electroreflectance in spectral region from 1.5 eV to 3 eV where photoluminescence bands usually take place. It was shown that there are few direct transitions in the spectral region were bulk silicon hasn't one. These direct transitions could be associated with photoluminescence bands.

1. Introduction

Porous silicon was under intensive investigation since 1990 due to peculiar optical properties. It is know that silicon nanoscaled crystalline wires are created during electrochemical or stain etching [1, 2, 3]. The main attention of the huge numbers of papers was attracted to the photoluminescence properties of the porous silicon [4, 5]. It could be mark out three photoluminescence bands:

3. Red-yellow photoluminescence band with maximum near 1.7-2 eV.
4. Green – near 2.6 eV.
5. Blue – 2.9 eV.

What is more, nature of the photoluminescence is under discussion till now, too. There were proposed a lot of models (quantum confinement, a-Si, SiH_x, SiO_x, at al) [5]. Knowing of the electronic structure of the nanosilicon crystalline wire can help in the understanding of the photoluminescence nature. It is known that electron is strongly localised in such nanosilicon wire what leads to changing of the electronic structure in comparison with bulk silicon. It was shown [4, 6, 7] that energy of the undirect gap is increasing with decreasing of the wire size. The fundamental edge could achieve 3 eV from 1.14 eV at the silicon crystalline wire size 1 nm [8]. But it was shown in last works [9, 10] that band edge remains undirect nature at the decreasing of the nanosilicon wire size.

It was shown too [11, 12] that oxygen strongly influence on the electronical properties of the silicon nanowires. Oxygen can create electron and hole states (traps) in the band gap when size of the wire achieved 3 or less nanometers [12]. Electroreflectance measuring of the anisotropically etched silicon showed direct

E. Buzaneva and P. Scharff (eds.), Frontiers of Multifunctional Nanosystems, 313–320.

transition in the spectral region of photoluminescence near 1.9 eV [13] where bulk silicon hasn't any transitions.

In present paper electroreflectance measuring were provided in the spectral region of porous silicon demonstrates photoluminescence properties (described above). The purpose was to established electronic structure of the nanosilicon wire and associated photoluminescence band with electroreflectance spectra.

2. Experiment

The porous nanoscaled silicon were prepared by the electrochemical and stain etching on the n- and p-type silicon wafer with {100} orientation and a resistivity 10-12 Ω·cm. Porous samples were electrochemically etched in HF(20%)-C_3H_7OH solution (1:1) [14] at different current density (from 1 to 120 mA·cm^{-2}). The depths of the pores in the silicon were scaled from few manometers till several micrometers. Stain etching was provided in the HNO_3: HF: H2O (3:1:5) 3-5 minutes with previous surface polishing in the HNO_3: HF (3:1) during 3-5 seconds [15].

The electroreflectance measuring was performed using standard electrolyte technique (0.1 N KCl water solution) at room temperature [16]. The modulation voltage was 1 V what was suited to the low field regime range. The spectra were measuring in the wide spectral region from 1.5 eV where red photoluminescence is usually observed till 4 eV where the bulk silicon has the direct transition (~ 3.4 eV). The transition energies were calculated from the electroreflectance spectra using three-points Aspnes methods for bipolar shape of signal [16] or as energy of the maximum of the unidirectional one.

3. Results and discussion

Fig. 1 and 2 show electroreflectance spectra of the porous silicon prepared by electrochemical etching with current density 1 and 120 mA·cm^{-2}. As one can see, spectra contain three direct transitions in the region where bulk silicon hasn't any one (near 1.9, 2.5 and 3.0 eV, see tab 1). Fig 1 and 2 didn't contain signal in the spectral region near 3.4 eV ($\Gamma'_{25} - \Gamma_{15}$ in the central of Brillouin zone). Thus, one can decide that electroreflectance spectra were measured only from thin layer of porous silicon samples that contained pores, nanosilicon crystalline wires, SiO_x and SiO_2 coat on the surface. Silicon wafer didn't influence on the spectra that are presented in Fig. 1, 2.

Presence of the electroreflectance spectra in the spectral region near 1.9 eV where bulk crystalline silicon hasn't any direct transition (fig. 1, 2) corroborate to existence of the electronic states in the nanosilicon wire. It was theoretically predicted that the presence of the SiO_x layer on the surface of the nanocrystalline silicon must leads to creation of the electron states in the band gap with energy that is close to red photoluminescence maximum (1.9 eV) [11, 12]. A Si=O double bonds is more likely to be formed and stabilized in the interface, since it requires neither a large deformation nor an excess element [12].

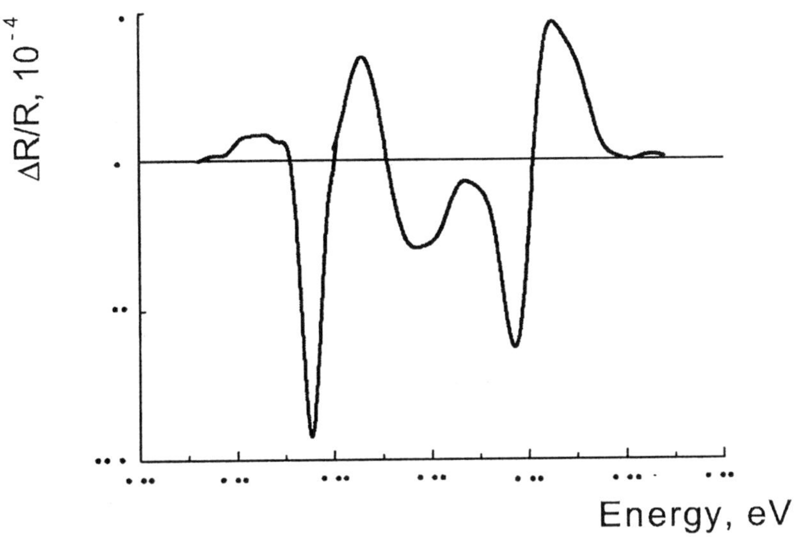

Figure 1. Electroreflectance spectrum of the porous nanoscaled silicon sample prepared by electrochemical etching with current density 2 mA·cm^{-2}.

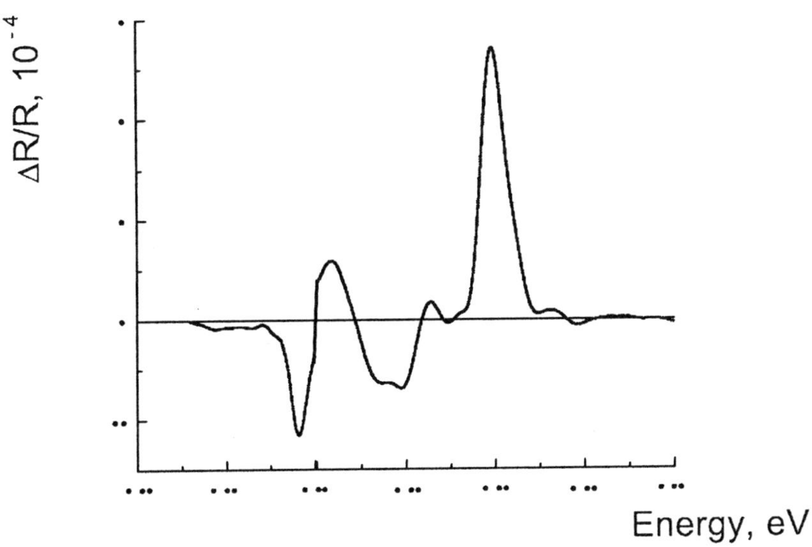

Figure 2. Electroreflectance spectrum of the porous nanoscaled silicon sample prepared by electrochemical etching with current density 120 mA·cm^{-2}.

Type of critical points is unknown. Thus in tab.1 collected the energies of direct transitions that were calculated for two-dimensional (2D), three-dimensional (3D) and for exiton type using three-points method [16]. As one can see the calculated energies of the direct transitions determined from different spectra presented in fig. 1 and 2 are almost the same (1.92 and 1.95 eV, respectively). Thus, energy of the transition is weakly depends on the current of electrochemical etching. In fig. 3 a presents electroreflectance spectrum obtained from the porous silicon samples that was prepared by stain etching. As one can see direct transitions registered in the spectral region near 1.95 eV, too. The calculated value of energy gives a little shift in the shorter wavelength region in comparison with fig.1 and 2.

TABLE 1. *Energies of the direct transitions calculated from electroreflectance spectra for different type of the critical points – three-dimensional, two-dimensional and exiton transition.*

Known photolumine-scence bands	Red, eV			Green, eV			Blue, eV		
Num	3D	2D	Ex	3D	2D	Ex	3D	2D	Ex
Fig. 1	1.92	1.91	1.93	-	-	-	3.0	3.0	3.0
Fig. 2	1.95	1.94	1.95	2.5	2.5	2.48	2.98	2.98	2.98
Fig. 3	2.04	2.04	2.04	2.64	2.64	2.65	-	-	-
Fig. 6	1.67	1.67	1.67	2.37	2.37	2.37	-	-	-

Such shifting from 1.91 to 2.04 eV could be connected with different value of the mechanical stress. It is known that porous silicon layer is under significant mechanical stress [5]. The porous silicon prepared by stain etching to must have less value of mechanical stress due to less roughness of the surface. Additional confirmation of smaller roughness of the porous silicon prepared by stain etching is existence of electroreflectance spectrum in the region near 3.4 eV, which can be suited to the silicon wafer (fig. 3 b).

All samples showed the same photoluminescence properties (fig. 4). The maximum of the photoluminescence spectra was place near 735 nm (1.69 eV). The energy of the direct transition (see tab 1) was near 1.95 eV. Such difference could be connected with exiton mechanism of the photoluminescence. The electron and hole is strongly localized on the traps that are connected with Si=O bonds and can easy formed exiton.

Electroreflectance spectra were observed in the region of the blue photoluminescence near 3.0 eV (see fig. 1, 2 and tab 1). The blue photoluminescence band is usually to ascribe to the SiO_x layer [17, 18] due to it was observed in the strongly oxygenated porous sample. Which kind of the bonds is related to the traps with energy near 3.0 eV is not established yet.

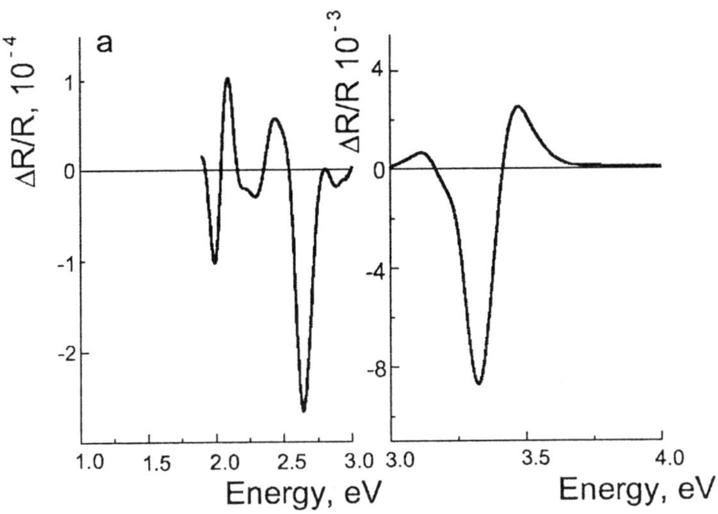

Figure 3. Electroreflectance spectrum of the porous nanoscaled silicon sample prepared by stain etching.

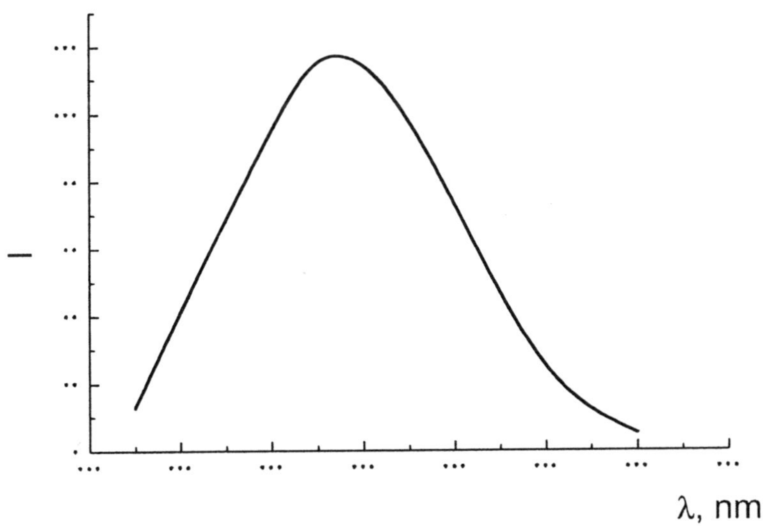

Figure 4. Typical photoluminescence spectrum of porous silicon that was prepared by electrochemical etching.

Electroreflectance spectra were observed in the spectral region near 2.5 eV, which related to the green band of the photoluminescence (fig. 2, 3; tab. 1). The transition energies calculated from spectra were changing from 2.5 eV to 2.65 eV. Such big difference of the values of energy could be connected to different value of mechanical stress of the porous silicon samples that were prepared by electrochemical and stain etching. Green photoluminescence band is often observed in the oxygenated porous silicon sample too [19].

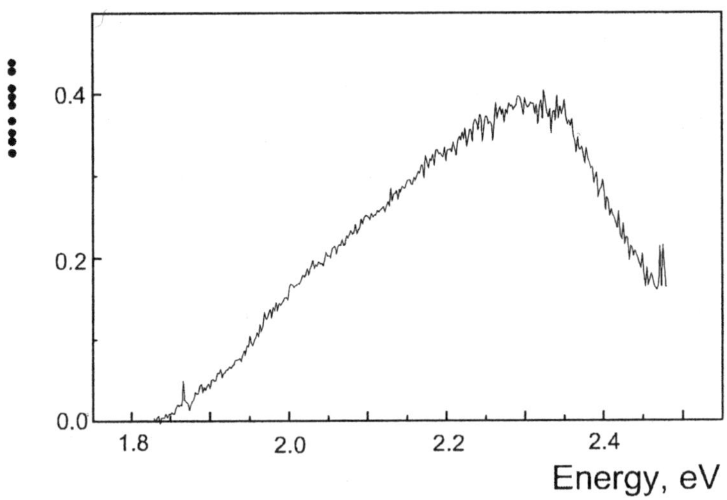

Figure 5. Photoluminescence spectrum of the sample prepared by deposition of the thin silicon films on the oxygenated silicon wafer with following annealing.

Similar photoluminescence band (fig. 5, maximum of photoluminescence spectra is placed near 2.31 eV) can be obtained in the sample that was prepared using following: silicon wafer was oxidising, then was deposited thin (near 10 nm) silicon film and process was repeated four-five times, then samples was annealing at 600°C during few hours (6 in our case). Such structure has strongly oxygenated silicon nanolayers. Electroreflectance spectra obtained from this sample are shown in fig. 6. As one can see, direct transitions there are not only in the spectral region near 3.4 eV what could be connected with silicon wafer but also in the region of green and red photoluminescence bands (2.37 eV and 1.67 eV, respectively). The energies of direct transitions calculated from the spectra presented in fig. 6 is shifted into long wavelength region in comparison with energies that were obtained for porous nanosilicon samples (see Tab. 1). Such shifting could be connected with smaller or stronger mechanical stress in different samples.

Thus, the existence SiO_x layer on the surface of the nanosilicon crystalline wire changed electronic properties of the silicon. Different bonds or different x in SiO_x layer created different electronic states in the band gap of the silicon. Energies of the states could be associated with maximums of the photoluminescence bands. It is interesting to note that value of state trap energies is higher than energies of the maximums of the photoluminescence bands, as usual (for example energy of direct transition – 2.37 eV and photoluminescence maximum – 2.31 eV for sample which spectra presented in fig. 5, 6). Thus can be concluded that most likely mechanism of the photoluminescence in such structures is exitonic one.

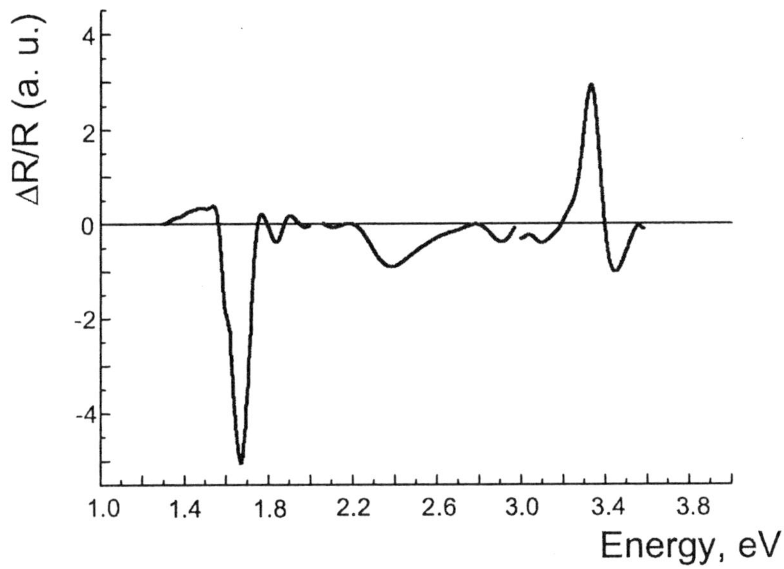

Figure 6. Electroreflectance spectra of the sample prepared by deposition of the thin silicon films on the oxygenated silicon wafer with following annealing.

Which bonds are suited to which electronic states are still under discussion. Theoretical calculations showed that Si=O double bond is most likely suits to the electronic states with energy close to 1.7 eV.

4. Conclusion

It was shown that existence of the SiO_x layer on the surface of the nanosilicon crystalline cluster (wire) changes silicon electronic properties. Electroreflectance spectra confirmed theoretical prediction that Si=O bond creates electronic states in the silicon band gap with energy near 1.7 eV. It was shown that there are electronic states with energy near 2.6 eV and 3.0 eV, which associates with known photoluminescence bands (green and blue, respectively). Mechanism of the photoluminescence has more likely exiton nature.

References

1. Canham L.T. (1990) Silicon quantum wire array fabrication by electrochemical and chemical dissolution of water. *Appl. Phys. Lett.* **57**, 1046-1048.
2. Lehmann V. and Foll H. (1990) Formation mechanism and properties of electrochemically etched trenches in n-type silicon. *Electrochem. Soc.* **137**, 653-658.
3. Starovoitov A., Bayliss S. (2000) Laser structuring of luminescent porous silicon during etching. *J. Por. Mater.* **7**, 267-371.
4. Svechnikov S.V., Sachenko A.V., Sukach G.A., Evstegneev A.M., Kaganovych E.B. (1994) Light emitting layers of porous silicon: receiving, properties and application. *Optoelectr. & Poluprov. Techn.* **27**, 3-29.
5. Venger E.F., Holiney R.Yu., Matveeva L.A. (2001) Study of the electronic states of the porous silicon using electroreflectance. *Optoelectr. & Poluprov. Techn.* **36**, 195-207.
6. Sanders G.D., Yia-Chung Chang (1992) Optical properties of free-standing silicon quantum wires. *Appl. Phys. Lett.* **60**, 2525-2527.
7. Bresler M.S., Yassievych I.N. (1993) Physical properties and photoluminescence of the porous silicon. *FTP.* **27**, 871-883.
8. Hao P.H., Hou X.Y., Zang F.L., Wang X. (1994) Energy band lineup at the porous-silicon/silicon heterointerface measured by electron spectroscopy. *Appl. Phys. Lett.* **64**, 3602-3604.
9. Ben-Chorin M., Averboukh B., Kovalv D., Polisski G. Koch. F. (1996) Influence of quantum confirement on the critical points of the band structure of Si. *Phys. Rev. Lett.* **77**, 763 – 766.
10. Wu X.L., Xiong S.J., Fan D.L., Gu Y., Bao X.M. (2000) Stabilized electronic state and its luminescence at the surface of oxygen-passivated porous silicon. *Phys. Rev. B.* **62**, R7759-R7762.
11. Kanemitsu Y. (1993) Visible photoluminescence from oxized Si nanometer-sized spheres: Exciton confinement on a spherical shell. *Phys. Rev. B.* **48**, 4883-4886.
12. Wolkin M.V., Jorne J., Fauchet P.M., Allan G., Delerue C. (1999) Electronic States and Luminescence in Porous Silicon Quantum Dots: The Role of Oxygen. *Phys. Rev. Lett.* **82**, 197 – 200.
13. Veneger E.F., Gorbach T.Ya., Matveeva L.A. (1999) Electroreflectance spectroscopy, electronic states and mechanism of the visible photoluminescence of the anisotropically etched silicon. *ZhETPh.* **5**, 1750-1761.
14. Gorbach T.Ya., Rudko G.Yu., Smertenko P.S., Svechnikov S.V., Valakh M.Ya., Bondarenko V.P., Dorofeev A.M. (1996) Simultaneous changes in the photoluminescence, infrared absorption and morphology of porous silioen during etching by HF. *Semicond. Sci. Thechnol.* **11**, 601-606.
15. Holiney R.Yu., Fendorenko L.L., Matveeva L.O., Strilchenko I.Yu., Venger E.F., Yusupov N.M. (2000) Electroreflectance of porous layers obtained by staine etching of laser modified silicon. *J. Phys. D: Appl. Phys.* **33**, 2875-2879.
16. Aspnes D.E. (1973) Third-derivative modulation spectroscopy with low-field electroreflectance. *Surf. Sci.* **37**, 418-442.
17. Tsybekov L. Vandyshev Yu.V., Fauchet P.M. (1994) Blue emmison in porous silicon: oxygen-related photoluminescence. *Phys. Rev. B.* **49**, 7821-7824
18. Haris C.I., Syvajarvi M., Bergmann J.P. (1994) Time-resolved decay of the blue emmision in porous silicon. *Appl. Phys. Lett.* **65**, 2451-2453.
19. Canham L.T. (1995) Luminescence band and their proposed origins in the highly porous silicon. *Phys. Status Solidi B.* **190**, 9-14.

SCANNING PROBE MICROSCOPY OF BIOMACROMOLECULES: NUCLEIC ACIDS, PROTEINS AND THEIR COMPLEXES

O.I. KISELYOVA, M.O. GALLYAMOV, N.S. NASIKAN and I.V. YAMINSKY
Chair of Polymer and Crystal Physics, M.V. Lomonosov Moscow State University Vorobyevy Gory, Moscow, 119899, Russia
O.V. KARPOVA, V.K. NOVIKOV
Department of Virology, M.V. Lomonosov Moscow State University Vorobyevy Gory, Moscow, 119899, Russia

1. Introduction

After successful imaging of DNA biomacromolecules using scanning probe microscopy (SPM) (Bustamante et al., 1992) much progress was achieved in the visualization of their different morphological features in air and liquid environments: cruciforms, R-loops, etc. SPM gives an opportunity for real-time studies of conformational changes of DNA molecules induced by chemical reagents - formation of torroidal and rod-like structures. Recently, SPM has greatly assisted in the measurements of electrical conductivity of individual DNA molecules (Kasumov et al., 2001). SPM opens new possibilities in the study of single macromolecule micromechanics: rigidity, strength of chemical bond and adhesion.

Though, it is only in few cases that SPM permits to observe inner structure of protein molecules, it can still be used in advanced experiments for chemical sensing (Friorini et al., 2001), measurements of intermolecular forces, antigen-antibody reactions. Atomic force microscopy (AFM) gives new advantages over other high-resolution techniques in the observation of dynamic processes involving protein molecules: aggregation, crystallization and thin film formation. AFM was applied for the *in-situ* observation of lysozyme crystal growth. The defect structure, kinetics of kinks and step movement was studied; kinetic parameters (step and kink velocities, nucleation rate, etc.) were defined (Chernov et al., 1999).

The present paper is devoted to AFM of plant viruses. Most plant viruses are well ordered complexes of biomacromolecules: nucleic acids and proteins. From the structural viewpoint virus particles have the simplest organization compared to other organisms. Quantitative information on their mechanical properties will help to verify biomechanical models developed for these "simple" biological systems.

In this work we image, study and compare two well-studied RNA-containing viruses belonging to two different families: tobacco mosaic virus (TMV) from tobamovirus family and potato virus X (PVX) from the potex virus group. Virus particles of both species are composed of RNA wrapped into protein molecules. TMV particles are rod-

E. Buzaneva and P. Scharff (eds.), Frontiers of Multifunctional Nanosystems, 321–330.

shaped, they look like rigid "sticks" (Knapp and Lewandowski, 2001), whereas PVX ones have elongated filamentous shape, are much softer and protein coverage is less dense. Here we make quantitative estimations of the rigidity of individual virus particles by means of AFM.

Traditionally when AFM is used for micro-mechanical measurements, the elasticity of materials is calculated on the basis of the information extracted from force curves (Radmacher et al., 1994). In our case, it was difficult to reliably maintain the probe over a single particle and we worked out and applied a different approach. The operation of AMF results in the 3D profile of the probe displacement, collected at constant probe-sample interaction force. The value of this force can be varied, which allows getting information about the elastic properties of the objects investigated. The probes impact on the sample leads to contact deformations (Drygin et al., 1998; Gallyamov et al., 2000), the value of which is determined by the applied force, elastic properties of the objects and the geometry of the contact area. For "soft" objects (polymer, biological ones) immobilized on a "rigid" substrate the consideration of probe-sample contact geometry allows to connect the dependency of deformation versus the applied force with elastic properties of the investigated objects.

2. Sample preparation

Purification of TMV and PVX particles was carried out as previously described by Drygin et al., (1998) and Atabekov et al., (2001) respectively. Mica, highly oriented pyrolitical graphite (HOPG) and aminopropylsilatran-treated mica (APS-mica) were used as substrates. In order to prepare APS-mica, freshly cleaved mica was incubated in the APS solution, rinsed in bidistilled water and dried (Shlyakhtenko et al., 1998). A 5-10 μl drop of the solution containing virus particles was placed onto the substrate, incubated for 5 minutes, rinsed with bidistilled water and dried. The average height of particles in each image was measured automatically using the software FemtoScan 001 (Filonov and Yaminsly, 1997).

3. Results and discussion

3.1. *IMAGING OF VIRUSES*

For AFM imaging TMV and PVX virus particles were applied on traditionally used substrates: graphite and mica. TMV particles adhered to both (Drygin et al., 1998), but the adsorption rate was higher on graphite surface (fig. 1 a). At the same time, TMV particles were more stable on mica substrate, which was used for contact mode experiments (data not shown). Interestingly, PVX particles did not adhere to mica and demonstrated no stability on graphite (Kiselyova et al., 2001). For their immobilization APS-treated mica (APS-mica) was applied. This substrate had previously been used in AFM experiments for nucleic acids immobilization (Shlyakhtenko et al., 1998). AFM images of PVX particles (fig. 1 b) show that their filamentous shape.

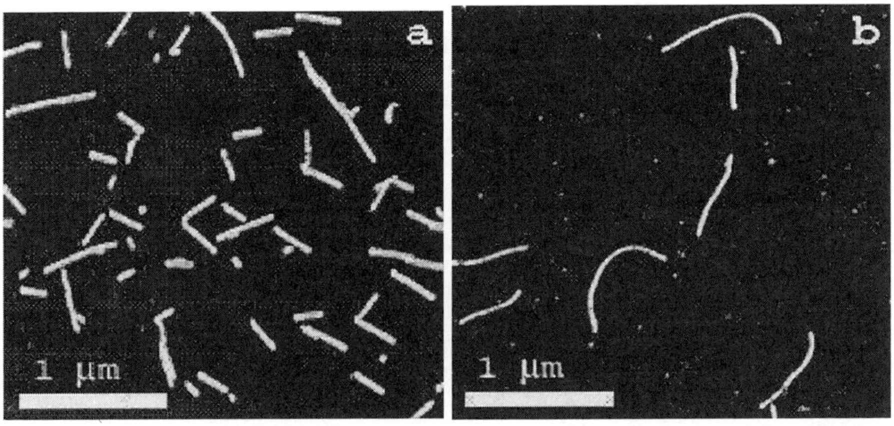

Figure 1. AFM images of (a) TMV particles on pure mica and (b) PVX particles on APS-mica obtained in contact mode

The height of particles revealed in tapping mode AFM images in was 13±2 nm, close to that provided by X-ray data reported by Varma et al., (1968) and particle's diameter measured by electron microscopy. Measurements of TMV particles height 18±2 nm gave good coincidence with conventional data as well (Knapp and Lewandowski, 2001). Distributions of particle's lengths revealed tail-to-tail aggregation effect (Drygin et al., 1998; Kiselyova et al., 2001).

In the next series of experiments PVX was subjected to RNase treatment. It is known that this procedure degrades the encapsidated RNA (Atabekov et al., 2001). Meanwhile, protein-protein interaction is so strong that it does not affect the integrity of virus particles. In AFM images RNase-treated particles looked the same and had the same dimensions as intact ones (fig. 2 a). Since TMV RNA is not destroyed by RNase treatment of TMV particles similar experiment was of no interest.

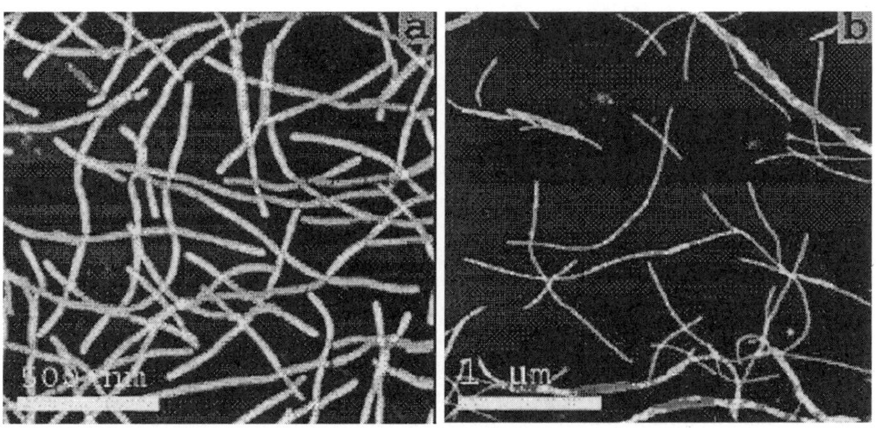

Figure 2. PVX particles subjected to treatments: (a) RNase treatment (b) – tripsin degradation.

Tripsin treatment of PVX degrades the 14-aminoasid N-terminus chain of PVX coat protein. This provokes particles side-to-side aggregation depicted in fig. 2 b.
Instability of PVX particles on HOPG yielded unexpected results. Intact virus particles were not observed, but considerably thinner structures (presumably viral RNA molecules) were revealed (fig. 3). RNA strands were seen as straight lines, their height was 0.55±0.05 nm, the width measured at half height – 8 nm. Previously, for RNA molecules, observed in AFM on freshly cleaved mica surface, these parameters were found to be in the intervals of 0.3-1.5 nm and 10-15 nm respectively (Drygin et al., 1998).

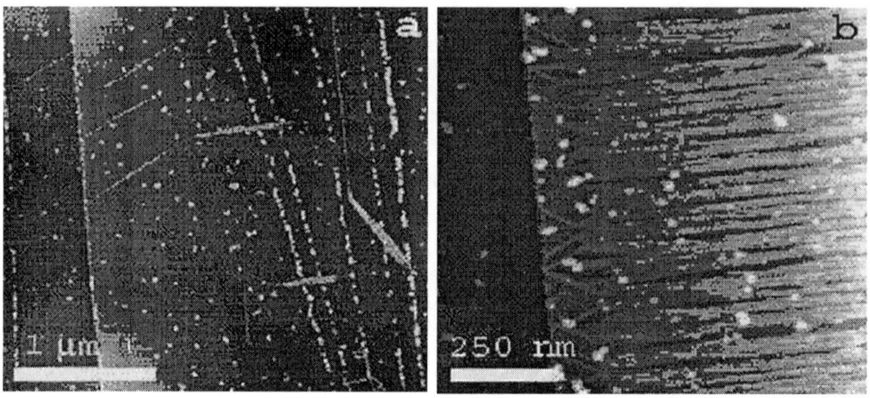

Figure 3. Decomposition of PVX particles on graphite substrate. Filaments are presumably RNA molecules unfolded and stretched on the crystalline lattice.

The apparent dimensions of RNA adsorbed on HOPG are closer to real RNA geometry, than those of mica-adsorbed RNA. We suppose that this difference is due to the well-known and widely discussed effect of thin water film, covering mica surface at ambient humidity (Guckenberger et al., 1994). The extended distribution of parameters for the RNA described by Drygin et al., (1998) can be presumably attributed to the fact that when adsorbed to mica, RNA has many bends and its secondary structure is not completely unfolded, whereas the molecules registered on HOPG are straightened.
The angles between adsorbed RNA molecules directions were multiple to 60°, which coincides with angles in HOPG crystalline lattice. This implies that substrate-RNA interaction was so strong that not only virus particles were decomposed, but the secondary structure of RNA was unfolded as well. RNA molecules tended to form flat parallel aggregates (fig. 3 b).

3.2. *CALCULATION OF CONTACT DEFORMATION*

Let's consider the contact area between an AFM probe tip and the investigated, specimen placed on a flat substrate surface. According to Hertz solution (Landau and Lifshitz, 1987), two bodies in contact, pressed by the force F, are deformed and brought towards each other by the distance h, the contact area no longer being a point but becoming a spot with a finite square S (fig 4.).

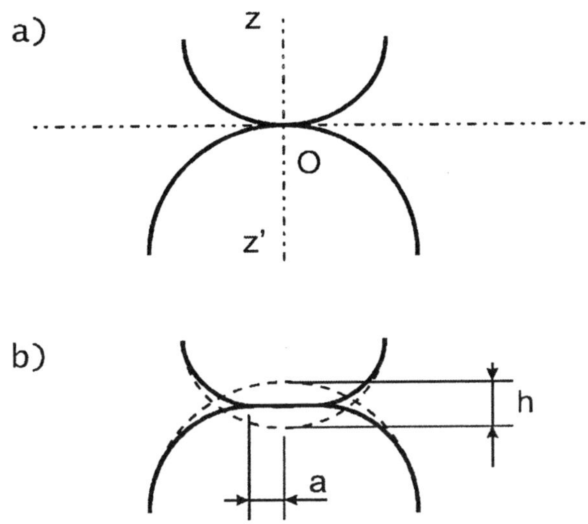

Figure 4. Contact deformation of two bodies. (a) – no compressing force, (b) compressing force applied.

The task analysis requires consideration of summarized curvature tensor of contact surfaces $\chi_{\alpha\beta}+\chi'_{\alpha\beta}$, which principle value *A* and *B* can be expressed in principle curvature radii of contact surfaces (Landau and Lifshitz, 1984). If the deformation is small compared to curvature radii, the contact area shape is an ellipse with semi axes *a* and *b*. These values and rapprochement *h*, are expressed in the implicit form using known task parameters: applied force *F* and contact geometry parameters *A* and *B*:

$$h=\frac{FD}{\pi}\int_0^\infty \frac{d\xi}{\sqrt{(a^2+\xi)(b^2+\xi)\xi}},$$
$$A=\frac{FD}{\pi}\int_0^\infty \frac{d\xi}{(a^2+\xi)\sqrt{(a^2+\xi)(b^2+\xi)\xi}},\qquad (1)$$
$$B=\frac{FD}{\pi}\int_0^\infty \frac{d\xi}{(b^2+\xi)\sqrt{(a^2+\xi)(b^2+\xi)\xi}}$$

D — being inverse effective elastic modulus of the contact area:

$$D=\frac{3}{4}\left(\frac{1-\sigma^2}{E}+\frac{1-\sigma'^2}{E'}\right),\qquad (2)$$

where *E* and *E'* are Young's moduli and σ and σ' - Poisson's ratios for the probe and the sample materials, respectively.

In certain practically important cases equations (1) can be simplified and solved analytically. Since virus particles investigated have cylindrical shape the model to be considered is the model of spherical tip and cylindrical sample.
For the contact of a spherical probe with (radius R) and a surface of a cylinder (radius R') parameters A and B are expressed in the following form

$$A = \frac{1}{2}\left(\frac{1}{R} + \frac{1}{R'}\right), \quad B = \frac{1}{2R}. \tag{3}$$

Unfortunately, in this case the parameters of interest cannot be explicitly expressed from (1). Numerical solution yields three equations to be solved sequentially:

$$\begin{aligned} \frac{b^2}{a^2} &= \frac{A}{B}\frac{K}{E} - \frac{A}{B} + \frac{K}{E}, \\ b^3 &= \frac{2FD}{\pi B}\frac{(K-E)}{\left(1 - \frac{a^2}{b^2}\right)} \\ h &= \frac{2FD}{\pi b}K, \end{aligned} \tag{4}$$

where $K \equiv K(\sqrt{1 - a^2/b^2})$, $E \equiv E(\sqrt{1 - a^2/b^2})$ — full elliptical integrals.
If the radius of the cylinder is less that the probe radius, (3) implies, that principle values of the summarized curvature tensor differ: $A > B$. Using (1), one can show that in this case $a < b$. If $a^2 << b^2$ equations (4) can be simplified, using a well known asymptotic for a full elliptical integral. Then, rapprochement h is expressed as follows:

$$h = \left(\frac{4}{\pi^2 C}\right)^{1/3} (C+1) B^{1/3} (FD)^{2/3} \tag{5}$$,

parameter C depending on the ellipse parameters a and b:

$$C = \ln\left(\frac{4b}{a}\right) - 1 = \frac{Bb^2}{Aa^2} \tag{6}$$

Formula (6) yields numerical value of b/a and C, provided B/A is known. Numerical solution shows that typical values of C for many tasks lie in between 1 and 3.

3.3. *EXPERIMENTAL DETERMINATION OF YOUNG'S MODULUS*

The difference of formula (5) from Hook's law is the non-linear dependence of the deformation h versus the applied force. This is explained by the fact that the size of the contact area also depends on the applied force.

We studied the deformation of virus particles at different values of the force applied from the probe (figure 5 (a) for TMV, (b) for PVX) and revealed that it fits the "two thirds" dependency (compare to (5),

$$h \sim (FD)^{2/3} f(R, R') \qquad (7),$$

where f is the function of probe and the sample radii R and R' respectively.
Generally, three terms contribute to the reduction of sample height in the AFM image due to contact deformation:

$$\ddot{A}h_{\acute{o}} = \ddot{A}h_{probe/sample} + \ddot{A}h_{sample/substrate} - \ddot{A}h_{probe/substrate}$$

Where the second term is practically equal to the first one for small radii of the sample.

Thus, the measured height of virus particles H, can be found from the following formula:

$$H = H_0 - 2(FD)^{2/3} f(R, R'),$$

where H_0 is the height (diameter) of a non-deformed particle.

Since the value of Poisson's ratios for the used materials $\sigma^2 << 1$, Young's modulus for silicon $E \sim 10^{11}$ Pa, and for biological objects E can be estimated as $E \sim 10^9$ Pa and less, the equation (2) can be simplified to:

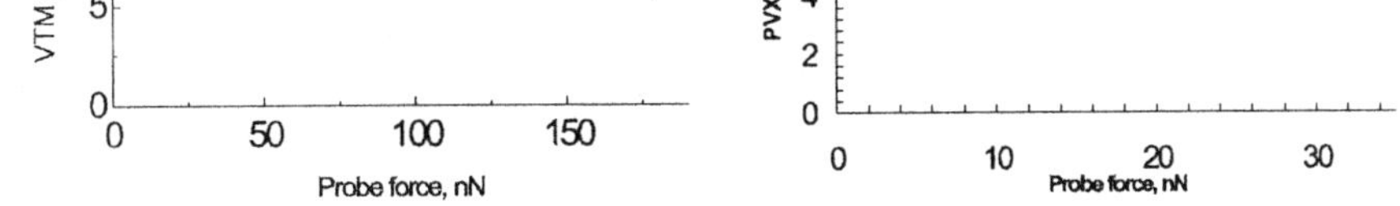

Figure 5. "Two thirds" dependency of particles' height versus the applied force.
(a) - TMV, (b) - PVX. In image (b) squares and continuous line correspond to intact particles, triangles and dotted line – to RNase-treated ones.

$$D \sim \frac{3}{4E} \tag{8}$$

The approximation of experimental points by theoretical curve was performed using the least meant square method. Experimental errors were determined as standard deviations of average h values obtained from several AFM images. The obtained parameters yielded the coefficient D, then Young's modulus was found to be $E \sim 3 \cdot 10^9$ Pa for TMV particles and $E \sim 8 \cdot 10^8$ Pa for PVX ones. Previously, Young's modulus was measured alternatively, providing a very close value of $E \sim 1.1 \cdot 10^9$ Pa (Falvo et al., 1997). Thus, the obtained value for PVX is almost by the order of magnitude less than that for TMV. This explains the difference in the appearance of two virus's particles adsorbed on the substrate: PVX particles are curved whereas TMV one remain practically straight.

For RNase-treated PVX particles the dependency of deformation versus the applied force was more drastic than for intact ones (fig. 5 b). Let's remind that in tapping mode images intact and RNase-treated particles had the same dimensions. The deformation was so essential that could no longer be described by the Hertz model. These data imply that though visually PVX particles remained unchanged, their rigidity was lost after RNA disruption.

3.4. *DEFORMATIONS OF VIRUS IN TAPPING MODE*

Though lateral tip-sample forces are very small in tapping mode, the deformation of soft objects occurs nevertheless. In tapping mode the contact between the probe tip and the sample is temporary and the contact deformations cannot be calculated according to the above-mentioned procedure. Here we compare the apparent heights of PVX particles obtained in contact and tapping regimes and study the influence of cantilever oscillations parameters on the obtained data. PVX was chosen because it is "softer" and deformations are easier to register.

Vertical dimensions of PVX particles imaged in contact mode, the contact force being minimized, was 11.0 nm, i.e. 20% less than in tapping mode (13.5 nm), presumably due to stronger deformation of particles in the contact regime.

It was demonstrated that the apparent height of PVX particles depended on the aspect ratio $\grave{A}/\grave{A}_0$, (loaded versus free oscillation amplitude). Namely, the height reduced as the aspect ratio was decreased and therefore the effect of the probe was strengthened (fig. 6 a).

Height reduction by 15% was registered at the aspect ratio value of 0.5. It is noteworthy, that the deformation of the particles was absolutely reversible. Interestingly, the amplitude of cantilever's free oscillations also influenced the apparent height. Two-times increase of the amplitude from 15 to 30 nm reduced the obtained height by almost 30 % (fig. 6 b). These data show that one has to be very careful about choosing tapping mode parameters when quantitative measurements are performed.

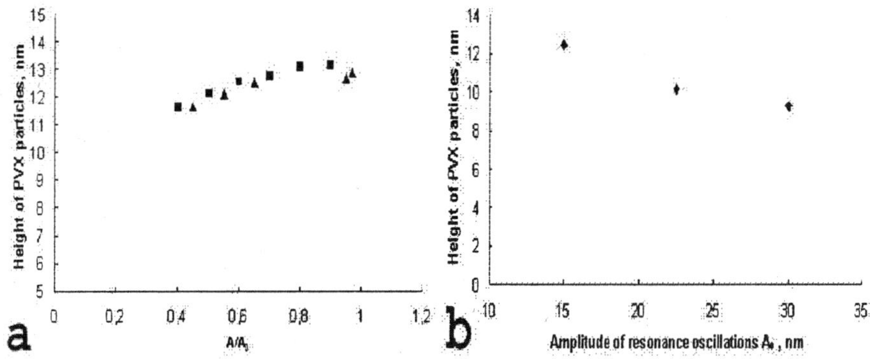

Figure. 6. The influence of tapping mode scanning parameters on PVX particles' height.
(a) height versus aspect ratio (A/A_0). Squares mark points obtained when reducing and triangles when increasing A/A_0.
(b) height versus free oscillations amplitude (A_0).

4. Conclusion

AFM can be successfully used as an imaging tool in descriptive virology. It observes virus morphology with subnanometer accuracy. It also presents direct data about mechanical properties of individual virus particles, which are not obtained in other high-resolution imaging methods. This is new information for biology and the way of its practical comprehension is to be developed. Still the demonstrated AFM method opens new opportunities in micromechanical measurements of different nanostructures in nanotechnological applications. Different virus particles may serve here as well-defined rigidity standards.

5. Acknowledgement

This work was supported by Russian Foundation for Basic Research (Grants no 00-07-90016, and 00-04-55020), and Russian Ministry of Science and Technology (Grant N 1.11.99, Programme "Atomic surface structures").

References

1. Atabekov, J.G., Rodionova, N.P., Karpova, O.V., Kozlovsky, S.V., Novikov, V.K., Arkhipenko, M.V. (2001), "Translational Activation of Encapsidated Potato Virus X RNA by Coat Protein Phosphorylation", Virology 286 (2), 466-474.
2. Bustamante, C., Vesenka, J., Tang, C.L., Rees, W., Guthold, M., and Keller, R. (1992), "Circular DNA Molecules Imaged in Air by Scanning Force Microscopy", Biochemistry, 31(1), 22-26.
3. Chernov, A.A., Rashkovich, L.N., Yaminsky, I.V., and Gvozdev, N.V. (1999), "Kink Kinetiks, Exchange Fluxes, 1D 'Nucleation' and Adsorption on the (010) Face of Orthorombic Lysozyme Crystals", Journal of Physics: Condensed Matter 11, 9969-9984.

4. Drygin, Yu.F., Bordunova O.A., Gallyamov M.O., and Yaminsky I.V. (1998), "Atomic Force Microscopy Examination of Tobacco Mosaic Virus and Virion RNA", FEBS Letters 425, 217–221.
5. Falvo, M.R., Washburn, S., Superfine, R., Finch, M., Brooks, F.P., Jr., Chi V., and Taylor R.M. (1997), "Manipilation of Individual Viruses: Friction and Mechanical Properties", Biophysical Journal 72, 1396-1403.
6. Filonov A.S., and Yaminsky, I.V. (1997), SPM control and image processing software,. Chichester: Advanced Technologies Center, Moscow.
7. Fiorini, M., McKendry, R., Cooper, M.A., Rayment, T., and Abell, Ch. (2001), "Chemical Force Microscopy with Active Enzymes", Biophysical Journal 80(5), 2471-2476.
8. Gallyamov, M.O., Drygin, Yu.F., and Yaminsky, I.V. (2000), "Atomic Force Microscopy Visualization of RNA and Ribonucleotides of Tobacco Mosaic Virus". Surface Investigation 15, 1127–1134.
9. Guckenberger, R., Heim, M., Cevc, G., Knapp, H.F., Wiegräbe, W., and Hillebrand, A. (1994), "Scanning Tunneling Microscopy of Insulators and Biological Specimens Based on Lateral Conductivity of Ultrathin Water Films", Science 266, 1538-1540.
10. Kasumov, A.Yu., Kociak, M., Guéron, S., Reulet, B., Volkov, V.T., Klinov, D.V., and Bouchiat, H. (2001), "Proximity-Induced Superconductivity in DNA", Science 291, 280-282.
11. Kiselyova, O.I., Nasikan, N.S., Yaminsky, I.V., and Novikov, V.K. (2001) "AFM Imaging of PVX Particles and PVX RNA", Physics of Low-Dimensional Structures 3/4, 167-174.
12. Knapp, E. and Lewandowski, D.J. (2001), "Tobacco Mosaic Virus Not Just a Single Component Virus Anymore", Molecular Plant Pathology, 2(3), 117-123.
13. Landau, L.D., and Lifshiz, E.M. (1987) Theoretical Physics, V. 7 (in Russian), Nauka, Moscow.
14. Radmacher, M., Fritz, M., Cleveland, J.P., Walters, D.A., and Hansma, P.K. (1994), "Imaging Adhesive Forces and Elasticity of Lisozyme Adsorbed on Mica with the Atomic Force Microscope", Langmuir 10, 3809-3814.
15. Shlyakhtenko, L.S., Potaman, V.N., Sinden, R.R. and Lyubchenko, Yu.L. (1998), "Structure and Dynamics of Supercoil-Stabilized DNA Crusiforms", Journal of Molecular Biology 280, 61-72.
16. Varma, A., Gibbs, A.J., Woods, R.D., and Finch, J.T. (1968), "Some Observations on the Structure of the Filamentous Particles of Several Plant Viruses", Journal of General Virology 2, 107-114.

APPLICATION OF ATOMIC FORCE MICROSCOPY IN PROTEIN AND DNA BIOCHIPS DEVELOPMENT

O.M. STUKALOV
Laboratory of Semiconductor Physics, Institute of Solid State and Semiconductor Physics NASB
P.Brovki Str. 17, 220072 Minsk, Belarus

1. Introduction

In recent years the biotechnology and analytical medicine have seen great advances in development of novel biochip technologies that open new horizons for identifying and quantifying biomolecules and diagnosing diseases. The hope is that the smaller/faster/cheaper approach that worked so well for computer-chip manufacturers will be applied to biotech as well. The goal is to use these tools to better understand the role played by genes in health and disease, to speed the drugs development, to improve diagnostic tools, and even to build handheld devices that produce parallel medical tests by one measurement (Cahill, 2001; Vo-Dinh et al, 2001; Weinberger et al, 2000).

Biochip is a special class of biosensors that has multiple transducer elements and is based often on integrated circuit microchips. The term ``biochip" comes from microelectronics where the word "chip" is used for a silicon-based substrate with fabricated miniaturized electronic circuits. However, over the last years, the term biochip has taken on a variety of meanings. A biochip is now generally defined as a material or a device that has an array of biosensors. In general, any device or component incorporating a two-dimensional array of reaction sites and having biological materials on a solid substrate has been referred to as a biochip. Biochips often involve both miniaturization, usually in microarray formats, and the possibility of low- cost mass production using computer-chip technologies.

Most of protein and DNA biochips consist from the solid substrate with the array of the microareas, each of which covered by thin films of different bioreceptors (e.g. an antibody, an enzyme, a protein, a nucleic acid) and a signal transducer. When the analyte interacts with the bioreceptor, the resulting complex has changed properties (e.g. mass, optical reflection, etc.) which is converted into a measurable electrical signal by the transducer. The most common types of analyte/bioreceptor complexes are based on interactions between native antibodies/antigens (Hyrata et al., 2001), both native and synthetic nucleic acid interactions through hybridization of complementary strands (Swanson et al., 2000; Caillat, 1999) and enzymatic reactions (Dempsey et al., 1997). The most prevalent signal transformation methods are the optical measurements, e.g. luminescence (Swanson et al., 2000; Caillat, 1999), absorption/reflection of light (Fritzsche et al., 2001), surface plasmon resonance (Green et al., 2000), etc.; electrochemical (Erdem et

E. Buzaneva and P. Scharff (eds.), Frontiers of Multifunctional Nanosystems, 331–340.

1999); mass and density-sensitive measurements, e.g. surface acoustic wave (Welsch et al., 1997; Kondoh et al., 1994), quartz microbalance (Spangler and Tyler, 1999), microcantilever (Fritz et al., 2000; Tamayo et al., 2001; Raiteri et al. (2001), etc; and even magnetoresistance (Baselt et al., 1998).

Scanning tunneling microscopy (STM) and atomic force microscopy (AFM, also called as scanning force microscopy, SFM) are the most recent inventions that produce revolutionary inroad in biomedical disciplines, including development of the biosensors and the biochips. Since invention of STM by Binnig and Roher in 1982 (Binnig and Roher, 1982) and AFM by Binnig, Quate and Gerber in 1986 (Binnig et al., 1986), these microscopies became standard in the physical sciences. Now there are a many of books and reviews specially focused on the STM/AFM techniques (Sarid, 1991; Weisendanger, 1994; Magonov and Wangbo, 1996), including applications in biological, biochemical and biophysical sciences (Ikai, 1996; Siedlecki and Marchant, 1998; Morris et al., 1999). The aim of this report to give short overview of possibilities of AFM technique, which may be useful in the development of DNA and protein biochips. Firstly, the operation principles and modes of the AFM are presented. Then usage of AFM for investigation of adsorption and adhesion of bioreceptor molecules on different surfaces is shown, because the strong immobilization of biomolecules is the key stage in the biochip design. And finally, application of AFM microcantilvers as transducers of bioinformation to electrical signals is reviewed. The review doesn't pretend on comprehension of extensively grown area of AFM in biochip technologies.

2. Operation principles of AFM

A wide range of AFM operating modes can be used in biomaterial interface examination both under ambient or aqueous conditions. But attention of this part is focused on some common modes including contact, tapping and force spectroscopy modes.

2.1. *VISUALIZATION MODES*

The Figure 1 illustrates schematic view of a AFM. The AFM probe is a flexible triangular or rectangular silicon microcantilever with sharp tip on the its end (Fig. 1, on the right). AFM users very often name the cantilever/tip assembly as "cantilever". The cantilever with tip (further will called simply "cantilever") is placed in contact with the sample. The forces acting between the tip and surface cause the cantilever to be bent. In most AFMs investigated sample is attached to the piezotube scanner (piezoscanner). The piezoscanner raster-scans the sample under the cantilever along X and Y axes from point to point, while the cantilever is unmovable (in other constructions the cantilever attached to the piezoscanner while sample is fixed). The change in surface topography changes a static bending of the cantilever at each point of the surface in the contact mode. In the tapping mode the cantilever is oscillated on the resonance frequency and the interaction forces change the amplitude and phase of cantilever oscillations. The feedback loop of the AFM maintains the static cantilever deflection (in the contact mode) or amplitude of the cantilever oscillations (in the tapping mode) at a constant value by approach/retract of the sample in Z direction in each point of the surface. Computer with special software

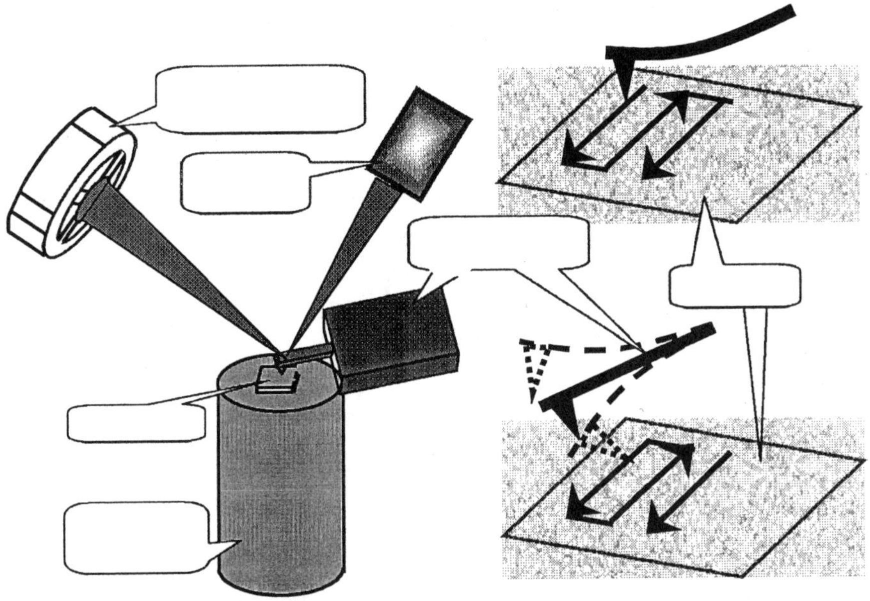

Figure 1. The schematic view of the AFM (on the left). A principle of the contact mode operation (on the right-top) and the tapping mode with oscillating cantilever (on the right-below).

controls all cantilever movements and collects image information. The AFM program transduces the voltage needed to elongate the piezoscanner along Z-axis to quantitative information about surface topography. The reflection optical technique is used for measurement of the nanoscale cantilever deflections in the most AFMs. In this technique semiconductor laser light is reflected off the back of the cantilever (see Figure 1) to the 4-segmented photodiode. The amount of the light reaching each of the photodiode segments is compared and used to generate a difference signal. The difference signal is used as the input to a digital feedback loop.

Contact mode operation is the simplest mode and easiest in use. It is suitable for imaging of hard inorganic surfaces, metals and often for high polymers. In most cases there are significant problems when contact mode is used for visualization of soft biological materials in ambient conditions. The fact is that there is the thin water adsorbate film on the any surface in air exists and formation of capillary between the tip and the surface causes significant forces, which are capable to destroy or shift bioobjects. Usage the contact mode in liquid or tapping mode both in air and liquid allows the capillary force to be avoided. Nevertheless, shear forces may also cause shift bioobjects in case the contact mode in liquid if the bioobjetcs have a weak adhesion to the substrate. In some cases these shifts in contact mode may be useful as it will be shown below.

2.2. *FORCE SPECTROSCOPY MODE*

In the force spectroscopy mode the AFM cantilever is approached to and retracted from the investigated surface in predetermined point. Cantilever's deflection is transduced to

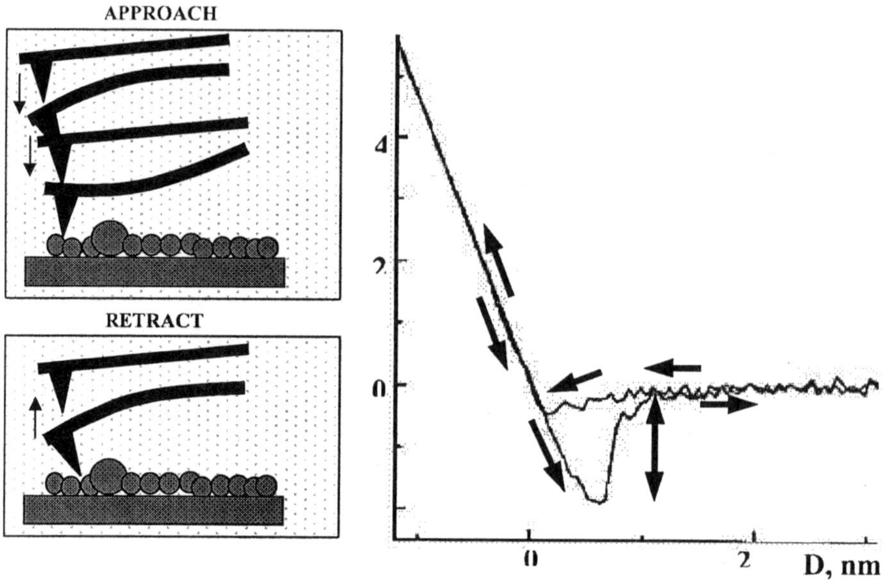

Figure. 2. Principle of force spectroscopy mode of AFM in liquid. The cantilever is approached to and retracted from the investigated surface. The dependence of interaction force versus passed distance (the force-distance curve) is recorded. Pull-off value on retract curve represents the adhesion force if measurements are conducted in vacuum or liquid, and the capillary force in case of measurements in ambient conditions.

the voltage signal, which may be converted to interaction force if cantilever stiffness (sometimes called as "force constant") is known. The dependence of force versus passed distance is recorded as force-distance curve (or simply force curve), as shown on Figure 2. The cantilever passes through attraction and repulsion states during approach and through hysteresis during retract. The lowest point on the force-distance curve corresponds to break of contact and the value of the hysteresis (often called as "pull-off value") represents the adhesion force between the tip and surface if measurements are carried out in liquid or vacuum environment and the capillary force in case of measurements in ambient conditions. The approach curve may be used to obtain local nanomechanical properties of soft material including bioobjects (Vinckier and Semenza, 1998). A procedure of interaction forces measurements, choice of the cantilever stiffness and processing of force-distance curves is described in details by Burnham et al.,(1998) and Burnham and Kulik (1999).

3. Contact and tapping modes of AFM for measurement of adhesion and absorption of biomolecules

Although in the most biological applications contact mode in air is not reliable, it may be successfully applied in some cases. For example, Pereira (2000) visualized single

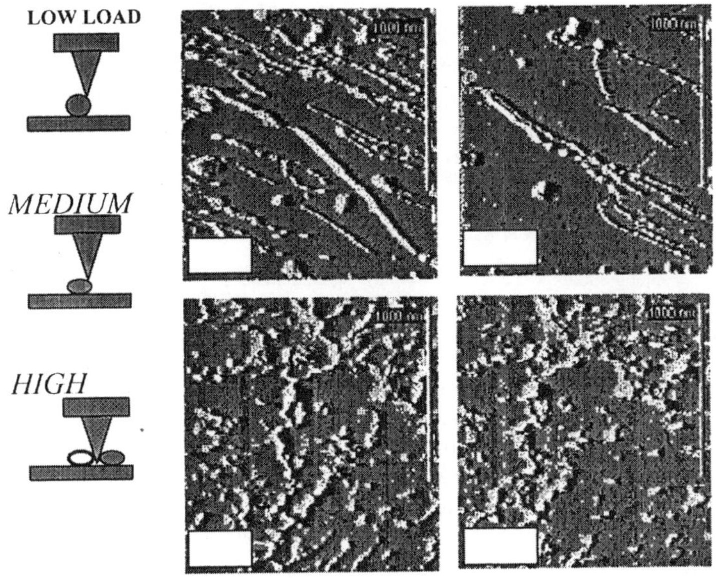

Figure. 3. Schematic view of the principle of measurement of the adhesion ability of the bioobjects using contact mode with consecutive increase of applied force (on the left). For example of application, the contact mode images of plasmid DNA adsorbed on the unmodified (upper two images) and hydrophobizated Si surface (low two images) are shown (see text). Scan size 1500 nm.

enzymes in the film and used the modified cantilever as nanobiosensor; Ridout et al. (1998) successfully investigated polysaccharide molecules.

Strong immobilization of biomolecules on the substrate surface with saving of its bioactivity is the very important task both in the biochips development and in preparation of biosamples for AFM investigation. The AFM allows the quantitative information about adhesion forces to be obtained. The most known mode for this is the chemical force microscopy, which will described below. I propose another method based on contact mode both in air and liquid. A principle of this method and example of its application for plasmid DNA deposited on Si unmodified and modified surfaces are presented on Figure 3. Normally the force between the AFM tip and the surface is kept as small as possible to prevent damage. If we will subsequently increase the tip-to-surface force during scanning of the same area, the shear force applied by the tip will increases too and at specific force value the bioobjects will start to displace (plasmid DNA in our case). This value of the force may be used to quantify adhesion ability of the bioobjects under interest to investigated surfaces. For example, I have compared the Si surfaces unmodified and hydrophobizated through silanization. If the load of the tip to be increased from 5 to 50 nN, the DNA molecules start to shift by the AFM tip in case of unmodified Si, while the molecules on hydrophobic surface not shifted but partially destroyed. Therefore it may be concluded that hydrophobization of Si surface improves adhesion of DNA molecules. Boyd et al. (2001) used the same method to measure adhesion between the bacterial cells and the stainless steel surface.

The tapping mode in air may be used to visualize both the bioreceptors on surfaces under interest to produce uniform stable films and the analyte biomolecules attached to

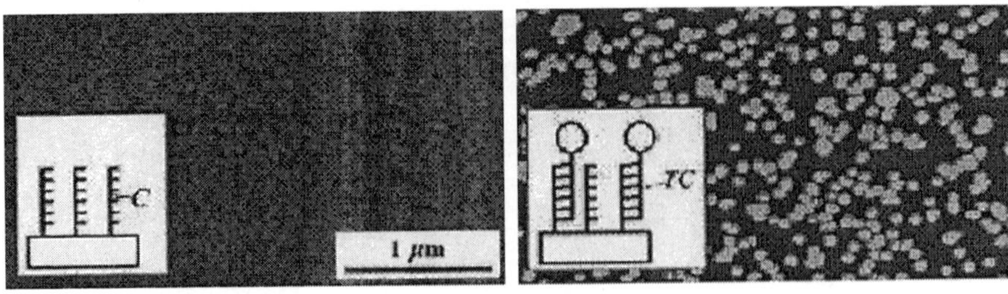

Figure. 4. Tapping mode AFM images of the single-stranded DNA molecules (capture molecules) immobilized on gold surface (on the left) and after incubation with 30-nm gold-labeled single-stranded DNA molecules (target molecules), which are complementary with the capture molecules. Reproduced from Csáki et al. (2001) with permission of authors.

bioreceptors thorough specific interactions to shows that biochip really work. In this case usage of optical methods in addition to AFM is preferable because typical scan areas of AFM lies between 1-100 μm and enough low to give information over the large areas (millimeters) of biochip surface. For example, Hirata et al. (2001) investigate lipid membranes with incorporated antibodies by the AFM and fluorescence microscopy. They received stable films with saving ability of specific interaction between immobilized antibodies and the analyte antigens. Patel et al. (1998) observed individual bioreceptor proteins on gold surface modified by self-assembled monolayers using tapping mode in air and used the AFM visual information about coverage density of proteins to characterize affinity of the modified surface to proteins absorption. Fritzsche with colleagues (Fritzsche et al., 2001; Csáki et al., 2001; Möller et al., 2000; Reichert et al., 2000) use AFM and optical microscopy to register hybridisation of complementary strands on DNA biochip surface. It their experiments single-stranded DNA molecules (the capture) are immobilized on the substrate surface. Other single-stranded DNA molecules are labeled by gold colloid nanoparticles (the target or the analyte). If the target molecules are complementary to the capture molecules, the hybridization reaction occurs and the label may be detected both by AFM (Figure 4) and optical microscopy in transmission or reflection mode. Usage the colloid label significantly simplifies and reduces the price of detection scheme of DNA biochips.

Ability to work in liquid environments is the most important peculiarity of AFM from biological point of view. For example, Fisher and Tjörnhage (2000) found, that exposing to air significantly changes surface structure of the lipid bilayer biosensor and only measurements in the tapping mode in buffer solution allow the real membrane structure with incorporated individual molecules to be observed. Liquid measurements allow the biological events to be investigated *in-situ* in real time. Chernov et al. (1999) use contact mode of AFM in liquid to explore the real time kinetics of the lysozyme single crystals.

4. Chemical force microscopy

The chemical force microscopy (ChFM) is powerful extension of AFM and now widely uses to measure the interaction forces between both chemicals and biomolecules. For biochips purposes ChFM allow the surface affinity to bioreceptor absorption to be

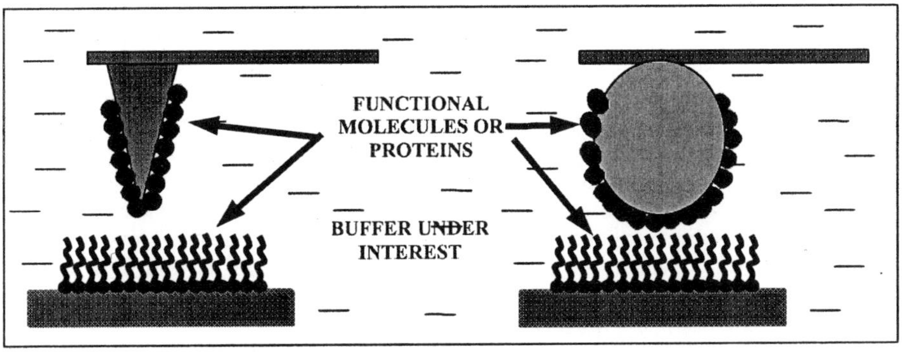

Figure. 5. Schematic view of chemical force microscopy with AFM tip (on the left) and tip-free cantilever with attached microbead (on the right) both coated by functional molecules.

measured directly. In ChFM the tip of AFM is coated by protein or other molecules as shown on Figure 5. Another extension of ChFM is usage of the tip-free cantilever with the microbead attached on its end coated by functional molecules (Figure 5, on the right). Commonly, ChFM works in liquid. The ChFM is used both in force spectroscopy modes to measure adhesion of the protein (Chen et al., 1997; Chen et al., 1998; Feldman et al., 1999; Vansteenkiste et al., 2000; Ros et al., 1999) or DNA (Lee et al., 1994; Noy et al., 1997) molecules on the tip or on the microbead (McGurk et al., 1999) to the modified surface and the visualization modes to recognize individual molecules (Ros et al. 1999; Pereira et al., 2000). A combination of force spectroscopy and visualization modes opens new possibilities to make nanocartography of surface forces. For example, Chen et al. (1997) produce 3-D maps of adhesion forces between bovine serum albumin (BSA) on the AFM tip and BSA and polystyrene on the substrate surface. These maps show that the adhesion force between BSA and polystyrene is much stronger then between BSA and BSA. In adhesion map mode AFM software measure force-distance curves in each point of surface simultaneously with topography in contact mode.

5. Microcantilever-based biosensors/biochips

Advances of the AFM techniques in biological sciences opens new area of the micromachined biochips based mainly on ability of the AFM microcantilever to sense weak interactions. Three types of the cantilever biosensors/biochips are published. First of these so-called Force Amplified Biological Sensor (FABS) uses the magnetic beads, and the molecules are probed by applying an external magnetic force (Baselt et al., 1996; Baselt et al., 1997). The second type of microcantilever biochip works like quartz microbalance sensor: absorption of the biomass onto surface of the vibrating microcantilever shifts its resonant frequency (Tamayo et al., 2001). And the third of them based on static deflection of microcantilever due to surface stress caused by abs orbed molecules (Fritz et al., 2000; Moulin et al., 2000). Raiteri et al. (2001) give a review that describes in details how micromachined cantilevers may be involved in biosensors and biochips.

A principle of second type biosensor on example of DNA detection is represented on Figure 6 (Fritz et al., 2000). Authors used two cantilevers, each of which is functionalized on one side with a different oligonucleotide base sequence (the capture). After injection of the solution of the oligonucleotide (the analyte) that is complementary to one of the capture, hybridization occurs on the cantilever and it starts to bend (Figure 6, on the left). The amount of bending of both cantilevers is measured by the laser-photodiode reflection method like in the conventional AFM. The authors found that the differential signal from two cantilevers is preferable to be used in this case. If then the solution of second oligonucleotide that is complementary to the one on second cantilever (not shown) will be injected, the second cantilever is bent as it is clearly shown on the Figure 6 (on the right). The same procedure used in case of proteins. The amount of cantilever bending lies in the range of tens of nanometers.

The usage of cantilevers has important advantages in that it does not require labeling, optical excitation, or external probes. The methodology is compatible with silicon technology and great parallelization into integrated devices will allow a new generation of DNA and protein biochips.

6. Conclusions

The advantages of atomic force microscopy in the development of the biochips have been shown. The AFM technique allows the bioreceptors immobilization procedure to be enhanced and absorption and adhesion of the DNA and protein molecules into biochip surface to be studied. The new generation of biosensor/biochips based on the microcantilever micromachining was demonstrated. These devices present feasibility, miniaturization, high sensitivity and parallelization, real-time monitoring and low cost. It expected that the micro- and nanomechanical technologies would produce significant hitch in the biochip technologies and find real applications in health industry.

Acknowledgements

Author thanks Dr. K.-S. Ha for stimulating of this work. Author also acknowledges Dr. Ch. Gerber and Dr. W. Fritzche for permission to use illustrations from their publications.

References

1. Baselt, D.R., Lee, G.U. and Colton, R.J. (1996), "Biosensor based on force microscope technology", *J. Vac. Sci. Technol. B* **14**(2), 789–793.
2. Baselt, D.R., Lee, G.U., Hansen, K.M., Chrisey, L.A. and Colton, R.J. (1997), "A High-Sensitivity Micromachined Biosensor", *Proc. IEEE* **85**(4), 672-680.
3. Baselt, D.R., Lee, G. U., Natesan, M., Metzger, S. W., Sheehan, P. E. and Colton, R. J. (1998), "A Biosensor Based on Magnetoresistance Technology", *Biosensors and Bioelectronics* **13**, 731-739
4. Binnig, G. and Rohrer, H. (1982), "Scanning tunneling microscopy", *Helv. Phys. Acta* **55**, 726-729.
5. Binnig, G., Quate, C.F. and Gerber, Ch. (1986), "Atomic force microscope", *Phys. Rev. Lett.* **56**, 930-933.
6. Boyd, R.D. and Verran, J. (2001), "A novel method for measuring bacteria-surface adhesion directly using the atomic force microscope", *Microbiology*, submitted.

7. Burnham, N.A., Kulik, A.J. and Gremaud, G. (1998), "Tip-Surface Interactions", in Procedures in Scanning Probe Microscopy (R.J. Colton et al., Eds.), pp. 565-84, John Wiley and Sons, Chichester, UK.
8. Burnham, N.A. and Kulik, A.J. (1999), "Surface forces and adhesion", in *Handbook of Micro/Nanotribology* (B. Bhushan, Ed.), pp. 247-2713, CRC Press, Boca Raton, Florida.
9. Cahill, D. (2001), "Protein and antibody arrays and their medical applications", *Journal of Immunological Methods* **250**, 81–91
10. Caillat, P., David, D., Belleville, M., Clerc, F., C. Massit, C., Revol-Cavalier, F., Peltie, P., Livache, T., Bidan, G., Roget, A. and Crapez, E. (1999), "Biochips on CMOS: an active matrix address array for DNA analysis", *Sensors and Actuators* B **61,** 154–162.
11. Chen, X., Davies, M.C., Roberts, C.J., Tendler, S.J.B. and Williams, P.M. (1997), "Recognition of protein adsorption onto polymer surfaces by scanning force microscopy and probe-surface adhesion measurements with protein-coated probes", *Langmuir* **13**, 4106-4111.
12. Chen, X., Patel, N., Davies, M.C., Roberts, C.J., Tendler, S.J.B., Williams, P.M., Davies, J., Dawkes, A.C. and Edwards, J.C. (1998), "Application of protein-coated scanning forcemicroscopy probes in measurements of surface affinity to protein adsorption", *Appl. Phys. A* 66, S631–S634.
13. Chernov, A.A., Rashkovich, L.N., I.V. Yaminsky, I.V. and Gvozdev, N.V. (1999), "Kink kinetiks, exchange fluxes, 1D 'nucleation' and adsorption on the (010) face of orthorombic lysozyme crystals", *J.Phys.: Condens. Matter.* **11**, 9969-9984.
14. Csáki, A., Möller, R., Köhler, J.M., Straube, W. and Fritzsche, W. (2001), "DNA monolayer on gold substrates characterized by nanoparticle labeling and scanning force microscopy ", *Nucleic Acids Research* 29, e81, 1-5.
15. Dempsey, E., Diamond, D., Smyth, M. R, Urban, G., Jobst, G., Moser, I., Verpoorte, E.M.J., Manz, A., Widmer, H. M., Rabenstein, K. and Freaney, R. (1997), "Design and development of a miniaturised total chemical anylysis system for on-line lactate and glucus monitoring in biological samples", *Analytica Chimica Acta* **346**, 341-349.
16. Erdem, A., Meric, B., Kerman, K. Dalbasti, T. and Ozsoz, M. (1999), "Detection of interaction between metal complex indicator and DNA by using electrochemical biosensor", *Electroanalysis* **11**, 1372-1376.
17. Feldman, K., Höhner, G, Spencer, N. D., Harder, P. and Grunze, M. (1999), "Probing resistance to protein adsorption of oligo(ethylene glycol)-terminated self-assembled monolayers by scanning force microscopy", *J. Am. Chem. Soc.* 121, 10134-10141.
18. Fisher, M.I. and Tjörnhage, T. (2000), "Structure and activity of lipid membrane biosensor surfaces studied with atomic force microscopy and a resonant mirror", *Biosensors and Bioelectronics* **15,** 463–471.
19. Fritz, J., Baller, M. K., Lang, H. P., Rothuizen, H., Vettiger, P., Meyer, E., Güntherodt, H.-J., Gerber, Ch. and Gimzewski, J. K. (2000), "Translating Biomolecular Recognition into Nanomechanics" *Science* **288**, 316-318.
20. Fritzsche, W., Csáki, A., Reichert, J., Möller, R. and Köhler, J.M. (2001), "New colloidal gold-based detection scheme for DNA chip technology", *European Microscopy and Analysis* **72**, 5-7.
21. Green, R. J., Frazier, R. A., Shakeshe, K. M., Davies, M. C., Roberts, C. J. and Tendler, S. J. B. (2000), "Surface plasmon resonance analysis of dynamic biological interactions with biomaterials", *Biomaterials* **21**, 1823-1835.
22. Hirata, Y., Laukkanen, M. L., Keina, K., Shigematsu, H., Aizawa, M. And Mizutani, F. (2001), "Microscopic characterization of Langmuir–Blodgett films incorporating biosynthetically lipid-tagged antibody", *Sensors and Actuators* B **3750,** 1–6.
23. Ikai, A. (1996), "STM and AFM of bio/organic molecules and structures", *Surface Science Reports* **26**, 261-332.
24. Kondoh, J., Matsui, Y., Shiokawa, S. and Wlodarski, W. (1994), "Enzyme-immobilized SH-SAW biosensor", *Sensors and Actuators* B **20**, 199-203.
25. Lee, G.U., Chrisey, L.A. and Colton, R.J. (1994), 'Direct measurement of the forces between complementary strands of DNA", *Science* **266**, 771-773.
26. Magonov, S. N. and Wangbo, M-H. (1996), *Surface analysis with STM and AFM*, VCH Publishers, Inc., New York, USA.
27. McGurk, S.L., Green, R.J., Sanders, G.H.W., Davies, M.C., Roberts, C.J., Tendler, S.J.B. and Williams, P.M. (1999), "Molecular interactions of biomolecules with surface-engineered interfaces using atomic force microscopy and surface plasmon resonance", *Langmuir* **15**, 5136-5140.

28. Möller, R., Csáki, A., Köhler, J.M. and Fritzsche, W. (2000), "DNA probes on chip surfaces studied by scanning force microscopy using specific binding of colloidal gold", *Nucleic Acids Research* 28, e91, 1-5.
29. Morris, V. J., Kirby, A. R. And Gunning, A. P. (1999), *Atomic Force Microscopy for Biologists*, Imperial College Press; London, UK.
30. Moulin, A.M., O'Shea, S.J. and Welland M.E. (2000), "Microcantilever-based biosensors", *Ultramicroscopy* **82**, 23-31.
31. Noy, A., Vezenov, D.V., Kayyem, J. F., Meade, T.J. and Lieber, C. M. (1997), "Stretching and breaking duplex DNA by chemical force microscopy", *Chemistry and Biology* 4, 519-527.
32. Patel, N., Davies, M.C., Heaton, R.J., Roberts, C.J., Tendler, S.J.B. and Williams, P.M. (1998), "A scanning probe microscopy study of the physisorption and chemisorption of protein molecules onto carboxylate terminated self-assembledmonolayers", *Appl. Phys. A* 66, S569–S574.
33. Pereira, R.S. (2000), "Detection of the absorption of glucose molecules by living cells using atomic force microscopy", *FEBS Letters* **475**, 43-46.
34. Raiteri, R., Grattarola, M., Butt, H.-J. and Skládal, P. (2001), "Micromechanical cantilever-based biosensors", *Sensors and Actuators B* **79**, 115-126.
35. Reichert, J., Csáki, A., Köhler, J.M. and Fritzsche, W. (2000), "Chip-Based Optical Detection of DNA Hybridization by Means of Nanobead Labeling", *Analytical Chemistry* 72, 6025-6029
36. Ridout, M.J., Brownsey, G.J., Gunning, A.P. and Morris, V.J. (1998), "Characterisation of the polysaccharide produced by *Acetobacter xylinum* strain CR1:4 by light scattering and atomic force microscopy", *International Journal of Biological Macromolecules* **23**, 287–293.
37. Ros, R., Schwesinger, F., Padeste, C., Plückthun, A., Anselmetti, D., Güntherodt, H.-J. and Tiefenauer, L. (1999), "SPM for functional identification of individual biomolecules", *Proceedings of SPIE* **3607**, 84-88.
38. Sarid, D. (1991), *Scanning force microscopy; with applications to electric, magnetic and atomic forces*, Oxford Publishing, New York, USA.
39. Siedlecki, Ch. and Marchant, R. (1998), "Atomic force microscopy for characterization of the biomaterial interface", *Biomaterials* **19**, 441-454.
40. Spangler, B.D. and Tyler, B. (1999), "Capture agents for a quartz crystal microbalance - continuous flow biosensor: functionalized self-assembled monolayers on gold", *Anal. Chim. Acta* **399**, 51-62.
41. Swanson, P., Gelbart, R., Atlas, E., Yang, L., Grogan, T., Butler, W. F., Ackley, D. E. and Sheldon, E. (2000), "A fully multiplexed CMOS biochip for DNA analysis", *Sensors and Actuators* B **64**, 22–30.
42. Tamayo J., Humphris, A.D.L., Malloy, A.M. and Miles, M.J. (2001), "Chemical sensors and biosensors in liquid environment based on microcantilevers with amplified quality factor", *Ultramicroscopy* **86**, 167-173.
43. Tamayo, J., Humphris, A. D. L., Malloy, A.M. and Miles, M.J. (2001), "Chemical sensors and biosensors in liquid environment based on microcantilevers with amplified quality factor" *Ultramicroscopy* **86**, 167-173.
44. Vansteenkiste, S.O., Corneillie, S.I., Schacht, E.H., Chen, X., Davies, M. C., Moens, M. and Van Vaeck, L. (2000), "Direct measurement of protein adhesion at biomaterial surfaces by scanning force microscopy", *Langmuir* 16, 3330-3336.
45. Vinckier, A. and Semenza, G. (1998), "Measuring elasticity of biological materials by atomic force microscopy", *FEBS Letters* **430**, 12-16.
46. Vo-Dinh, T., Cullum, B. M. and Stokes, D. L. (2001), "Nanosensors and biochips: frontiers in biomolecular diagnostics", *Sensors and Actuators* B **74**, 2-11.
47. Weinberger, S.R., Morris, T.S. and Pawlak, M. (2000), "Recent trends in protein biochip technology", *Pharmagenomics Review*, 1. http://www.ashley-pub.com/html/display_article.asp?ArticleID=3052
48. Weisendanger, R. (1994), *Scanning probe microscopy and spectroscopy*, Cambridge University Press, Cambridge, UK.
49. Welsch, W., Klein, C., Öksüzoglu, R.M., Schickfus, M.V. and Hunklinger, S. (1997), "Immunosensing with surface acoustic wave sensors", *Sensors and Actuators* A **62**

PECULIARITIES OF TH.TERRESTRIS SPORES SURFACE ULTRASTRUCTURE INVESTIGATED BY AFM

E.N. GROMOZOVA[1], P.M. LYTVYN[2], V.S. PODGORSKY[1]
[1]Institute of Microbiology and Virology, NAS, Ukraine,
154, Zabolotnogo st., Kyiv 252143, Ukraine, e-mail: gren@carrier.kiev.ua
[2]Institute of Semiconductor Physics, NAS, Ukraine,
45, pr. Nauky, Kyiv 03028, Ukraine, e-mail: plyt@isp.kiev.ua

1. Introduction

Properties of different micro-organisms are intensively investigated because they display several vital functions, and, among them, the most interesting are: protecting the cytoplasm from an environment, providing the cell with a rigid structure, determining the cell shape, controlling cell adhesion and cell aggregation processes. These processes at the macrolevel have their consequences both in natural environment and in biotechnologies.

Investigations of a composition and properties of a surface of both dormant spores and those in the course of germination for various kinds of fungi was done by many researchers (Gerin et al., 1993; Dute et al., 1989; Ellis, 1978). However, the problem whether there are differences in their structure at varied directions of formecreating processes remains open.

The atomic force microscopy (AFM) as a new method for subnanometer surface features investigations was intensively developed during last decade. The method has a large potential not only in 3-D surface mapping but also in investigations of surface physico-chemical properties.

One of the features of the given method of research consists in original state of sample preparations. The samples are not exposed to any processing, i.e. the method (practically not destructive) allows to investigate the behaviour of the biological systems in natural conditions; this is especially valuable at study of the morphogenetic processes. All this distinctions are combined with high 3-D resolution.

2. Experimental procedure

The surface ultrastructure of the samples was studied by means of a Digital Instruments Nanoscope IIIa atomic force microscope. The measurements were performed at ambient conditions in tapping mode to increase sensitivity and resolution as well as to decrease contact force between sample surface and AFM tips during surface topography measurements. The scanning was taken by a 100 μm G-scanner using etched silicon tip with a nominal radius of curvature 10 nm.

Spores were cultivated at various conditions and then positioned on fresh cleaved mica surface or on glass. Samples were placed at AFM for measurements after drying during few minutes.

E. Buzaneva and P. Scharff (eds.), Frontiers of Multifunctional Nanosystems, 341–346.

3. Results and discussion

As it could be seen from a fig.1, the spores of Thielavia terrestris, washed off from agar, are oval (the average size is 3.8 × 1.9 μm). A surface of spores in inoculum is rather smooth (the ledges of a surface do not exceed 2 nm) according to the AFM topography data.

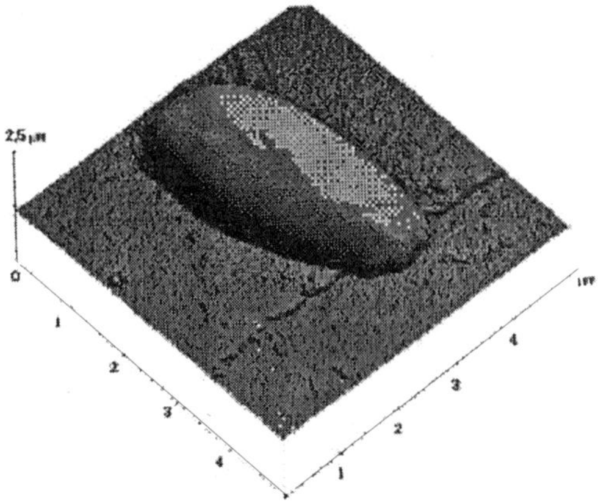

Figure 1. 3-D AFM image of a Th.terrestris spore.

Nevertheless, in the AFM phase registration mode, which is highly sensitive for adhesive and viscoelastic, properties of surface there are areas with significant phase contrast. The areas form a certain granular structure. Such areas are homogeneously distributed over surface (the average size is 50 nm), but the chains of separate grains of length 100-200 nm (fig.2.) are observed also.

The surface large-scale roughs can lead to insignificant changes in size and distribution of grains in these areas, but general tendency does not vary. The properties described differ from the so-called "rodlet" structures of protein nature marked by many authors as for dormant spores of a lot of fungi (Beever et al., 1978; Beever et al., 1979; Claverie-Martin et al., 1986). Though by means of AFM it was proved only for the spores of Phanerochaete chrysosporium (Dufrene et al., 1999).

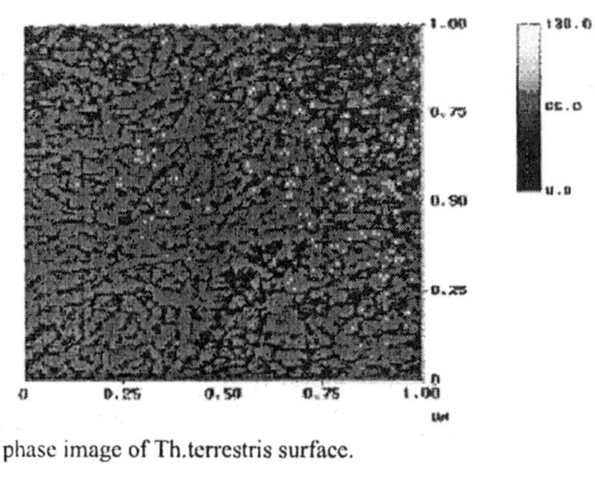

Figure2. AFM phase image of Th.terrestris surface.

During an incubation process of the spores in Chapek liquid mix culture their swelling was observed. The sizes became 4.86 × 2.7 μm after cultivation during 8 hours. The structure of a surface depends on the conditions of fungi development strategy. On the surface of spores with "pellet's" development strategy the local areas with significant phase contrast were observed as fibril structures (width about 10 nm and length up to 260 nm). Together with a fibril structures in a field of scanning the granular formations with lateral sizes approximately 100 õ 60 nm are seen. The granular formations appear in surface topography as hillocks with height of 15-20 nm. Density of such formations is about 20 un/microns (fig.3.).

The density of the fibril structures decreased after cultivation during 10 hours but their length and thickness were increased (up to 1000 nm and 10-20 nm, accordingly). The density of granular formations decreased (up to 5 un/microns) thus some of them were enlarged up to 300-400 nm (fig.4,a). The fibril layer of the spore's surface in this period is developed as can be observed in AFM topography mode (fig.4, b).

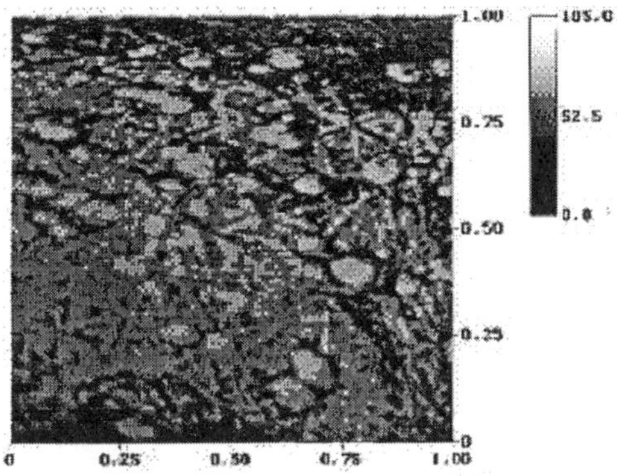

Figure3. AFM phase image of the Th.terrestris spores after cultivation during 8 hours. Spores have "pellet's" development strategy.

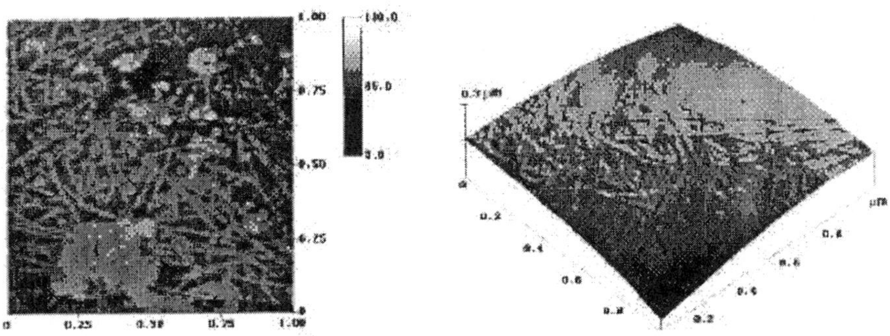

(a) (b)

Figure 4. The surface of Th.terrestris spores with "pellet's" development strategy after cultivation during 10 hours: phase image (a); 3-D topography image (b).

For dispersed development strategy of the spores the fibril structures were absents after cultivation during 8-10 hours. There are granular structures (size 30-80 nm) with less density in comparison to the initial surface.

Fibril structures occurred at nogerminated spores only after cultivation at 18h (length 50-200 nm and width about 20 nm) (fig.5,a) and distinctly were traced as long, twisting strings after 20h (length about 700 nm, width ~ 15 nm) (fig.5,b). In both tests the separate large granular formations with the size range from 260 nm up to 500 nm were observed.

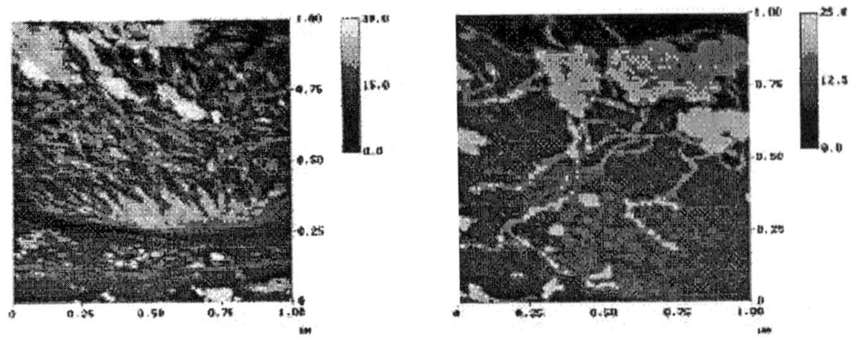

(a) (b)

Figure 5. The phase image of Th.terrestris spores with "dispersed" development strategy after cultivation during 18 hours (a) and 20 hours (b).

Thus, fibril structures were revealed at pellet type of growth much earlier than at dispersed type and in much larger quantity. The granular areas were integrated and settled down on the surface; sometimes they were completely replaced by the fibril structures. It is necessary to note, that the similar structural changes were observed in the pellet type strategy caused not only by the reaction of environment, but also by others morphogen factors (for example low temperature).

One of the distinct features of the different development strategies is the variety in degree of adhesive and viscoelastic properties of the spore's surfaces (fig.6).

It is must be emphasized, that in case of dispersed strategy, decreasing of AFM phase shift signal value are more essential. That testifies both decreasing of surface adhesive forces and viscoelasticity of surface. Obviously, viscoelastic component prevails in this case. During the development the spores became less rigid, the wall gains plasticity. It is necessary to specify, that on background of general decrease of interaction forces in local surface areas there can be an increase of adhesion forces as is shown in work of Gerin (Dufrene et al., 1999). The fact, that in dispersed type this decrease is more obvious testifies both shorter period of the spore's germination, and the greater elasticity of their surface connected to smaller quantity of structural elements.

Thus, the above-presented results allow ascertaining the changes of ultra structure and molecular forces of interaction of the Th.terrestris spores surfaces. They depend not only on the process of swelling, but also on the morphogenes (of pellet and dispersed types of mycelia). The fibril structures on the sizes, which we revealed and the description corresponds to those that were specified for Pen.roqueforti (Hwang et al.,

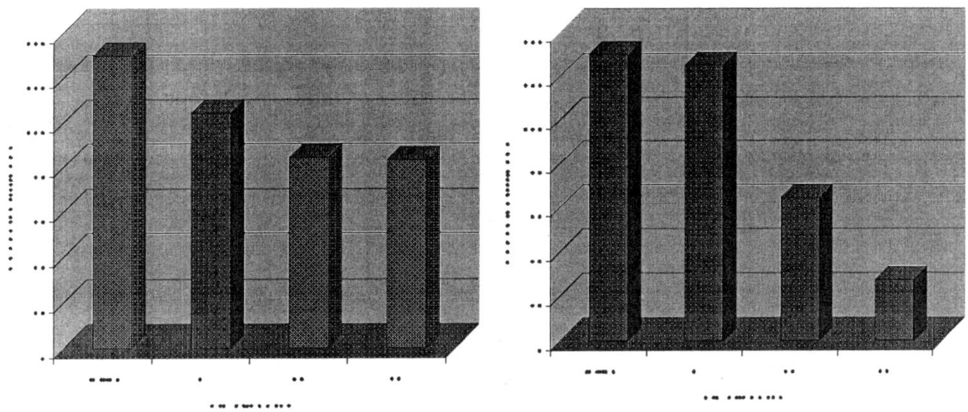

(a) (b)

Figure.6. Dependence of AFM phase signal shift value on the cultivation period for spores of "pellet's" (a) and "dispersed" (b) development strategies which reflects the changes of adhesive and viscoelastic properties of spores surfaces.

1977). Though we do not establish the nature of this fibrous layer, it is possible to assume that this is of the polysaccharide origin. According to literary data polysaccharides can be presented as well organized fibril or mucous gel covering germination tube and sprouting spore. The data, received by us, correspond with results of the electron-microscopic study of the sprouting spore Syncephalastrum racemosus (Hobot et al., 981). They have shown that in the process of the spore swelling rodlet-protein layer collapses and polysaccharide appears on a surface. Gerin et al., (1993) assume, that polysaccharides which take part in the processes of the growing of the pellet form should have the same origin as polysaccharides that appear during the spore germination. It is known that the fungi walls are built with microphibrilles and amorphous matrix that is immersed in them. Chitin or cellulose more often represents the skeleton part. The substances of matrix are glucans, galactans and heteropolysaccharides which are conjugated with proteins (Elinov, 1989). Probably, the changes, which we observed in the structure of walls of the swelling spores, are a prototype of the cellular walls of the hyphae of the pellet and dispersed types of the mycelia.

All we have stated above is coordinated with our earlier hypothesis about conformity of the pellet form to adverse conditions of the cultivation. According to the opinion of Feofilova (Feofilova, 1994) the tregalose during nonspecific stress (this carbohydrates prevails in the spores of the majority of the fungi (Sussman, 1966)) transforms into glucans (chitin and glucosamin). This is inherent to the pellet form making it different from the dispersed form. And especially important is that these structural changes are observed, as it was shown by our investigations by means of AFM, at the early stages of cultivation of spores in a lag-phase. Feofilova, Nemtsev, Terioshina (Feofilova et al., 1998) have shown that the ratio of chitin and glucan in the cellular walls of Aspergillus niger does not vary during growth, though it depends on composition of nutritious environment. It means, that the composition of the structural polysaccharides of the cellular walls is determined at the earliest stages of development (on our opinion is in a lag-phase).

The found fact that structural differences of the spores surfaces depend on conditions of swelling, extend modern ideas about biosynthesis of cellular walls of fungi. As known, the cellular wall is formed in the region of apex during whole period of the hyphae growth de novo. We consider that the program of this synthesis, as well as a level of the predecessors of chitin and glucan, is defined before occurrence of the apex during the stage of swelling. This fact is very important for the problem of the morphogenes of fungi. Retrieved data has the great practical interest in connection with the purpose of obtaining the cellular walls of certain composition in the biotechnology processes.

References

1. Beever R.E., Dempsey G.P. (1978), "Function of rodlets on the surface of fungal conidiospores", *Nature.* **273**, 608-610.
2. Beever R.E., Redgwell R.J., Dempsey G.P. (1979), "Purification and chemical characterization of the rodlet layer of Neurospora crassa conidiospores", *J. Bacteriol.* **140**, N.8, 1063-1070.
3. Claverie-Martin F., Diaz-Torres V.R., Georghegan M.J. (1986), "Chemical composition and electron microscopy of the rodlet layer of Aspergillus nidulans conidiospores", *Curr. Microbiol.* No.14, 221-225.
4. Dufrene Y.F., Boonaert C.J.P., Gerin P.A., Asther V., Rouxhet P.G. (1999), "Direct probing of the surface ultrastructure and Molecular interections of dormant and germinating spores of Phanerochaete chrysosporium", *J. of Bacteriol.* **181**, No. 17, 5350-5354.
5. Dute R.R., Weete J.D., Rushing A.E. (1989), "Ultrastructure of dormant and germinating conidia of Asp.ochraceus", *Mycologia.* **81**, No 5, 772-782.
6. Elinov N.P. (1989) *Khimicheskaia microbiologiia*, Moscow, Vysshaia Shkola.
7. Ellis D.H. (1978), "Fine structure of spore germination in Epicoccum purpurascens", *Trans. Br. Mycol .Soc.* **70**, 170-173.
8. Feofilova E.P. (1994), "Biokhimicheskaya adaptatsyia gribov k temperaturnomu stressu", *Microbiologia*, **63**, No5, 757-776.
9. Gerin P.A., Dufrene Y., Bellon-Fontaine M., Asther M., Rouxhet P.G. (1993), "Surface properties of conidiospores of Phanerochaete chrysosporium and their relevance to pellet formation",*J. of Bacteriology.* **175**, No 16, 5135-5144.
10. Hobot J.A., Gull K. (1981), "Changes in the organization of surface rodlets during germination of Syncephalastrum racemosum", *Protoplasma.* **107**, No.3, 339-343.
11. Hwang D.H., Chabot J.F., Hood L.F., Kinsella J.E. (1977), "Ultrastructural changes in Pen. Roqueforti during germination and growth", *J. Food Biochem.* **1**, No. 1, 3-14.
12. Nemtsev D.V., Kozlov V.P., Terioshina V.M., Memorskaia F.S., Feofilova E.P. (1998), "Izmeneniya v sostave strukturnykh komponentov kletochnoj stenki Aspergillus niger v zavisimosti ot uslovij kul'tivirovaniya", *Pricladnaya biokhimia i mikrobiologia.* **34**, No. 1, 95-98.

RELAXATION OF NANOSTRUCTURED MOLECULAR MATERIALS UNDER THE INFLUENCE OF SOLVENT VAPORS

Y. SHIRSHOV, , B. SNOPOK, O. RENGEVYCH
Institute of Semiconductor Physics, National Academy of Sciences of Ukraine
45 Prospect Nauki, Kiev 03028, Ukraine; Phone: +(38044) 265-1827
V. KALCHENKO
Institute of Organic Chemistry, National Academy of Sciences of Ukraine
2 Murmanskaya St., Kiev, Ukraine
A. COLEMAN
Institute of Biology and Chemistry of Proteins, CNRS, Lyon, France

Abstract.

In recent years the self-assembling processes in some materials have become very important in connection with development of the nanometer-scale molecular technology. Calyxarenes are the molecules of just such a type. The possibility of thin evaporated and Langmuir–Blodgett calyxarene films swelling under the action of solvent molecules has been demonstrated previously. Some models of separate molecules complexing were proposed and theoretical models were developed to explain the effect of structural relaxation under the influence of solvent vapor. Here we present the results of our investigation of thin thermally deposited films of some calyxarenes and their derivatives treated in solvent vapors. The efficient modern methods, such as surface plasmon resonance, atomic force microscopy and quartz crystal microbalance, were used. Dramatic changes in optical constants and film morphology were observed. The numerical data analysis was performed and physico-chemical model was proposed to explain the results obtained and predict some new effects. Our results could be used in research and development of specially tuned sorbents for electronic nose – a new and very promising kind of information devices.

1. Introduction

The materials demonstrating different sorption sensitivities to different classes of molecules serve as a basis for the modern line of multichannel analyzers of individual gas contaminants or chemical image recognition in complex mixtures. Among numerous physical effects, adsorption on surface or in the bulk of thin molecular layers is one of

E. Buzaneva and P. Scharff (eds.), Frontiers of Multifunctional Nanosystems, 347–368.

the most often used. Molecular adsorption on a sensitive layer is usually detected with either optical (Slavík et al., 2001; Goepel, 1994; Malmsten, 1994; www.analytical-microsystems.de) or gravimetric techniques (O'Sullivan et al., 1999; Fredriksson et al., 1998;. DiNatale et al., 2000; Lucklum et al., 2000). Besides, some other effects are often used for this, such as changes in conductivity (Szczurek et al., 1998; Dickert et al., 2001; Gardner et al., 1994; Hatfield et al., 1994), permittivity (Steiner et al., 1995; Torsi et al., 2000) or other physical parameters of a thin layer.

Evidently a number of molecules adsorbed by a sorbent with pores in its bulk is determined by sorbent structure – presence of specific voids in its bulk or features of its surface relief. These details of solid or polymer structure measure from several fractions of a nanometer to several tens of nanometers, i.e., they are typical nanostructured materials. Calyxarenes, resorcinarenes and cavitands are classical examples of nanometer cavities (Nelli et al., 1993; Hartmann et al., 1996; Dickert et al., 1993; Janshoff et al., 2000). Where individual molecules in a gas phase or solution are concerned, then both size and shape of nanocavities are determined solely by the stereochemical features of cell-like molecular structures.

Another example of nanostructured materials (if not so determined from the viewpoint of microcavity size and shape) exists in polymers (Varadan et al., 1999; Dodd et al., 1992.). They are extensively used when developing sensitive sorbents with nanopores tuned to molecules-analytes using imprinting technique. In this case layer polymerization is realized in the presence of molecules-analytes that are removed from the bulk by prolong washing in a pure solvent. As a result, a random system of microcavities is formed whose shape and size are close to the molecule-analyte parameters. A particular advantage of this technique lies in the possibility to form an internal surface of a microcavity that chemically complements the molecule-analyte surface (Kriz et al., 1997).

The principal advantages of this class of sensitive elements are a possibility to modify polymer backbone using some specific peripheral groups of preset chemical functionality, as well as the ease of fabrication and measurement . So the required sensitivity can be obtained for a limited number of molecules only. They may be donor of acceptor ones, depending on the type of the polymer backbone and peripheral substituents that increase or decrease the charge carrier concentration. The composite sensitive elements of conducting polymer sensors are polymers containing intercalated conducting particles (Hawkins et al., 1999). They use the effects of coating structure modification resulting from interaction with a gas environment. Their selectivity that is due to gas molecules "solubility" in the film bulk also is limited (Hsuing et al., 1999). The conductibg polymer sensors demonstrate a nonlinear response and dependence on moisture presence in a sample.

The composites from polymers and conducting nanoparticle ensembles (carbon black, fullerenes, colloidal gold) represent a particular line in formation of gas-sensitive layers tuned to definite features of an analyte (Krummel et al., 1999; Carotta et al., 2001; Guadarrama et al., 2000). The following two features of nanoparticles are principally used in this case: (i) formation of disordered network for charge carrier transport through jumping over the percolation level (Kukla et al., 1996; Zee et al., 1999; Bidan , 1992) and (ii) formation of self-assembled molecular ensembles on colloidal gold or fullerene surfaces for control over the adsorption capacitance and selectivity of gravimetric sensors (Shih et al., 2001; Schlupp et al., 2000).

All the substances-cavitands, in particular, calyxarenes, except polymer layers (that are not true solids) were tacitly considered as carriers of the principal features of solids, namely, constant volume and shape, stable crystalline structure (usually polycrystallites separated with amorphous interlayers). Presence of microcavities in the material bulk makes it possible to sorb a definite number of molecules-analytes that form complexes with substituted calyxarenes (Kunsagi-Mate et al., 2001; Gutsche et al., 1989). It seems natural that, along with microcavities, a rich variety of cavities of different shapes and much bigger volume exist in the bulk of such sorbents. These cavities provide unhindered delivery of molecules. The nature of such cavities depends on the film deposition technique and the rates of deposition and solvent removal. These cavities should result in a nonspecific sorption that is not related to the host–guest complexing.

The above model for a calyxarene sorbing layer does not basically assume any substantial change in film thickness under molecular adsorption. This adsorption should rather increase the refractive index leaving thickness practically unchanged since both film morphology and structure have been set by the technique of film deposition onto the substrate. Therefore a possibility of appearance of adsorption parameters instability during the adsorption–desorption cycles should not cause anxiety. One always can bring sorbent to its initial state by heating it in vacuum or in atmosphere with a pure carrier.

However, some recently obtained facts require revision of the above notion of the films of calyxarenes, resorcinarenes and other organic substances. These facts indicate at essential modification of a solid (in a sense that was defined above) molecular layer under action of the solvent vapors. Revealing of such modification has become possible due to application of new experimental techniques, namely, surface plasmon resonance (SPR) and atomic force microscopy(AFM). Very valuable information is given also by ellipsometry and low-energy neutron diffraction.

It has been shown earlier (Nabok et al., 1995; Shirshov et al., 1997) that thicknesses of thermally deposited tert-butylcalyx arene films measured by ellipsometry can grow by up to 20% their initial values under the effect of toluene and chloroform. The quartz crystal microbalance (QCM) technique has shown that the Langmuir–Blodgett (LB) films of substituted calyx arene can absorb up to 20 benzene molecules per calyxarene molecule. In this case torus-shaped calyxarene micelles of micron size may appear on the film surface.

The above serves as evidence that self-organization processes may occur in thin calyxarene films exposed to organic solvent vapors. This fact deserves more detailed consideration since it is related to the problem of stability of such promising materials operation as sensitive layers in gas analyzers and devices of the electronic-nose-type.

Here we make an attempt to investigate the structural changes of thin calyxarene films treated in organic solvent vapors, as well as to make a forecast concerning the use of such materials in the electronic-nose-type systems intended for gas analysis.

2. Materials and Instruments

Calyxarenes (C[n]A) used in this study were synthesized as the tert-butylcalyx[n]arene (n = 4, 6, 8) and its derivatives at the Institute of Organic Chemistry of the National Academy of Sciences of Ukraine and Laboratory of Industrial Chemistry of the Lyon Claude Bernard University. These compounds, of crown-like shape, were obtained as fine powders. The last stage of synthesis was recrystallization from toluene solution.

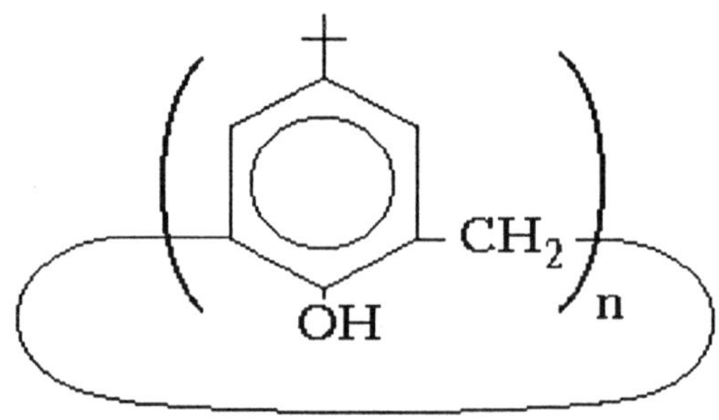

Figure 1 Picture of tert-buthil calyx[n]arene molecules

Substrates for calyxarene deposition were silicon plates or polished glass slides (15×20×2 mm) covered with thin (50 nm) gold films prepared by thermal evaporation onto substrate from Knudsen cell in a vacuum of 10^{-3} Pa. The films had a constant thickness in the 50–100 nm range over the whole surface.

3. Methods

Structural relaxation of amorphous and polycrystalline materials is usually accompanied with changes in their optical parameters (thickness, refractive index), as well as electronic (electron spectra) and dielectric (Shirshov et al., 1990) properties. To measure concurrently thickness and refractive index of thin calyxarene films is rather difficult. This is so because ellipsometry that is usually applied for this purpose does not allow measurement performance in the kinetic mode. This difficulty can be got round by applying a new technique for study of the optical constants of thin films on gold surfaces with surface plasmon-polariton resonance (SPR) (Kostioukevich et al., 1995). The PPR technique consists in using a thin gold film as a resonator to excite the plasmon-polariton oscillations in the case when plasmon wave vector $\boldsymbol{k}_p$ coincides with the longitudinal component of wave vector of the incident light wave $\boldsymbol{k}_x = (2\pi/\lambda)\sin\theta_i$. Under this condition light reflection peaks sharply. This peak can be registered quickly and very accurately by varying either angle of incidence or light wavelength (Fig.1). $\boldsymbol{k}_p$ strongly depends on the ambience and dielectric film parameters. That is why the SPR technique is very convenient when dealing with thin films on gold surfaces. Another advantage of the SPR technique is that plasmons are excited on the substrate inner side, while the surface to be studied remains free and open to adsorption.

Figure 2 general view of the QCM experimental plant

SPR measurements were carried out using a device SPR-4m developed at the Institute of Semiconductor Physics of the National Academy of Sciences of Ukraine (Kiev, Ukraine). The general view of the device and gas system is shown in Fig.2. The experimental cell volume was about 100 μl.

Film surface morphology was studied with an atomic force microscope Nanoscope III (Santa Barbara, USA). The measurements were performed in the contact and tapper modes. The latter one turned out to be preferable since, due to insufficient material (thermally sputtered organic layers) hardness, organic material often adhered to the probe in the contact mode.

4. QCM Devices and el Ectronics

The adsorption properties were studied with a multichannel (8, 16 channels) quartz resonance device whose base frequency was 10 MHz. A multisensor QCM-based analyzer involves the following units: (i) a flow-type chamber with sensor array; (ii) a quartz oscillator unit; (iii) a unit for frequency measurement with RS232 interface based on a dedicated microprocessor; (iv) a gas mixture generator; (v) a computerized system to collect and process information. To provide vapor-gas mixtures stability (if the sample could be supplied in liquid state only), a mechanical agitator maintained a constant speed of the gas carrier (air) that was passed through the sample volume. The test oscillators were connected to the frequency meter in series; time for measurement was 1 s and the frequency reading accuracy was 1 Hz. We did not use an additional thermal stabilization of the system, as well as control over gas carrier humidity and common pressure.

The AT-cut crystals were used as physical transducers. They operated in the shear vibration mode; the fundamental frequency was 10 MHz. The measuring procedure involved the following stages: air circulation until the transducer frequency is stabilized (±5 Hz); vapor-gas mixture circulation; refinement stage using pure air circulation.

Calixarene films were prepared by thermal sputtering onto the metal electrodes of QCM (on one side). The transducers with deposited sensitive layers were placed in the measuring plant and exposed to ethanol vapor (in standard conditions) until a reproducible response of the whole sensor array was achieved. This required from 10 to 15 cycles. Such a pre-starting procedure enabled to stabilize parameters of thin films. After this the sensitive element array was used to perform the planned measurements with some kind of perfumes.

5. SPR measurements

The order of SPR experiment included:

1) measurement of the initial SPR curve in air;
2) inlet of solvent vapor and observation of SPR minimal shift due to adsorption;
3) measurement of SPR curve after the continuous state has been obtained (first adsorption);
4) replacement of the ambience by pure air and observation of SPR minimum recovery (first recovery);
5) measurement of the whole SPR curve after the continuous state has been obtained.

Usually after three adsorption- recovery cycles the process became renewable.
Both thickness and refractive index of calyxarene films were extracted from the whole SPR curve using special fitting program (Beketov et al., 1998) that permitted to fit from one to four parameters simultaneously. We note that the system investigated included at least five optical parameters - three for gold and two for calyxarene films. So we used the step-by-step fitting procedure to obtain all the five parameters of the reflected system.

Typical view of SPR curves for calyxarene films before and after toluene adsorption is shown in Fig.3. The problem was to extract all the data for two investigated layers, namely, gold and calyxarene ones. The rough evaluation of gold film parameters may be obtained on the sample without calyxarene. The data for initial gold film parameters are such: refraction coefficient, n_{Au} = 0.075; extinction coefficient, k_{Au} = 3.382; thickness, d_{Au} = 51.453 nm.

Table 1. *Data calculated from SPR curve measurement.*

#	Experimental condition	n_{Au}	K_{Au}	d_{Au}, nm	n_{Cx}	d_{Cx}
1	Initilal film c[8]A	0.125	3.043	61.233	1.502	51.16
2	I toluene adsorption	0.125	3.043	64.78	1.488	60.03
3	I recovery	0.125	3.043	62.32	1.457	59.61
4	II adsorption	0.125	3.043	65.64	1.474	62.88
5	II recovery	0.125	3.043	63.81	1.464	60.91
6	III adsorption	0.125	3.043	64.61	1.468	60.1
7	III recovery	0.125	3.043	66.72	1.568	50.153

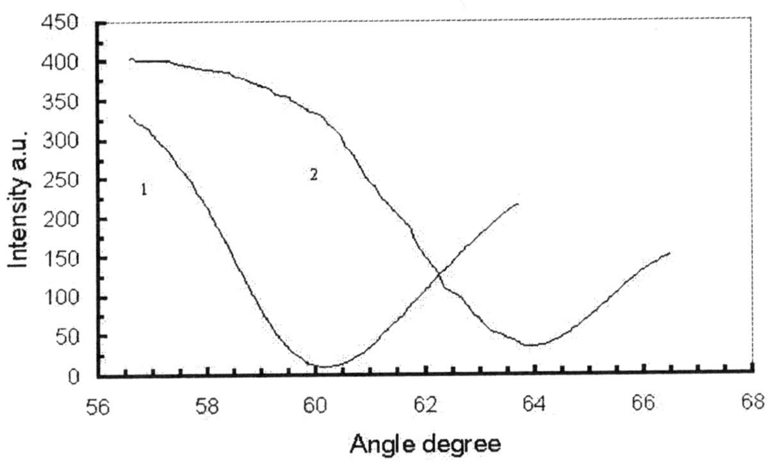

Figure 3 Typical view of SPR curves for Calix[8]arene film before 1(left) and after 2(right) adsorption of toluene

Using these values, the curves 1, 2 (Fig.3) and fitting algorithm, we obtained all the data for gold and calyxarene films before and after adsorption. Toluene vapors adsorption leads to SPR curve shift (Fig.3, curve 2). One can see that adsorption of toluene leads to d_{Cx} increase and n_{Cx} decrease. The adsorption also changes both n_{Cx} and d_{Cx} (rows 3–6). Similar behavior was demonstrated also by the optical data of C[4]A in contact with chloroform vapor. Fig.4 shows the calyxarene refractive index n_{Cx} versus film thickness d_{Cx} curves.

Kinetics of SPR curves changing at adsorption and recovery. One can see from Fig.3 that adsorption effect is accompanied by a SPR curve shift along the angle axis. The curve form change is too small to be detected visually; it is, however, sufficient for simultaneous extraction of two parameters, namely, thickness d and refractive index n. One can show that optical response at thin film formation on a solid surface under external reflection of light is proportional to the product nd (Rengevych et al., 1999).

It accounts, to a first approximation, for a mass of the substance that has been added to the film or extracted from it (Shirshov et al., 2001). Therefore the SPR curve shift may be considered as a measure of the adsorbed substance. The device SPR-4m makes it possible to monitor the curve position on the angle axis, thus enabling one to register rather fast kinetics of sample mass changing.

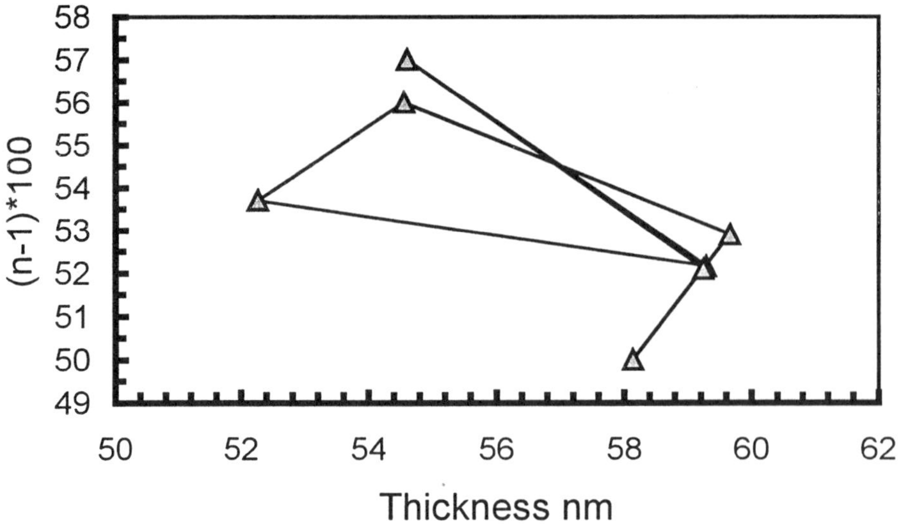

Figure 4 . Correlation between refractive index and film thickness after adsorption (right down) and recovery (upper left points)

Shown in Figure.5 are the typical kinetic curves obtained at repeated adsorption and desorption. The nonmonotone character of these curves is practically the same for all the host–guest pairs studied.

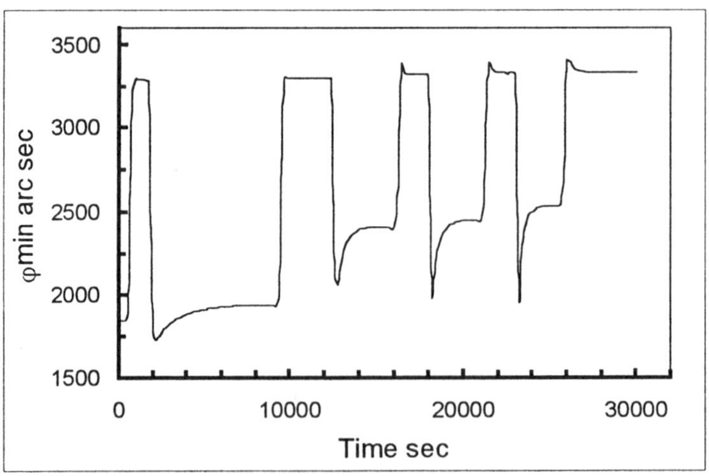

Figure. 5. Kinetics of the SPR-curve minimum angle φ_{min} change during the adsorption- recovery cycles for calix[8]arene in the toluene vapors.

6. Measurements with planar polarization interferometer

The nonmonotone adsorption kinetics was supported by SPR study of ethanol adsorption onto thin calyx [6] arene layers.

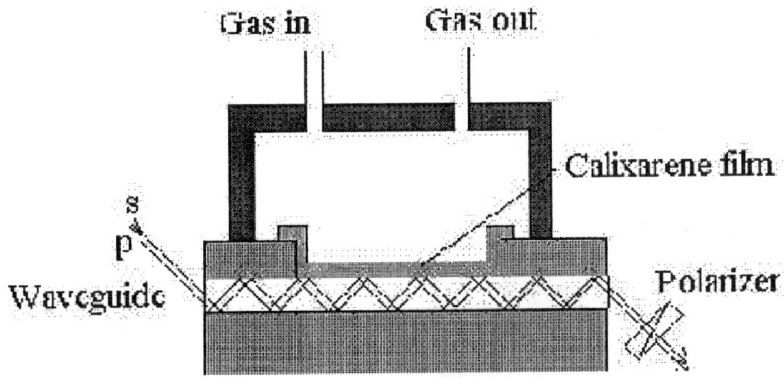

Figure 6 Scematic drow of wavequiding chip and signal change after adsorption-recovery

The PPI technique is described in detail in (Shirshov et al., 1998). Its essence is briefly shown in Fig.6. The *s*- and *∂*-polarized light beams are concurrently introduced

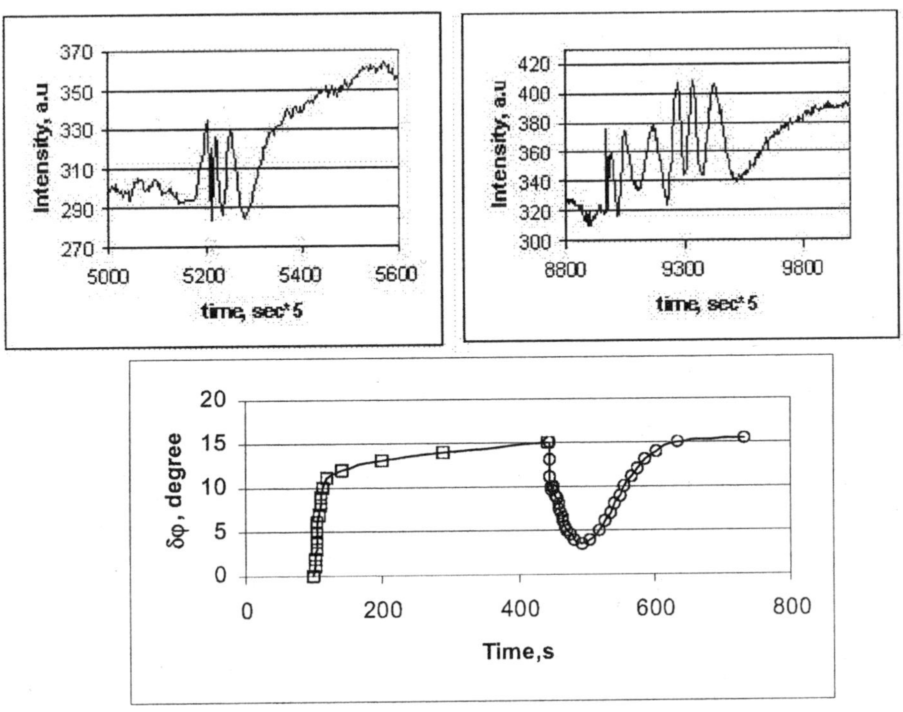

Figure.7. Kinetic of PPI-output signal after ethanol adsorption on C[6]A and recovery

into a light guide through the input facility. At some distance the light guide cladding is removed. This provides access for sensitive layer sputtering. The *s*- and *p*-polarized light beams interact with ambience in different ways. As a result, an elliptically polarized signal is formed at the light guide outlet. Its phase is determined by the refractive index and thickness of a gas-sensitive layer. Thus it becomes possible to examine with high accuracy the sensitive layer changes due to adsorption. They manifest themselves in fixation of the interference band number with a photosensitive detector placed after a linear polarizer.

Shown in Figure.7 is a typical PPR signal taken at adsorption (b) and desorption (c). The phase difference changes given in Figure.7 are monotone at adsorption and

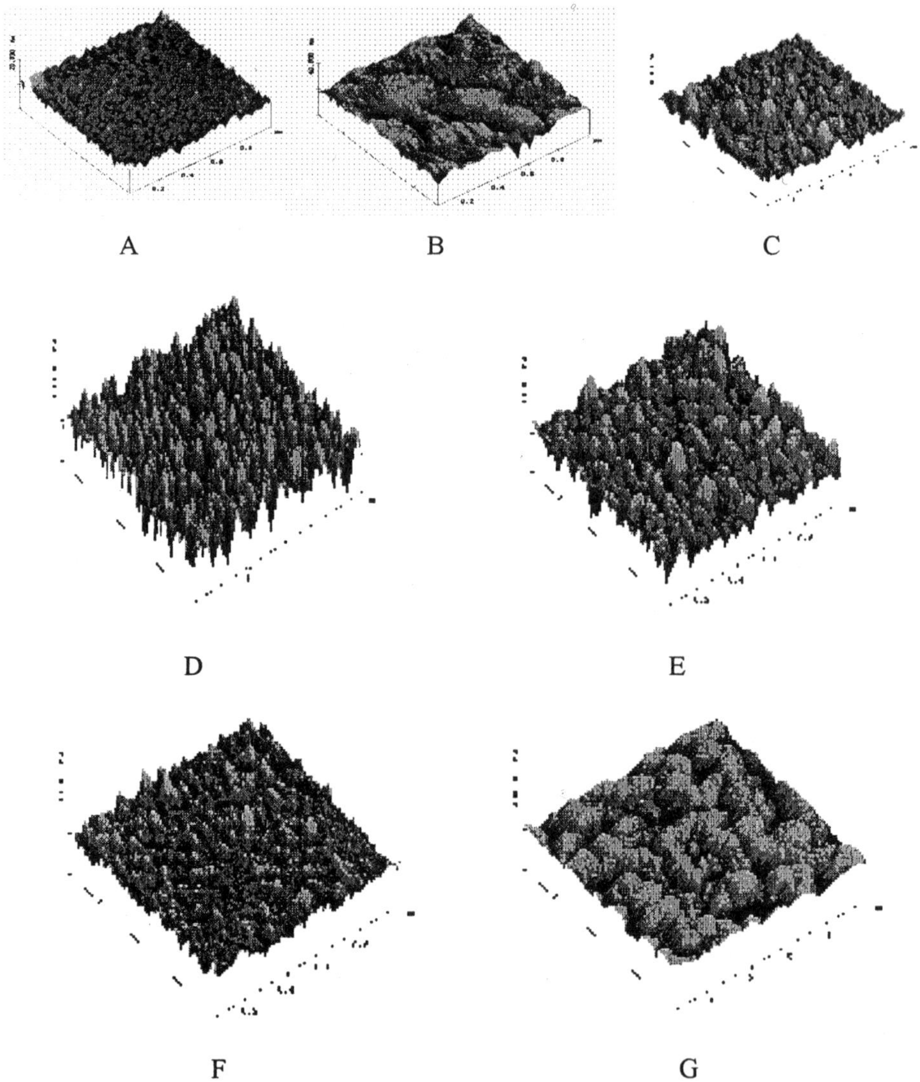

Figure 8 AFM pictures of calixarene films

nonmonotone at desorption. The signal value might be related to the changes in film thickness and refractive index in the following way. A phase difference change of 12 π for *s*- and *ð*-polarizations corresponds to thickness increase of about 4 nm (the initial thickness value being 100 nm) or refractive index change of 0.006.

One can see from Fig.7 that ethanol adsorption onto a calyx/6/arene (C[6]A) film results in a monotone increase of either thermally sputtered layer thickness (of about 4%) or refractive index (of about the same value). Contrary to this, ethanol desorption is accompanied with a more complex change in the PPR response. At first the signal returns to a value that is close to the initial one, and then unexpectedly returns to a higher value. The above experiment demonstrates an unusual behavior of calyx/6/arene films during the ethanol saturated vapor adsorption–desorption cycles

7. Atomic force microscopy

When studying the calyxarene films morphology with atomic force microscopy, we used the tapping mode. The films 50–100 nm thick have been sputtered onto a cleaned polished silicon surface. Treatment of samples in saturated vapors of chloroform or toluene has been performed by placing them in a sealed container with a free surface of the evaporated liquid. We used the following two types of treatment: (i) cyclic exposure to molecular vapors (for 10s) and pure air (for 5 min.) and (ii) long-term (for 15, 30 and 60 min.) treatment. The surface areas at three points of the sample surface were scanned; their sizes were from 1×1 to 10×10 μm. Shown in Fig.8 are the patterns of tert-butyl[4,6,8]calyxarene films taken both in the initial state and after various exposures to saturated vapors of chloroform and toluene molecules. è òîëóîëà. Fig.8a demonstrates the starting film surface whatever the type of molecules. It should be noted that the statistical parameters of the starting films are close to those for gold films on a smooth surface. These are randomly located hill-shaped objects. They may be characterized by a root-mean square height of irregularities and grain size in the substrate plane. The device Nanoscope III is equipped with software that enables one to obtain the values of these parameters. They are given in Table 2 along with comments on the visually observed general patterns.

Shown in Fig.8b, c are the Ñ/4/À surfaces after exposure to saturated vapors of chloroform and toluene, respectively. (Such an extreme action has been chosen because cycling and treatment for shorter times resulted in slight changes of surface morphology only.) One can see that this treatment leads to appearance of grooves (whose maximal depth is 15–18 nm) at a protruding surface that form a plateau. The groove depth is about 20% the total film thickness (100 nm). According to a rough estimation, the ratio between the groove and plateau areas is about 1/4. Contrary to the above, the effect of toluene molecules (Fig.8c) is in formation of a set of randomly placed "hills" whose maximal height is 30–35 nm, i.e., twice the depth of grooves at the previous pattern. Ñ/6/À demonstrates higher lability. In this case changes in surface morphology appear even at more "soft" treatments, such as cycling in solvent vapors. This cycling results in appearance of a random hilly pattern. The hills are rather high (maximal height of 30–35 nm), with bases of 80–100 nm. This means that the pyramid side makes the angle of about 30° with the substrate plane. It should be noted that long-term treatment

Table 2

NN	Materi al	Treatment conditions	Mean roughness	Mean grain plane size	Commentary
1	C[4}A	none	3-4 nm	20-50 nm	Random hills
2	C[6]A	none	3-4 nm	25-55 nm	Random hills
3	C[8]A	none	3-5 nm	25-50 nm	Random hills
4	C[4}A	Chloroform cycled	10-15	200-400 nm	Platoes separated narrow ravine
5	C[4]A	Toluene cycled	15-20 nm	500-700 nm	Random hills
6	C[6]A	Cloforofm cycled	30-40 nm	200-300 nm	Random hills
7	C[6]A	Toluene cycled	8-10 nm	80-100 nm	Random hills
8	C[6]A	Chloform treated	4-5 nm	60-70 nm	Random hills
9	C[6]A	Toluene treated	40-50 nm	600-800 nm	Systems of titled crystals column-like
10	C[8]A	Chloform treated 15 min			System of lines under 45 degree in the plane
11	C[8]A	Chloform treated 30 min			System of lines under 45 degree in the plane
12	C[8]A	Chloform treated 60 min	100 nm	1000-1200	Part of band consisted of cubic crystals array
13	C[8]A	Toluene treated 15 min	4-6 nm	600-800 nm	Parallel band distanced at 800-1000 nm
14	C[8]A	Toluene treated 15 min	50-60 nm	100 nm	System of columns separated narrow gaps
15	C[8]A	Toluene treated 60 min	140-180 nm	200 nm	System of tilted columns

without cycling results in a more gently sloping relief. In this case the pyramid side makes the angle of about only 15° with the substrate plane due to smaller (7–10 nm) hill heights.

Exposure to toluene vapors demonstrates more interesting and dramatic behavior – see Fig.8f, g. Cycling results in a shallow (80–100 nm) and not so high (8–10 nm) relief. Contrary to this, long-term exposure leads to appearance of pronounced self-organized "mountain systems" of a peculiar closed or half-closed form. Vivid imagination could distinguish here "ridge" orientations making angles of 90 and 45° with each other. More close inspection (taking into account scale distortion) reveals that these "ridges" are made from thin plane plates that make the angle of about 30° with the substrate plane. These formations are rather high (40–55 nm); the "ridge" widths are from 800 up to 1000 nm.

Ñ/8/À is a material that demonstrates the highest structural lability. Even short-term exposure to chloroform vapors results in appearance of a self-organized set of lines that are an aggregation of ridges – see Fig.8h. These ridges are formed due to drift of the neighboring material that starts from movement of the surface molecular layers toward the condensation centers. The material is taken out of the "inter-ridge" areas from depths no more than 10–15 nm (this is about 10–15% the total film thickness).

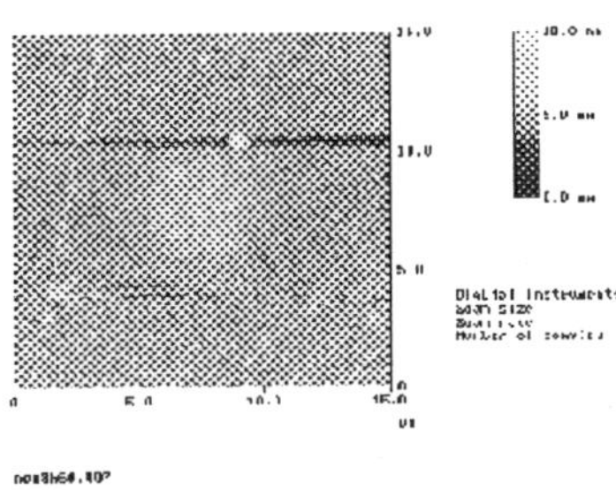

H

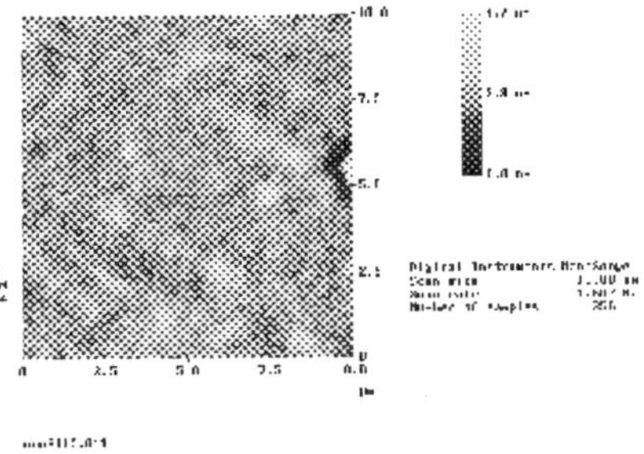

I

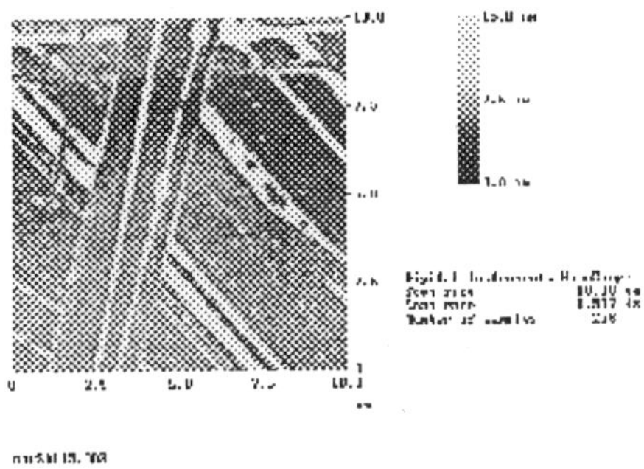

J

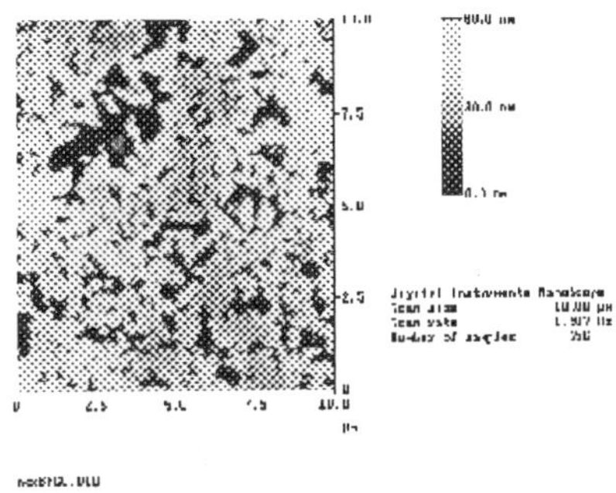

K

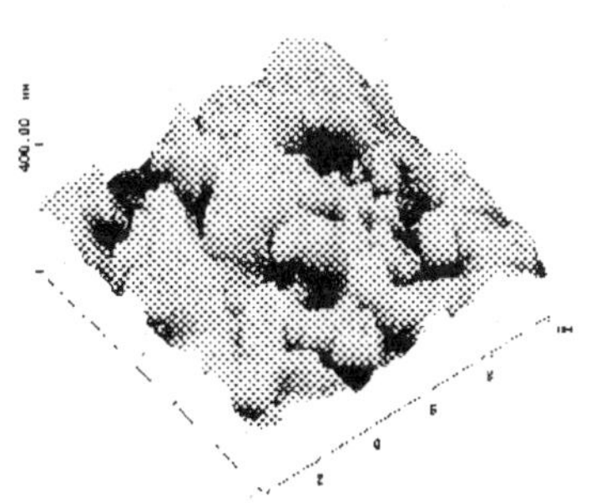

L

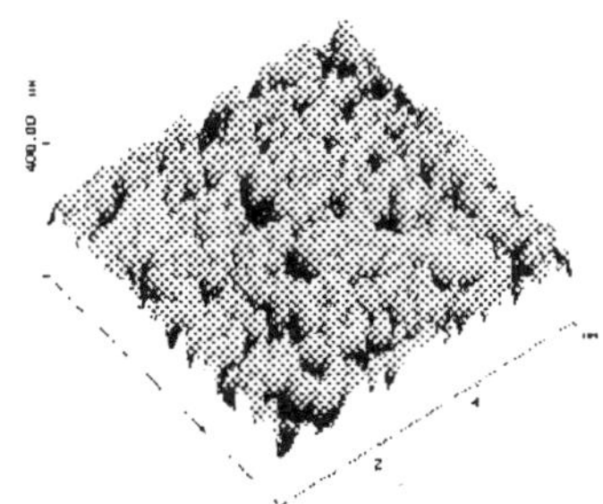

M

Figure 8 Continue

Increasing of treatment duration favors the formation growth. In this case a pronounced set of bands is formed. They make the angle of 45° with each other in the surface plane. Further increase of the time of exposure to chloroform vapors makes individual bands to coalesce into crystallites with right angles (fig.8i). Making the vertical and horizontal scales equal enables one to conclude that these crystallites are parallelepipeds that are prolate along the surface. Their thicknesses lie between 100 and 150 nm, while the parallelepiped area is 500×500 nm.

Effect of toluene on the C/8/A film somewhat differs from that of chloroform. Relief formation (after several tens of seconds) starts as appearance of small (2 to 5 nm) ridges on the mean surface. These ridges are oriented as parallel line segments that sometime intersect at right angle. Then formation of individual plane hills begins. Their form is close to square measured 500×500 nm. When the time of exposure to toluene grows up to 15 min., then individual "wells" (whose depth is up to 60 nm) appear. The material moves from them to the vertical prisms with square (side of about 120–150 nm) bases and height of about 60 nm. The area of wells is about 1/4 of the total area. Increase of the time of exposure to toluene is accompanied with growth of the vertical prisms. Their height increases up to 100–120 nm; the ratio between their area and the total area grows up to 0.5. The base form remains square with side of about 150 nm. All results obtained by AFM are summarised in the Table 2.

8. Quartz crystal microbalance data

The experimental frequency variation for the array of sensitive elements versus time curves taken for three perfumes are shown in Fig.9. An analysis of the sensor array response made it possible to note the following principal regularities in the kinetic curves behavior: (i) for all the sample types these curves are monotone with flattening out at about 1000 s; (ii) both absolute values of frequency variation and its growth rate for individual sensors are determined by the sample type; (iii) for sensors with sensitive coatings based on C[6]A, C[8]A the signal magnitudes are maximal, contrary to sensors modified with C[4]A (Fig.10). Usually the kinetic curve has at least two different portions, namely, fast (until about 500 s) and slow. Typical responses of sensors after change of gas flow from pure air to gaseous sample have two different parts – fast and slow components of relaxation curve during the experiment (about 20 min). It is necessary to stress that irreproducible and intricate kinetic behavior of the response curve during the first three to five (after evaporation) measurement cycles has been observed. After that the response stays stable and can be used for correct data processing.

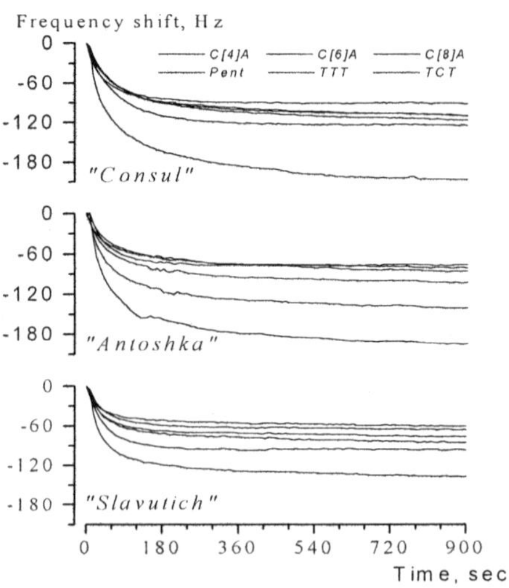

Figure.9 Thypical frequency change after starting exposition in perfumes vapor

9. Discussion

The data obtained by SPR demonstrated change of effective film thickness during adsorption and desorption of small toluene and chloroform molecules. Significant change of solid-state volume is quite uncommon effect when relatively small molecules are adsorbed. As a rule, we can observe this effect for either the polymer substances or biological objects. It is known as swelling. For solids the same effect takes place only in the case of outside irradiation by particles with quite high energy (implantation), or at high temperatures in the laminated structure substances (intercalation). The effect of metal swelling due to the action of high-power radioactive irradiation is also known. It seems likely that the last two effects cannot be used to explain the observed thickness changes of the calyxarene films under the action of solvent vapors, since they were treated neither thermally nor by irradiation. So we deal with the effect similar to reversible swelling of polymer films.

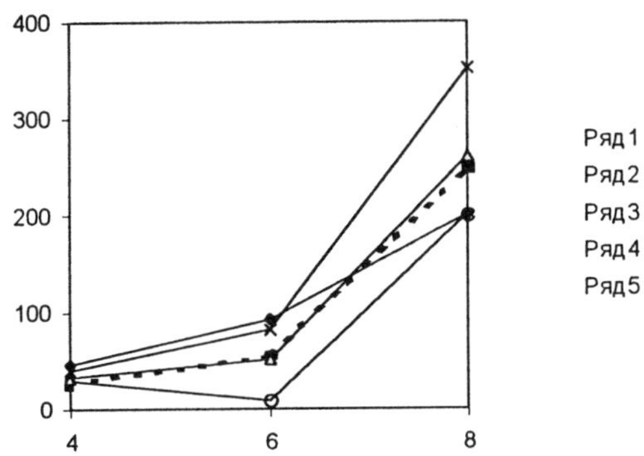

Figure.10 Comparison of responses tertbuthyl[4,6,8]arenes on the action different perfumes (1-3), ethanol (4) and water (5).

We use classical theory of refraction to analyze a relationship between the refractive index and thickness of calyx[n]arene films. It is known (Feinman, 1963) that

$$R_a = 3\frac{n_a^2 - 1}{n_a^2 + 2} - 3\frac{n_i^2 - 1}{n_i^2 + 2}\frac{d_i}{d_a}$$

where a is the polarizability, N is the volume concentration of molecules, and n is the refractive index. We assume that refraction is additive and the surface molecular concentration of calyx[n]arene does not change after the solvent adsorption. Then we can determine a relation between layer thickness, its refraction index and the additional adsorbent refraction R_a, where i and a correspond to film parameters before and after adsorption, respectively. The equation takes into account increasing of film thickness and decreasing of calyxarene concentration. Since additional refraction R_a is caused by toluene presence, it can be determined as $R_a = N_t\alpha_t$, where index t indicates toluene and the α_t is toluene polarizability. We can calculate the toluene concentration in the sample under the adsorption-recovery processes. Figure. 4 shows the data for calyx[4]arene. Note that the reversible (after three adsorption–recovery cycles) values of N_a are equal to 3.4×10^{20} and 1.5×10^{20} for adsorption (point 6 in Figure. 4) and recovery (point 7 in Figure. 4), respectively.

In order to determine the volume concentration of calyx [4] arene molecules from the refraction measurements, we have to know the value of calyxarene polarizability. Toluene polarizability is 2.35×10^{-22} cm^{-3} (CRC Handbook, 1993-94). A calyx[4]arene molecule consists of four toluene molecules, four fragments of CH_2 and OH. By assuming that calyxarene polarizability is a sum of those for its ingredients, we can evaluate the calyxarene polarizability: 1×10^{-21} cm^{-3}. Calculated concentration of absorbed concentration of toluene molecules are shown in Figure. 11.

The volume concentration of calyxarene molecules is about 4.68×10^{20} cm^{-3} for the initial sample whereas the maximal concentration of adsorbed toluene molecules is 3.4×10^{20} cm^{-3}. The authors assume that such an agreement between the values indicates at the adsorption of toluene molecules into calyxarene cavities.

The equilibrium system state after recovery (point 3 in Figure. 3) is not identical to the initial system state (point 1 in Figure. 3). Data in table 1 (rows 1, 3) also provide support for this view. After three adsorption-recovery cycles kinetics of φ_{min} change becomes reversible (points 8, 9 in Fig. 3). As the adsorption film thickness increases, the refractive index decreases and vice versa in the case of recovery. Note, that the system returns to the initial parameters values (d = 50 nm, n = 1.5) after several adsorption-recovery cycles. This suggests that some film ordering mechanism exists for the films deposited under stretched state.

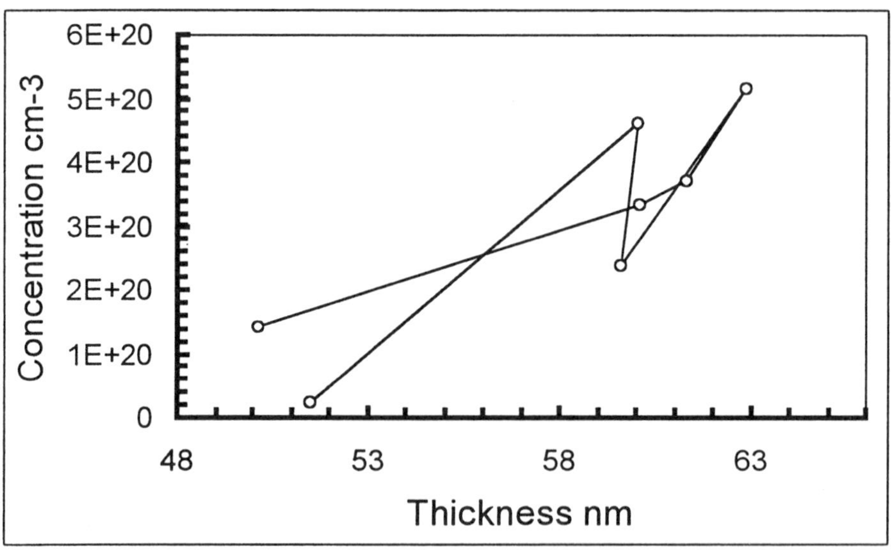

Figure 11. Correlation between C[6]a film thickness and conceuntrattion of absorbed molecules of toluene

We propose to take the first three adsorption- recovery cycles as a structural relaxation of thermally deposited films to the stationary state. Actually during the adsorption- recovery cycles the film surface became matted, hence film roughness increases and film thickness is not well defined. So, the data obtained from SPR measurement are the effective ones. In this way we have to relate the first stage of calyxarene film structure relaxation to the formation of microcrystals or some other ordered structures in the initially strained film.

Probably, together with microcrystals formation a guest-host complex is also formed at the first stage of structural rearrangement. This leads to changes of both refraction and film thickness. The process has to remain after formation of a stationary configuration of calyxarene film (after 3 adsorption- recovery cycles). The agreement in order of magnitude between the adsorbed toluene molecules amount and the number of calyxarene cavities supports this suggestion.

Concerning non-monotone kinetic shown on the figure.5 and 7d it would said follow. Figure. 5 shows kinetics of the SPR-curve minimal angle φ_{min} change during the adsorption- recovery cycles for calyx[8]arene in the toluene vapors. First toluene adsorption leads to sharp (for several seconds) and monotone increase of φ_{min}. The system recovery leads to non-monotone φ_{min} change. At first φ_{min} becomes smaller than that in the initial state for several seconds and then relaxes slowly approximately to the initial value.

The equilibrium system state after recovery (point 3 in Figure. 5) is not identical to the initial system state (point 1 in Figure.5). Data in Table 1 (rows 1, 3) also provide support for this view. After three adsorption–recovery cycles kinetics of φ_{min} change becomes reversible (points 8, 9 in Figure. 5). As the adsorption film thickness increases, the refractive index decreases and vise versa in the case of recovery. Note that the system returns to the initial parameters values (d = 50 nm, n = 1.5) after several adsorption-recovery cycles. This suggests that some film ordering mechanism exists for the films deposited under stretched condition.

Actually the nontrivial behavior of C[6]A films was observed in experiments with planar polarization interferometer (Figure. 7). The reverse branch of adsorption–desorption cycle shows clearly defined minima and practically full return to the state after adsorption. (We can note however that absolute changes of film thickness in the case of ethanol are much smaller than in the cases of toluene or chloroform).

One of possible explanation may be as follows. The first three adsorption–recovery cycles lead to structural relaxation of thermally deposited films to the stationary state. Actually during the adsorption–recovery cycles the film surface becomes matted, hence film roughness increases and film thickness is not well defined. So, the data obtained from SPR measurement are the effective ones. In this way we have to relate the first stage of calyxarene film structure relaxation to the formation of microcrystals or some other ordered structures in the initially strained film.

Probably, together with microcrystals formation a guest–host complex is also formed at the first stage of structural rearrangement. This leads to changes of both refraction and film thickness. The process has to remain after formation of a stationary configuration of calyxarene film (after three adsorption–recovery cycles). The agreement in order of magnitude between the adsorbed toluene molecules amount and the number of calyxarene cavities supports this suggestion.

The above facts serve as evidence that structural relaxation of solid organic films of the calyxarene type does occur. As a result, the surface/volume ratio for the film material increases. This fact should make a favorable effect on the adsorption capacitance of a sensing layer. Concurrently with it, the kinetics of sorbent response to the sorbate pressure front is to become substantially faster. It goes without saying that these are the positive results of calyxarene films cycling in the solvent vapors. At the same time the sorbent parameters might be treated as variable, since each new contact with an analyte can lead to further structural relaxation. Taking into account considerable potentialities of the discussed materials for various applications in the future, one should conclude that further investigations and search for the ways to overcome the discussed difficulties are needed.

On the other hand, treatments of stressed films of complex chemical substances in the solvent vapors could be used for oriented crystallization and formation of layers with the required surface/volume ration that would be optimal for certain guest–host pairs.

Examination of Figure. 8c, d shows that surface morphology of Ñ/4/À films does not considerably change; the films do not demonstrate formation of edged crystallites.

Similar calculations have been performed for some other calyxarenes. It was found, in particular, that the adsorption capacitance of 6 and 8 calyxarenes essentially exceeds that of Ñ/4/À (see Figure.10). Therefore in this case an assumption that the 1:1 complexes are formed at adsorption of guest molecules should be supplemented with an evident statement that vapor molecules can be sorbed not only in specific microcavities but also in other pores that are certain to be present in thermally sputtered stressed films of organic molecules. Such pores are not chemically determined and can house molecules according to the mechanisms of weak molecular interaction – from dispersion mechanism to condensation in pores. This statement is induced by the results

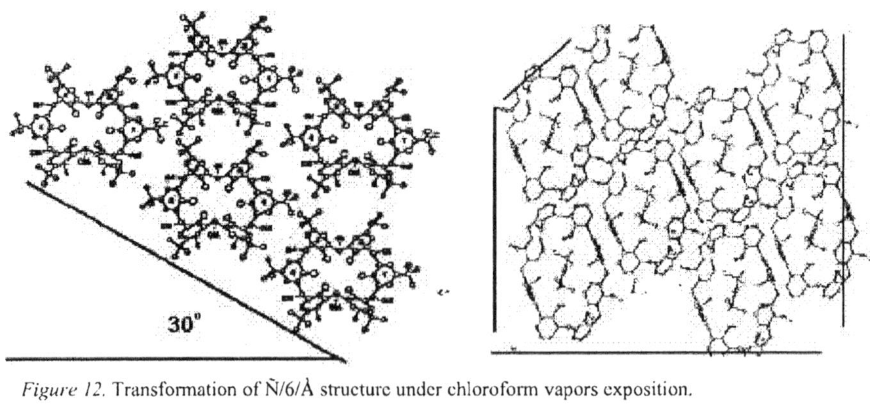

Figure 12. Transformation of Ñ/6/À structure under chloroform vapors exposition.

of atomic force microscopy studies. As discussed earlier, in the limiting case the adsorption–desorption cycles and long-term treatments form from amorphous sediment systems of microcrystals whose height is twice the initial thickness. At the same time areas appear that are free of the film material. Such a process of typical crystallization has to be accompanied with considerable transfer of calyxarene molecules. A driving force for such mass transport seems to be stresses resulting in a corresponding chemical potential gradient. In this case adsorption of a considerable amount of solvent molecules serves as lubricant. Before being fixed in a specific micropore, a guest molecule experiences a stage of condensation at the micrograin boundaries, thus making their orientation easier.

The Ñ[6]À complexes with some molecules are known to crystallize in the space group *P*1. It corresponds to an inclined parallelepiped one angle of which is close to 110° (Vicens et al., 1991). One can see from Fig.8 that Ñ/6/À treatments in toluene and chloroform vapors during maximal times of exposure lead to appearance of objects whose symmetry is close to hexagonal. The assumed microstructure of such objects is presented in Fig.12.

Contrary to the previous case, Ñ/8/À films exposed to toluene vapors crystallize into microcrystals of cubic symmetry. It follows from (Vicens et al., 1991) that Ñ/8/À molecules can form 1:2 complexes with chloroform and toluene. In this case a crystal structure may appear that is prolate along one of the hexagonal axes, and four molecules can form a square subunit. Bearing this in mind, one could assume that formation of the objects shown in Fig.8i, j at interaction between the “guest–host” molecules occurs due to growth of crystallites whose unit cell is formed by four joined 1:2 complexes.

The results obtained could elucidate kinetics of the mechanism for microcrystallites formation.

10. Conclusion

Simultaneous investigations of thickness and refractivity index of thin thermally deposited films of some calyxarenes during treating in solvent vapors have demonstrated increasing of the first and decreasing of the second parameter by adsorption and opposite change by desorption. Refractivity index increasing after adsorption is not surprising and was quantitative explained as enhancing of electron density of solid. But nearly reproducible changing of film thickness was unexpectedly reached +/-20% for C[6]A and C[8]A films. Such effect can't be explained in a simple way taking into account that C[n]A is the kind of solid covalent binding matter. The key for comprehension of this situation is producing by investigations of treated films surface morphology in nanoscale region. The facts taken as a whole suggest that interaction of thermally deposited calyx[n]arene with saturated vapors of dissolvent (toluene, chloroform) leads to both reversible and irreversible changes in the refractive index and film thickness. Irreversible changes can be related to the reconstruction of film microstructure whereas reversible changes are a consequence of guest-host complex formation typical for calyxarene.

Together with guest–host complex a microcrystals formation is also formed at the first stage of structural rearrangement. In the limiting case of the solvent vapor treatment systems of microcrystals forms from amorphous sediment. Such a process of typical crystallization has to be accompanied with considerable transfer of calyxarene molecules. A driving force for such mass transport seems to be stresses resulting in a corresponding chemical potential gradient. In this case adsorption of a considerable amount of solvent molecules serves as lubricant. Before being fixed in a specific micropore, a guest molecule experiences a stage of condensation at the micrograin boundaries, thus making their orientation easier.

11. Acknowledgments

The authors are grateful to E. Matsas and S. Zynio for their help in preparation of samples and instruments, O. Telezhnikova and I. Kruglenko for performing measurements.

The work was partially supported by the Scientific & Technical Programs of Ukraine "Sensorics" and "Certification", as well as the INCO–Copernicus Grant N IC15-CT96-0818

References

1. Beketov G.V.,Shirshov Yu.M., Shinkarenko E.V., Chegel V.I., 1998. Surface plasmon resonance spectroscopy: prospects of superstrate refractive index variation for separate extraction of molecular layer parameters. Sensors and Actuators B48, 425-432.

2. Bidan G., 1992. Electroconducting conjugated polymers: new sensitive matrices to build up chemical or electrochemical sensors. A review. Sensors and Actuators B6, 45-56.

3. Carotta M.C., Martinelli G., Crema L., .Malagu, Merli M., Ghiotti G., Traversa E., 2001. Nanostructured thick-film gas sensors for atmospheric pollutant monitoring. Sensors and Actuators B 76, 336-342.

4. CRC Handbook of chemistry and physics, CRC Press Boca Raton, 74-th Ed. 1993-94.

5. Dickert F.L., Bruckdorfer Th., Feigl H. et al., 1993. Supramolecular detection of solvent vapours with QMB and SAW devices. Sensors and Actuators Â13-14, 297-301.

6. Dickert F.L., Halikias K., Hayden O., Piu L., Sikorski R., 2001. Sensors based on fingerprints of neutral and ionic analytes in polymeric materials. Sensors and Actuators B 76, 295-298.

7. DiNatale C., Paolesse R., Macagnano A., Mantini A., Mari P., D'Amico A., 2000. Qualitative structure-sensitivity relationship in porphyrins base QMB chrmical sensors. Sensors and Actuators B 68, 319-323.

8. Dodd G.H., Bartlett P.N., Gardner J.W.,1992. Odours, the stimulus for an electronic nose, in Sensors & Sensory Systems for an Electronic Nose (eds. J.W.Gardner and P.N.Bartlett). NATO ASI Series E: Applied Science 212, pp. 1-12.

9. Feinman R., Leighton R., Sands M., 1963. The Feinman lectures on physics, Edison-Wesley Publishing Company. Inc, London, p. 64.

10. Fredriksson C., Kichiman S., Rodahl M., Kazemo B., 1998. The piezoelectric quartz crystal mass and dissipation sensor. Lengmuir 14, 248-251.

11. Gardner J.W., Pearce T.C., Friel S. et al.,1994. A multisensor system for beer flavour monitoring using an array of conducting polymers and predictive classifiers. Sensors and Actuators B18-19, 240-243.

12. Goepel W., 1994. New materials and transducers for chemical sensors. Sensors and Actuators B B18-19, 1-21.

13. Guadarrama A., Rodrigues-Mendez M.L., de Saja J.A., Rios J.L., Olias J.V., 2000. Array of sensors based on conducted polymers for the quality control of the aroma of the virgin olive oil. Sensors and Actuators B 69, 276-282.

14. Gutsche C.,D. In: J.F.Stoddart (Ed.), 1989. Calixarenes: Monographs in supramolecular chemistry, Royal Sosiety of Chemistry, Cambridge, pp.90-97.

15. Hartmann J., Hauptmann P., Levi,S, Dalcanale E., 1996. Chemical sensing with cavitands: influence of cavity shape and dimensions on the detection of solvent vapors. Sens. Actuators B35, 154-157.

16. Hatfield J.F., Neaves P., Hiks P.J. et al., 1994. Towards an integrated electronic nose using conducting polymer sensors. Sensors and Actuators B18-19, 221-228.

17. Hawkins T.M., Travers P.J.,1999. Counter-ion influence on the response of conducting polymer gas sensors. Proceedings of 6-th International Symposium ISOEN99, Tuebingen, pp.113-116.

18. Hsuing C.M., Munoz B., Li J., Vicic M, Koper O., 1999. Relationship between the physical structure of a polymer composite based electronic nose and its response. Proceedings of 6-th International Symposium ISOEN99, Tuebingen, pp.150-153.

19. Janshoff A., Steinem C., Michalke A., Henke C., Galla H.-J., 2000. Monofunctionalized b-cyclodextrins as sensor elements for detection of small molecules. Sensors and Actuators B 70, 243-253.

20. Kostioukevich S.A., Shirshov Yu.M., Matsas E.P., Stronski A.V., Subbota Yu.V. Chegel V.I., Shepelyavi P.E., 1995. Application of surface plasmon resonance for the invesyigation of ultrathin metal films. Proc.SPIE v.2648, pp.144-151.

21. Kriz D., Ramstrom O., Mosbach K., 1997 (June). Molecular imprinting: new possibilities for sensor technology. Anal.Chem.News&Features, pp.345a-349a.

22. Krummel Ch., Krage R., Reichl D., Gauglitz G., 1999. VOC detection by means a new polymer-based optical transducer. Proc.of 9-th Intern. Trade Fair and Conference for Sensors, Transducers and Systems, Numberg vol.1 pp. 139-143.

23. Kukla A.L., Shirshov Yu.M., Piletsky S.A., 1996. Ammonia sensors based on sensitive polyaniline films. Sensors and Actuators B37, 135-140.

24. Kunsagi-Mate S., Nagy G., Kollar L., 2001. Calculation of energy compexes host-guest Investigation of calixarene (host) and neutral benzotrifluoride (guest). Sensors and Actuators B76, 545-550.

25. Lucklum R., Hauptmann P., 2000. Response of quartz-crystal microbalance: mass sensitivity, viscoelacticity and acousctic amplification. Sensors and Actuators B 70, 30-36.

26. Malmsten M., 1994. Ellipsometry studies of protein layers adsorbed at hydrophobic surfaces. Journ. of colloid and interface science 166, 333-342.

27. Nabok A.V., Lavrik N.V., Kazantseva Z.I.,Nesterenko B.A., Markovskiy L.N., Kalchenko V.I., Shivaniuk A.N., 1995. Thin Solid Films 259, 244-247.

28. Nelli P., Dalcanale E., Faglia G. et al., 1993. Cavitends as selective materials for QMB sensors for nitrobenzene and other aromatic vapours. Sensors and Actuators B13-14, 302-304.

29. O'Sullivan C.K., Guilbault G.G., 1999. Commercial quartz crystal microbalances - theory and application. Biosensors and Bioelectronics 14, 663-670.

30. Rengevych O.V., Shirshov Yu.M., Ushenin Yu.V., Beketov A.G., 1999. Separate determination of thickness and optical parameters by surface plasmon resonance: accuracy consideration Semiconductor Physics, Quantum Electronics & Optoelectronocs 2, 28-35.

31. Schlupp M., Heil C., Koch A., Muller-Albrecht J., Schramm U., Bargon J., 2000. Metal colloids as sensor-active materials for the detection of valatile compounds and solvents. Sensors and Actuators B71, 9-12.

32. Shih J.-S., Chao Y.-Ch., Sung M.-F., Gau G.-J., Chiou Ch.-S., 2001. Piezoelectric crystal mnembrane chemical sensors based on fullerene C60. Sensors and Actuators B76, 347-353.

33. Shirshov Y. M., Snopok B.A., Samoylov A.V., Kiyanovskij A.P., Venger E. F., Nabok A.V., Ray A. K., 2001. Analysis of the reponse of planar polarization interferometer to molecular layer formation: fibrinogen adsorption on silicon nitride surface. Biosensors and Bioelectronics 16, 381-390.

34. Shirshov Yu.M., Svechnikov S.V., Kiyanovskii A.P., Ushenin Yu.V., Venger E.F. and Merker R., 1998. A sensor based on the planar-polarization interferometer. Sensors & Actuators A68, 384-387.

35. Shirshov Yu.M., Gerasimenko V.S., Kunitskij I.E., Kutichina Zh.Ya., Nazarov A.N., 1990. Thermally stimulated depolarization of phosphosilicate glasses. Ukrainian Physical Journ. 35, 544-549.

36. Shirshov Yu.M., Zynio S.A., Matsas E.P., Beketov G.V.,.Prokhorovich A.V, Vtnger E.F., Markovskiy L.N., Kalchenko V.I., Soloviov A.V., Merker R. 1997. Supramol. Sci. 4, 491 - 494.

37. Slavík R., Homola J.,Tyroký J., Brynda E., 2001. Novel spectral fiber optic sensor based on surface plasmon resonance. Sensors and Actuators B: Chemical 74, 106-111.

38. Steiner F.-P., Hirlemann A., Cornila C., Noetzel G., Bacchtold M., Korvink J.G., Gopel W., Baltes H., 1995. Polymer coated capacitive microintegrated gas sensor. Proc.of the 8-th Int.Conf. on Solid State Sensors and Actuators, Eurosensors IX, Sockholm, pp.814-817.

39. Szczurek A., Szekowka P., Lichnerski B.W., 1998. Volatile organic compounds mixtures guantification with TGS sensor array. Proc.of the 12-th European Conference on Solid-State Transducers, Southhampton, pp. 841-844.

40. Torsi L., Dodabalapur A., Sabbatini L. and Zambonin P.G., 2000. Multi-parameter gas sensors based on organic thin-film-transistors. Sensors and Actuators B: Chemical 67, 312-316.

41. Varadan V.K., Gardner J.W., 1999. Smart tongues and smart noses, in Smart Structures and MEMS (ed. V.K. Varadan) . Proceedings of SPIE 3673, 67-76.

42. Vicens, J.; Bohmer, V., 1991. Calixarenes: A Versatile Class of Macrocyclic Compounds; Kluwer, Dordrecht, .www.analytical-microsystems.de

43. Zee F., Judy J.W., 1999. Miniaturization of a carbon black/polymer array electronic nose. Proceedings of 6-th International Symposium ISOEN99, Tuebingen, pp.120-123.

BIOSPECIFIC INTERACTIONS ON THE OPTICAL TRANSDUCER SURFACE – THE BASE OF INFECTION DIAGNOSTICS.

N.F. STARODUB, L.V. PIROGOVA, V.P. ARTYUKH,
V.N. STARODUB
A.V. Palladin Institute of Biochemistry of the Ukrainian National Academy of Sciences, Department of Biochemistry of Sensory and Regulatory Systems 9 Leontovicha Str., 01030 Kiev, Ukraine

Abstract.

An immune sensor based on the surface plasmon resonance (SPR) was developed for express diagnostics of bovine viral leucosis (BVL). Sensor was used for the detection of level of antibodies against bovine leukaemia virus (BLV) in the blood serum. Industrial manufactured BLV antigen for screening test in the agar gel radial immune diffusion (RID) required the additional purification for the use in immune sensor analysis. It was shown that the immune sensor analysis is more sensitive, rapid and simple in the comparison with the traditional RID test. It was stated that the developed immune sensor can be used for performance of screening of BVL at farms and at the minimal dilution of the serum in 1:500.

1. Introduction

BVL is oncological disease of blood and rather of the hemolymphopoetical system. It is characterised by abnormality maturation process of blood cells, which is accompanied by the intensive generation mainly young form cells (Balida et al., 1977). All vertebrate mammalians, birds, fishes subjected to fall leucosis sick. Especially this disease extended among livestock of many countries. Although animals can become infected with BVL at any age, tumour (lymphosarcomas) is seen typically in animals over 3 years old. Infections are usually subclinical; only 30-70% of infected cattle develop persistent lymphocytes, and 0.1-10% develop tumours. Signs depend on the site of the tumours and may include digestive disturbances, neurological manifestations, weight loss, weakness or general debility (`Angelino et al., 1998).

Epidemiological and End Results Program of USA were used to perform a descriptive epidemiological study of leukaemia in Iowa. Data were also collected on livestock population patterns in the state and a survey was performed to locate cases of lymphosarcoma within the cattle population. Ecological relationships between human leukaemia, livestock populations and bovine lymphosarcoma were investigated. There is a high positive correlation between acute lymphoid leukaemia in males and cattle density. This relationship is greater for dairy cattle than for beef cattle. Consequently, it can predicate that leukaemia is potential problem from both public health and economic perspectives (Donham et al., 1980). The feature of the cattle leucosis is duration period of disease without noticeable pathology symptoms. Virus of the type C is retrovirus

E. Buzaneva and P. Scharff (eds.), Frontiers of Multifunctional Nanosystems, 369–376.

family from an oncovirus rod. It is agent of this type of leukaemia. Leucovirus, except for revertase, contains six proteins, of which are superficial (an envelope protein) glycoprotein (gp51) and inside protein (p21) (Matthews, 1979).

Prophylaxis and elimination are complicated in connection with very wide extent this disease. The main cause of an extension is inclusion of ill animal without free occurrence of the leucosis symptoms in healthy herd. Therefore an improvement of the express diagnostic methods for detection of early phase of bovine leucosis is extremely significant and urgent target.

More then 20 various variants from haematological, histological (Balida et al., 1977) and up to use of polymerise chain reaction (PCR) (Fechner et al., 1996; Blankenstein et al., 1998) as well as different variants of immunological methods, for example, classical fluorescence (Ferrer et al., 1972), immune diffusion (Miller et al., 1972), radioimmune analysis (Franz et al., 1986), immunoblot (Fechner et al., 1996) were developed. The immunological methods were based on the detection of antibodies in blood serum, which will be induced by the agent of disease. Usually diagnosis of the oncoviral infection in Ukraine is realised with the use of the RID. This method is simple and inexpensive but that later has low sensitivity as well as it needs a long time for realization. At present, analysis by enzyme linked immune assay (ELISA) method becomes more full-scale applicated (Nguyen et al., 1993; Carli et al., 1999). At the comparison of immune diffusion, ELISA method and western blot it was stated that ELISA method has the number of merits, in particular, it has high specificity and sensitivity, but faults too (Dolz et al., 1999). First of all, in order to realize ELISA method it is necessary to have laboratory conditions, as well as high-qualified personnel. Moreover, this method demands more six hours for realization only, without take into consideration of the time transportation samples. The price of analysis by ELISA method is comparatively high, too. At present, the express and all the more field method for diagnosis of BVL is absent. Therefore, urgent problem is development of new, sensitive, specific and simultaneously simple and fast methods of BVL diagnostics for all stages of the pathology evolution. The creation of the instrumental analytical method on the basis of biosensor principles is one of the alternative approaches. The objective of the presented paper is development of the immune sensor on the basis of phenomenon of the SPR under the principles give an account of Homola et al., (1999) for the express diagnosis of BVL. SPR immune sensor will be able to register level of appropriate antiviral antibodies, which were induced by means of BVL in blood serum. To this end, the BLV antigen would be immobilized on transducer surface and use for binding of specific antibodies.

2. Material and methods.

As the optical transducer it was used the thin gold film, which was made between two mediums. It is glass and test solution, which have different refraction indexes. The polarized laser beam came in medium (special glass prism) with angle at higher of the critical angle and was subjected total inner reflecting. In the circumstances electromagnetic component of the light ray penetrated outside of gold field. Under definite angle incidence the oscillating vector of light wave was coincided with the vector oscillating of free electrons in area layer of gold film. That affects caused excitation of the surface plasmons that was accompanied by consumption part of the beam energy, therefore declined intensity of the reflected light. That effect was denominated as SPR phenomenon. The values of a resonance angle very responses to emergence of the additional layer immobilization of biological material. Thickness of the spraying gold layer on the glass plate was 20 nm. The glass gilt of plate was interconnected with prism by means of immersion liquid (polyphenil ester) with the refraction index of 1.6. Antigen of BLV was received from supernatant fluid of standard culture of FLK cells. BLV antigen was preliminary cleared and concentrated. BLV antigen was received from firm "Leikopol"- Poltava. Supernatant fluid from culture of FLK-cells was subjected to centrifugation and treatment with Triton X 100. The concentrated BLV antigen was used for RID analysis. In immune sensor analysis it was used BLV antigen, which was additional purified with the help of ultrafiltration and sorption on porous glass. In experiment it was utilized 10-fold dilution (in distillate water) of the purified antigen. Cow serums were received from distinct farms of Poltava region (Ukraine). All serums were checked by RID analysis and conditionally distribute on three groups: RID – positive, RID- negative and RID – doubtful, i.e. serums, which are not entail emergence of clear precipitation line in gel. For SPR immune sensor analysis it was prepared different dilutions serums in 0.05-M phosphate buffer, pH 7.4, which include 0.14 M NaCl (PBS). Serums from the immunised cows were received after vaccination those by preparations, which were manufactured by firm "Leikopol". 1% solution of the bovine serum albumin (BSA) from Sigma was prepared and used for blockage of free places on transducer surface. Testing of serums was carried out in the following way. At first, resonant angle was registered after injection in the measuring cell (10 μl) of distillated water. Next, into the measuring cell the solution of BLV antigen was added and it was kept that twenty minutes at room temperature. After that, measuring cell was washed in order to remove antibodies and again to registrate resonant angle. For the elimination of non-specific interaction of the serum components with biological layer on transducer surface 1% of BSA solution was introduced in the measuring cell for 10 min at room temperature. After that, measuring cell was washed with the help of PBS and angle shift was recorded. Later on, transducer surface was sequentially treated with the antiserum solutions at the following dilutions: 1: 20000; 1:10000; 1:5000; 1:1000 and 1:500. Each of samples was incubated for 10 min. Every time the measuring cell was washed with PBS contained 0.1% of Tween-20 to delete unbound components and after that the response of immune sensor was determined. Obtained results were treated by statistically.

3. Results and discussion.

At the beginning of our work it was registered the level of the sorption of the BLV antigen on the transducer surface. In this case we used 10-fold of diluted antigen. Shift of the resonant angle was reached up to 4500-6000 sec. After treatment of the transducer surface by BSA RID-positive antiserum (500-fold dilution) was injected into the measuring cell. It turned out, that resonant angle did not changed. The absence of response is resulted in low concentration of the antigen proteins of gp51 and p24 in the prepared BLV antigen. Therefore, this BLV antigen needs in additional purification to remove non-specific elements and to achieve increasing concentration of gp51 and p24 proteins on the transducer surface. After procedure of antigen filtration through porous glass it was stated that 10-fold diluted solution provided the stable physical sorption of biological molecules on the transducer surface, which did not destroy at the washing of the measuring cell and at its storage. Immobilisation of antigen accompanied layer formation, which resulted in change of resonant angle in the range of 4000-4500 sec. The treatment of the transducer surface by BSA did not effect on the value of resonant angle. Evidently, quantity of free sites on the transducer surface was absent in generally or was minimal and antigen concentration was optimal for formation of dense layer. At the injection of the serum from leukaemia cows in the measuring cell SPR response was correlated with degree of the serum dilution (*Figure. 1*). The prepared BLV antigen allowed to us to determine specific antibodies in blood serum after dilution in 10000 fold.

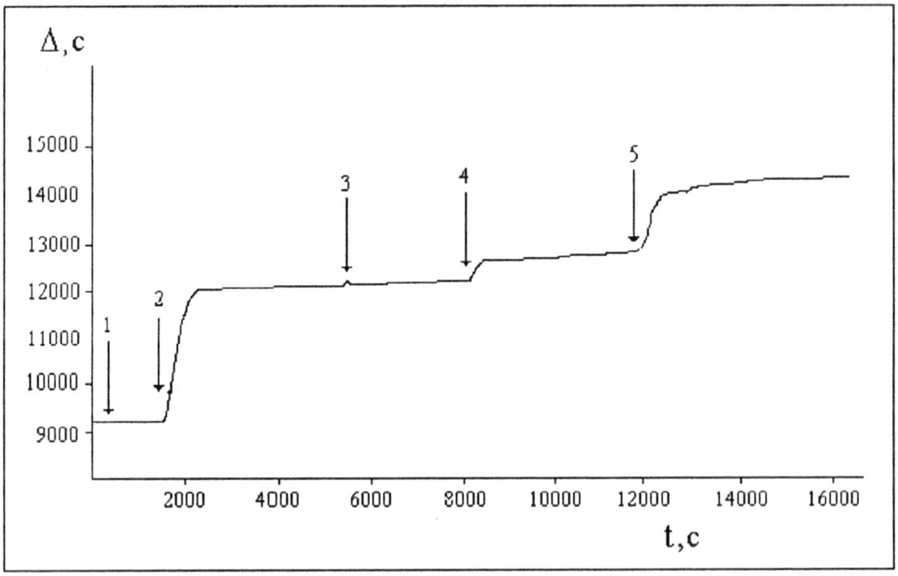

Figure. 1 Sensor output after addition of biological substances in the measuring cell 1 – 5 distillated water, antigen, BSA and antiserum next dilutions From 1: 10000 to 1: 1000, respectively.

It is necessary to pay attention that at this dilution of serum from ill cows RID – analysis given positive results too. Next stage of investigation was the examination of series of cow blood serums from different farms of Poltava region. We analysed blood serums with next dilutions: 1:20000, 1:10000, 1:5000, 1:5000, 1:2000, 1:1000, and 1:500. Magnitude of angle shift had the inverse negative relationship to the degree of serum dilution, i.e. the response was proportional quantity of specific antibodies in samples. It was investigated 25 RID-positive, 18 RID-negative samples and 15 samples, which was conditionally denominated as RID-positive. In *Figure 2* it is presented the

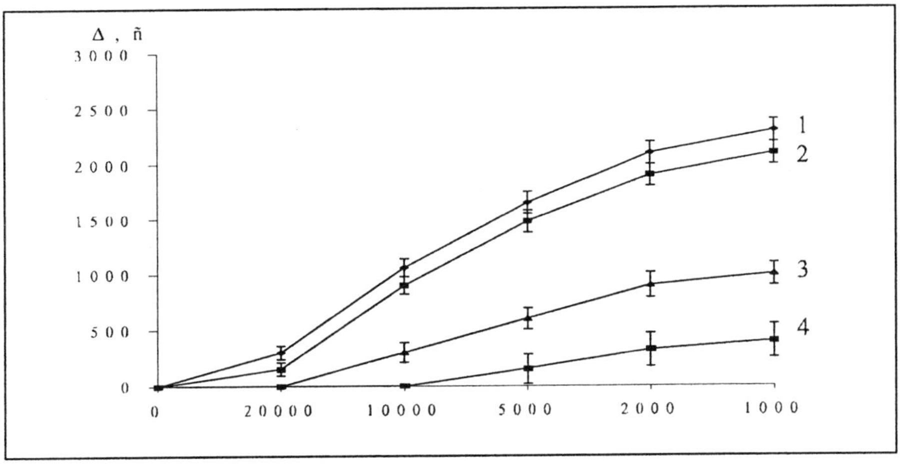

Figure. 2. The shift of resonance angle of the SPR-based immune sensor at the different dilution of cattle serums with various extent of the RID-reply. 1 – 4 -RID (++), RID (+), RID (- +), RID (-), respectively. Abscissa – serum dilution; ordinate - shift of the resonance angle.

average magnitude data, which were received. The 10000-fold diluted serum from ill cows were able to shift resonant angle in the range of 700-1000 sec. RID-doubtful serum samples were demonstrated the presence of the specific antibodies at 1000–500

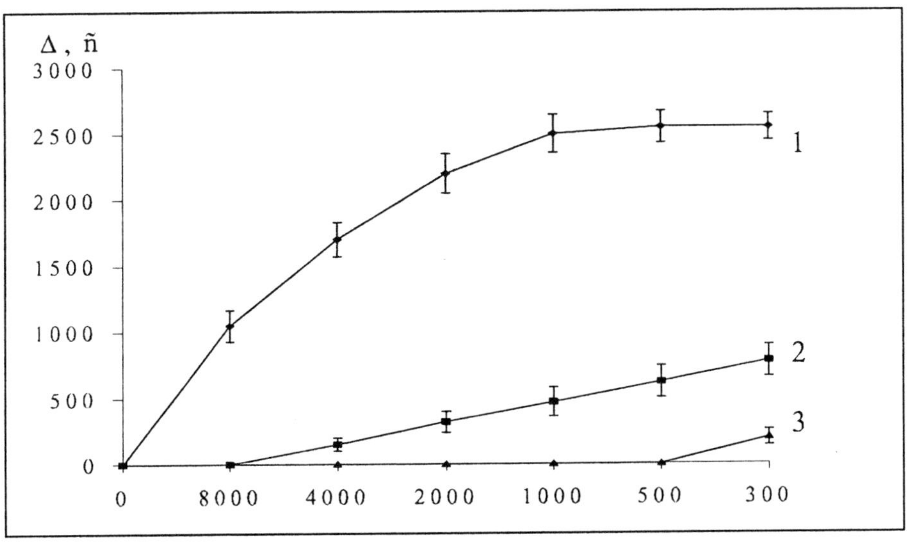

Figure 3. The shift of resonance angle of the SPR-based immune sensor at the different dilution of serums from different cow groups. 1-3 RID - positive, RID – negative and healthy caws, respectively. Abscissa – serum dilution; Ordinate - shift of the resonance angle.

fold dilutions only. The maximal discrepancy between SPR-device and RID-analysis was stated at the investigation of RID-negative samples. In this case it was shown that the shift of resonant angle was within of 100 – 600 sec. Moreover, SPR immune sensor demonstrated existence of resonant angle shift even at 500- fold dilutions when RID analysis was negative. Evidently, SPR immune sensor is more sensitive then RID analysis.

As a control samples it was used cow serums from herds without leukaemia disease (*Figure 3*). The serums caused shift of the resonant angle at 1:300 fold dilution only. Obviously that is result in non-specific interactions, for high concentration of protein components.

It turned out, that serum from vaccinated cows may cause specific signal of the immune sensor even at 1:20000-fold of dilution and this signal has strongly correlated with fact vaccination (*Figure 4*). It is necessary to separately note that common times, which is required for SPR immune sensor analysis equals to 40 min, including time for immobilisation of antigen on transducer surface, blocking free binding sites and washing of the measuring cell. For direct testing procedure of the serums it is necessary to spend no more than 10 min. It is possible to use removable elements - preliminary treated appropriate glass plates. Furthermore traditional RID-analysis demands laboratory condition only and three a days for its fulfilment.

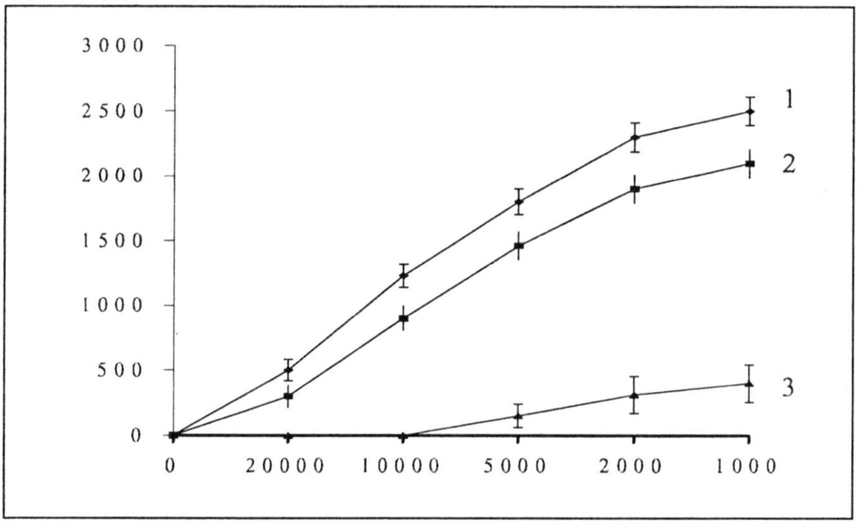

Figure 4. The values of the immune sensor resonance angle at the investigation of serums from RID – positive and vaccinated cows. 1 – 3 serums from vaccinated, RID – positive and RID – negative cows, respectively.

Immune sensor analysis may be fulfilled with the field version of equipment as the compact SPR-device. Such SPR- device with above mentioned properties gives chance to realize fast and inexpensive screening of the leukaemia immediately on stockyards or quarantine stations.

4. Conclusion.

Immune sensor analysis on the basis of the SPR phenomenon corresponds to fast and handy method for diagnostic of cattle leucosis at the determination of level of the anti-leucosis antibodies in serum. Serial prepared BLV antigen, which is produced by biotechnological plants in Ukraine and used for the wide screening in case of employment of RID analysis, needs in additional purification. Immobilisation of the purified BLV antigen on transducer surface of SPR allows to us to determine level of the specific antibodies in blood serum of the ill and vaccinated animals. Immune sensor analysis exceeds sensitivity inherent to one of traditional RID. Minimal serum dilution for the express determination of specific antibodies in the cow serums with the help of SPR immune sensor should be equal no less then 1:5.00.

The presented experimental material is only commencement of the investigations in branch of the using immune sensor analysis for leukaemia diagnostics.

Thereupon it arises many questions, which demand separate and detailed study. In particular, the problems connected with differentiation of situation, which is induced by the leukaemia virus and vaccination, should be clarified. Probability it is arisen by cross-reaction between different types of antibodies, which may be presented in blood at the different organism state. For diagnostics of BVL of cattle would be very interested to use milk or urine. It is necessary to fulfil the comparative examination of the sensitivity SPR- immune sensor and ELISA and others methods.

References

1. Angelino, J.L., Garcia, M. and Birgel, E.H. (1998), "Productive and reproductive performance in cattle infected with bovine leukosis virus", *Dairy Res.* **65**, 693-695.
2. Balida, V.and Ferrer, J.F. (1977),"Expression of the bovine leukemia virus and its internal antigen in blood lymphocytes ", *Proc. Soc. Exp. Biol. Med.* **156**, 388-391.
3. Blankenstein, P. and Fechner, H. (1998)," Possibilities and limitations for use of the polymerase chain reaction (PCR) in the diagnosis of bovine leukemia virus (BLV) infection in cattle", *Dtsch Tierarztl Wochenschr.* **105** , 408-412.
4. Carli, K.T., Sen, A., Batmaz, H. and Minbay, A . (1999), "Detection of IgG antibody to bovine leukaemia virus in urine and serum by two enzyme immunoassays", *Lett. Appl. Mucrobil.* **28**, 416-418.].
5. Donham, K.J., Berg, J.W. and Sawin, R.S. (1980), "Epidemiologic relationships of the bovine population and human leukemia in Iowa", *Am. J. Epidemiol.* **112**, 80-92.
6. Dolz ,G. and Moreno, E.(1999), "Comparison of agar gel immunodiffusion test, enzyme-linked immunosorbent assay and western blotting for the detection of BLV antibodies", *Zentralbl Veterinarmed [B]*. 46, 551-558.
7. Ferrer, J.F., Avila,L. and Stock, N.D. (1972), " Serological detection of type C viruses found in bovine cultures", *Cancer Res.* **32,** 1864-1870.
8. Fechner, H., Kurg, A., Geue L., Blankenstein P., Mewes G., Ebner D. and Beier D. (1996), "Evaluation of polymerase chain reaction (PCR) application in diagnosis of bovine leukaemia virus (BLV) infection in naturally infected cattle", *Zentralbl Veterinarmed* [B]. **43**, 621-630.

9. Franz J., Hampl J., Hofirek B., Skrobak F., Svoboda I.and Granatova M. (1986), "Radioimmunologic detection of antibodies to bovine leukemia virus ", *Vet Med (Praha)*. **31**, 459-468.
10. Homola, J., Yee, S. S. and Gauglitz, G. (1999), " Surface plasmon resonance sensor: review", Sensors and Actuator B. **54**, 3-15.
11. Miller, J. and Olson, C. (1972), " Precipitation abtibody to an internal antigen of theC-type virus associated with bovine lymphosaroma", J. Nat. Cancer. Inst. **49**, 1459 – 1469.
12. Matthews, R.E.F.(1979), "Classification and nomenclature of viruses", *Intervirology*, **12**, 128-296.
13. Nguyen, V.K., and Maes, R.F. (1993) "Evaluation of an enzyme-linked immunosorbent assays for detection of antibodies to bovine leukemia virus in serum and milk", *J Clin Microbiol*. 979-998.

THIN-FILM BIOTERMOSENSORS

A. N. SHMYRYEVA[1], N.F.STARODUB[2]
1-National Technical University of Ukraine "KPI"
Faculty of electronics, Microelectronics Dept., corpus 12,
Pr. Pobeda 37, 03056 Kiev, Ukraine
2- Institute of biochemistry, Kiev, Ukraine

Abstract.

The thin films materials on the basis of nickel, yttrium, gadolinium, ytterbium, and cerium were evaporated and researched. Thin-film temperature-sensitive elements and heat flows sensors, manufactured according to microelectronic technology have linear metrological characteristics within wide temperature range of -150...+500°C, speed of response lesser than 1sec, positive temperature coefficient is $6{,}4\cdot10^{-3}$ 1/degrees, dimensions are (1 –1,5) x(1 –3)x0,5 mm. Strontium oxide entering in quantity of 5-20 % to CeO_2 composition lead to falling conductivity with minimum at 15 weight % of SrO and with following conductivity rise. For this composition it was obtained the maximum temperature resistance coefficient is TRC=$8\cdot10^{-2}$ 1/degree. Oxide cerium entering has also led to energy state altering of semiconductor surface and adsorption activity. The films of CeO_2 –10 % ZnO composition has the biggest temperature sensibility of $5.58\cdot10^{-2}$ 1/degree.

1. Introduction

Thermobiosensors provide the heat energy transformation that is excreted as a result of different biochemical reactions (material – enzyme, antibody – antigen, virus – crate and etc.) to electrical signal on his surface. The main problems in this field are the necessity of high level microminiaturization, high response, localization of heat objection process, increasing of thermo sensibility. This paper is dedicated of the researches in the field of film materials on the basis of nickel composition. Some natural phenomena in the case of introduction of such metals as yttrium, gadolinium, ytterbium, cerium were defined more precisely too.

2. Experiment

Among of known chemical elements the nickel has the highest temperature resistance coefficient of $6.7\cdot10^{-3}$ 1/degree in volume material. This value has altered in the wide range of $(1.2 - 6.7)\cdot10^{-3}$ 1/degree in dependence on pure material, impurity presence, with accounting of high nickel activity as absorbent as was noted by Meixner et al. (1989). The nickel high corrosion hardness in air and same chemical aggression ambient and resistance to oxidation at heating to 600-800 ^{0}C are more important

E. Buzaneva and P. Scharff (eds.), Frontiers of Multifunctional Nanosystems, 377–382.

properties. But the known nickel applications for thin film thermistor development are limited of temperature resistance coefficient equalled of $3.8 \cdot 10^{-3}$ 1/degree and working temperature of 250 ^{0}C receiving that is connected with phase transition. In metal alloys the charge carriers concentration doesn't depend on temperature. So the alloys electro-conductance temperature dependence is entirely determined of charge carrier mobility dependence on temperature.

The 1% addition of yttrium is efficient for increase of alloy refractoriness; at the same time fine-grain structure of the film after long-duration heating is retained. The introduction into the composition gadolinium's nickel in prescribed quantities has ensured a solution of the problem aimed at increasing of surface resistance and thermosensibility of the film. At 3 % gadolinium intake the specific resistance increased from 2.4 to 3.4 Ω/□ at simultaneous increasing of relation's resistance at 100 ^{0}C to resistance at 0 ^{0}C, videlicet from 1.34 to 1.44. The graduated characteristics has clear-cut linear dependence in the temperature range from –70 to 400 ^{0}C. The alloying of ytterbium nickel provides an improvement in plasticity of the film. The incorporation of ytterbium made it possible to increase TCR by 25-30 % under the extension of operating temperature interval up to -150...+600°C.

The principal relationships of electrophysical and thermophysical properties of synthesised materials due to technological processes parameters of vacuum electronic-beam and magnetron sputtering have been determined, namely: speed of deposition (Fig. 1), substrate temperature, temperature and time of annealing in vacuum. The temperature resistance coefficient dependence studying on substrate temperature demonstrated that substrate temperature rise from room to 300 ^{0}C has not given significant temperature resistance coefficient altering. Its maximum is observed at temperature of 200 ^{0}C (of $4.4 \cdot 10^{-3}$ 1/degree). However, at temperature higher of 180 ^{0}C deposited nickel film had very big mechanical strains up to needle structure formation. The optimum temperature value at film nickel evaporation is of 150-170 ^{0}C. At annealing immediately after film process evaporation in vacuum facility the surface resistance increased with insertion on saturation part (Fig. 2).

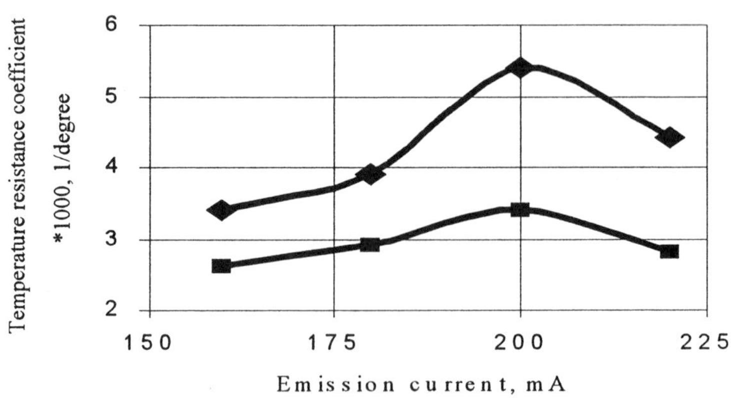

Figure 1.: Ni-Gd film thermosensibility dependence on electron beam emission for various film thick d, where: 1- d=0.4 μm; 2 – d=0.2 μm.

This work is dedicated of thin film system studying based on CeO_2-SrO, CeO_2-ZnO, CeO_2-EuO. The thin films of indicated systems are formed by vacuum evaporation from original powder-like materials in special proportions of composition components ("flash" method). This method supplies the film formation of specified stoichiometric composition with amorphous structure. Having evaporating on substrates that have the built-in positive surface charge from silicon oxide the films has close-grained structure and semiconductor properties.

The CeO_2, EuO and couple CeO_2/EuO were used as source materials. The substrate temperature during the evaporation was 160°C and 200^0C. The light-doses of evaporated materials reached the tungsten filament with definite frequency supplied for the keeping of stoichiometric composition of double systems. The contact system was formed by titanium and nickel system evaporating throw a mask.

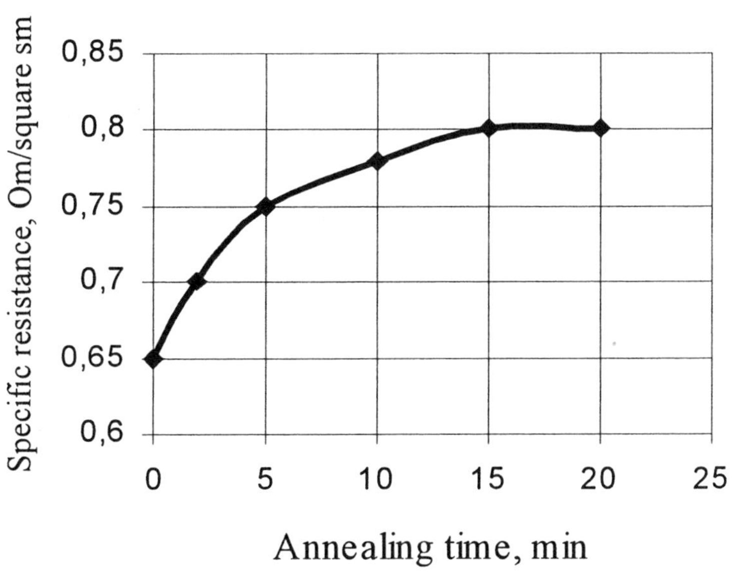

Figure 2. : Ni-Gd film surface resistance dependence on annealing time.

From electrical properties the oxides based on CeO_2 is wide band gape semiconductors. Strontium oxide entering in quantity of 5-20 % to CeO_2 composition lead to falling conductivity with minimum at 15 weight % of SrO and with following conductivity rise. For this composition were obtained the maximum temperature resistance coefficient TRC=$8 \cdot 10^{-2}$ 1/degree.

The films of CeO_2 –10 % ZnO composition had the biggest temperature sensibility of $5.58 \cdot 10^{-2}$ 1/degree. These films were obtained by flash evaporation of powder materials too. On inducing of zinc oxide we have got the changing of energy states semiconductor surface, conductance actions, absorption activity, photosensitivity (Table 1.) as was shown by Shmyryeva et al. (1998).

Table 1.

Composition	Voltage-current characteristic β - coefficient unlinierity	Temperature (T) influence. Function of temperature coefficient resistance dependence, R_0, T_0-original values	Illumination effect $I_{photocurrent}/I_{dark}$	Moisture effect $I_{moisture}/I_{dry}$
CeO_2	$I=AU^{\beta}$ 1.04	$R=R_0 exp(T_0/T)$ $2.63 \cdot 10^{-2}$	0.6	1.8
$ZnO+CeO_2$ (10% 90%)	I=RU	$R=R_0-AT$ $5.58 \cdot 10^{-2}$	0.1	1.0
$ZnO+CeO_2$ (20% 80%)	$I=AU^{\beta}$ 1.09	$R=R_0 exp(T_0/T)$ $3.5 \cdot 10^{-2}$	1.0	2.1
$ZnO+CeO_2$ (50% 50%)	$I=AU^{\beta}$ 1.37	$R=R_0 exp(T_0/T)$ $2.25 \cdot 10^{-2}$	2.35	1.7

3. Results

The ZnO (10 %) – CeO_2 (90 %) films have the most thermosensivity of $5.58 \cdot 10^{-2}$ $1/^0C$ and linearity dependence with negative temperature coefficient what have permitted to create the thin-films temperature sensors. These sensors different from another due to own high speed response (less than 0.5 sec), unique sensibility, manufacturability and linearity of metrology characteristics. For another system CeO_2-ZnO compositions the temperature dependencies are exponential what are explained by thermal stimulated tunnelling through parabolic potential barrier. Current-voltage characteristics of films CeO_2/EuO were investigated. They have unlinear character and could be approximated by power law dependence (fig.3).

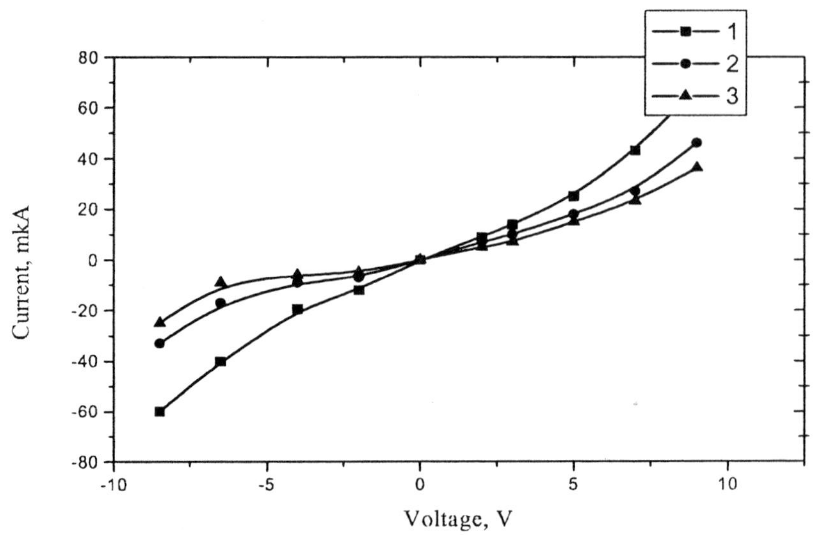

Figure 3. Current-voltage characteristics of EuO-CeO_2 films at substrate temperature of 200 ^0C under different content of EuO: 1- 10 % EuO; 2 –40 %EuO; 3- 60 % EuO.

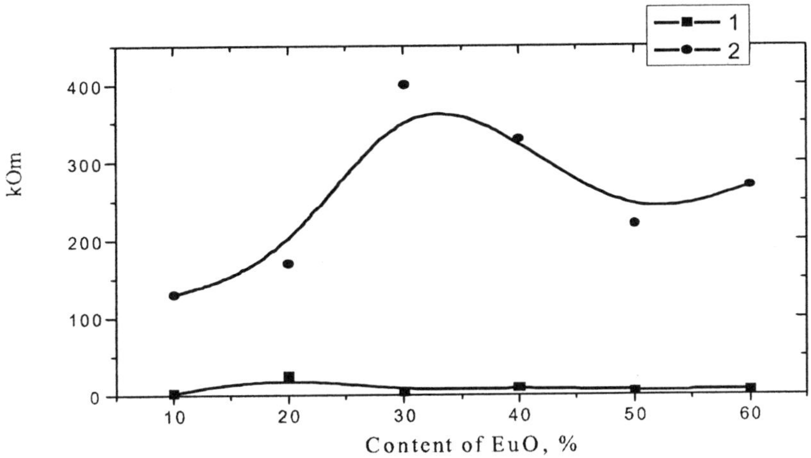

Figure 4. Film's resistance dependence on EuO content at different substrate temperature: 1–T=160 ^{0}C; 2–200^{0}C

The films have amorphous structure. The rare-earth oxide films have been the materials with prevailing electron conductivity mechanism as was shown by Jayaraj et al. (1989). Ion conductivity presence and it has less then of 0.25 % from general one. We can assume that in this case we deals with current limiting of space charge. The presence of two parts on I-V characteristics can be explained by system non uniformity, for example, by stoichiometric composition local derivation, different impurities, dielectric interlayers between grains. There are some mechanisms which form the current, consequently the current dependence will have sophisticated character and to differ for various materials. At substrate temperature gain the deposited films transfer from amorphous station to polycrystalline. In this case polycrystalline potential barrier gains that leads to conduction decreasing, films resistance increases (fig.4). It is also confirmed by the fact that the temperature resistance coefficient of the samples

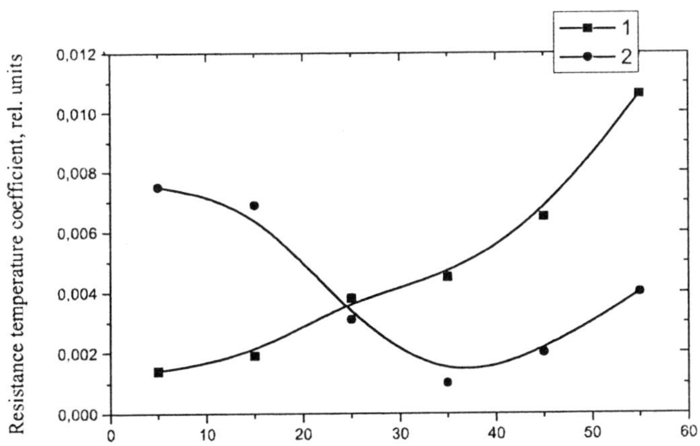

Figure 5. The temperature resistance coefficient dependence on EuO content in EuO-CeO$_2$ films at substrate temperature of: 1 – 160 ^{0}C; 2 – 200 ^{0}C.

deposited on the wafers with the temperature of 200 ^0C has less value than for the samples with substrate temperature of 160 ^0C (fig. 3). It was obtained that double-composition EuO-CeO_2 oxide films with EuO contain of 30-40 weight % demonstrate the extreme properties: a minimal duck conductivity is of $10^{-10}\ \Omega^{-1}m^{-1}$, a maximal photo-sensibility and non linearity of current-voltage characteristics is of 1.48, minimal temperature resistance coefficient is of 10^{-3} 1/degree

4. Construction and recommendation

As a result of ours investigation we propose thin film binary oxide systems based on cerium oxide ZnO (10 %) – CeO_2 (90 %) è CeO_2 (85 %) – SrO (15%) for development of stabile high sensitive temperature and heat flux sensors, temperature biosensors for measurements different biology matters concentration and analyse of organic and water phase. The working principle of sensitive element, which is made by differential pair scheme, based on heat flux measurements that segregated at control substance reaction with appropriate reagent (enzyme, antibody, alive hutch and etc.)

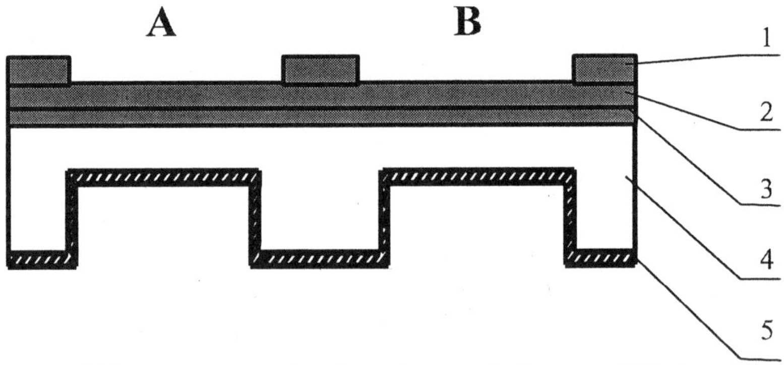

Figure 6. Temperature biology sensor construction: 1- metal contact; 2- thermo sensibility layer; 3- dielectric layer; 4- substrate; 5-enzyme.

This devise is characterised by high sensitivity, precision, uniformly, universality, approachability, and very small samples dimension, high response, multifunctional.

It was obtained the possibility to determine the glucose concentration in the blood from 1.5 to 20 mM/l at biosensor sensibility 0.2 mM/l, determine time 40 s and less that is in according with international standard of express diagnostics of glucose content in the blood.

References

1. Jayaraj, V.K. and Vallabhan, C.P.G. (1989), "Dielectric and optical properties of europium oxide films", *Thin Solid Films,* vol. 177, pp. 59-67.
2. Meixner, H. and Fleischer, M. (1996), "Thin film gas sensors based on metal oxides", *Electronic Ceramics and Applications,* Electro ceramics 5 International conference, University of Aveiro, Portugal, September 2-4, pp.71-80.
3. Shmyryeva, A. N., Dushejko, M. G., Scurtul, K. D. and Semikina T. V. (1998) "New thin films oxide materials CeO_2 – ZnO for sensoring", *Sensors springtime in Odessa.* Workshop, 29 –30 May, pp. 70 - 71.

POROUS SILICON AS TRANSDUCER FOR IMMUNE SENSORS: FROM THEORY TO PRACTICE

V.M. STARODUB
Dept.of Biochemistry of Sensory and Regulatory Systems of the Institute of Biochemistry of the Ukrainian Academy of Sciences, Kiev, Ukraine[*]

Abstract

Porous silicon (PS) attracts more and more attention of the investigators. It has already found its application not only in optoelectronic elements, such as light-emitting diodes, photodetectors, etc., but also in biochips, which can be implanted into human organism, and biological sensors capable of measuring a number of substances. PS has recently been shown to possess visible photoluminescence (PhL) and electroluminescence. The nature of the PhL is still under discussion but quantum-dimensional theory seems to prevail as a possible explanation of the PhL mechanism. In this work a new application of PS, namely its promising use for the creation of immune sensors, is discussed. Laboratory prototypes of immune sensors based on the PS PhL were developed for the registration of myoglobin in blood serum and biological components in the air that can cause allergic reaction in people. The results obtained with the help of the sensor approach were compared with the data of the ELISA method. The difference between them was not statistically significant but sensor analysis was much shorter, simpler and cheaper than the ELISA method. The concentration of biological compounds in the air was determined also with an immune sensor based on the optical fibres in combination with enhanced chemical luminescence. The immune sensor based on the PS PhL showed sensitivity nearly to the optical fibre sensor but it was simpler in the construction and operation. Moreover, it allowed direct registration (without need of any label) of specific interactions between immune components. The obtained results show that PS and especially its ability to emit visible PhL can be successfully used for the development of new types of biological sensors. The elaborated models of the immune sensors on the basis of PS PhL are promising for their further improvement and production of principally new instrumental devices for the monitoring of the air contamination by biological components of different nature and diagnostics of the state of the human organism. Moreover, achievements of this study intrigue and stimulate further investigations on the use of the PS for enzyme-, DNA- or other types of biological sensors.

Key words: porous silicon, photoluminescence, biosensors, myoglobin, allergic compounds, biological contaminants, fibre optics.

E. Buzaneva and P. Scharff (eds.), Frontiers of Multifunctional Nanosystems, 383–398.

1. Introduction

Investigations in the field of biosensors are intensively developing in the last 30 years. As result, during this time a lot of prototypes of these instrumental devices were proposed for medical diagnostics, environmental monitoring and control of biotechnological processes. These devices are capable of determining as biological molecules as single cells, and also different organic and non-organic substances, low and high molecular weight organic components, enzymes and their substrates, antibodies and antigens, oligonucleotides and lectins, different forms of lipids, saccharides, proteins, aminoacids and many others [1-5].

Biosensors differ in the transducer used, which can be of different nature. They can be based on electrochemical, optical, mechanical, calorimetric, magnetic, etc. transducers and use such effects as fluorescence, chemical luminescence, light absorption and scattering, evanescent wave, non-emitting energy transfer, surface plasmon resonance, PhL of PS, etc. Biosensors differ also in the biological sensitive element, which can be comprised of enzymes, immune components, nucleic acids, lectins, receptors, cells, tissues and even whole organisms.

Presently number of possible biosensor prototypes can be estimated as more than 400 [1-5]. Nevertheless, only limited number of biosensors have found their practical application. Enzyme sensor for glucose belongs to those, which are now practically in use. As a reason for this can be a very large demand for such a sensor. At the same time, rather low practical application of biosensors can be explained by the fact that biosensors are only at the beginning of their development. Nevertheless, already now at the creation of biosensors it is necessary to turn to principally new approaches, which use nano-technology, stable selective and sensitive elements as well as apply multi-parametrical analysis to provide highly selective, sensitive and wide screening.

In this paper we will discuss use of new types of transducers, namely porous silicon, for the creation of biosensors.

2. Porous silicon and immune sensors

Nowadays PS attracts more and more attention of the investigators. It has already found its application not only in optoelectronic elements, such as light-emitting diodes [6,7], optical filters [8], etc., but also in biochips [9,10], which can be implanted into human organism, and biological sensors capable of measuring a number of substances [11-17]. PS has recently been shown to possess visible PhL and electroluminescence [18, 19]. The nature of the PhL is still under discussion but quantum-dimensional theory seems to prevail as a possible explanation of the PhL mechanism [18]. In this work a new application of PS, namely its promising use for the creation of immune sensors, is discussed.

2.1. *PREPARATION OF PS*

PS was obtained using the technology according to which monocrystalline silicon was irradiated with a laser beam followed by a chemical treatment [20]. The scheme of the PS preparation is shown in *Figure 1.*

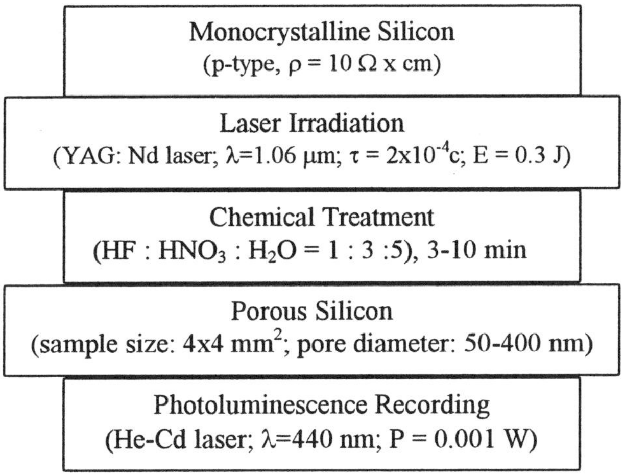

Figure 1. Process of PS formation.

Monocrystalline silicon (p-type, ρ = 10 Ω•cm) was irradiated with focused laser beam in the regime of surface scanning. For this purpose we used YAG laser with the wavelength of 1.06 μm, impulse duration of $2x10^{-4}$ sec and energy of 0.3 J. Then, silicon surface was treated during 3-10 min with the mixture of HF, HNO_3 and H_2O (the volume ratio: 1 : 3 : 5 respectively). The obtained PS wafer was cut into single pieces with the size of 4 x 4 mm.

2.2. *MORPHOLOGY OF THE PS SURFACE*

The surface morphology was investigated with atom force microscope (AFM). The surface contains pores with the depth from 10 to 200 nm and the diameter (d) from 50 to

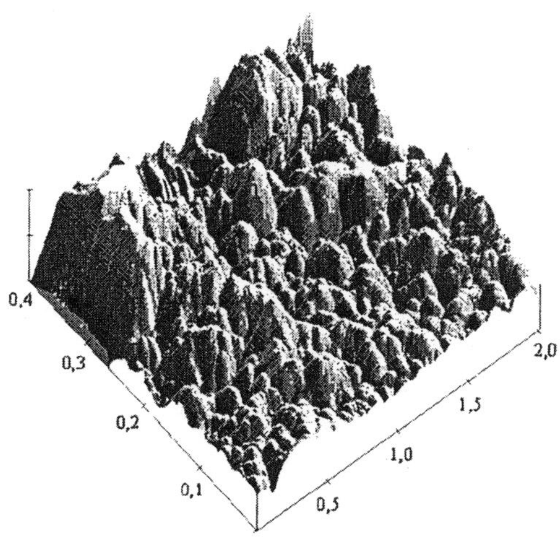

Figure 2. PS surface under AFM

400 nm and silicium crystals with the width from 50 to 400 nm. The AFM image of the PS surface is presented in *Figure 2*. According to the existing classification of PS surfaces on microporous (d < 2 nm), mesoporous (2 nm < d < 50 nm) and macroporous (d > 50 nm) [21] as well as taken into account the results of our AFM investigations, it can be concluded that the obtained PS samples are macroporous.

2.3. *PhL of the PS*

The prepared PS was capable of emitting visible PhL at the excitation by He-Cd laser beam with the wavelength of 440 nm and energy of 0.001 W. PhL was characterised by a maximum at 650 nm.

Increase in the energy density (from 11.5 to 33.5 J/cm^2) of laser beam during PS preparation lead to the rise in the PhL intensity. That is why the laser beam was focused in such a way that its density was on the level of 33.5 J/cm^2. These data are in agreement with the results presented in [22].

The spontaneous decay of PS PhL at the temperature of 300 K is described by a 'stretch exponent' and it lasts during ≥ 200 μsec (*Figure 3*).

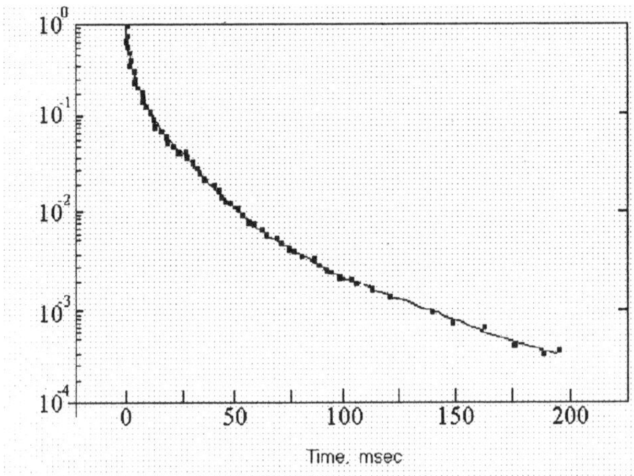

Figure 3. Spontaneous decay of the PS PhL.

To register PS PhL a measuring set-up was assembled. It consisted of a number of elements among which were He-Cd laser, standard monochromator UM-2, photomultiplier FEU-83 and personal computer (*Figure 4*).

The structural peculiarities of the surface and main characteristics of the PS PhL were studied at the contact of the PS sample with different environments. It was shown that at the immersion of the PS sample into distilled water, buffer solution (20 mmol/L, pH 7.3), buffered sodium chloride solution as well as solutions of proteins and different antigens or antibodies over a wide range of concentrations (from 10 to 1000 mg/L) the intensity and shape of the PhL spectra remained almost immutable after 2 h of measurements. At the same time, it was discovered that formation of the specific immune complex on the PS surface caused extinguishing of the PS PhL (*Figure 5*).

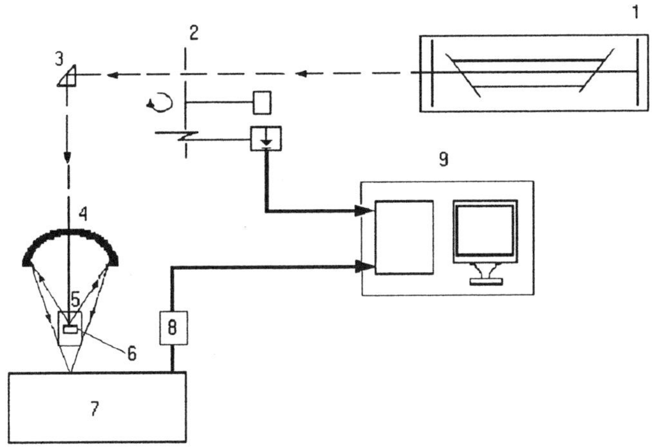

Figure 4. Block-scheme of the set-up for PS PhL measurement:
1 - laser, 2 - modulator, 3 - prism, 4 - Al mirror, 5 - measuring quartz cell, 6 – PS sample, 7 - monochromator, 8 - photomultiplier and 9 - personal computer.

In our experiments it was shown that PS PhL decreased up to zero level when during two hours the specific immune complexes between myoglobin and its specific antibodies (monoclonal IgG) and specific biological components from the air and their specific antibodies (polyclonal IgG) were formed on the surface of PS sample.

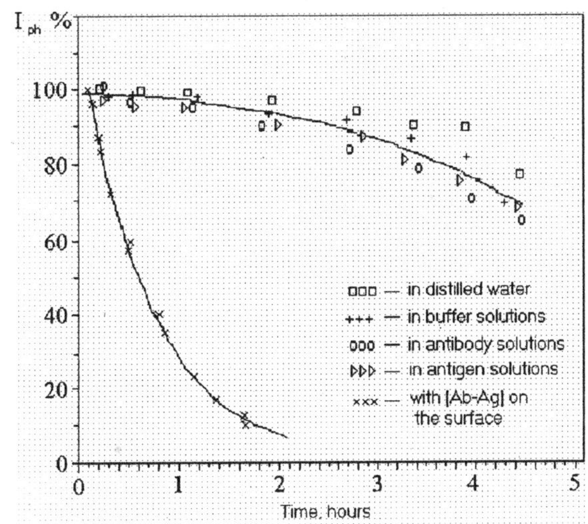

Figure 5. PS PhL as a function of time under contact of PS with different environments: distilled water, buffer solutions, separately with the solutions of antibodies or antigens and at the formation of specific immune complex on the surface.

We assume that extinguishing of PS PhL may be explained by release of hydrogen from silicon bonds and its subsequent capture by the immune complex. Torn silicium bonds are known to intensify the non-radiactive channel of recombination [18]. This assumption is based on the fact that the presence of non-radiactive channel of recombination on the PS surface is the key factor affecting the intensity of PS PhL in the frame of any existing theories to explain the nature of PS PhL.

2.4. *IMMUNE SENSOR FOR THE MEASUREMENT OF MYOGLOBIN*

The discovered effect was for the first time used for the development of laboratory prototype of immune sensor based on the PS PhL for the determination of myoglobin in blood serum [20, 23]. Antibodies or antigens were immobilised on the PS surface by spontaneous physical sorption.

The AFM image of the PS surface with immobilised antibodies to myoglobin is presented in *Figure 6*. It can be seen that the PS surface is uniformly covered with a net of biological molecules.

The AFM image of the PS surface with the specific immune complex between myoglobin and its specific antibodies is presented in *Figure 7*.

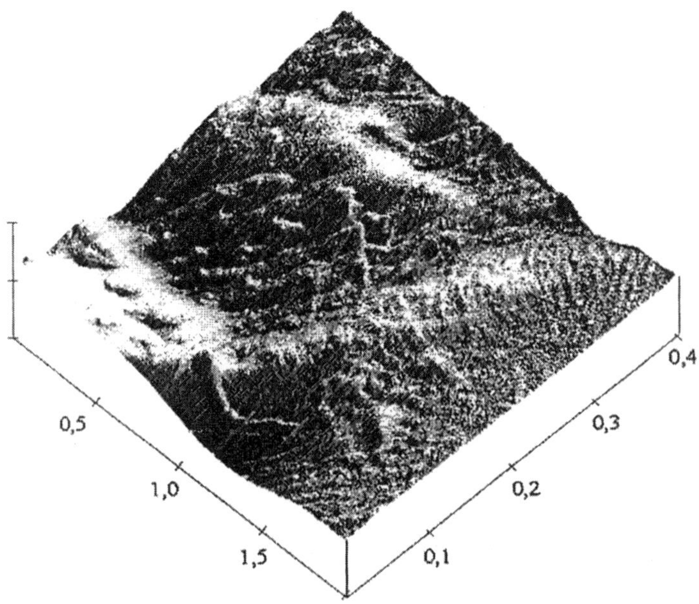

Figure 6. AFM of the PS surface after immobilisation of antibodies to myoglobin.

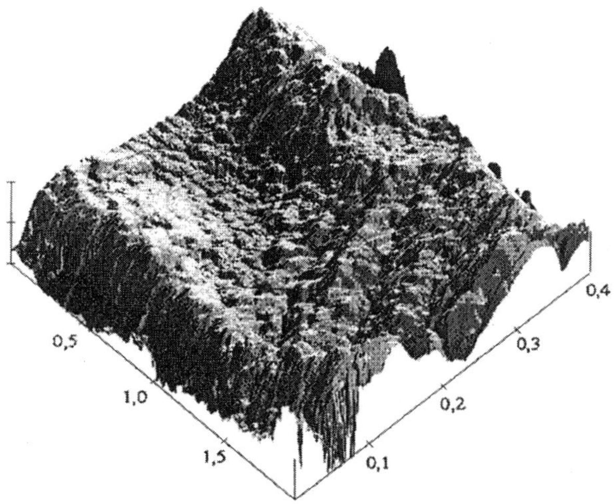

Figure 7. AFM of PS surface with the specific immune complex between myoglobin and its specific antibodies on the surface.

The degree of the PhL intensity reduction depends on the time of immune complex formation on the PS surface and on the concentration of the immune components in the tested solution.

The optimal concentration of antibodies to be deposited on the PS surface and optimal time of their immobilisation were studied. It was found that the optimal concentration of antibodies to be deposited on the PS surface was 100 mg/L *(Figure 8)*, while optimum time of their deposition was equal to 60 min *(Figure 9)*.

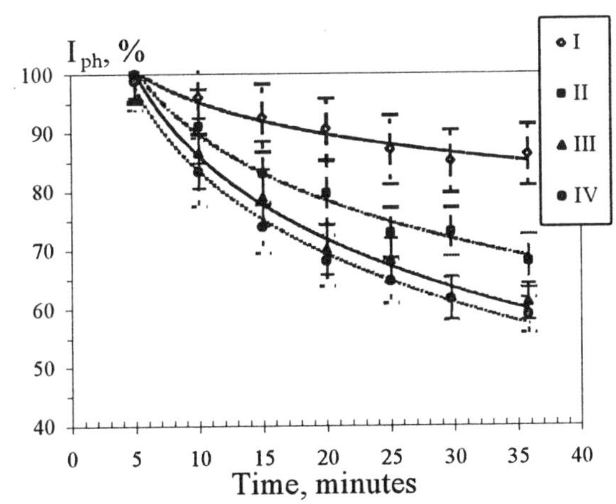

Figure 8. PhL intensity as a function of time under the formation of the specific immune complex on the PS surface. Concentration of antibodies to myoglobin in immobilisation solution: 1 – 4, II – 20, III – 100 and IV – 500 mg/L.

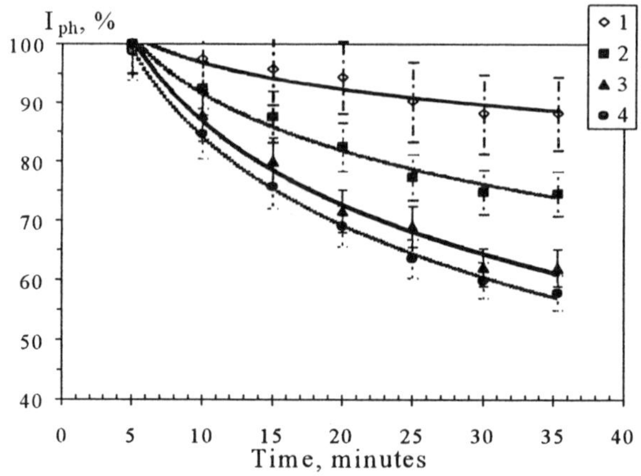

Figure 9. PhL intensity as a function of time under the formation of the specific immune complex on the PS surface. Immobilisation time of antibodies to myoglobin on the PS surface : 1 – 15, 2 – 30, 3 – 60 and 90 min.

In the following experiments it was shown that minimum concentration of myoglobin, which could be registered with the help of the proposed immune sensor, was 10 μg/L. The PhL intensity changed linearly in the range of myoglobin concentrations in buffer solution from 0.01 to 1 mg/L *(Figure 10)*.

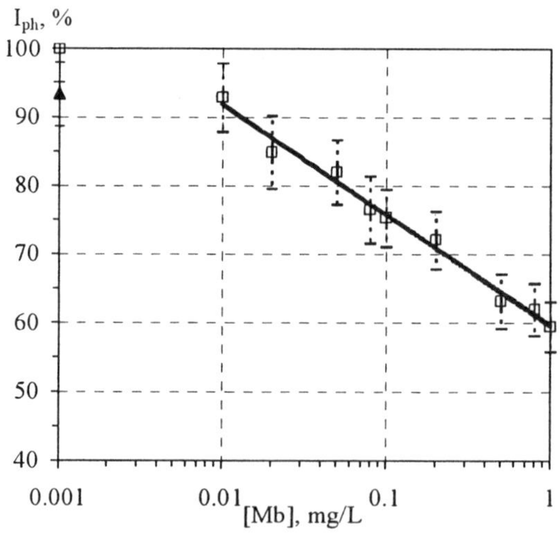

Figure 10. PhL intensity as a function of myoglobin concentration in buffer solution. Concentration of antibodies to be immobilised - 100 mg/L, time of their immobilisation -60 min; time of immune complex formation on the PS surface - 15 min.

To examine the work of the developed immune sensor in real conditions we have analysed samples of blood serum from 3 healthy persons. These samples were mixed and subsequently diluted 10 and 100 times. Followed experiments showed that at the immersion of the PS sample with the specific antibodies immobilised on the surface, into the sample of non-diluted blood serum approximately 30% reduction of the PhL intensity was observed *(Figure 11)*. At the same time for both diluted samples the level of the PhL intensity almost did not change. Such differences in the intensity of the PS PhL can be explained by the fact that the blood serum of a healthy person may contain myoglobin in concentrations up to 100-150 μg/L.

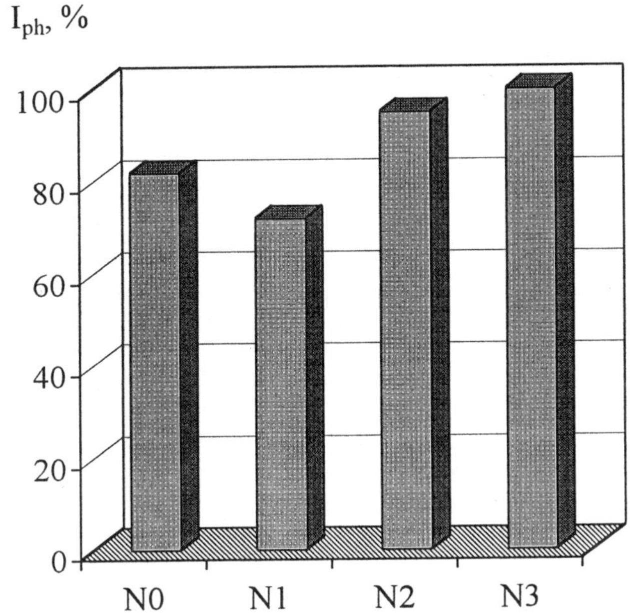

Figure 11. The intensity of the PS PhL at the immersion of PS with antibodies to myoglobin into buffer solution containing myoglobin (0.1 mg/L) (N0) and blood serum: undiluted (N1); diluted 1 : 10 (N2) and 1 : 100 (N3).

To verify this assumption we added different quantities of myoglobin (from 1 μg/L to 10 mg/L) to the samples of preliminary diluted (1 : 10) blood serum. The obtained results (*Figure 12*) confirmed made assumption. Moreover, these data showed that the PhL intensity linearly decreases in the range of myoglobin concentrations in blood serum from 0.01 to 1mg/L. This conclusion is very important because it testifies that the developed immune sensor can be also used for the detection of myoglobin in the blood serum of patients with myocardial infarction, when the myoglobin concentration can vary in the range of 10-100 μg/L.

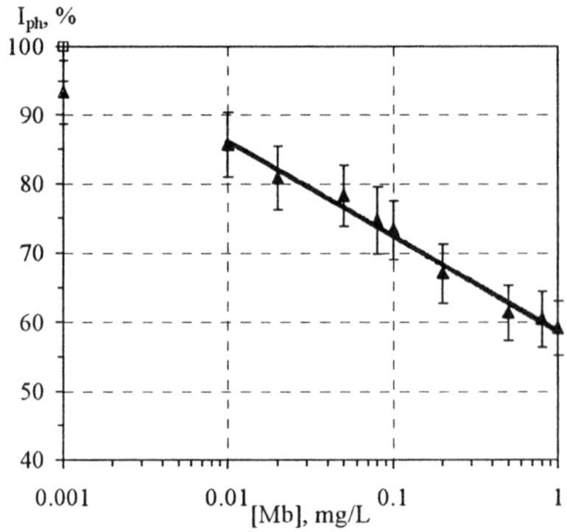

Figure 12. PhL intensity as a function of myoglobin concentration in blood serum diluted 1 : 10. . Concentration of antibodies to be immobilised - 100 mg/L, time of their immobilisation - 60 min; time of immune complex formation on the PS surface - 15 min.

For the analysis of the myoglobin content in real samples we used blood serums from healthy people (I group), patients with the initial stages of myocardial infarction (II group) and patients with acute heart failure (III group). In total 15 persons, 5 persons from each group, were investigated. The samples of blood serum from patients suffering from the heart failure were diluted 1 : 10 and the samples from healthy persons remained undiluted. The results obtained with the help of the sensor approach were compared with the data of the ELISA method. The averages of myoglobin concentrations are given in Table 1.

Table 1. *Myoglobin concentration in blood serum of healthy people and patients with myocardial infarction*

Concentration of myoglobin, μg/L					
ELISA-method					
Group type	1	2	3	4	5
I	14.3 ± 0.58	19.7 ± 1.53	22.0 ± 1.43	22.6 ± 1.52	29.3 ± 1.27
II	21.1 ± 1.13	15.8 ± 0.93	17.5 ± 0.82	17.9 ± 0.92	22.7 ± 1.33
III	61.2 ± 2.55	42.4 ± 2.04	51.8 ± 2.32	68.9 ± 2.93	45.6 ± 1.86
Optical immune sensor based on PS PhL					
I	15.3 ± 1.82	18.6 ± 1.67	21.4 ± 2.07	23.3 ± 2.39	31.2 ± 3.28
II	20.6 ± 1.87	16.3 ± 1.57	18.3 ± 1.67	17.3 ± 1.82	21.8 ± 2.24
III	58.7 ± 5.82	40.6 ± 3.87	54.3 ± 5.16	71.6 ± 6.79	43.3 ± 4.41

The obtained results testify that the difference between the data obtained by the ELISA-method and the developed immune sensor is not statistically significant. Unfortunately it was stated that the immune sensor is a little bit less precise than the ELISA-method. This probably can be explained by the fact that PS samples from different series of wafers were used. The precision of the sensor approach can be increased via improvement of the production process. At the same time, it is important to mention that the sensor analysis is much shorter, simpler and cheaper than the ELISA method.

The overall time of the analysis by immune sensor is about 80 min including duration of all preliminary steps. It can be substantially shortened through the decrease of duration of antibody-antigen interaction and in advanced performed preparation of selective layer on the transducer surface as well as through the fulfilment of analysis in a kinetic regime.

To study operational stability of the developed immune sensor the surface of PS was after each cycle of measurements treated with solution of HCl (0.1 N) or acetate buffer (50 mmmol/L, pH 2.2) during 5 min. It was then washed with phosphate buffer (50 mmol/L, pH 7.3) and distilled water, after which the PS PhL was measured. It was found that after the first cycle of measurements the level of the PhL intensity was only 50% of the initial level. After this, the samples used in the first cycle of the measurements were placed into myoglobin solution (100 mg/L) for 60 min and then they were again treated with acid medium according to the procedure presented above. After the second treatment of the PS surface with solution of low pH the intensity of the PhL reached zero level. According to our opinion the most possible explanation of such an effect is the direct damage of the PS surface as result of the influence of solutions with low pH. Thus, the PS samples are not suitable for multiple use. Nevertheless, this cannot be a problem at the mass production of the sensors because of wide availability of silicon, its low cost and simplicity of PS obtaining.

2.5. *IMMUNE SENSOR BASED ON THE PS PhL FOR THE MONITORING OF BIOLOGICAL COMPONENTS IN THE AIR*

The developed immune sensor was used for the monitoring of biological compounds in the air of the biotechnological plant producing amino acid lysine, which is situated in the Kiev region [24-26].

To determine quantity of lysine producents and their protein components in the air specific polyclonal antibodies were obtained. For this purpose the dried cells of lysine producents were suspended in 5 M saccharose solution, then they were frozen and de-frozen several time. After this centrifugation (500 g, 30 min) and dialysis against 50 mmol Tris-HCl buffer were performed. Obtained components were used for the immunisation of rabbits. After two months antiserum was obtained from immunised rabbits. The obtained antiserum was then used to prepare affinity purified specific antibodies. The obtained antibodies showed no cross reactivity with bovine serum albumin, myoglobin, haemoglobin and casein.

Air sampling was performed according to the filtration technique with the help of air sampler constructed for in field application and manufactured by "Polytechmed" (Ukraine). The air was collected at 1.2 m^3/h during half an hour. For the desorption of collected particles from the filter into solution, filters were put for 15 min into the solution of 20 ml of $NaHCO_3$ buffer (50 mmol/L, pH 7.3). The obtained extract was centrifuged (500 g for 30 min) and concentrated up to 2 ml with the help of vacuum evaporator.

The monitoring of lysine-producents and their protein components was curried out in 4 working rooms of the biotechnological plant: fermentation, lysine concentration, drying and packing of finished product as well as in six places outside the plant: on the territory of the plant, in the vicinity of the chimney; at a distance equal to 500 m from the chimney; in three different places of residential area at distances equal to 2, 4 and 6 km from the chimney. The air sampling was performed in 2 days, which were characterised by different climatic conditions: 1) at the presence of wind and 2) at its absence. The level of specific biological components in air was determined on the basis of calibration curves constructed preliminary for each of individual methods. The total concentration of proteins in the air was determined by the standard Lowry method.
The study of the sensitivity of the immune sensor based on the PS PhL towards lysine producents and their components showed that it equals 100 μg/L (330 ng/m^3).

The developed immune sensor was tested for monitoring biological components in the air of working rooms of a biotechnological plant producing lysine and in the ambient environment. The obtained results were confirmed by the data of the ELISA method and an immune sensor based on the optical fibres in combination with enhanced chemical luminescence, that was for the first time designed to monitor air quality *(Table 2 and 3)*.

Table 2. *Concentration of biological components in working rooms of biotechnological plant.*

#	Place of air sampling	Concentration of total protein (mg/m^3)	Concentration of specific biological substances (mg/m^3), determined with:		
			sensor based on PS PhL	Optical fibre sensor	ELISA-method
1.	Woork room of fermentation	0.15 ± 0.02	0.032 ± 0.008	0.035±0.005	0.033±0.004
2.	Woork room of lysine concentration	0.25 ± 0.04	0.049 ± 0.010	0.052 ± 0.007	0.053±0.010
3.	Woork room of lysine drying	1.60 ± 0.17	0.540 ± 0.080	0.520 ± 0.030	0.510±0.030
4.	Woork room of lysine packing	3.50 ± 0.35	0.840 ± 0.090	0.830 ± 0.060	0.820±0.010

Table 3. *Concentration of specific biological components in the ambient air.*

#	Place of air sampling, distance to the chimney	Concentration of specific biological components ($\mu g/m^3$)			
		1-st day, moderate wind		2-d day, almost no wind	
		sensor based on the PS PhL	optical fibre sensor	sensor based on the PS PhL	optical fibre sensor
1.	Nearby	0.50±0.07	0.52±0.04	3.33±0.37	3.37±0.31
2.	500 m	0.50±0.08	0.48±0.06	2.17±0.18	2.19±0.16
3.	2 km	0.33±0.05	0.36±0.03	1.50±0.12	1.47±0.09
4.	4 km	0.33±0.06	0.35±0.03	1.33±0.16	1.34±0.13
5.	6 km	<0.33	0.29±0.03	0.83±0.09	0.81±0.06

It was shown that at the process of lysine production air pollution in those working rooms where the initial stages of synthesis took place was minimal in comparison with the air contamination of the rooms at later production stages. The value of the total protein content in the air increased 10 times moving from the beginning to the end of the lysine synthesis and the subsequent packing of finished product. For every working room investigated the concentration of total proteins in the air was 4 times higher than the concentration of specific biological components. This can be explained by the presence of rather high concentrations of naturally occurring plant or animal proteins. Such a high difference in the concentrations of total and specific biological components leads to the conclusion that traditional biochemical approaches for the measurement of total protein quantities are not suitable for the monitoring of specific biological components coming from a specific biotechnological unit.

Having investigated the ambient air it was established that due to the variety of weather conditions a rather wide area should be monitored for the presence of biological pollutants.

The results obtained with the use of different approaches were statistically identical. Nevertheless the accuracy of used methods decreased in row: ELISA-method, optical fibre sensor and sensor based on the PS PhL. The lower accuracy of the last approach is probably connected (as also in case of myoglobin determination) with the application of the PS samples from wafers of different series.

The immune sensor based on optical fibre and enhanced chemical luminescence has been already applied for the detection of a number of biochemical substances [27-29]. The construction of the sensor for the detection of biological components in the air was performed as described in these papers. The sensitivity of the optical fibre sensor towards specific biological pollutants was about 20 μg/L, or 66 ng/m^3, with the overall time of the analysis equal to 45 min.

Immune chemical analysis was accomplished in a competitive or "sandwich" way when one of immune components was labelled with horse radish peroxidase. To measure the activity of this enzyme buffered solution (tris-HCl buffer, 50 mmol/L with pH 8.1) containing hydrogen peroxide, luminol and *p*-iodphenol in the concentration of 2, 0.05 and 0.06 mmol/L, respectively, was used

Although the immune sensor based on the Ph PS showed lower sensitivity in comparison with the optical fibre sensor, it was simpler in the construction and operation. Moreover, it allowed direct registration (without need of any label) of specific interactions between immune components.

3. Conclusion

As a result of work completed phenomena of the PhL intensity reduction under the formation of a specific immune complex on the PS surface was discovered. It was used for the construction of laboratory prototypes of immune sensors for the determination of myoglobin in human blood serum as well as cells of lysine producents and their components in the air. The proposed method turned to be simple, fast, sensitive, with further possibilities to shorten its duration. The obtained results as well as the prevalence of Si, its low cost and the simplicity of the PS production led to the conclusion that the elaborated models of the immune sensors on the basis of PS PhL are promising for their further improvement and production of principally new instrumental devices. Their practical application would provide a possibility to monitor air contamination by biological components of different nature and would allow fast, straightforward, and sensitive diagnostics of the state of the human organism.

The work of the developed sensor was compared with the ELISA method and an immune sensor based on the optical fibres in combination with enhanced chemical luminescence. It showed sensitivity nearly to the optical fibre sensor but it was simpler in the construction and operation. Moreover, it allowed direct registration (without need of any label) of specific interactions between immune components.

The obtained results show that PS and especially its ability to emit visible PhL can be successfully used for the development of new types of biological sensors. Moreover, achievements of this study intrigue and stimulate further investigations on the use of the PS for enzyme-, DNA- or other types of biological sensors.

4. Acknowledgements

This work was partially supported by the INCO-Copernicus Foundation (grant ICA2-CT-2001-10007).

5. References

1. Turner A.P.F., Karube I., Wilson G.S. (1989) "Biosensors: Fundamentals and Applications", Oxford University, Press., Oxford, UK, 800.
2. Starodub, N.F. (1990) "Electrochemical biosensors and biochemical diagnostics", *Lecture notes of the Int. sensors center of biocybernetics of the Acad. of Sci. of the Socialist. Countries*, **3**, 173-202.
3. Starodub, N.F. (1996) "New variants of electrochemical and optical biosensors: achievements and main modern directions of development", *Lecture notes of the ICB seminars. Biomeasurements. Electrochemical measurements of biochemical quantities. Ed. P. Bergveld and W.Torbicz. April 1995*, 159-181.
4. Starodub, N.F., Starodub, V.M. (2000) "Immune sensors: sources of origination, achievements and perspectives", *Ukr. Biochem. J.* **72**, 147-164.
5. Starodub, V.M., Starodub, N.F. (2001) "Electrochemical and optical biosensors: origin of development, achievements and perspectives of practical application", *NATO Proc. "Novel Processes and Control Technologies in the Food Industry", Ed. F. Bozoglu et al.*, 63-94.
6. Lackwood, D.J. (1998) "Light emission in silicon", *Semicon. And Semimetals*, **49**, 1-35.
7. Mehra, R.M., Agarwal, V., Mathur, P.C. (1995) "Development and characterization of porous silicon (a review)", *Phys. Stat. Sol.*, **9**, 75-79.
8. Mattei, G., Marucci, A., Yakovlev, V.A. (1998) "Splitting of porous silicon microcavity mode due to the interaction with Si-H vibrations", *Mat. Sci. and Engineer, B*, **51**, 158-161.
9. Canham, L.T. (1997) "Biomedical applications of porous silicon", *Properties of porous silicon. EMIS Data Reviews Series*, **18**, 371-376.
10. Canham, L.T., Reeves, C.L., Wallis, D.J. (1997) "Silicon as an active biomaterial", *Mat. Res. Soc. Symp. Proc.*, **452**, 579-589.
11. Sorli, B., Garcia, M., Benhida, A. (1999) "Porous silicon layer used as a humidity sensor", *Proc. Europ. Mat. Conf. E-MRS 1999 Spring Meeting"*, 1-8.
12. Rittersma, Z.M., Benecke, W. (1999) "A novel capacitive porous silicon humidity sensor with integrated thermo- and refresh resistors gas sensors", *Proc 13-th Europ. Conf. Solid-State Transducers Eurosensors XIII, Sept. 12-15, 1999, the Hague, the Netherlands*, 371-374.
13. Thust, M. et al. (1996) "Porous silicon as a substrate material for potentiometric biosensors", *Meas. Sci. Technol.*, 7, 26-29.
14. Bogue, R.W. (1997) "Novel porous silicon biosensor", *Biosens. And Bioel.*, **12**, xxvii-xxix.
15. Lin, V.S.-Y. et al. (1997) 'A porous silicon-based optical interferometric biosensor", *Science*, **278, 31,** 840-842.
16. Kelly, M.T., Bocarsly, A.B. (1998) "Mechanisms of photoluminescent quenching of oxidized porous silicon. Applications to chemical sensing", *Coordination Chem. Rev.*, **171**, 251-259.
17. Foucaran, A. et al. (1997) "Porous silicon layers used for gas sensor applications", *The Solid Films*, **297**, 317-320.
18. Canham, L.T. (1990) "Silicon quantum wire array fabrication by electrochemical and chemical dissolution of wafers", *Appl. Phys. Lett.*, **57**, 1046-1048.
19. Koshida, N., Koyama, H. (1991) "Efficient visible photoluminescence from p-Si", *Japanese J. Appl. Phys.*, **30**, L1221-L1223.
20. Starodub, N.F. et al. (1996) "Use of the silicon crystals photoluminescence to control immunocomplex formation" *Sensors and Actuators B*,. **35-36**, 44-47.
21. Schoening, M.J. et al. (1997) "Miniaturization of potentiometric sensors using porous silicon microtechnology", *Electrochim. Acta.*, **42, 20-22,** 3185-2193.
22. Svechnikov, S.V. et al. (1994) "Photoluminescence of porous silicon obtained at the use of laser irradiation", *Ukrain. Phys. J.*, **39, 6,** 704-705.
23. Starodub, N.F., et al. (1999) " Use of the silicon crystals photoluminescence to control immunocomplex formation", *Sensors and Actuators B*, **35-36**, 44-47.
24. Starodub, V.M., Starodub, N.F. (1999) "Optical immune sensors for monitoring protein substances in the air" *Book of abstracts of the 13th European Conference on Solid-State Transducers, September 12-15, the Hague, the Netherlands*, **87-88.**

25. Starodub, V.M. et al. (1998) "Optical sensors for medical diagnostics and environmental monitoring", *Lecture Notes of the ICB Sseminars "Measurements and Microtechnology for Determination of (Bio)chemical Quantities in Medical and Environmental Monitoring", March, 1998, Warsaw, Poland,* 192-205.
26. Starodub, V. Urge-Vorsatz, D., Starodub, N. (1999) "A new method based on a bioaffinic sensor utilizing the photoluminescence of porous silicon to monitor protein substances in the air", *Proc. E-MRS'99 Spring Meeting "Microcrystalline & Nanocrystalline Semiconductors",* **I-I,** 13.
27. Starodub, N.F. et al. (1992) "Optoimmunosensor for analysis of specific and non-specific classes of immunoglobulins", *Sensors and Actuators* **B7**, 371-375.
28. Starodub, N.F., Arenkov, P.J., Starodub, A.N., Berezin, V.A. (1993) "Fiber optic immunosensors for detection of some drugs", *Sensors and Actuators* **B13-14**, 728-731.
29. Starodub, N.F., Arenkov, P.J., Atarodub, A.N., Berezin, V.A. (1994) "Fiber optic immunosensors based on enhanced chemiluminescence and their application to determine different antigens", *Sensors and Actuators* **B18-19**, 161-165.

A POROUS SILICON MICROCAVITY AS AN OPTICAL AND ELECTRICAL MULTIPATRAMETRIC CHEMICAL SENSOR

Z. GABURRO, L. PAVESI
INFM and Dipartimento di Fisica Universita' di Trento, ITALY
C. BARATTO, G. FAGLIA, G. SBERVEGLIERI
INFM and Dipartimento di Chimica e Fisica Universita' di Brescia, ITALY

Abstract

The lack of selectivity of chemical sensors encourages the investigations of single sensors with different mechanism of interaction with the gases. In fact, the wider the spectrum of information that can be extracted from one sensor, the more the sensor can be made selective by the analysis of the response pattern.

We experimentally demonstrate that porous silicon optical microcavities (PSM) can be effectively used as multi-parametric gas sensors. Semiconductor microcavities are formed by a layer structure as a Fabry-Perot filter: a central active medium is embedded between two dielectric multilayer mirrors. The parameters monitored in our sensors are the DC electrical conductance, the photoluminescence (PL) intensity and the spectral position of the resonance cavity peak. Measurements are carried out at room temperature a) in static air saturated with various solvents and hydrocarbons, and b) under controlled flux of a mixture of dry air and gaseous traces. The examined species under controlled flux are NO_2 (1-21 ppm), relative humidity (20% - 80%), ozone (200ppb) and ethanol (500 to 30000 ppm). The PSM were produced by electrochemical etching of p type (6-9 Ω cm) and p+ type (0.005-0.02 Ω cm) silicon wafers. For electrical measurements, gold contacts were evaporated on the top surface of p+ PSM.

We show that the spectral position of the peak depends on the refractive index of the gas, whereas the luminescence intensity depends mainly on its low frequency dielectric constant. This indicates that independent information is available through these parameters. Moreover, the dynamics of the response of the parameters is different. Finally, in the case of ethanol, the time response of the resonance peak is discussed in detail and compared with simulated gradual filling of the cavity. The simultaneous measurement of the peak intensity and the resonance position can be used to monitor the penetration depth of the gas inside the cavity.

1. Introduction

The use of porous silicon (PS) as a sensor material is suggested by its large surface versus volume ratio. Typical values are in the range of 500 m^2/cm^3, and depend on the dopant

E. Buzaneva and P. Scharff (eds.), Frontiers of Multifunctional Nanosystems, 399–412.

level and on the substrate. Chemistry of PS is therefore determinant to the electronic, optical and electrical properties [1].

Demonstrated transduction mechanisms are electrical [2-8], electrochemical [9], and optical [3,10-14]. For example, the electrical conductivity (G) of PS changes with the environment, and variations have been reported to depend exponentially on the dipole moment of the organic solvent, and on the gas pressure [3]. An interesting application has been suggested by the response of G to NO_2, a dangerous air pollutant. It has been found that the presence of NO_2 has an anomalously large impact on the G of PS ($\Delta G/G$ in excess of 5 have been measured at 100 ppb, where G is the electrical conductance of the sample) [15]. Also the photoluminescence (PL) of PS, which is normally quenched by surrounding gases [16], is a very interesting transduction mechanism, since PS has large PL efficiency at room temperature, thus low power excitation is required for measurements. An effective and practical sensing parameter based on PL variations is the integrated PL intensity. Other optical sensing parameters are reflectivity [17-19], and optical waveguiding [20].

An additional sensing parameter is the effective refractive index of the PS. In fact, PS is a structure composed of crystalline silicon alternated with voids (pores). Since the structures of the PS considered here are sized much less that optical wavelengths, the structure has an effective refractive index, which depends both on the refractive index of Si and of the gas which fills the pores. In order to monitor refractive index of PS samples, a convenient way is to fabricate the samples as suitably thin single layers [14,21], or stacks of thin layers such as a PS microcavity (PSM), which exhibits well resolved Fabry-Perot interference fringes [22-25]. In the PL spectrum of a PSM, a narrow and large intensity peak is observed at the cavity resonance wavelength, whose spectral position depends on the optical thickness, hence on the refractive index, of the cavity. At the same spectral position, a corresponding dip in the reflectivity spectrum is also observed. If the refractive index of the environmental gas changes, the Fabry-Perot fringes shift.

In this work, we propose that PSM's might be used as multiparametric sensors. Simultaneous monitoring of several parameters in single devices is required for gas selectivity. We show that in PSM's at least three parameters are available: the electrical conductivity, the integrated PL intensity and the resonance peak wavelength. The three parameters differently respond to physical properties of gases. The conductivity is affected in complex way by chemical and electrical interaction. The PL intensity is also affected by a combination of the gas characteristics, but the molecular electric dipole is one of the dominant agents in PL quenching. The cavity peak wavelength, on the other hand, is more easily linked to the refractive index of the PS layers. Although we have not yet tested the three sensing parameters simultaneously, our measurements show that the multiparametric idea is expected to work in PSM's.

2. Optical response in static saturated vapor

We have first chosen the simplest configuration to measure the optical response. The PSM's for this study were produced by electrochemical etching on p-type silicon wafers (6-9 Ω cm). The choice of the substrate doping was motivated by the fact that luminescence emission is much higher for lightly doped substrates than for heavily doped ones [24]. The parameters for the fabrication were selected to have the initial cavity peak in the yellow-green region of the spectrum [24]. This way, the peak shifts towards the red portion of the visible spectrum. The red shift of the high-reflectivity results in a color change of the PSM's, which is clearly visible with naked eye. This effect allows a first approximate but immediate evaluation of the presence of a chemical. In addition, the reflectivity shift does not cause the exit the PSM resonance from the luminescence band after immersion in organic solvents.

We have characterized the optical properties under saturated vapor of various solvents in sealed transparent containers acquiring the PL spectra under argon laser excitation (488 nm wavelength). The measurements were performed at room temperature (20 ^{o}C).

Marked red shifts of the cavity peak, as well as large changes of the integrated PL intensity were observed. Figure 1 and 2 show the observed peak shifts, whereas the effect of the solvents on integrated PL intensity is shown in Figure 3.

The substitution of air with solvent in the pores changes the value of refractive index of the PS layers, resulting in the red shift (Figure 1 and 2). In the case of total filling of the pores the shift can be easily calculated with the Bruggeman effective medium approximation [27]. The result of this calculation are shown together with the experimental data in Figure 2. In this calculation the refractive index values of PS have been determined from reflectivity spectra of PS single layers obtained with the same etching parameters used for the PSM, and variations of these values with wavelength have been included. This latter effect turned out to be very important, as shown for comparison in Figure 2, and in particular explains why we found a pore filling close to 100% with every solvent. This contrast with the results reported in [17] for Bragg reflectors, where the authors found only a partial filling of the pores almost independent from the solvent used.

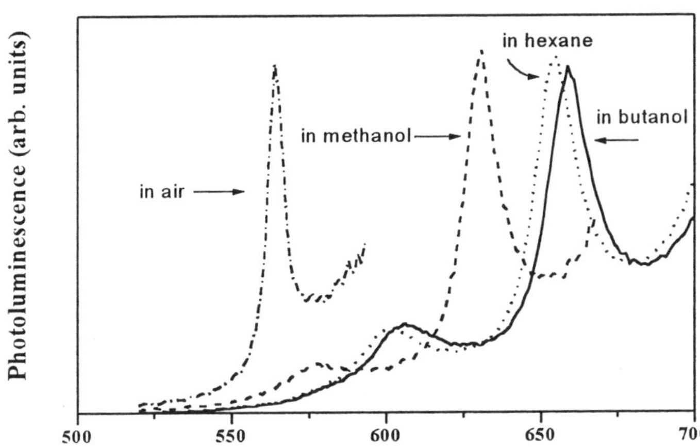

Figure 1. PL spectra of a PSM initially centered at 570 nm after immersion in different organic solvents. Intensities are not in scale. Dash-dotted line: in air. Dashed line: in methanol. Dotted line: in hexane. Solid line: in butanol.

Measurements performed in saturated vapors gave the same results obtained for the liquid solvents within the experimental error, confirming that the red shift is due to capillary condensation into the pores [17]. For the lower values of refractive index the relative shift is slightly higher than that calculated. This probably depends on the fact that refractive index of PS produced as single layer is slightly different than the index of PS produced in multilayered structures with the same etching parameters. In spite of approximations used the agreement with experimental measurements is very good.

As opposed to the peak shift, calculation of PL intensities in different organic solvents is much more complicate, for several reasons. Dependence of intensity with low frequency dielectric constant ε has been reported [16]. According to the model in [16] PL is strongly quenched by high values of ε, because the Coulomb interaction experienced by the electron-hole pairs is diminished, affecting the radiative recombination rate.

Results confirm the trend observed in [16]. The quenching effect is much less pronounced in our case. It is possible that the time elapsed between production on PS and measurements is important, because we used sampled aged in air for about two weeks. Results are however reproducible within the experimental error for PSM's produced with the same etching parameters. The intensity is enhanced with respect to air for almost every solvent because in the case of a PSM the absorbed radiation depends on the stop-band position, which moves towards the red after immersion. Furthermore the emission enhancement due to the cavity effect depends on the peak position with respect to the emission maximum, which in turn, depends partially on the solvent [16].

Reversibility of the process has been tested. PL variations and peak shift are totally reversible when the solvent is a hydrocarbon. Samples are partially damaged when dried from polar solvents, but the reversibility is strongly improved if they are rinsed in

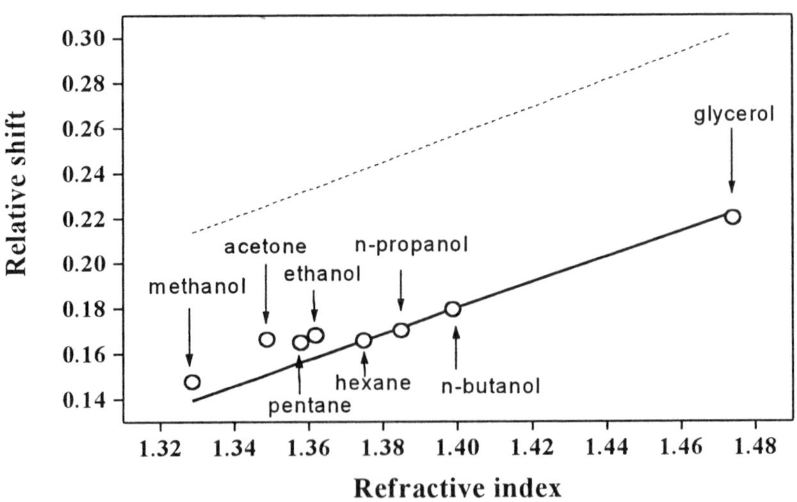

Figure 2. Relative peak shift for total filling versus refractive index for a PSM initially centered at 570 nm. Dashed line: calculated shift without inclusion of variation of porous silicon refractive index with wavelength. Solid line: calculated shift with this effect included. Open circles: experimental points.

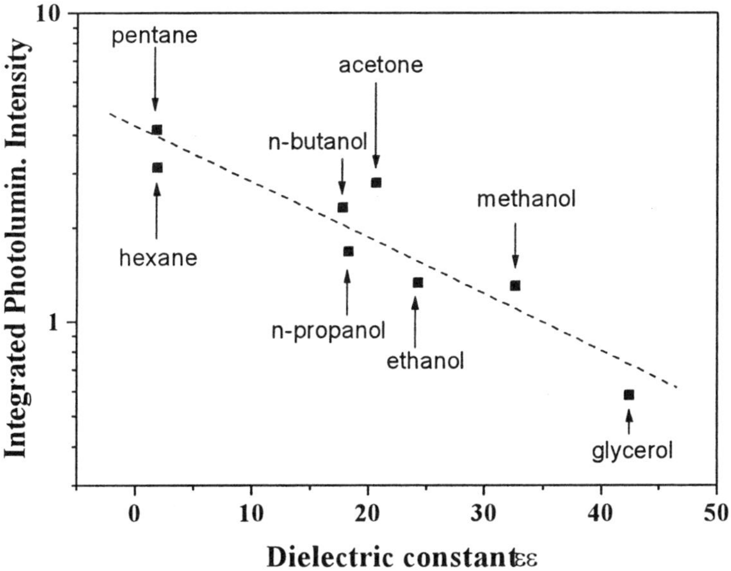

Figure 3. Integrated relative intensity (550-850 nm) as a function of the dielectric constant of the solvent for a PSM. The data are normalized with respect to the emission in air. The dashed line is only a guide for the eyes.

pentane before drying. The degradation is less pronounced for measurements in saturated vapors, and it is supposed to be negligible for low concentrations.

Variations of both peak position and luminescence intensity can be coupled easily in order to determine the concentration and the nature of the chemical solvent (gas discrimination). For example, ethanol and pentane have very similar values of refractive index and therefore similar relative shift of the resonance peak, whereas their dielectric function is very different, and luminescence emission is much higher in pentane than in ethanol.

3. Optical response to different concentrations in controlled flux

A more sophisticated flux-controlling system was used to characterize the response to different gas concentrations. We have tested the response to NO_2, water and ethanol. PL spectra have been acquired under controlled flux of gases (0.3 liters/minute). Dry air, NO_2, and ethanol up to 1000 ppm came from certified bottles, while for higher concentrations of ethanol and for relative humidity the flow-through saturation method was used. The carrier (dry air) flowed through a condenser kept at T=20 °C, placed after a bubbler (T=40 °C). At the outlet of the condenser the flow was saturated with vapors. The saturated vapor so far obtained was diluted with synthetic air in the desired percentage. Similarly to the measurements in saturated vapor, the PSM's for this study were also produced by electrochemical etching on p-type silicon wafers (6-9 Ω cm) to optimize the PL efficiency. Samples were 6 months old.

Samples were at room temperature (20 °C) at all times. To measure the PL intensity and the PSM's peak position, we have used argon laser excitation (488 nm wavelength), a dispersion grating and a CCD detector to acquire PL spectra. The acquisition time was 2 seconds for a full spectrum. The laser intensity was constantly monitored during the measurements, and spectra were afterwards normalized.

In the case of NO_2, we have first conditioned the high resistivity PSM's for 1 hour in dry air. We have used a slightly larger concentration than required for practical applications to facilitate the experiment. We decided to use 21 ppm, which was the highest available in our experimental setup. This required to run the measurement in dry air. Results are shown in Figure 4. Further characterization is required to test the response in realistic conditions (100 ppb to 1 ppm).

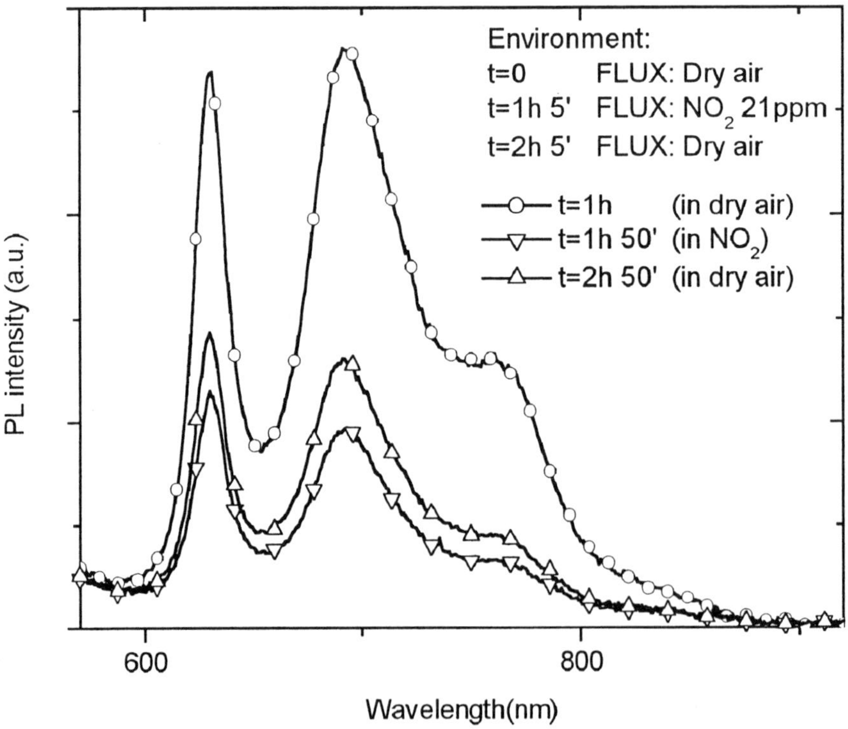

Figure 4. PL spectra of a high resistivity PSM, under controlled flux (0.3 l/m), of dry air (circles), dry air + NO_2 (21 ppm, downward triangles), and back to dry air (upward triangles). Timing of switching of flux, as well as of spectra acquisition, is indicated in the plot (room temperature).

Change of relative humidity at T=20°C in the range from dry to saturation produces a reversible modification of both PL intensity and cavity peak position. Figure 5 reports behavior of the peak wavelength at ~630 nm (cavity resonance peak) when switching from dry air to different values of humid air and then again to dry air. The peak shift is of about 10 nm when passing from dry saturation.

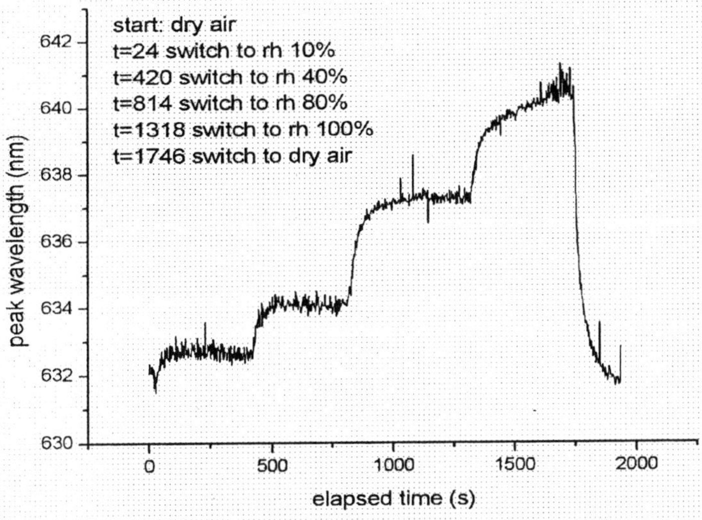

Figure 5. Behavior of the peak wavelength at ~630 nm (cavity resonance peak) when switching from dry air to different values of humid air and then again to dry air at 20°C.

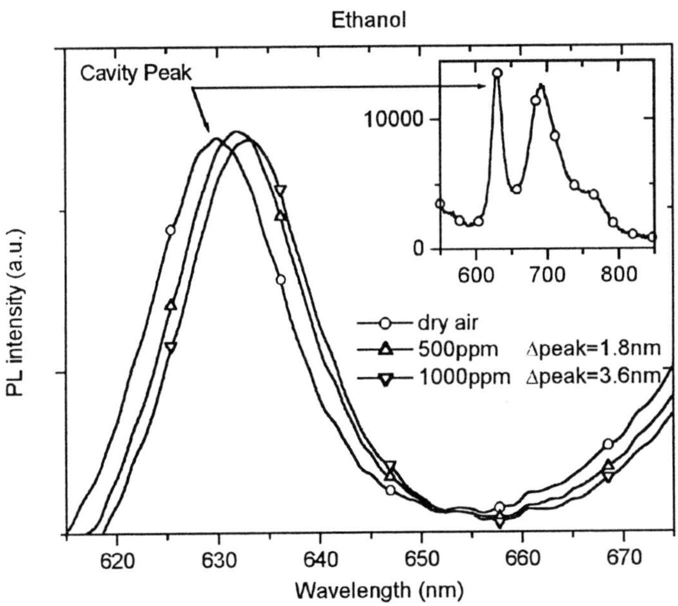

Figure 6. PL spectra of a high resistivity PSM, under controlled flux of air containing ethanol (0.3 l/m), at room temperature. Spectra have been acquired 20 minutes after the switching of the flux with the indicated concentration. Dry air (circles), dry air + ethanol (500 ppm: upward triangles, 1000 ppm: downward triangles).

In the case of ethanol, we have first conditioned the PSM in dry air for 1 hour. We have then exposed the PSM to 500 ppm of ethanol for 20 minutes, monitoring the PL spectra every 2 seconds. We then switched the concentration to 1000 ppm and repeated as above. The last spectra in dry air, 500 ppm, and 1000 ppm of ethanol are shown in Figure 6. This way, we have verified that 10 minutes were enough to reach stable spectra. Thus, we have afterwards applied a controlled flux increasing the ethanol concentration every 10 minutes, starting from 1000 ppm and finishing at saturated vapor (approximately corresponding to 30,000 ppm). Figure 7 shows sample spectra acquire d during this second measurement. The amount of the total shift of Figure 7 was smaller than what reported in Figure 2 since the samples had different porosity. For the measurement in Figure 7, we have used an aged sample, to improve reversibility.

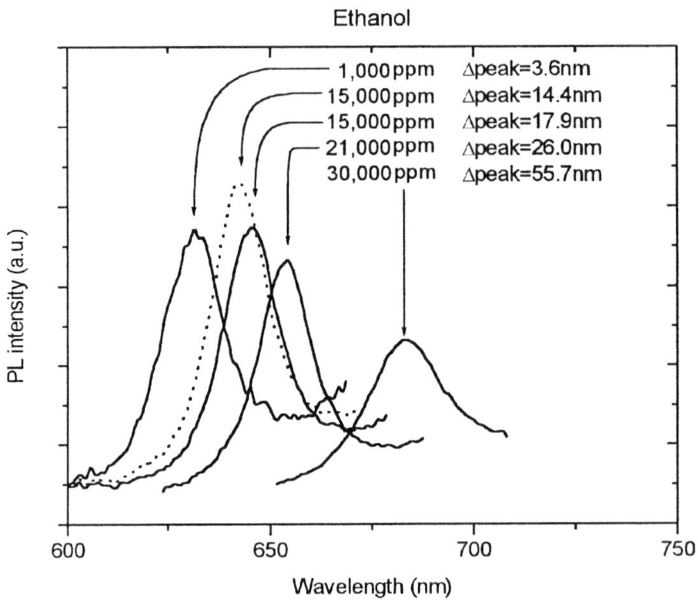

Figure 7. PL spectra of a high resistivity PSM, under controlled flux of air containing ethanol (0.3 l/m), at room temperature. Dry air + ethanol (1,000, 15,000, 21,000 and 30,000 ppm) after 10 minutes. Dashed line: dry air + ethanol, 15,000 ppm, 30 seconds after the switching of the flux.

4. Dynamics of optical response in controlled flux

The results of the controlled flux setup with ethanol (Figure 7, comparison the 2 curve at 15,000 ppm taken after different exposure times) suggested that a characterization of the dynamic response of the sensors was also possible. We show this idea discussing the dynamic response to ethanol of a high resistivity PSM (6-9 Ω cm). The DBR's high-index layers were fabricated using an etching current density of 5 mA/cm^2 for 33 s, whereas the low-index layers using 50 mA/cm^2 for 5 s. The cavity was formed with 6-periods DBR's. The spacer was fabricated forcing 25 mA/cm^2 for 35 s. The sample was 6 months old.

The sample was conditioned in dry air for 2 hours. Then, data acquisition started (t=0 s) under the flux of dry air. The flux was changed at time t=80 s to air containing saturated vapor of ethanol, and back to dry air only at t=634 s. The wavelength of the cavity peak and the PL intensity at that wavelength are reported in Figure 8 as a function of time.

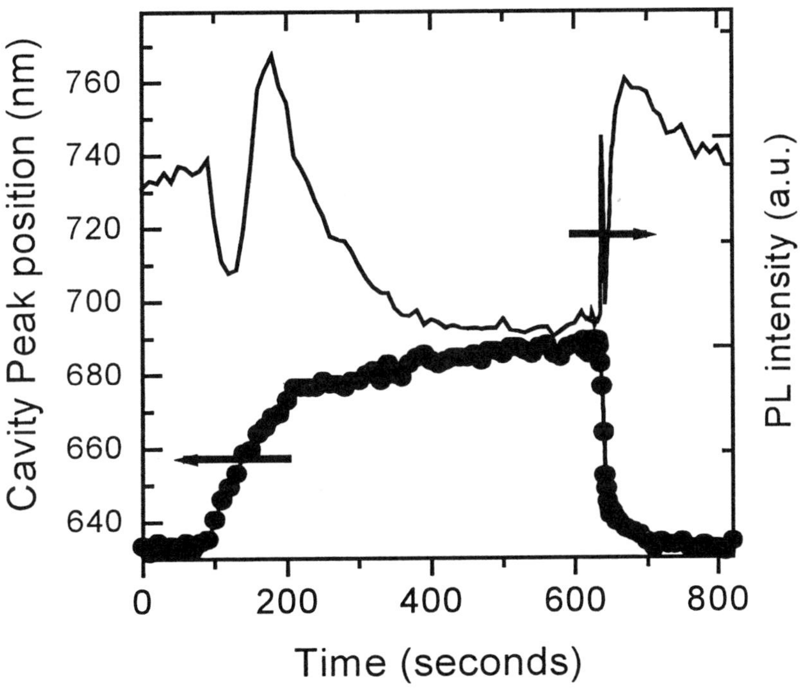

Figure 8. Wavelength position of the resonance peak of the cavity (line with solid circles, left y-axis), and PL intensity of the peak (solid line, right y-axis), as a function of time, of the porous silicon microcavity during exposure to ethanol. The sample is kept under constant flux (0.3 liters/minute) of dry air (from t=0 to t=80 s), air saturated with ethanol (t=80 to 634 s), and back to dry air (t=634 to 820 s).

To understand the origin of the oscillations in the PL intensity, we observe that, using both quantities of Figure 8, the time dependence can be eliminated (Figure 9, top plot), and it is possible to approach the study of the ethanol penetration with a quasi-static model. The hollow circles refer to the ethanol filling, whereas the recovery after the switching back to dry air is reported with solid circles.

To simulate the quasi-static filling, we first point out that the cavity exchanges the gas with the outside only at the interface air/1st DBR. The filling of the cavity with

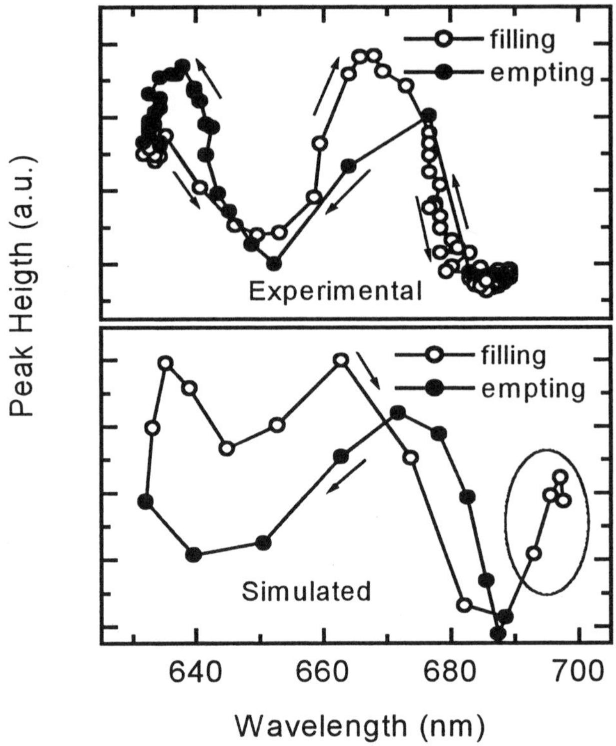

Figure 9. Comparison between experimental (top) and simulated (bottom) filling (hollow circles) and empting (solid circles) of the cavity with ethanol. The experimental data are the PL intensity of the resonance peak of the cavity versus its wavelength position. The simulated data show the depth of resonance peak of the reflectivity spectrum versus its wavelength at 14 progressive levels of ethanol filling. In the simulation of filling, the air (whose dielectric constant is assumed $n_{air}=1$) is replaced by a mixture of air and ethanol with effective dielectric constant $\varepsilon=\varepsilon_{air}+0.4\varepsilon_{eth}-\varepsilon_{air})$, and vice versa for the empting.

experimental filling (top plot, hollow circles) can be compared with the simulated filling (bottom plot, hollow circles).

The comparison points out the following differences. First, the experimental PL peak does not increase at the beginning as expected from the simulation, probably due to fast diffusion to deeper layers before completing the filling of upper layers. The fact that the initial filling is fast is consistent with the observations of Gao *et al.* [21]; the filling probably slows down after the ethanol starts condensing in the pores, making the abrupt empty/full picture closer to reality. Second, the final increase, predicted by the simulation (circled 4 points at longer wavelength side), is not observed in the experiment. This can be due to a) capping of air at the bottom of the cavity, which prevents further diffusion of ethanol into the last layers, b) onset of PL quenching, which starts to be significant on time scales of the order of 1 minute.

To simulate the recovery, we have assumed and taken into account the air capping, and eliminated the last 4 points of the filling simulation. We have then simulated the emptying with the same routine for the filling, switching the dielectric constants (Figure

9, solid markers). The simulation of the recovery works fairly at the beginning, but becomes too crude after emptying few layers in proximity of the air/cavity interface, since the ethanol starts diffusing from the bottom of the cavity. The signature of the diffusion is the final increase of the PL observed in the experimental plot.

5. Electrical response in controlled flux

As third sets of measurements, we have started the electrical characterization under controlled flux. Two gold contacts have been evaporated (electrode thickness was about 50 nm) on the top surface of low resistivity PSM's (0.001 Ω cm substrate). We have started the electrical characterization with low resistivity substrates because low resistivity is expected to have a larger electrical response [15]. We used a volt amperometric technique at constant bias in which the sensor was biased at constant voltage (1V) and the current output was read by a picoammeter.

We have exposed low resistivity PSM's to 1, 2 and 4 ppm of NO_2, under several humidity conditions, to reproduce known performances. Details of the measurement timing are shown in Figure 10.

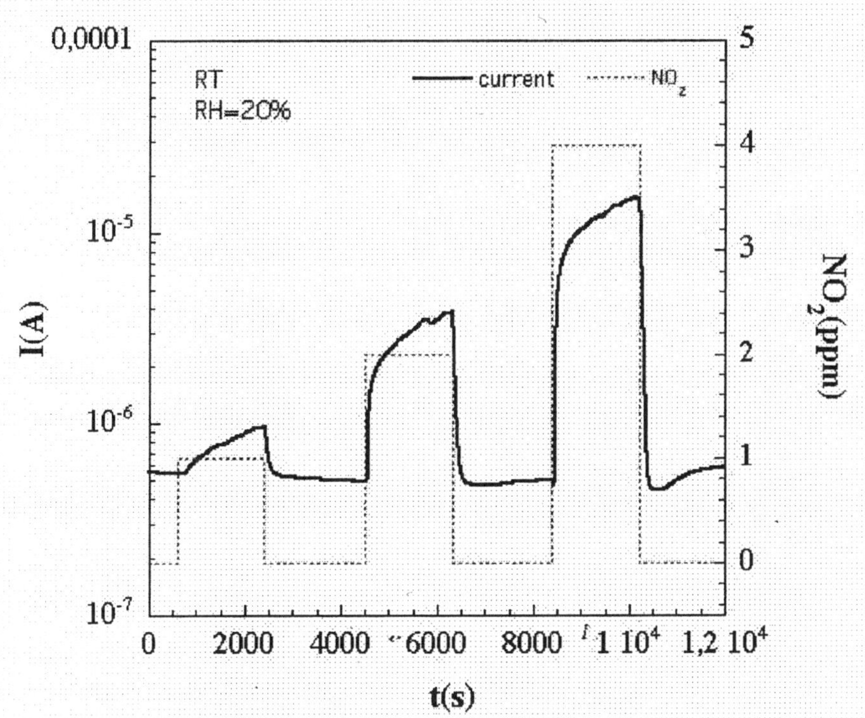

Figure 10. Electrical current through a low resistivity PSM under controlled flux (0.3 l/m) of humid air (20%) containing 1, 2 or 4 ppm of NO_2. The dotted line represents the NO2 concentrations referring to the right scale, while solid line represents device current referring to the left scale.

The behavior of the PSM with ozone is quite different from that with NO_2. Only electrical measurements have been performed so far with ozone. We observed a response much smaller than that to NO_2 with 200 ppb of ozone, in a different way from the well-known sensitivity of all semiconducting gas sensors towards this oxidizing gas. The lack of sensitivity to ozone is beneficial for NO2 detection, since ozone and NO_2 are usually encountered together in practical applications (pollutant sensing).

The electrical conductance increases with relative humidity as shown in Figure 11. The remarkable difference between the response of the cavity peak position (Figure 5) and the conductance (Figure 11) is the kinetic of the response, that is much faster for the peak shift than in the case of conductance.

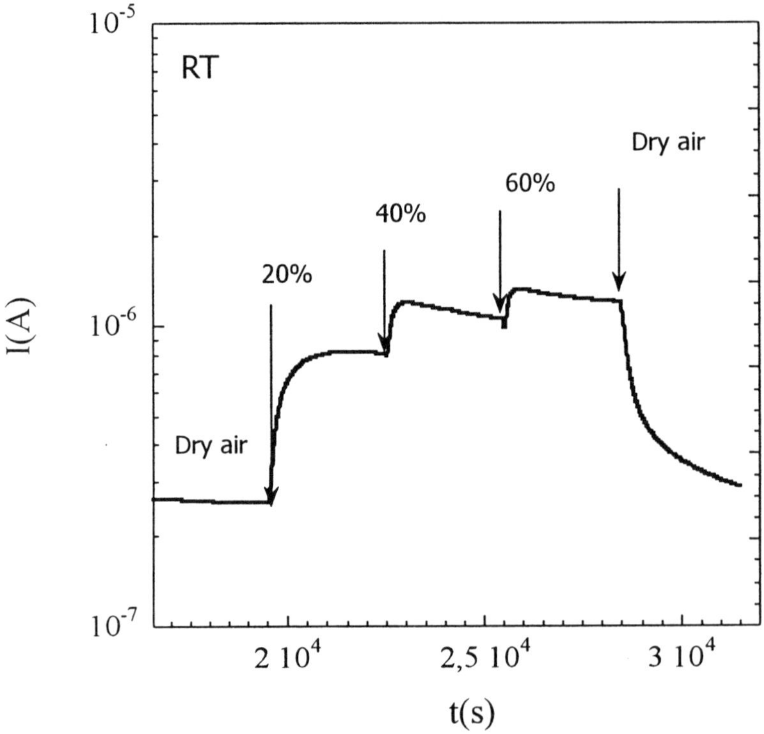

Figure 11. Dynamical variation of current flowing through a PSM biased at 1 V when switching from dry air to different values of humid air and then again to dry air at 20°C.

6. Conclusions

In porous silicon optical microcavities a) the electrical conductance b) the photoluminescence intensity and c) the cavity peak wavelength are properties which change with the environment, and thus allow gas sensing. Although to date we have not yet performed a simultaneous monitoring of the three sensing parameters, their simultaneous monitoring might allow gas discrimination.

The optical response in static air saturated with vapor of several solvents and hydrocarbons shows that the position of the cavity resonance depends on the refractive index, and the photoluminescence intensity depends mainly on molecular dipole moment. The effects of the gas on the photoluminescence intensity, however, have complex origin and might depend also on other properties of the gas.

Preliminary measurement of the optical response in controlled flux shows that, in the case of ethanol and water, there is a *monotonic* relationship between the concentration of the gas and the relative shift of the position of the resonance of the cavity. This property allows a one-to-one correspondence between measured shift and gas concentration. Compared to electrical conductivity and photoluminescence intensity, the response of the peak position is fast, since it is triggered by the *presence* of the gas in the porous layers. On the contrary, the electrical conductivity and photoluminescence intensity changes under the gas electrical or chemical *action*. Therefore, time resolved measurements are best obtained using the peak position as sensing parameter. As a sample application, we demonstrate that the shift of the peak allows tracking of the status of gas penetration and the exposure to the gas in time.

We have also shown that the PSM's electrical conductivity is affected by the presence of NO_2, and the effect is observable at concentrations of the order required by practical applications.

7. Acknowledgements

We acknowledge the support of INFM – progetto PAIS SMOG 2001, and the support of Provincia Autonoma di Trento.

References

1. A. Halimaoui, in: L.T. Canham (Ed.), Properties of Porous Silicon, IEE INSPEC, The Institution of Electrical Engineers, London, 1997, p. 12.
2. R. C. Anderson, R. S. Muller, and C. W. Tobias, Sens. Actuators A21-A23, 835 (1990).
3. M. Ben-Chorin, A. Kux, and I. Schechter, Appl. Phys. Lett. 64, 481 (1993).
4. A. Foucaran, F. P. Delannoy, A. Giani, A. Sackda, P. Combette, and A. Boyer, Thin Solid Films 297, 317 (1997).
5. P. G. Han, H. Wong, M. C. Poon, and N. Wang, J. Vac. Sci. Technol. A17, 1832 (1999).
6. I. Schechter, M. Ben-Chorin, and A. Kux, Anal. Chem. 67, 3727 (1995).
7. A. Motohashi, M. Kawakami, H. Aoyagi, A. Kinoshita, and A. Satou, Jpn. J. Appl. Phys. Part 1 34, 5840 (1995).
8. A. Motohashi, M. Ruike, M. Kawakami, H. Aoyagi, A. Kinoshita, and A. Satou, Jpn. J. Appl. Phys. Part 1 35, 4253 (1996).
9. T. Taliercio, M. Dilhan, E. Massone, A. M. Gue, B. Fraisse,and A. Fou-caran, Thin Solid Films 255, 310 (1995).
10. J. L. Coffer, S. C. Lilley, R. A. Martin, and L. A. Files-Sesler, J. Appl. Phys. 74, 2094 (1993).
11. J. Harper and M. J. Sailor, Anal. Chem. 68, 3713 (1996).
12. J. H. Song and M. J. Sailor, J. Am. Chem. Soc. 119, 7381 (1997).
13. S. Zangooie, R. Bjorklund, and H. Arwin, Sens. Actuators B43, 168 (1997).
14. C. L. Curtis, V. V. Doan, G. M. Credo, and M. J. Sailor, J. Electrochem. Soc. 140, 3492 (1993).
15. C. Baratto, G. Faglia, E. Comini, G. Sberveglieri, A. Taroni, V. La Ferrara, L. Quercia, G. Di Francia, Sens. Actuators B77, 62 (2001).
16. S. Fellah, R. B. Wehrspohn, N. Gabouze, F. Ozanam, J.-N. Chazaviel, J. Lumin., 80, 109 (1999).

17. P. A. Snow, E. K. Squire, P. St. J. Russell, and L. T. Canham, J. Appl. Phys 86, 1781 (1999).
18. V.S.Y. Lin, K. Motesharei K. P. S.Dancil, M. J. Sailor, M. R. Ghadiri, Science, 278, 840, (1997).
19. S. Zangooie, R. Jansoon, and H. Arwin J. Vac. Sci. Techol. A 16, 2901, (1998).
20. H. F. Arrand et. al. J. Lumin., 80, 119 (1999).
21. J. Gao, T. Gao, and M. J. Sailor, Appl. Phys. Lett. 77, 901 (2000).
22. L. Pavesi, Riv. Nuovo Cimento 20, 1 (1997).
23. S. Chan, and P. M. Fauchet, Appl. Phys. Lett. 75, 274 (1999).
24. V. Mulloni, and L. Pavesi, Appl. Phys. Lett. 76, 2523 (2000).
25. V. Mulloni, C. Mazzoleni, and L. Pavesi, Semicon. Sci. and Techn. 14, 1052 (1999).
26. Z. Gaburro, N. Daldosso, L. Pavesi, G. Faglia, C. Baratto, G. Sberveglieri, Appl. Phys. Lett. 78, 3744 (2001).
27. D. A. G. Bruggeman, Ann. Phys. (Leipzig), 24, 636 (1935).

COMPOSITE SILICON-BASED PHOTONIC CRYSTALS AS LIGHT EMISSION AND SENSOR ELEMENTS

L.A. KARACHEVTSEVA
Institute of Semiconductor Physics of NAS of Ukraine
Prospect Nauki 45, Kyiv 03 028, Ukraine

1. Introduction

A new class of composite materials named photonic crystals are a potential technology for modification of the absorption spectrum of electromagnetic waves in semiconductor devices due to photonic band gap formation. The light-assisted method of n-Si electrochemical etching allows the formation of macroporous photonic crystals with a high ratio of pore depth to pore diameter and with strong periodicity. Research into photonic crystals has mainly been directed towards optical and microwave applications. However, one of the main features of the photonic band gap is the decay wave formation and the transition to Joule thermal energy. Anomalous coefficients of the light absorption due to decay wave formation allow the development novel thermal and photo detectors on the basis of macroporous silicon.

Microporous silicon with pore sizes about 10 nm has already been proposed for the development of silicon-on-insulator (SOI) structures in integrated circuit technology (Theunissen, 1972). Observation of the room-temperature photoluminescence and electroluminescence has demonstrated that these are practical and efficient PS-based emitters (Canham, 1990). During the 90s, macroporous silicon with pore sizes greater than 50 nm also received dynamic development (Lehmann et all, 1990). Macroporous n-Si was formed as 2D structures and possessed a photonic band gap in which the optoelectronic applications of silicon, the dominant material for electronics, are widened.

Optical transmittance of 2D periodical photonic structures was investigated theoretically and experimentally (Robertson et all, 1992). Especially, transverse and longitudinal electromagnetic wave components were measured and calculated for 2D macroporous silicon in direction perpendicular to pores (Grunnihg et all, 1996). Optical transmittance of 2D periodic structures was investigated for parallel to rods (out-of-plane) direction theoretically only [9-11]. By the use of the plane wave method Maradudin et all (1994) have calculated the dispersion curves (photonic bands) of electromagnetic wave propagation in a structure consisting an infinite array of parallel, long triangular lattice of air cylinders. Sigalas et all (1998) studied the transmission of electromagnetic waves propagating in 2D photonic crystals for out-of-plane incident

E. Buzaneva and P. Scharff (eds.), Frontiers of Multifunctional Nanosystems, 413–426.

angle as high as 85^0. There is a full calculated photonic band gap for both E- and H-polarization for dielectric constant ratio higher than 12.25. So the aim of our work is experimental investigation of optical transmittance by 2D macroporous silicon structures in direction parallel to pores.

Electrophysical parameters of porous silicon have been investigated since the beginning of the 80s. It has been found that porous silicon layers are highly resistive, with resistances exceeding $10^5\,\Omega\cdot$cm for high levels of porosity (Zimin , 2000). A model has been proposed, based on the formation of regions that are depleted by charge carriers due to several factors: capture of carriers by traps on pore surfaces, leaving of extrinsic atoms in an electrolyte and passivation of the impurity atoms by hydrogen. The resistivity of the macroporous silicon structures of *p*-type conductivity exceeded the resistivity of the initial silicon by 1.6-15 times and was determined by the transport of charge carriers through the low resistance regions of the silicon matrix. Over the last few years, research into *n*-type macroporous silicon structures has undergone significant development. Thus, the next aim of our investigation was to study the electron conductivity, concentration and mobility in 2D macroporous silicon structures on *n*-Si substrates, taking into account the effect of macropore size and concentration.

Multi-element IR-detectors of a new type – uncooled bolometric, pyroelectric and thermoelectric matrices have been actively developed during recent years (Rogalsky, 1999). These are highly attractive for broad usage in general industry, medicine, ecological applications, transport and so on. The important characteristics required of microbolometric matrices are a high temperature coefficient of electrical resistance, low excess noise and good radiation absorption in the working spectral region. The structures based on amorphous (Malyarov et all, 1999) or polycrystalline silicon (Piotrovsky et all, 1998) satisfy such requirements. In this work the results of research into macroporous silicon structures made by photoanodic etching are given. Strong absorption of electromagnetic radiation was measured on macroporous silicon structures with the microporous layers removed so that apparently the initial silicon substrate[11]. Thus the bolometric characteristics of the macroporous silicon were estimated. The ohmic indium contact to macroporous silicon structures was work up, and the temperature dependencies of conductivity, concentration and mobility of electrons, and spectra of excess noise were measured.

The mechanism of macropore formation in silicon by electrochemical etching in a solution of hydrofluoric acid is connected to the generation and transfer of non-equilibrium holes to the n-Si electrochemically treated surface as a result of optical band-to-band electron-hole generation or collision ionization in strong electrical fields. The as-prepared macroporous silicon structures consist of microporous layers on the macropore walls. Such a complex macropore structure with nanocrystals within the micropore layer might be the important element of optoelectronic devices due to emitting and waveguiding (photonic) functions. Microporous silicon layers on macropore walls have their own features: formation in a direction perpendicular to [100], with the total surface area equal to the area of the macropore surface. So photoluminescence of micropore-macropore silicon were presented in this work. The local chemical states and the structure of microporous layers on macropore walls were investigated both before and after KOH etching using Fourier transform infrared spectroscopy and scanning electron microscopy.

2. Fabrication of macroporous silicon structures

The mechanism of macropore formation in silicon by electrochemical etching in a solution of hydrofluoric acid is connected to the generation and transfer of non-equilibrium holes to the n-Si electrochemically treated surface as a result of optical band-to-band electron-hole generation or collision ionization in strong electrical fields. Regimes of the electrochemical pore formation in the macroporous silicon structures were investigated for a linear change of applied voltage (Karachevtseva et all, 1998). This is important for cylindrical macroporous etching and for the control of their diameters. The relationship between the parameters of the electrochemical process (intensity of illumination, current density, applied voltage) was analysed from the point of view of the diffusion-drift theory. Dependencies of the nonequilibrium hole distribution through the depth of the n-Si samples with various values of hole lifetime and the electricalfield strength were calculated for the macropore formation regimes. The reduction of the hole lifetime and the electrical field strength leads to an essential decrease of hole concentration along the silicon anode thickness were estimated. The experimental dependencies of the hole concentration p_0 and current density j_p during macropore formation correspond to the diffusion-drift model of the nonequilibrium hole transfer for an anode thickness exceeding the length of hole diffusion and for comparatively large macropore radiuses (Figure 1 a,b). Macropores with diameters D_p = 1-15 μm, depths h_p = 50-250 μm and concentrations $N_p = 10^5$-10^6 cm^{-2} were formed due to the generation and transfer of nonequilibrium holes to the n-Si electrochemically treated surface and using the developed regimes of the electrochemical process stabilization.

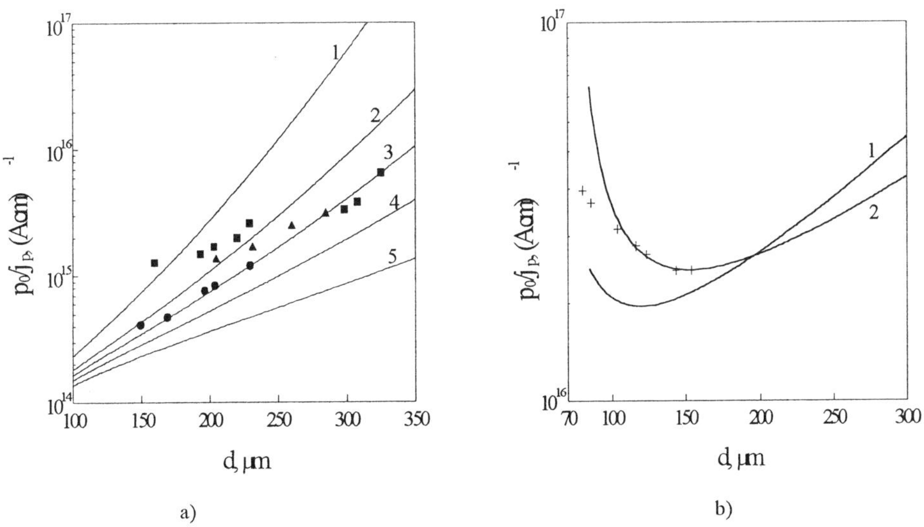

Figure 1. Experimental (signs) and theoretical dependencies of p_0/j_p on silicon anode depth during pore formation for U = const (1a) and U = ad + b, a = 3.2.10^{-3} V/cm, b = -2.7 10^{-2} V (1b). 1a: hole lifetime τ_p, 10^{-7}: 1 - 3, 2 - 5, 3 - 7, 4 - 10, 5 - 20; 1b: hole lifetime τ_p, 10^{-7}: 1 - 7, 2 - 10.

The starting material consisted of n-type silicon with [100] orientation and 2-5 Ω.cm resistivity. The initial n-Si plates were chemically polished in a 1:3 HF and HNO_3 water solution and anisotropically etched in 10% solution of KOH in water to allow pit formation (the periodic structures were anisotropically etched after the photolithography procedure). Ohmic contacts were provided by rubbing some In-Ga eutectic alloy on the edge of the silicon plate and by a platinum electrode in the electrolyte. The sample was mounted in an electrochemical cell and connected to a potentiostat by the 3-electrode scheme. The applied voltage was measured relative to a platinum wire near the sample surface. The electrolyte was 5 weight percent hydrofluoric acid in water. During the electrochemical etching process the sample was illuminated by radiation from a 100 W tungsten lamp. Additional anisotropy etching in 10% KOH solution was used for removing the products of the electrochemical process from the macropore walls.

3. Optical transmittance of 2D periodical photonic structures

Optical transmittance of 2D periodic structures was investigated for parallel to rods (out-of-plane) direction theoretically. By the use of the plane wave method Maradudin have calculated the dispersion curves (photonic band structure) of electromagnetic wave propagation in a structure consisting an infinite array of parallel, long triangular lattice of air cylinders. Accodingly to Maradudin results in two-dimensional periodic structure with period a absolute photonic band gap forms at $\lambda = 1.5\lambda_a$ (λ_a - is the optical period of the photonic structure, equaled to $(a-D_p)\varepsilon^{1/2}+D_p)$ by 2D photonic structure with $\varepsilon = 13$ and the cylinder volume V = 0.84. Soucoulis calculations predicted photonic band gap formation common for H- and E-polarization of electromagnetic irradiation at $\lambda \approx 1.2\lambda_a$ at $\varepsilon = 12.3$ and V = 0.73. We measured absorption bands at $\lambda = 1\text{-}1.5\lambda_a$ on periodical macroporous silicon structures. Obtained results correlate qualitatively with calculations. Absolute band gap below the first band was not measured. So macroporous silicon structures are effective media for long electromagnetic waves.

Absorption spectra of macroporous silicon with different pore diameters and concentration have common features at $\lambda \leq \lambda_a$. There is an essential reduction in transmittance of electromagnetic radiation with a shortening of the wavelength. Recalculated absorption coefficient is equaled 300-500 cm^{-1}. Really, for earlier measurements at $\lambda = 1.15$ μm and $\lambda = 3.39$ μm of He-Ne laser optical transmittance of the macroporous structure was $3\text{-}6\cdot10^{-3}$ apparently homogeneous material (Karachevtseva et all, 2000). At short wavelengths $\lambda < D_p$ the optical transmittance slightly grows. Transmission spectra contain steps or oscillations (Figure 2a-c). The step frequency in the long-wave part of the spectrum is proportional to the distance between the pores ($\Delta\nu^l \sim a-D_p$), but in the short-wave region it is proportional to the diameter of the pores ($\Delta\nu^s \sim D_p$), as in Figure 3a,b. The frequency intervals of the steps are related via $\Delta\nu^s/\Delta\nu^l \approx 4$ for the commensurable quantity $a-D_p$ and D_p. Transmission spectra of macroporous silicon as well as the formation of the steps can be explained by a model of directed and radiated optical modes. Macroporous silicon is a short wave-guide structure with a large contrast in dielectric constants of the pores and the silicon matrix. The full electromagnetic field of optical wave-guides can be represented as the sum of fields of directed modes, transferring energy of radiation over a infinite distance in a spatially stable regime, and fields of radiation in a spatially nonstable regime. For short waveguides, the second component plays a vital role

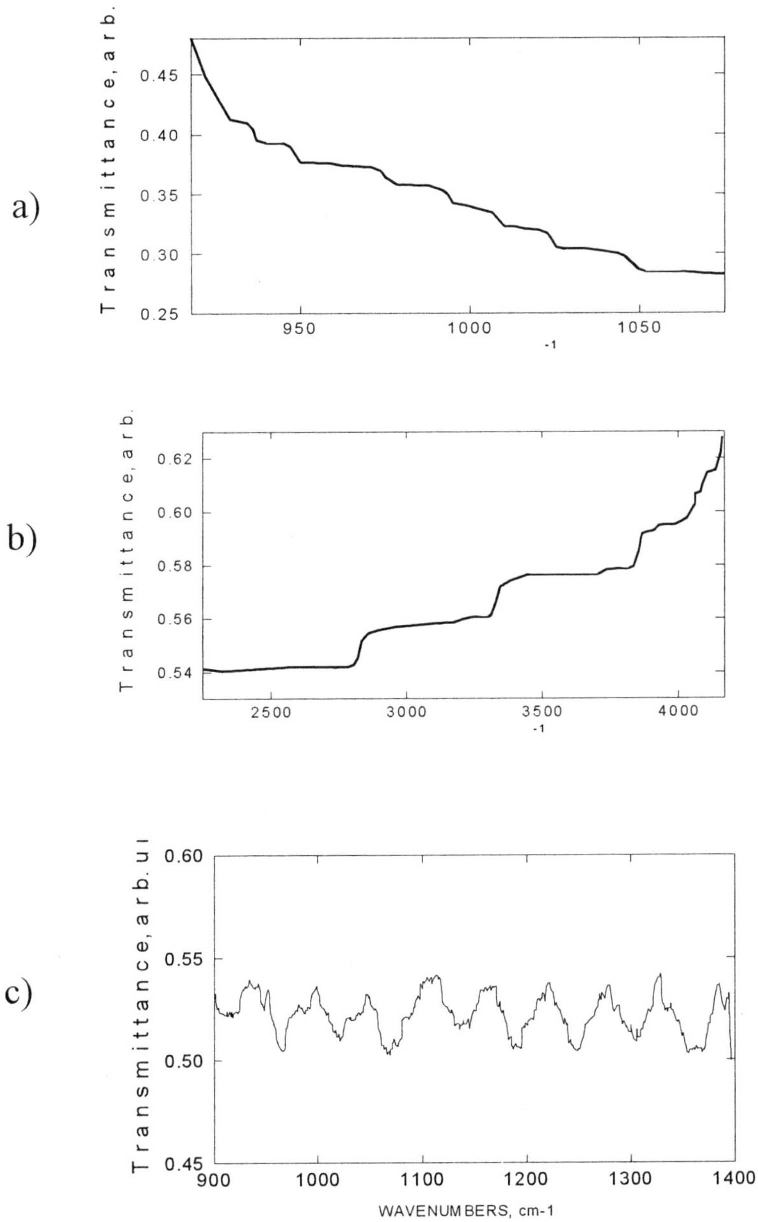

Figure 2. Long-wave steps (a), short-wave steps (b) and oscillations (c) in the transtission spectrum of macroporous silicon structure.

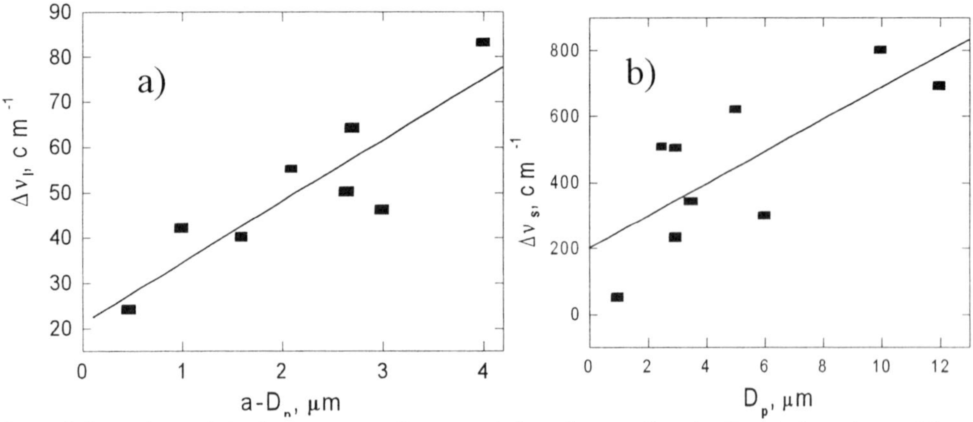

Figure 3. Dependency of the short-wave step frequency $\Delta\nu^s$ on the pore diameter D_p (a); dependency of the long-wave step frequency $\Delta\nu^l$ on the distance between pores a-D_p (b).

In the short-wave spectral region the directed optical modes form on macropores, because the step growth of transmittance takes place for $\lambda \leq D_p$. So macropores present themselves as wave-guides of a metal type (the walls of the macropores are enriched with electrons). For normal incidence of electromagnetic irradiation on macroporous structures there might be sliding radiation incidence on macropore walls. It is known, that for sliding incidence of radiation on conductive surface the absorption coefficient for TM (p) waves is 10^3-10^4 times greater of absorption coefficient for TE (s) waves. So TE modes are formed on conductive macropore walls with order n = 1, because the first directed mode is formed for $\lambda = D_p$. In the middle region $D_p < \lambda < \lambda_a$ it is possible the directed mode formation for the silicon wave-guides and decay mode formation for the macropores. Such modes are formed in the same medium (the silicon matrix) and So transmission spectra of macroporous silicon as well as the steps were explained by a model of directed and decay optical modes formed by macroporous silicon as a short

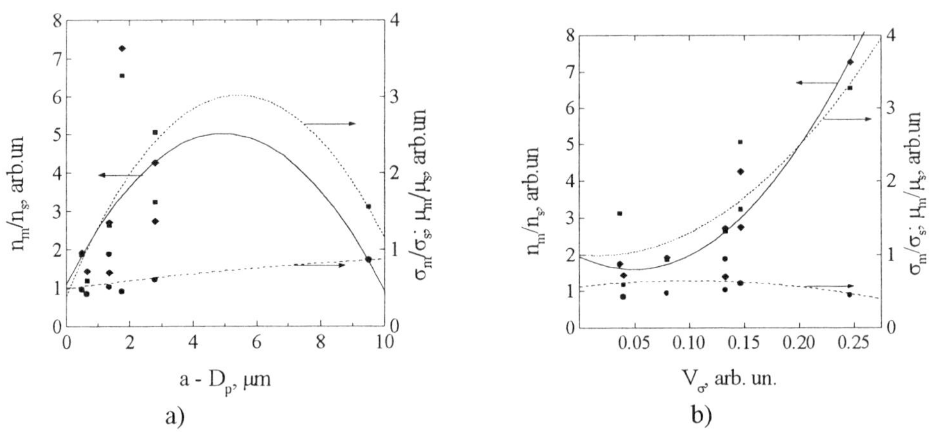

Figure 4. Relative conductivity of macroporous silicon structures (•) and a matrix of macroporous layers (♦) for different values of the macropore volume (a); relative electron concentration (•) and mobility (♦) of macroporous silicon structures for different values of the macropore volume (b).

waveguide structure with a large contrast in dielectric constants of the pore and the silicon matrix. So optical transmittance by composite macroporous silicon structures with conductive pore walls might have characteristics of metal waveguides. As well as optical transmittance by composite macroporous silicon structures with conductive pore walls might have characteristics of metal waveguides. As well as optical transmittance by composite macroporous silicon structures with isolated pore walls might have characteristics of dielectric waveguides.As well as optical transmittance by composite macroporous silicon structures with isolated pore walls might have characteristics of dielectric waveguides. Investigation the photonic gap, as well as wave guide band formation will permitt to determine the decay wave energy dissipation in composite photonic crystals of the macroporous silicon.

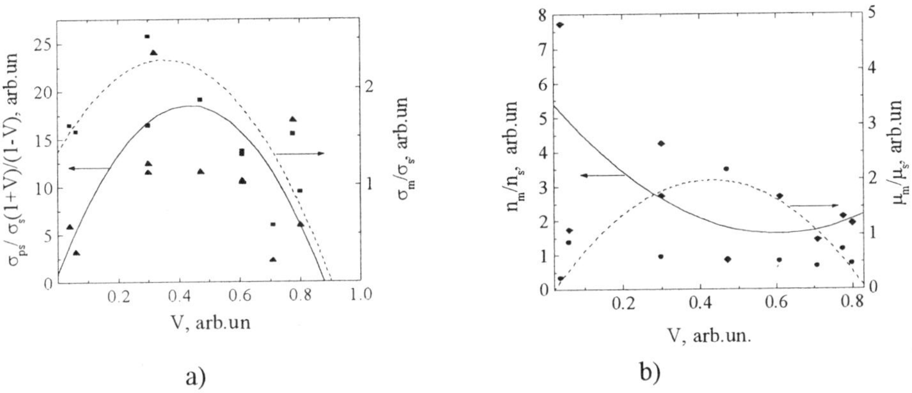

Figure 5. Dependencies of electron conductivity (•), concentration (Δ) and mobility (♦) in macroporous silicon structures on distance between macropores (a) and on volume of regions enriched by electrons (b).

4. Electron transport properties in macroporous silicon with the conductive pore surface

Research into photonic crystals has mainly been directed towards optical and microwave applications. However, one of the main features of the photonic band gap is the decay wave formation and the transition to Joule thermal energy. Anomalous coefficients of the light absorption due to decay wave formation allow the development novel thermal and photo detectors on the basis of macroporous silicon. Thus the bolometric characteristics of the macroporous silicon were estimated. The ohmic indium contact to macroporous silicon structures was work up, and the temperature dependencies of conductivity, concentration and mobility of electrons, and spectra of excess noise were measured.

A method of creation of an ohmic contact without an intermediate dielectric layer was realized for the formation of metal-silicon and metal-macroporous silicon low resistance contacts (Hroderic, 1978). Thermal annealing of the "In - n-Si" and "In – macroporous n-Si" structures was performed at temperature of 200-500°C. With a change in annealing temperature from 200°C up to 400°C the transient contact resistance decreased 2 – 2.5 times and reached an optimal value of $R_0 \approx$ 4-10 $\Omega \cdot cm^2$, that is, less than the transient resistance of an Al-porous silicon contact, formed at a temperature of T = 400°C.

We investigated electron transport properties in macroporous silicon with the conductive pore surface. For this purpose silicon structures were anisotropically etched in potassium hydroxide (KOH) after electrochemical treatment, and a high concentration of donor type OH^- bonds takes place on the macropore wall surface.

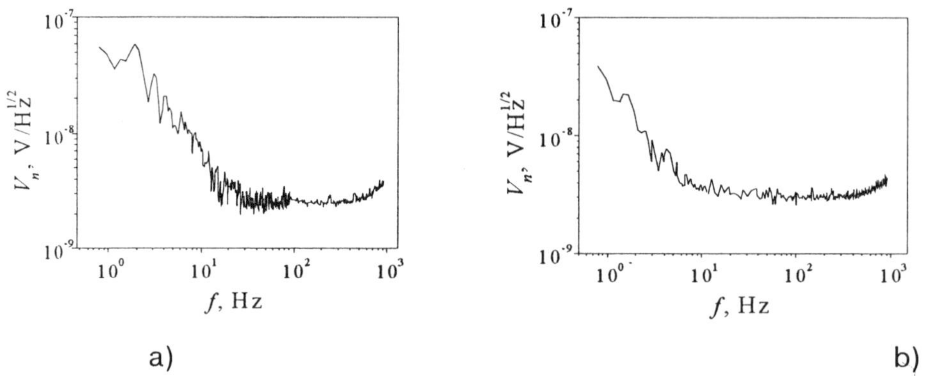

Figure 6. Noise voltage V_n as a function of frequency for silicon substrate with $\rho = 2\ \Omega\cdot$cm (a) and macroporous silicon structure (b).

Figure 4a shows the dependencies of the relative electron conductivity of the macroporous silicon structures on the macropore volume V. The results of the measurement of the electron conductivity on the macroporous layers were recalculated, taking into consideration the two-layer structure, including the measured parameter σ_m, and the silicon substrate parameters σ_s. The matrix conductivity σ_{ps}^M of the macroporous layer was determined according to the theory of effective media for cylindrical inclusions. For a comparative analysis of the electron conductivity in the matrix of the macroporous structures the value σ_{ps}^M was considered relatively to the appropriate silicon substrate parameter σ_s. The conductivity of electrons in the two-layer silicon structures, and also in a matrix of macroporous layers grow with an increase of the macropore volume. These parameters reach a maximum for V = 0.3 - 0.4, and then decrease (Figure 4a). The electron concentration in the two-layer structures grows with an increase of the macropore volume and reaches a maximum for V = 0.3 - 0.4, as electron conductivity (Figure 4b). Thus the electron mobility decrease monotonically. An explanation of the experimental results was reached after analysis of the dependencies of electron conductivity, concentration and mobility on the distance between pore a - D_p (a – average lattice constant of the macroporous structure). Figure 5a shows such dependencies with a strong maximum in the electron conductivity and concentration at $a - D_p \approx 2$ μm due to regions with thickness $\Delta D \approx 1$ μm, enriched by electrons, formed around pores. Figure 5b shows the dependencies of electron conductivity, concentration and mobility on the volume of the regions V_σ in macropore structures, reached by electrons. Figure 5b shows the pronounced growth of electron conductivity and concentration over the volume V_σ due to regions around pores of thickness $\Delta D \approx 1$ μm, enriched by electrons. electron conductivity in the matrix of the macroporous structures the value $\sigma_{ps}{}^{M}$ was considered relatively to the appropriate silicon substrate parameter σ_s.

5. Bolometric characteristics of macroporous silicon structures

The temperature dependencies of conductivity, concentration and mobility of macroporous silicon structures on n-Si substrates in the range 77-300K repeat the temperature curves of the corresponding parameters for crystalline silicon. The index p in the electron mobility - temperature relationship $\mu_p \sim T^p$ is high for n-type silicon compared with other major electronic semiconductor materials. The value p = -2.6 does not correspond to the theoretical value p = -1.5 due to various phonon scattering mechanisms. For example, two-phonon scattering results in a cubic relationship; $\mu_n \sim T^{-3}$. For impurity ionization the free charge concentration remains constant, and the temperature dependence of conductivity is determined by the temperature dependence of mobility, as in our case. The diapasons of the relative temperature changes σ^{77}/σ^{300}, n^{77}/n^{300} and μ^{77}/μ^{300} for macroporous silicon structures and crystal n-Si are comparable. The temperature coefficient of resistance ranges from 0.4 - 1% at room temperature to 1.5 - 4% at 77K.

Figure 6 shows the dependencies of excess noise on frequency for the silicon substrate (a) and for macroporous silicon (b). The noise level is comparable for the initial substrate and for macroporous silicon. White noise is mainly due to Johnson one. 1/f noise is visible at frequencies < 10 Hz. Absolute noise values are $(2\text{-}5).10^{-9}$ $VHz^{-1/2}$ for frequencies > 10 Hz, which is more than 10 times lower than the noise level in amorphous and α-Si. For an average time constant of 10 ms, voltage responsitivity of 10^3 W/K, and temperature coefficient of resistance of 1%, the detectivity of thermal bolometric elements on the base of macroporous silicon might be equaled $5\cdot10^9$ $cmHz^{1/2}W^{-1}$.

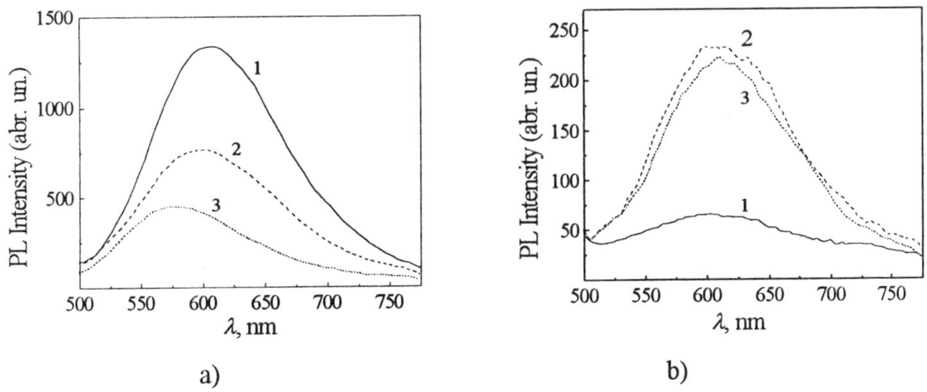

Figure 7a. PL spectra of macroporous silicon structures formed by anodization at various currentdensities through macropores; (1) j_p = 40 mA/cm^2, sample N1; (2) j_p = 46 mA/cm^2, sample N2, and (3) j_p = 54 mA/cm^2, sample N3 from Table 1.

Figure 7b. PL spectra of macroporous silicon structures; (1) sample N4, (2) sample N5, and (3) back surface of the mechanically treated sample N4 from Table 1.

So macroporous silicon structures are a promising material for the fabrication of thermal detectors due to their bolometric characteristics, which are better than those of amorphous or polycrystalline silicon. After addition anode oxidation and silica layer formation on the pore wall the photoconductivity signal in macroporous silicon structure is increased apparently the silicon crystal one due to high absorption coefficients.

In general, high-conductivity coatings on macropore walls increase the light absorption and the energy dissipation in the spectral ranges of both photonic and waveguide bands. On the other hand, low-conductivity coatings form potential barriers on the pore walls as a result of an amplification of the photovoltaic effect. So surface layers of aluminium with silicon, boron and silicon carbides on macropore walls of the about 100 nm thickness might be formed by a method of gas transport reactions.

6. Photoluminescence in complex micropore-macropore silicon structures

Photoluminescence (PL) spectroscopy was performed at room temperature by using a photoluminescence-lifetime measurement system with a streak scope. The excitation source was a nitrogen laser pulse, and PL was measured by the photon-counting method. The chemical states of the micropore-macropore structures were identified using Transmission Fourier Infrared (FTIR) Spectroscopy. For the infrared analysis the back surface of the samples was mechanically polished to make specimens of 80 μm thickness. To subtract the background absorption of the Si substrate, the same silicon wafer of 80 μm thickness was used as a reference. The microstructure of the outer

TABLE 1. *Characteristics of macroporous silicon samples*

Sample number	Resisti-vity . □.cm)	Current density, j (mA/cm^2)	Macro-pore diame-ter, D_p, (□m)	Pore con-centra-tion, N_p.10^{-5} (cm^{-2})	Pore volume, V_P (arb.un.)	Pore current density, j_P (mA/cm^2)	icPS ick-ess m)	MicPS volume (arb.un.)
1	4.5	4.4	3	16	0.11	40	700	0.13
2	4.5	9.7	4	20	0.19	46	250	0.06
3	4.5	18	4	26	0.33	54	150	0.05
4	2	11	3	34	0.24	51	300	0.11
5	2	11	5	12	0.21	58	150	0.03

surfaces and cleaved cross-sections of the micropore-macropore silicon structures were studied by scanning electron microscopy (SEM) using a JEOL JSM-6100 operating at 15kV.

Orange photoluminescence, peaking near 600 nm with 10 μW/cm^2 intensity was measured on macroporous silicon with nanocrystals within a microporous layer (Figure 7a). The process of macropore formation is self-controlled and is determined by the

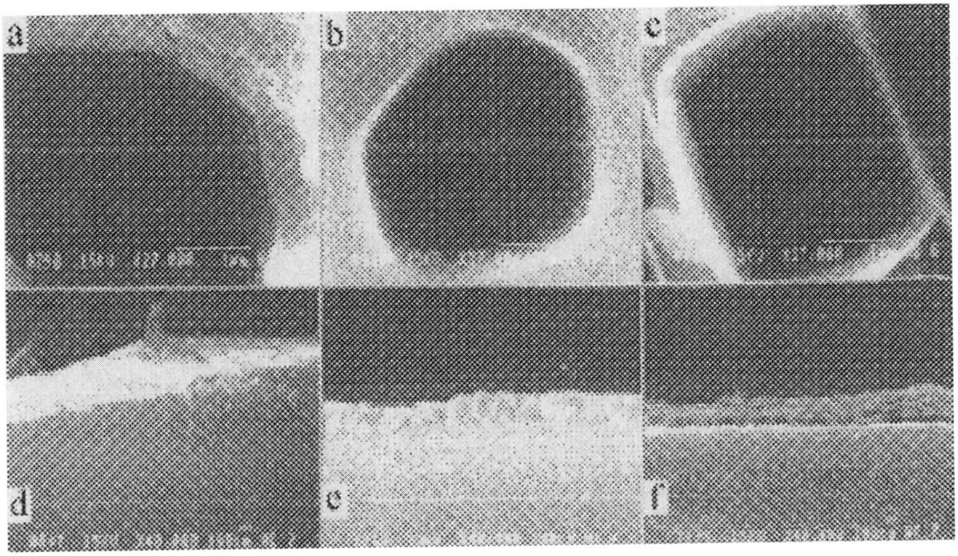

Figure 8. SEM images of macropore outer surfaces (a)-(c) and cross sections of macropore walls (d)-(f); (a) and (d) sample N1, (b) and (e) sample N2, (c) and (f) sample N3 from table 1.

current density $j_{ps} \approx 20$ mA/cm^2 through the pores. However, for a silicon anode with deep pits or pores a surface charge of holes accumulates at the tips for current densities $j_p > j_{ps}$. Regimes of the microporous layer formation corresponding to the condition $j_p \geq j_{ps}$ were realized in our case (Table 1). PL intensities, lifetimes and peak widths were comparable to PL parameters of flat microporous layers (Petrova-Kox et all, 1992; Cullis et all, 1994; Tsai et all, 1993). After KOH treatment there was no measurable PL. Figure 7b shows that orange PL was observed from macropore walls for silicon structures having microporous layers. This is confirmed by the observation of the same PL peak from the mechanically polished back surface of the macropore structures.

According to SEM results, the thickness of the microporous layers on macropores was 100-700 nm (Figure 8). Microporous layers with a tree-like structure (samples N1 and N4 from Table 1) were formed in a direction perpendicular to [100], as in figure 3a, e, whereas higher current densities j_p give thinner microporous layers without the tree-like strucutre (Figure 8b, c, f, h). After KOH treatment, the microporous layers on macropore surfaces grew thinner, and the macropore walls were roughened, in a similar manner to NaOH treatment of the flat microporous layers.

FTIR spectroscopy data (Figure 9) confirmed that the microporous layers on the macropore walls contain hydrides, oxides and organic species as in the flat microporous layers (Almeida et all, 1990; Anderson et all, 1993; Prokes et all, 1995). However, the peaks on the FTIR spectra are stronger relative to the plane microporous layers. This is due to the fact that the total surface area of the microporous layers equals the surface area of the macropore walls. A strong peak at 465 cm^{-1} was observed, due to the rocking motion of the oxygen atoms perpendicular to the Si-O-Si plane. Such a peak is characteristic of bulk SiO_2 and thermal SiO_2 films on silicon plates. A peak at 465 cm^{-1} was not observed in the flat microporous silicon. The observation of this peak in macroporous silicon structures may be due to the large macropore surface area leading to a large total surface area within the microporous layer. Figure 9b showed that the 465 cm^{-1} peak intensity decreased strongly after KOH.

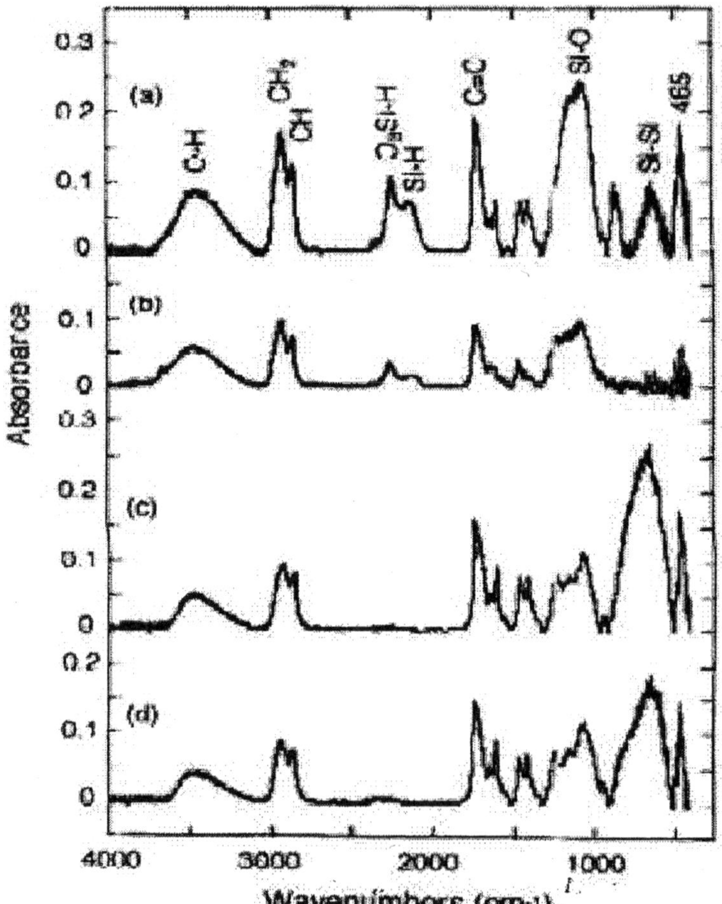

Figure 9. FTIR spectra of samples N4 and N5 from table 1, treated with 10% KOH for different time; (a) 0 s, (b) 10 s, sample N4; (c) 5 s, (d) 20 s, sample N5.

Samples with lower PL intensities corresponded to higher current densities during macropore formation (Figure 7a, Table 1). A blue shift was observed after the PL intensity decrease. According to our SEM and FTIR results, higher current densities gave thinner micPS with large porosity, and hydride peaks became negligible. Reduction of hydrogen bonds on the microporous surfaces formed dangling bonds which acted as nonradiative recombination centres. So microporous layers on macropore walls showed a lower PL intensity due to the thinness of the PS layers and the small hydride concentration. A blue shift in the PL peak is reported (Cullis et all, 1999) to be evidence of the reduction in Si particle size.

So composite macropore structure with nanocrystals might be the important element of optoelectronic devices due to emitting and waveguiding (photonic) functions. And gas-phase reactions can be realised at unusually low temperatures due to the interaction of silicon with highly reactive gaseous reagents. Extremely new results of the processes of the SiC nanowire and nanotube formation showed the growth on the silicon macropore surface in carbon media the spiral wires and tubes of the about 100 nm diameter. So coatings with nanoparticles on macropores might be effective emitting elements in photonic macroporous silicon structures.

7. Conclusions

The light-assisted method of electrochemical etching of n-Si allows the formation of cylindrical macropores with a high ratio of pore depth to pore diameter and with strong periodicity. Such structures are highly promising for modification of the electromagnetic wave spectra in semiconductor devices. The method of macropore formation on low resistance silicon plates was improved, and the regimes of electrochemical pore formation in macroporous silicon structures correspond to the diffusion-drift model of the nonequilibrium hole transfer.

Transmission spectra of macroporous silicon were explained by a model of directed and decay optical modes formed by macroporous silicon as a short wave-guide structure with a large contrast in dielectric constants of the macropore and the silicon matrix. Anomalous coefficients of the light absorption due to decay wave formation allow the development novel semiconductor devices on the basis of composite macroporous silicon.

Macroporous silicon structures are a promising material for the fabrication of thermal detectors due to their bolometric characteristics, which are better than those of amorphous or polycrystalline silicon. Conductive and isolated coatings improve bolometric characteristics and amplificate the photovoltaic effect in composite macroporous silicon.

Micropore layers of thickness 100 - 700 nm on macropore walls of two-dimensional silicon structures were formed for the high regime of the anodization current density during macropore formation. Photoluminescence in composite macropore silicon structures with nanocrystals of microporous layers was firstly measured. So coatings with Si and Si-content nanoparticles on macropores might be effective emitting elements in photonic macroporous silicon structures.

References

1. Almeida, M., Pantano, C.G. (1990), Structural invstigation of silica gel films by infrared spectroscopy, J. Appl. Phys. 68, 4225-4232.

2. Anderson, R. C., Muller, R.S., Tobias, C.W. (1993), Chemical surface modification of porous silicon, *J. Electrochem. Soc.* **140**, 1393-1396.
3. Canham, L. T. (1990), Silicon quantum wire array fabrication by electrochemical and chemical dissolution of wafers, *Appl. Phys.Lett.* **57**(10), 1046-1048.
4. Cullis, G., Chanham, L. T., Williams, G. M., Smith, P. W., and Dosser, O. D. (1994), Correlationof the structure and optical properties of luminescent, highly oxidized porous silicon, *J. Apple. Phys.* **75**, 493-501.
5. Grunning, G.,and Lehmann, V. (1996), Fabrication of 2D infrared photonic crystals in macroporous silicon, article in *Photonic Band Gap Materials*, Eds. C. M. Soukoulis, Kluwer Academic Publishers, 437-444, .
6. Hroderic, E.H. (1978), *Metal-Semiconductor Contacts*, Clarendon, Oxford.
7. Karachevtseva, L. A., Litvinenko, O. A., Malovichko, E. A. (1998), Stabilization of Electrochemical Formation of Macropores in *n*-Si, *J.Theor.and Experim.Chem.*(Rus.) **34**(5), 314-318.
8. Karachevtseva, L.A., Lytvynenko, O.A., and Stronska, E.J. (2000), Development and optical characteristics of the macroporous silicon structures, *Semiconductor Physics, Quantum Electronics & Optoelectronics* **3**, 22-25.
9. Lehmann,V., Foll, H. (1990), Formation Mechanism and Properties of Electrochemically Etched Trenches in n-Type Silicon, *J. Elektrochem. Soc.* **137**(2), 653-659.
10. Malyarov, V.G., Khrebtov, I.A., Kulikov, Yu.V. (1999), Comparative investigations of bolometric properties of thin-film structures based on vanadium dioxide and amorphous hydrated silicon, *Applied Physics* (Rus.) **2**, 86-96.
11. Maradudin, A., McGurn, R. (1994), Out of plane propagation of electromagnetic waves in a two-dimensional periodic dielectric medium, *J. Mod. Opt.* **41** (2), 275-284.
12. Petrova-Koch,V., Muschik, T., Kux, A., Meyer, B.K., and Koch, F. (1992), Rapid thermal-oxidized porous Si, *Appl. Phys. Lett.* **61**, 943-951.
13. Piotrovski, J., and Dobransky, L. (1998), Micromachined thermal devices - detectors and emitters, *Opto-Electronics Rev.* **6**, 69-78.
14. Prokes, S.M., and Carlos, W.E. (1995), Oxygen defect center red room temperature photoluminescence from freshly etched and oxidized porous silicon, *J. Appl. Phys.* 78, 2671-2674.
15. Robertson, M., Arjavalingam, G. (1992), Measurement of photonic band structure in a two-dimensional periodic dielectric array, *Phys. Rev. Lett.* **68**(13), 2023-2026.
16. Rogalsky, A. (1999), *Infrared Detectors*, Electrocomponent Science Monographs, V.10, Gordon and Breach Publishers.
17. Sigalas, M.M., Bismas, R., Ho, K.M., and Soukoulis K.(1998), Theoretical investigation off-plane propagation of electromagnetic waves in two-dimensonal photonic crystals, *Phys. Rev. B*, **58**(11), 6791-6794 .
18. Theunissen, M. I. I. (1972), Etch channel formation during anodic dissolution of n-type silicon in aqueos hydrofluoric acid, *J. Electrochem.Soc.* **119**(3), 351-359.
19. Tsai, C., Li, K.-H., Sarathy, J., Shin, S., and Campbell, J.C. (1991), Thermal treatment studies of the photoluminescence intensity of porous silicon", *Appl. Phys. Lett.* **59**, 2814-2816.
20. Turner, D. R. (1958), Electropolish silicon in hydrofluoric acid solutions, *J. Electrochem. Soc.* **105**(7), 402-408.
21. Zimin, S.P. (2000), Classification of porous silicon electrical properties, *Phizika i Tekhnika Poluprov.* **34**(3), 359-363.
22. Zimin, S.P. Mosquitoes E.P. (1998), Influence of a short-term annealing on conductivity of porous silicon and transient resistance of a contact aluminium - porous silicon, *J. Technical Physics Letters* (Russ.) **24**, 45-51.

OPTICAL TRANSMISSION OF MACROPOROUS SILICON

A.D.REMENYUK, E.V.ASTROVA, I.G.LANG, L.I.KOROVIN, V.B.SHUMAN

194021, Ioffe Physico Technical Institute, Polytechnicheskaya 26, St.Petersburg, Russia

Macroporous silicon (MaSi) has regular structure, its macropores have high aspect ratio of the pore length to the diameter and can propagate through the total wafer thickness. The structure consists of an array of long parallel pores of circular or square cross-section, characterized by a dielectric constant $\varepsilon_a = 1$, embedded in a medium whose dielectric constant is $\varepsilon_b = 12$ and form periodic structure of high contrast. The intersections of the axes of the pores with a perpendicular plane form one of the two-dimensional Bravais lattice, usually a square or triangular lattice. The material is promising for application in various fields, such as photonics, vacuum electronics, x-ray focusing, etc. In spite of the fact that 10 years have passed since the technique of electrochemical etching of macropores in n-Si under backside illumination has been proposed [1], its technology and properties have not been studied much. Unlike microporous Si, macroporous Si can have low porosity (<10%) and in many respects retains the properties of a single crystal. The MaSi is compatible with conventional processes of Si device fabrication and the field of its application expands [2 - 4].

The optical properties of MaSi excite a grate deal of interest recently in the case, when electromagnetic waves propagate in a plane perpendicular to the pore axes. Due to periodicity of the structure the dispersion curves for the light propagation form bands that are separated by gaps at points of symmetry in the corresponding two- dimensional first Brillouin zones. This is a case of so called photonic crystals.

Recently it has been shown that MaSi can be used as a short-pass filter in the visible wave range if the light propagates along the through pores [5] . The short-pass filter is that which transmits light of short and absorb long wavelengths. They are an important part of many imaging systems. They can be used by themselves or in combination with narrowband filters to transmit isolated portions of the spectrum, often in the ultraviolet (UV). For example, in UV studies of the sun, short-pass filters can eliminate unwanted visible light [6]. Such filters are normally made up of a multilayer thin film coating on a glass substrate. But for short enough cutoff wavelength, the materials of the filter itself – both the substrate and the thin film layers – limit the filter performance. One way around this problem would be to create a short-pass filter in which the only transmissive medium is vacuum or air. Macroporous silicon seems to be more flexible and tunable material. So the pores are the apertures of small diameter of large length in an absorbing medium, that leads to light diffraction at the entrance of the pores.

Here we report the results of more detailed theoretical and experimental studies of MaSi spectral characteristics for this case.

E. Buzaneva and P. Scharff (eds.), Frontiers of Multifunctional Nanosystems, 427–430.

1. Sample preparation.

The samples under the study were fabricated by photoelectrochemical etching [7]. The starting material was single crystal (100)-oriented Czochralski grown n-type silicon with a resistivity ρ=15 Ω·cm. A standard photolithographic process was used to form pits on the polished surface of the silicon. Deep pores were etched electrochemically in 4 % HF aqueous solution for 300 to 450 min under backside illumination under the voltage 4V and a current density of j=5 mA/cm^2 was used. The pore diameter was altered by oxidation of the samples and subsequent etching of the oxide. The through pores formed a triangular lattice with a pitch d = 12 μm. The pore radii R were 2,5…4,3 μm and the lengths L were 120…208 μm varying from a sample to sample. In order to obtain through pores the remaining after electro-chemical etching nonporous substrate was removed from the sample backside by mechanical polishing. Typical samples, which we have studied, are shown on Fig. 1.

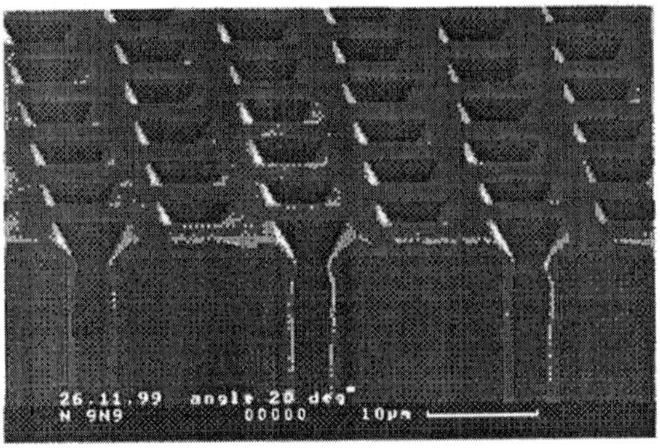

Figure. 1. Scanning microscope image of macroporous silicon

2. Results and discussion.

Optical transmission measurements were performed using a SF-26 spectrometer fitted with photomultiplier as a detector. The sample was placed normally to the wafer surface, i.e. the beam was parallel to pore axes. Fig.2 shows the transmission spectra of four samples. The figures near the curves indicate the pore radii and length in μm. The transmittance strongly depends on the pore radius, whereas the influence of the pore length is much more weak. The spectral region of sharp change of transmission shows the shift to the long wavelengths, when pore diameter increases and the pore length decreases. The absolute transparency is rather low, the reason of that is not quite clear now.

The figures show the pore radii and the pore length in μm for each of 4 samples. The general expression for calculation spectra was obtained from general diffraction theory Landau [8], taking into account only the diffraction at the entrance of the pores and transmittance of the pores, which is limited by absorbance of pore walls:

$$\widetilde{T}_C(\lambda)=\frac{4\pi}{\sqrt{3}d^2}\int_0^R r_i dr_i\left\{\left[\int_0^\beta dx\cos xJ_0\left(\sqrt{ax}\right)\right]^2+\left[\int_0^\beta dx\sin xJ_0\left(\sqrt{ax}\right)\right]^2\right\} \tag{1}$$

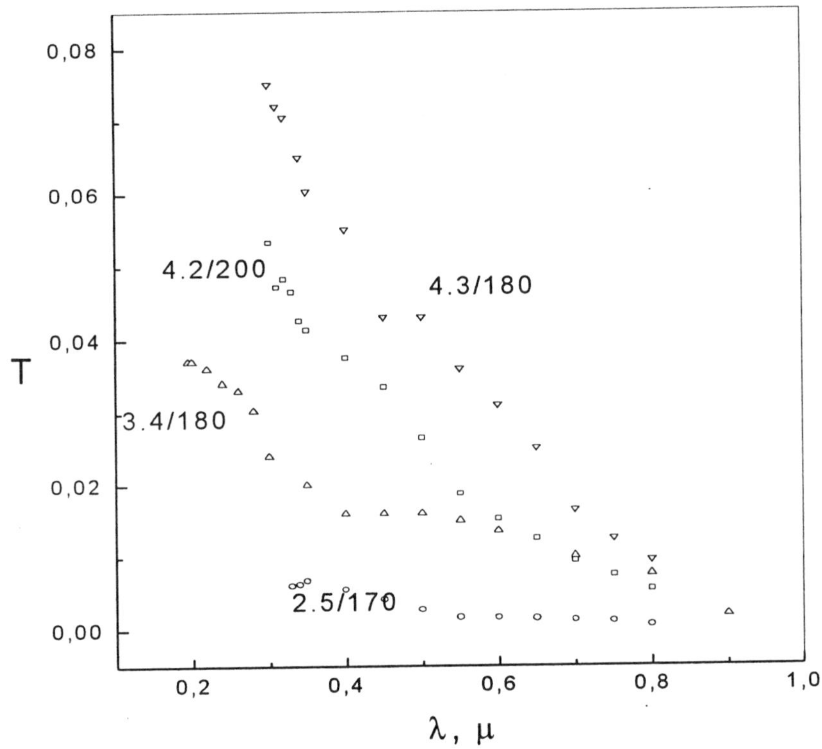

Figure.2. The transmission spectra of four samples of MaSi

$a=\dfrac{4\pi r_i^2}{\lambda L}$, $x = kr^2/2L$, $k = 2\pi/\lambda$, J_0 – Bessel function, $\beta=\dfrac{\pi R^2}{\lambda L}$; λ - light wavelength, T_C – the sample transmission, r_i – the pore radius.

It does not consider the diffraction at the pore exit and the interaction of the light, transmitted by each pore, nevertheless it is more accurate, than simplified expression used in [5].

Fig. 3 shows the comparison of the calculated dependence with the experimental data for the sample with R = 4.3 μm and length L = 180 μm. A good agreement of experimental and calculated results is observed at the short wavelengths. The difference at λ > 0.65 μm can be explained by light diffraction at the pore exit, which has not been taken into consideration by expression (1). Light divergence from each pore is higher for long wavelengths, so that at β<< 1 the significant part of light beam is out of light receiver, that looks like a decrease of the sample transmission.

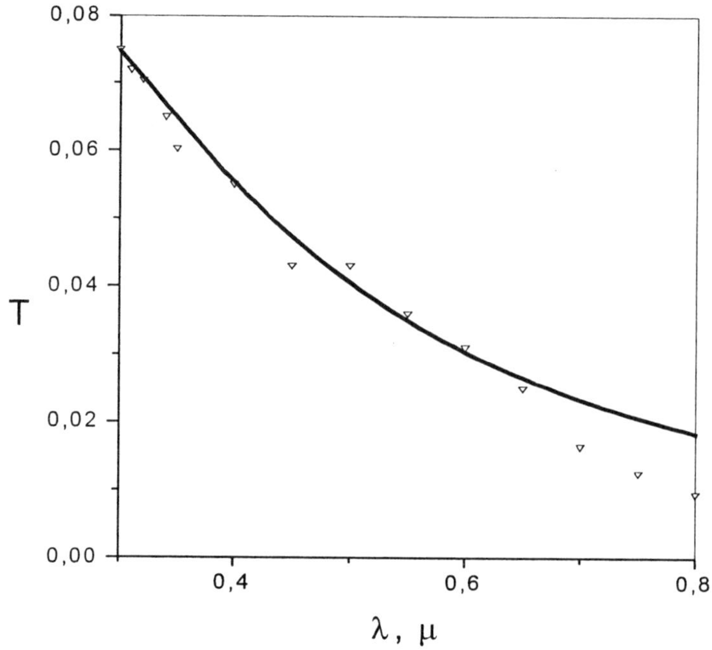

Figure. 3. The results of calculation according to (1) (dot – dashed curve) in comparison with experimental data (points) for the sample with R = 4.3 μm and L=180 μm

To conclude, our results confirm the short-pass optical characteristics of MaSi in the visible wave range and suggest the improved model for interpretation of the experimental data.

3. Acknowledgment

The authors would like to thank the Russian state program "Physics of nanostructures" and St. Petersburg Scientific Center program "Low dimensional structures" for financial support.

References

1. V.Lehmann and H.Foll. J. Electrochem. Soc. **137,** 653, (1990).
2. A.Birner, R.B.Wehrpohn, U.M.Gosele, K.Bush. Advanced Materials, 13, 377, (2001).
3. V.Lehman, U.Gruning, Thin. Sol.Films, 297, 13 (1997).
4. C.P.Beetz, R.Boerstler, J.Steinbeck, B.Lemieux, D.R.Winn. Nuclear Insstruments and Methods in Physics Research, A422, 443, (2000).
5. V.Lehman, R.Stengl, and H Reisinger, R.Detemple, W.Theiss, Appl. Phys. Lett., 78, 589, (2001).
6. J.Wallace. Laser Focus World, April 2001, p.36
7. E.V.Astrova, A.D.Remenyuk, A.G.Tkachenko, I.L.Shulpina, Technical Physics Letters, 26, 1087 (2000).
8. L.D.Landau, E.M.Lifshits, Field theory, Moscow,1960 (in Russian).

MAGNETORESISTIVE SENSORS AND MEMORY

K.-M.H. LENSSEN
Philips Research Laboratories,
Prof. Holstlaan 4, NL-5656 AA Eindhoven, The Netherlands

Abstract:

An introduction to magnetoresistance (AMR, GMR and TMR) effects is given, as well as an overview of their application in sensors and Magnetic Random Access Memory (MRAM). Different areas of sensor applications are discussed, including automotive sensors, current sensors and input devices. Presently, there is a large interest in the development of MRAM, since this new type of solid-state memory may be able to replace several types of conventional memories. Because MRAM promises to be non-volatile, fast and dense, it could enable e.g. "instant-on" computers and portable devices with a much longer standby time. The MRAM technology also offers possibilities for the realization of compact, high-sensitivity, low-cost, multi-analyte biosensors.

Key words: anisotropic magnetoresistance / AMR / giant magnetoresistance / GMR / tunneling magnetoresistance / TMR / spin-tunnel junctions / multilayers / magnetic sensors / sensor applications / automotive sensors / magnetic random access memory / MRAM / magnetoresistive biosensors

1. Introduction

Magnetoresistive (MR) materials show a change in electrical resistance R as a function of the applied magnetic field H, which makes them an interesting class of materials for a large variety of field sensors. Moreover, they can be applied in a new type of solid-state memory: Magnetic Random Access Memory (MRAM).

There exist many different magnetoresistance effects, listed in the following short overview:

1. *Magnetoresistance*

   The classical magnetoresistance effect due to the Lorentz force is usually relatively

E. Buzaneva and P. Scharff (eds.), Frontiers of Multifunctional Nanosystems, 431–452.

small, but in semiconductors and two-dimensional electron gases it can be used for sensor applications [1].

2. *Anisotropic magnetoresistance (AMR)*
   The anisotropic magnetoresistance effect is observed in thin ferromagnetic films.
3. *Giant magnetoresistance (GMR)*
   This effect is found in inhomogeneous magnetic material systems, like metallic multilayers and granular systems.
4. *Tunneling magnetoresistance (TMR)*
   The TMR effect occurs in spin-tunnel junctions, magnetic multilayers comprising a tunnel barrier.
5. *Colossal magnetoresistance (CMR)*
   The CMR effect, which is found in perovskites, was called "colossal" because the relative resistance ratio is even larger than for the giant magnetoresistance effect.
6. *Infinite magnetoresistance ("IMR")*
   This list of superlatives could be completed by the superconducting transition in superconductors, since this provides a real "infinite" magnetoresistance ratio according to the conventions used in the other magnetoresistance effects.

The scope of this paper is limited to the magnetoresistance effects that are found in magnetic thin films: AMR, GMR and TMR, since these are at present most interesting for modern sensor and memory applications. First these MR effects will be presented shortly. After that, it is discussed how magnetic sensors can be realized based on these phenomena and the broad range of applications of such sensors is illustrated. Further, an introduction is given to the upcoming, potentially very important application area of MRAM. Finally, the opportunity to realize magnetoresistive biosensors based on the device technology developed for MRAM is shortly discussed.

2. Magnetoresistance effects in magnetic thin films

2.1. *ANISOTROPIC MAGNETORESISTANCE (AMR)*

The AMR effect was discovered already in 1857 by W. Thomson (the later Lord Kelvin) [2] and is observed in ferromagnetic materials, particularly in thin films. It is the phenomenon that the resistivity ρ depends on the angle φ between the magnetization M and the current I (Figure 1). The typical relative resistance change (the MR ratio $\Delta\rho/\rho$) is in the order of 2% for thin-film materials that are applied in sensors. The angle dependence of the resistivity is given by

$$\rho(\varphi) \approx \rho(90°) + \Delta\rho \cos^2 \varphi. \qquad (1)$$

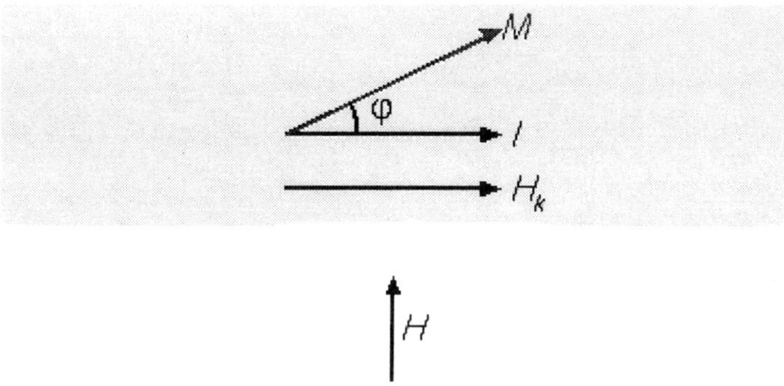

Figure 1. Schematic view of the relevant quantities in the AMR effect (M is the magnetization, I is the current, H_k is the anisotropy field and φ is the angle between magnetization and current directions)

2.2. *GIANT MAGNETORESISTANCE (GMR)*

In 1988 a more pronounced magnetoresistance effect was discovered [3], which was much larger than the known AMR effect, and therefore it was baptized giant magnetoresistance (GMR). It is the phenomenon that the resistance of a material depends on the angle between magnetization directions at different locations in the material. This definition implies that the GMR effect occurs in magnetically inhomogeneous material systems, which can be granular systems [4,5] or discontinuous multilayers [6], but for applications mainly multilayers are considered. In this case the resistance of the multilayer depends on the mutual angle between the magnetization directions of different ferromagnetic layers in the multilayer. If the direction of at least one of the magnetizations can be changed by an applied magnetic field, we have (in principle) the basic element for a magnetic sensor.

The GMR effect is generally explained by spin-dependent electron scattering. This principle is schematically illustrated in Figure 2. In a simplified two-current model a distinction is made between spin-up and spin-down electrons. In the case that the ferromagnetic layers have parallel magnetization directions, only one type of electrons is scattered strongly, resulting in a relatively low resistivity. For an antiparallel configuration of the magnetization directions both types of electrons are scattered appreciably and thus the resistivity is high.

The resistivity ρ as a function of the angle ϑ between the magnetization directions is described by

$$\rho(\vartheta) \approx \rho(0°) + \Delta\rho\,(1-\cos\vartheta)/2. \tag{2}$$

The quantity $\Delta\rho/\rho(0°)$ is called the GMR ratio. While for practical AMR sensors the MR ratio is typically about 2%, the GMR ratio of modern GMR materials is several times larger (~5%-15%). Equation (2) shows that the angle dependence of the GMR - effect has a period of 360°, in contrast to the AMR effect which has a period of only 180°.

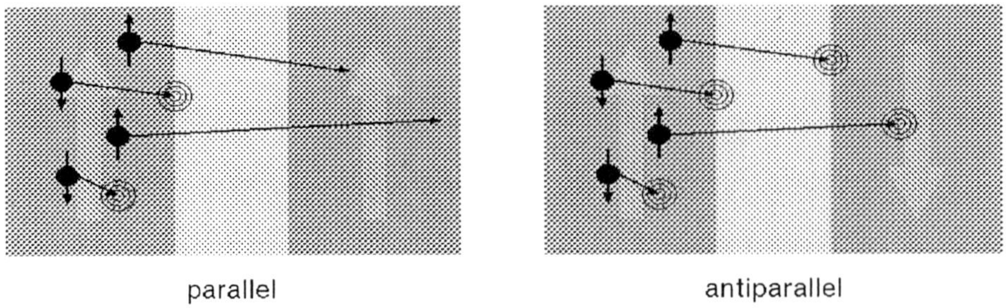

Figure 2. Illustration of spin-dependent electron scattering in GMR multilayers

2.3. *TUNNELING MAGNETORESISTANCE (TMR)*

The tunneling magnetoresistance effect was already discovered in 1975 [7], but it was limited to low temperatures for a long time. Only in 1995 commercial mass applications became feasible when a sizable TMR effect was observed at room temperature [8].

From the application side, the TMR effect could be viewed as a variant of the GMR effect, with a larger resistance and a larger MR ratio (up to ≈50%), although the physical principle is different. Instead of the metallic spacer layer (usually copper) in GMR multilayers, in spin-tunnel junctions the ferromagnetic layers are separated by a tunnel barrier (usually aluminum oxide). The tunnel probability of an electron with a certain spin depends on the relative orientation of the magnetization directions. Similar as for GMR, antiparallel magnetizations result in a large resistance, while parallel magnetizations correspond with a relatively low resistance. While in GMR devices the current is generally applied in the plane of the film (CIP), in TMR devices the electrons have to tunnel through the barrier and therefore the current has to be perpendicular to the plane of the layers (CPP). This requires extra fabrication steps. The angle dependence is the same as for the GMR effect.

3. MR sensors and applications

3.1 *AMR SENSORS*

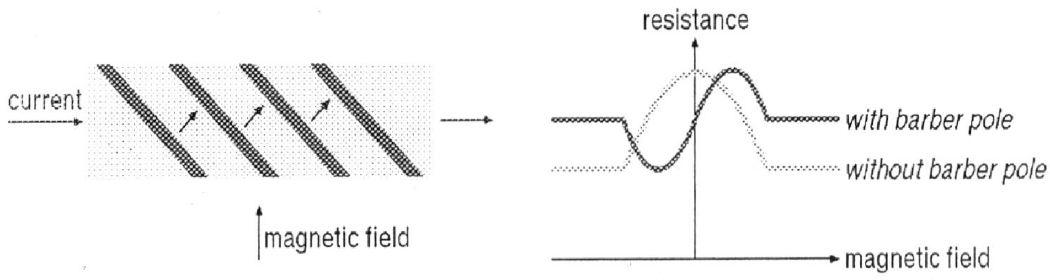

Figure 3. Illustration of linearization by means of a barber-pole structure

Because in AMR materials the resistance change takes place in a small magnetic-field range, this phenomenon is very interesting for high-sensitivity sensor applications, like for example recording heads [9,10] or earth magnetic field measurements [11,12]. However, since the magnetoresistance curve is symmetrical around zero field (light grey curve in Figure 3), the sensitivity is low in just this region. Therefore, linearization is desired, and an elegant solution for this is the so-called barber pole [13]. This structure consists of a series of well-conducting metal (e.g. Au or Al) stripes, which are placed under an angle of ±45° with respect to the current direction (see the left-hand side of Figure 3). The result is that the current will flow as long as possible through the barber-pole stripe and then cross the AMR material towards the next stripe via the shortest route. Effectively, the current direction is thus rotated over 45° and the magnetoresistance curve is shifted over 45°, which leads to a steep and linear graph around zero field (dark curve in Figure 3).

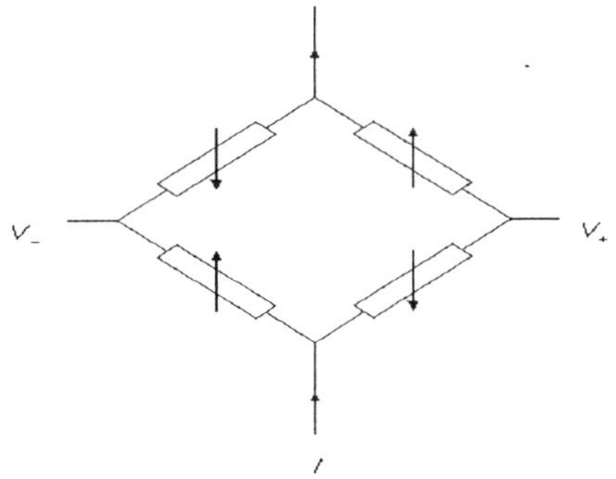

Figure 4. Schematic view of a Wheatstone-bridge configuration

For magnetic field sensors it is often desirable to eliminate the influence of temperature variations and to realize a bipolar output by the use of a Wheatstone-bridge configuration. In the bridge the MR elements in two of the bridge branches should have an opposite response to a magnetic field than the other two elements (see Figure 4). This can easily be achieved by placing the barber poles under –45° and +45° on the two pairs of AMR elements.

3.2 *GMR SENSORS*

3.2.1 *Types of GMR sensor materials*

There are several kinds of GMR multilayers; the basic types are shown in Figure 5.

The oldest type is the antiferromagnetically (AF) coupled multilayer (Figure 5a). It consists of alternating ferromagnetic and nonmagnetic layers, of which the thicknesses are chosen such that the exchange coupling between the magnetic layers is antiparallel. An external magnetic field will rotate all magnetization directions towards a parallel configuration, and thus reduce the resistance. This effect is independent of the direction of the magnetic field (neglecting anisotropy fields which are usually relatively small).

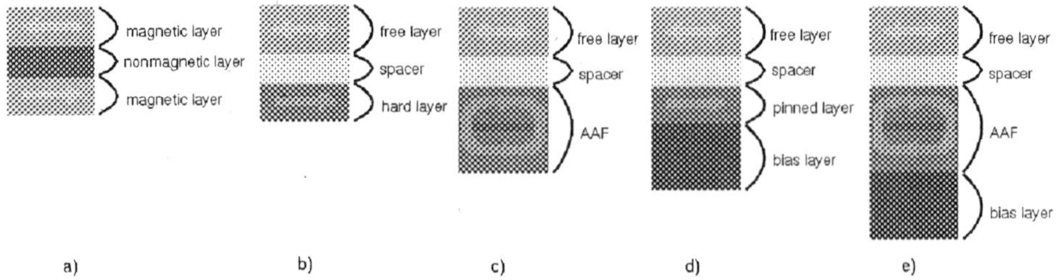

Figure 5. Schematic representation of the basic elements of different types of GMR multilayers: a) antiferromagnetically coupled multilayer, b) hard—soft multilayer, c) hard—soft multilayer with AAF, d) exchange-biased spin valve, e) spin valve with exchange-biased AAF

The other four types of multilayers try to keep the magnetization direction of one magnetic layer (or set of magnetic layers) fixed over the field operation range, while another one is made sensitive to external magnetic fields. Antiparallel coupling is not required and usually any coupling between fixed and free layers is even undesirable.

In hard—soft multilayers (Figure 5b) this is accomplished by using two kinds of ferromagnetic materials with different coercivities. The hard layer, with a relatively high coercivity, does not rotate significantly for not too large fields, while the magnetization of the soft layer (with a low coercivity) easily rotates in an applied field. An issue is to realize the combination of a really hard layer with a really soft layer in a single stack. Besides complications of the material structure, there is also a magnetostatic effect in sensor elements: the stray field from the hard layer acts on the soft layer and thus makes it less sensitive. A solution for the latter problem is the use of an Artificial AntiFerromagnet (AAF), also called Synthetic AntiFerromagnet (SAF or SyAF), as the hard layer (Figure 5c) [14]. Such an AAF consists of two (or more) ferromagnetic layers that are exchange-coupled antiferromagnetically by a thin nonmagnetic layer (this is similar to a multilayer of the type of Figure 5a [15]), in which the thicknesses of the magnetic layers are chosen such that the net magnetic moment is (close to) zero. In this way there is virtually no stray field, and moreover, the exerted torque on the AAF by an external field is small because of the vanishing magnetic moment, which strongly improves the rigidity of the hard layer.

The major disadvantage of hard—soft multilayers remains, however, that for finite magnetic fields the magnetization of the hard layer also changes direction and this limits the useful operation range. At a certain value of the field the magnetization of the hard layer can flip completely into the direction of the external field; this flipping gives rise to a jump from one branch of the magnetoresistance curve to another.

A completely different type of GMR multilayer is the exchange-biased spin valve (Figure 5d) [16]. In this kind of multilayer the magnetization direction of one ferromagnetic layer is pinned via the exchange-biasing effect [17—19], usually by an adjacent antiferromagnetic layer. A review of exchange-biased spin valves can be found in [20]. In contrast to hard—soft multilayers the output signal of exchange-biased spin valves in their operation range is not affected by exposure to large magnetic fields. Because of the good linearity and the very high sensitivity in the low-field range [21], this is the preferred type of GMR material for recording applications [10]. An example of a magnetoresistance curve of a simple exchange-biased spin valve is shown in Figure 6. In the past the thermal stability of the exchange-biasing effect was poor, but

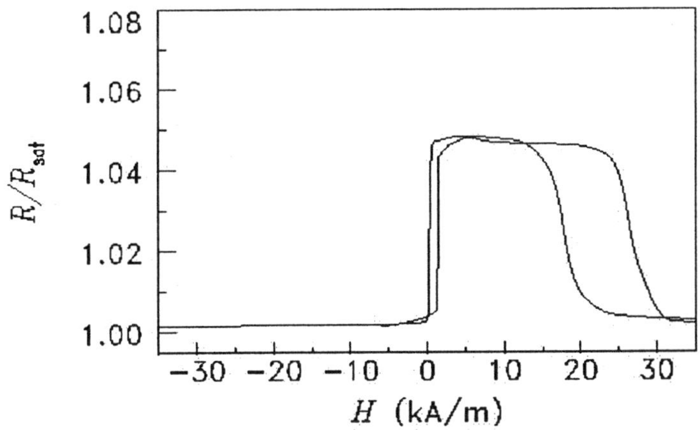

Figure 6. Magnetoresistance curve of a simple exchange-biased spin valve

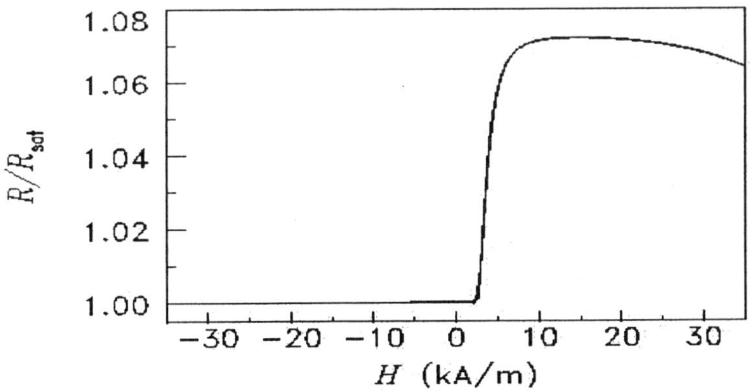

Figure 7. Magnetoresistance curve of a spin valve with exchange-biased artificial antiferromagnet

with the discovery of new materials like Ni-Mn, Ir-Mn and Pt-Mn this is no longer a limitation for practical applications. For automotive and industrial sensor applications the magnetic field range over which the GMR effect is nonzero is, however, too small. This has recently been solved by using an AAF as the pinned layer (Figure 5e) [22]. The magnetoresistance curve of a spin valve with exchange-biased AAF is presented in Figure 7. The sensor material of this last type that we developed will be discussed in more detail in the next part of this paper.

It should be pointed out that Figure 5 only indicates the basic structures. In reality the shown building blocks are sometimes repeated (in particular in the case of AF-coupled multilayers) and usually a buffer layer and a cap layer have to be added to the stack, respectively to induce the desired microstructure and for protection. Of course many variations on the shown basic structures exist. For example, for types b—e the layer with fixed magnetization is positioned at the bottom of the stack, but it is also possible to have stacks with the fixed layer at the top or with two fixed layers (both at the bottom and the top) and one free layer.

A more detailed overview of GMR sensors can be found in [23].

3.2.2 *Wheatstone-bridge GMR sensors*

In our research at Philips we focussed on spin valves with exchange-biased AAF

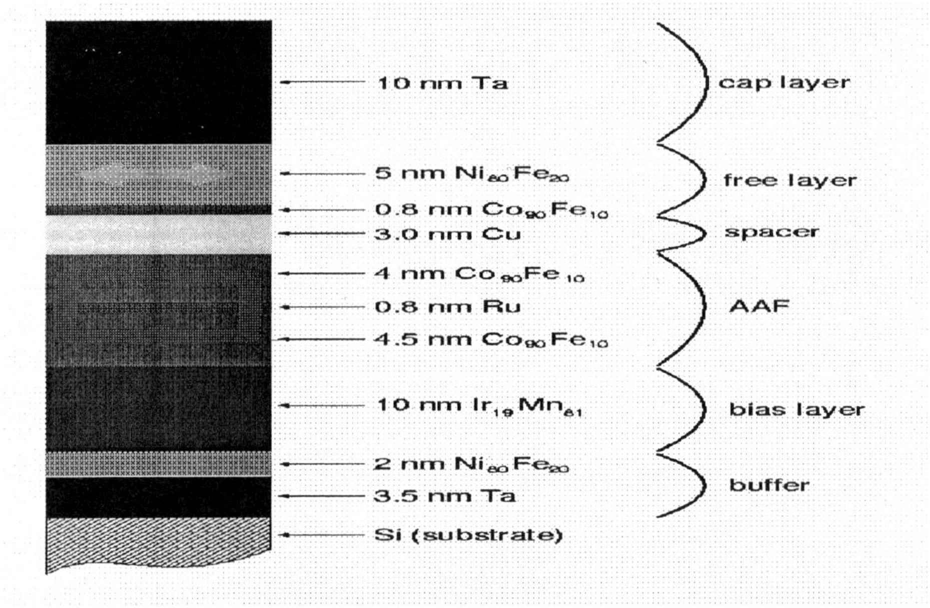

Figure 8. Schematic composition of our GMR multilayer

[22], since these material systems have the best properties for robust GMR sensors with Wheatstone-bridge configuration for automotive and industrial applications.

A typical example of the stack composition is 3.5 nm Ta/2.0 nm $Ni_{80}Fe_{20}$/10.0 nm $Ir_{19}Mn_{81}$/4.5 nm $Co_{90}Fe_{10}$/0.8 nm Ru/4.0 nm $Co_{90}Fe_{10}$/3.0 nm Cu/0.8 nm $Co_{90}Fe_{10}$/5.0 nm $Ni_{80}Fe_{20}$/10.0 nm Ta (Figure 8). Care has been taken to avoid interfaces between easily mixing materials (so e.g. no Ni near Cu). Ir-Mn has been chosen as the exchange-biasing material because of its good robustness against temperature (blocking temperature ≈290°C) [24]. The films are deposited using dc magnetron sputtering, while a magnetic field of about 20 kA/m is applied in the plane of the film.

This GMR material system shows a nice step-like magnetoresistance curve, with a GMR ratio of typically ≈6% (see also Figure 7). It can withstand very large magnetic fields without changes of the properties, as has been checked up to 200 kA/m. At 200°C the GMR ratio is only approximately halved, which is still larger than for the AMR effect at room temperature [25]. This offers possibilities for GMR sensors that still can operate at 200°C or more, which is higher than for any other magnetic microsensor.

Using this GMR material system we have realized sensor elements with dimensions down to 10 μm, but this is certainly not the lower limit.

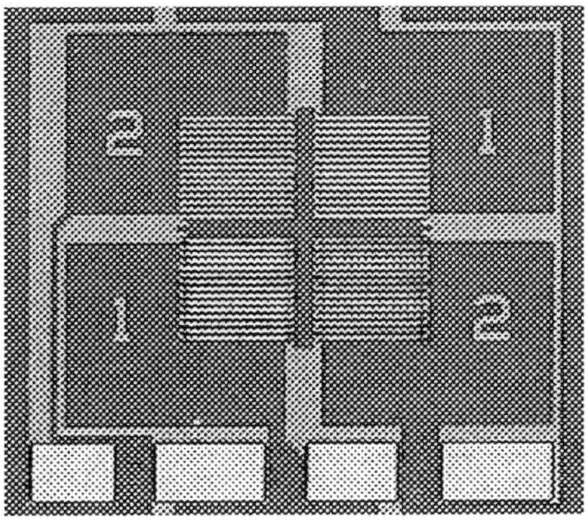

Figure 9. Photograph of a GMR sensor chip with full Wheatstone-bridge configuration (active area 0.5 mm x 0.5 mm)

For a Wheatstone bridge it is necessary to make elements with opposite output signals on one substrate; in the case of exchange-biased spin valves this requires pairs of elements with opposite exchange-biasing directions and therefore opposite responses to an applied magnetic field. Because in particular for automotive and industrial sensor applications robustness with respect to temperature and field amplitude is essential, this is not trivial to achieve. At the Philips laboratories several solutions have been developed, including a double-deposition method [23] and localized heating in a magnetic field. In this way we succeeded to fabricate full Wheatstone bridges without affecting the favorable sensor properties of our material (Figure 9). The output curve of one of the first prototype GMR sensors is shown in Figure 10 (dashed curve).

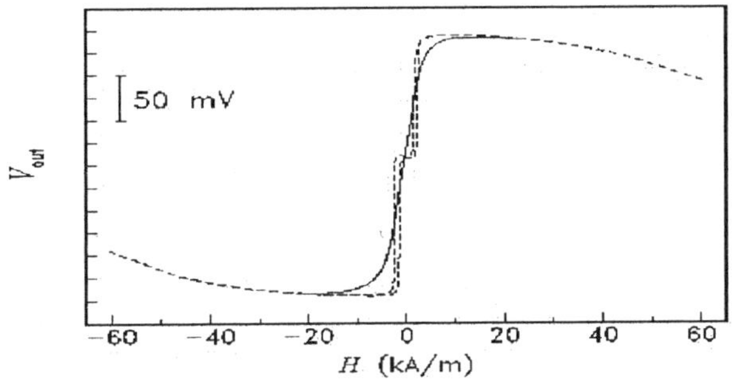

Figure 10. Output voltage of a prototype of a full GMR Wheatstone bridge as a function of magnetic field with 5 V input (dashed line); idem with a small bias magnet (solid line)

The narrow plateau around zero field is due to a small offset field in the magnetoresistance curve of the material, caused by slight coupling between the free and pinned layers. The operation range is several times larger than for existing GMR sensors. This allows sensor configurations with significantly better signal-to-noise ratios and larger immunity against disturbing fields.
For most automotive applications the low-field range is not important, but, if necessary, the small hysteresis can be eliminated by adding a small permanent bias magnet, so that a smooth output curve is obtained (solid line in Figure 10). This also makes the narrow plateau disappear. Although the prototype sensor was not optimized

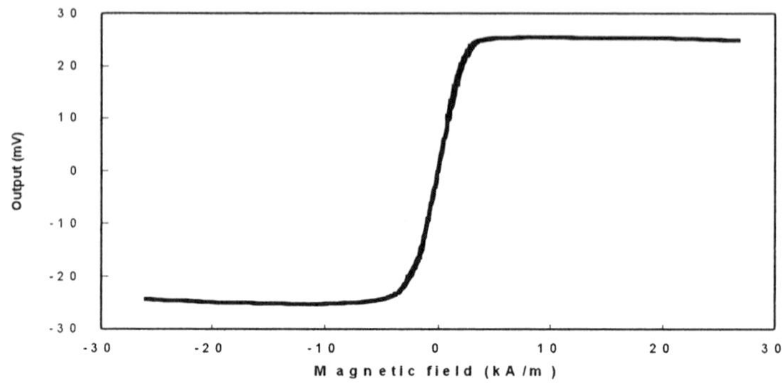

Figure 11. Output characteristics of a robust GMR sensor with a full Wheatstone-bridge configuration [26]

for low-field sensitivity at all, it already provides a larger sensitivity than typical AMR sensors that are presently used for automotive applications. By further optimization, recently sensitivities of 10—20 (mV/V)/(kA/m) and a hysteresis smaller than 0.1 kA/m have been achieved without the need for a bias magnet (Figure 11) [26].

The GMR sensors are also very suitable for angle measurements. The output signal as a function of the angle of the magnetic field (e.g. from a permanent magnet that is attached to the object to be measured) is shown in Figure 12 for an input voltage of 5 V.

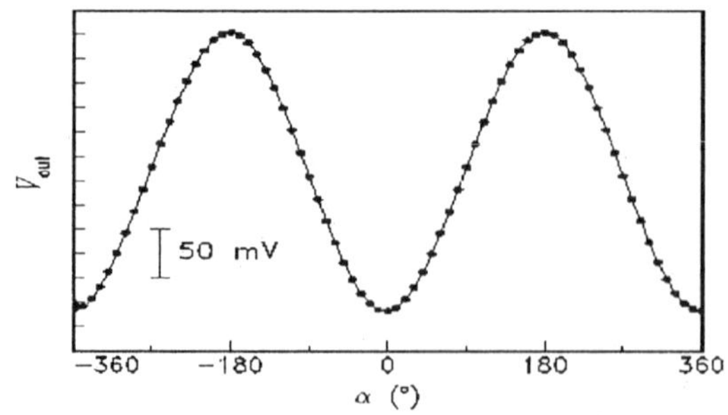

Figure 12. Output voltage of a GMR Wheatstone bridge in an angle-sensor geometry, using a field of 16 kA/m (solid line); idem after exposure to 200 kA/m (dots)

The output is independent of the field amplitude over a very large field interval; it gives useful sinusoidal signals up to fields of about 40 kA/m. As can also be seen in Figure 12 the original output curve is reproduced without changes after rotation in a field of 200 kA/m (dots). This is an essential advantage over the present commercial GMR angle kA/m (dots). This is an essential advantage over the present commercial GMR angle sensors, which can already be irreversibly damaged by a field of a typical permanent magnet.

3.3 *SENSOR APPLICATIONS*

3.3.1 *Introduction*

The three main application areas of magnetoresistance are (1) magnetic recording (e.g. HDD heads and tape read heads [10]), (2) magnetic field sensors and (3) magnetic random access memory (MRAM). In this section the wide range of magnetic field sensor applications will be illustrated, while the following section is devoted to MRAM.

3.3.2 *Automotive magnetic sensors*

In automotive and industrial environments the main basic applications for (G)MR sensors are position, angle and rotation-speed sensing.

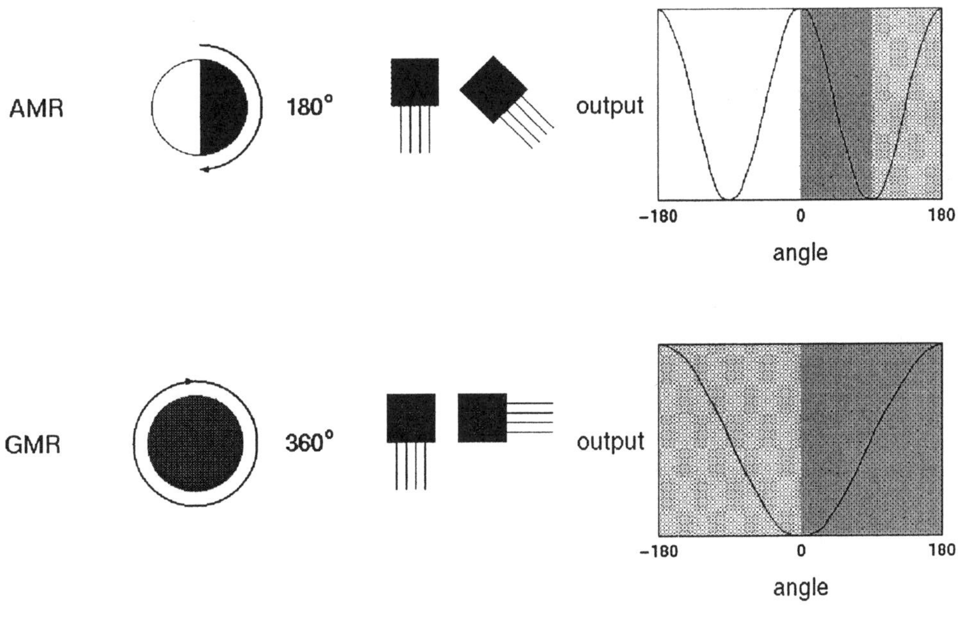

Figure 13. Schematic view of analog angle sensing by means of two mutually rotated AMR (at the top) or GMR (at the bottom) sensors

As already discussed above, GMR sensors are very suitable for analog angle sensing because of their intrinsic angle dependence. This is schematically illustrated in Figure 10. Because of its 180° period an AMR sensor (see equation (1)) can measure a range of 90° unambiguously (dark grey area in Figure 13). With two AMR sensors that are mutually rotated over 45° this range can be doubled (dark grey plus light grey area).

Since the angle dependence of the GMR effect has a period of 360° (see equation (2)) it is possible to cover the full revolution of 360° with two GMR sensors that are mutually rotated over 90°. Such a contactless angle sensor can be used in automotive applications to determine many relevant quantities, like throttle-valve position, chassis height and pedal position.

Another wide-spread application is rotation-speed sensing, which is used in e.g. anti-lock braking systems (ABS). The principle is illustrated in Figure 14. The sensor is positioned between a (soft)magnetic gear wheel and a permanent magnet, with its sensitivity direction in the tangential direction. If a tooth of the wheel is exactly opposite the sensor (lower left in Figure 14), the field lines from the magnet cross the sensor perpendicularly (in which direction it is not sensitive) and the sensor gives null output. If, however, a tooth is approaching from above (lower right in Figure 14), the field lines are deviated upwards, resulting in a component along the sensitive direction of the sensor and the output signal is (e.g.) positive. In an analogous way, for a tooth below the sensor the output signal is negative. In this way a rotating wheel gives an output signal of positive and negative pulses (approximating a sinusoidal signal), as indicated in the middle of Figure 14. The number of peaks per time interval is a measure for the rotation speed.

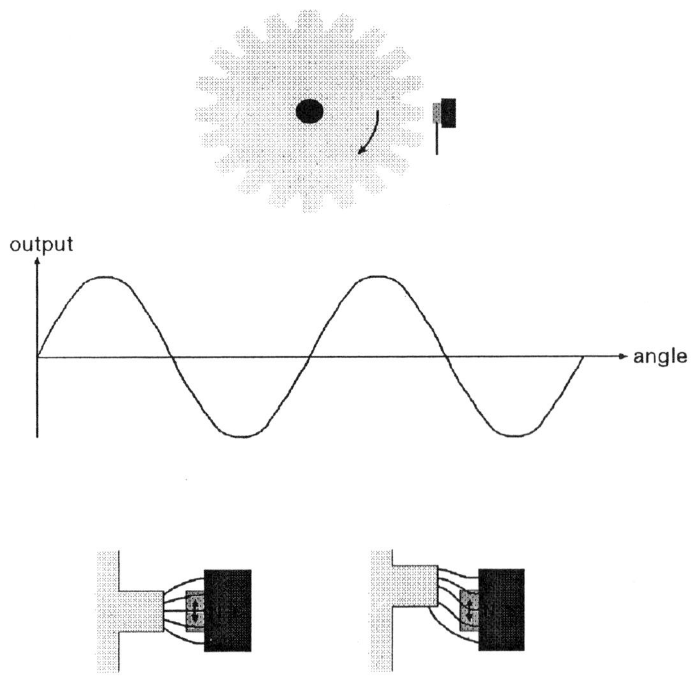

Figure 14. Schematic view of rotation-speed sensing

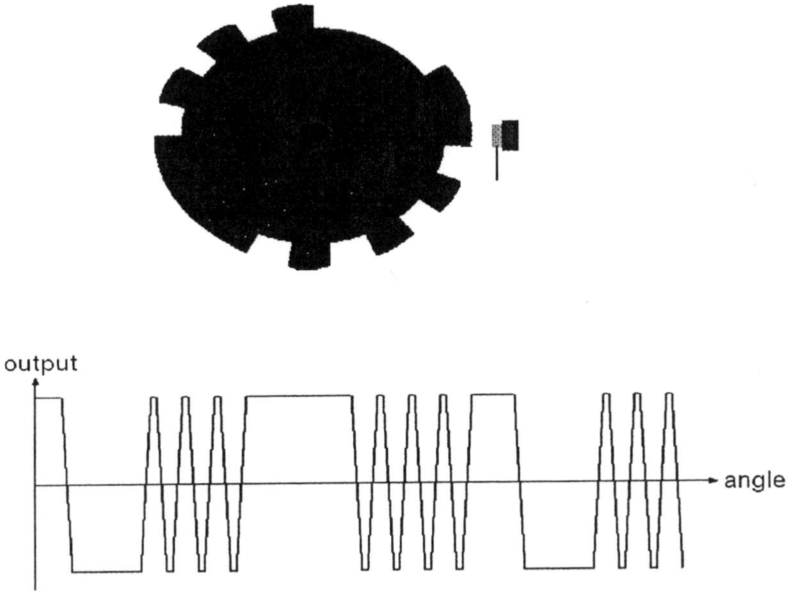

Figure 15. Schematic view of digital position sensing

A third application is digital position sensing, which can be seen as a special variant of rotation speed sensing (see Figure 15). In this case the moment at which a tooth passes the sensor has to be detected very accurately and reproducibly. This is, for example, used for crankshaft-position sensing. The required specifications for this application are very high. See also [27] for more information about the application of AMR sensors for position sensing and how to exploit the possibility of MR sensors to measure field direction independent of field strength.

3.3.3 *Current sensors*

Magnetoresistive sensors are very suitable to measure a current in a contactless way. A current through a conductor induces a magnetic field, and by measuring the amplitude of this field (at one or several locations) the amplitude of the current can be calculated (Figure 16).

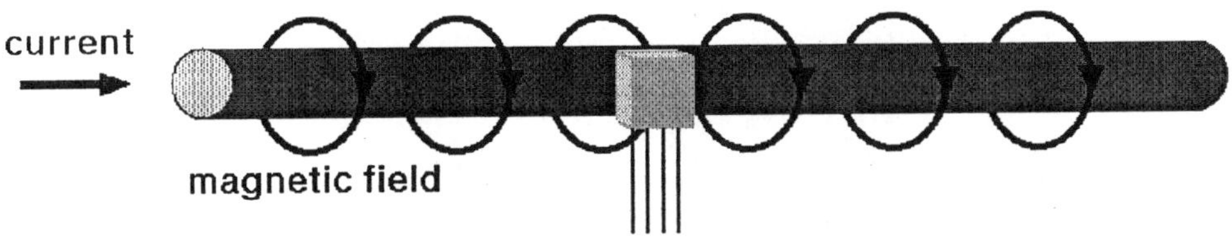

Figure 16. Illustration of the principle of non-intrusive current sensing

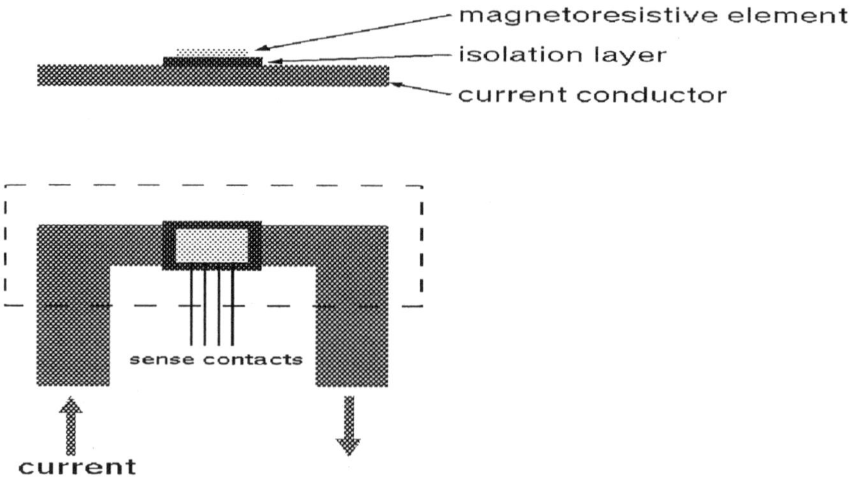

Figure 17. Schematic view of a current-sensor chip

This non-intrusive current sensing is of importance in situations where direct contact would be dangerous (for example, because of high voltages) or where intersecting the current is undesirable (for example, in control systems at industrial plants or air conditioning systems).

The fact that magnetoresistive sensors are thin-film devices allows the realization of current sensor chips, in which both the current-carrying conductor and the MR element are integrated (Figure 17).

3.3.4 *Input devices*

An example of an application area where the use of magnetoresistance sensors is not obvious but nevertheless brings clear benefits is the field of input devices for computers or consumer electronics.

Figure 18. A track ball

A track ball is a well-known type of input device, a.o. for computers (Figure 18). An annoying problem is that the ball can start to slip, if it becomes greasy or dirty and thus the read-out of its position becomes corrupted. This problem can be avoided if the read-out can be done in a contactless way. This can be achieved by implementing a regular, (soft)magnetic pattern on the track ball and sense its rotation around two axes using magnetoresistive sensors with bias magnets [28] (see Figure 19). Actually, this can be considered as a kind of two-dimensional generalization of the rotation-speed sensing concept. Of course, this idea can also be used in a computer mouse.

As stated already earlier, AMR sensors are also very suitable for use in an electronic compass. This enables the realization of a input device that can be used “in the air”.

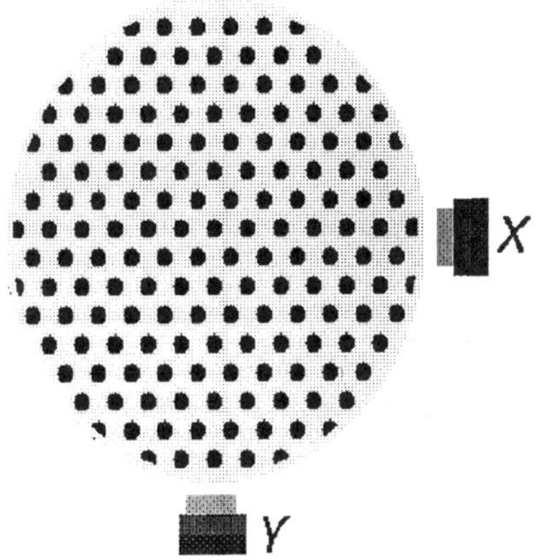

Figure 19. Principle of contactless read-out of a trackball

Figure 20. Use of a compass based on AMR sensors for cursor control

Using three mutually orthogonally mounted AMR sensors, the direction and amplitude of the local magnetic field can be measured (note that it is irrelevant whether this is the undisturbed earth magnetic field or the earth magnetic field distorted by the presence of e.g. iron objects). The relative rotation of this three-dimensional compass in any direction can be monitored by means of the sensor output signals. This can be used to control the cursor movement on a display [29] (Figure 20).

3.3.5 *Other sensor applications*

The range of applications of magnetoresistive sensors is seemingly unlimited; besides the already mentioned examples, e.g. in brushless motors, coin detection, traffic detection, mobile phones [30], mine detection, etc. etc. As the most recent development can be mentioned the use of GMR sensors in pick-up elements for string instruments, in particular electrical guitars [31].

4. Magnetic Random Access Memory

4.1 *INTRODUCTION*

Magnetic Random Access Memory (MRAM) is often presented as the ideal, all-purpose solid-state memory of the future. Indeed, it is expected that MRAM may provide a combination of properties that at present are only found distributed over different types of memories (SRAM, EEPROM, Flash, DRAM, etc.). This is certainly a very attractive perspective; if two (or more) different memory technologies could be replaced by a single memory technology (c.q. MRAM), the development costs for a next generation of a product comprising embedded memory could decrease dramatically. However, for this to come true, it is important that the specifications of MRAM will not merely be a compromise, only suitable for certain niche markets, but that they will really be able to compete with the (future) state-of-the-art “conventional memories”.

4.2 *Magnetic RAM principles*

An MRAM is a memory (1) in which the stored data are represented by magnetization directions and (2) in which the read-out is done by a resistance measurement (for an overview of MRAM see, for example, Refs. [32—35]). This definition implies that its operation relies on a magnetoresistance phenomenon. Early MRAMs were based on the anisotropic magnetoresistance (AMR) effect. Since the amplitude of the AMR effect in thin films is typically less than 3%, the use of AMR-based MRAMs was limited mainly to military and space applications.

The discovery of the larger giant magnetoresistance effect [3] changed this situation. The GMR effect allowed the realization of smaller elements with a higher resistance and a larger MR effect (5 à 15%), and therefore a higher output signal. This enabled, in principle, the realization of MRAMs for general applications

A breakthrough in the field of magnetic tunnel junctions around 1995 [8] improved the perspectives of MRAM even further, when a large tunnel magnetoresistance (TMR) effect was demonstrated at room temperature. Since then TMR effects with amplitudes up to ≈50% have been shown, but because of the strong dependence of the TMR effect on the bias voltage, the useable resistance change in practical applications is at present around 30%.

In general, both GMR and TMR result in a low resistance if the magnetization directions in the multilayer are parallel, and in a high resistance when the magnetizations are oriented antiparallel. In TMR multilayers the sense current has to be applied perpendicular to the layer planes (CPP) because the electrons have to tunnel through the barrier layer; in GMR devices the sense current usually flows in the plane of the layers (CIP), although a CPP configuration might provide a larger MR effect, since the resistance perpendicular to the planes of these all-metallic multilayers is very small. Nevertheless, supported by the rapidly continuing miniaturization, the possibility to base MRAMs on CPP GMR recently got serious attention [36].

For the representation of the bits there exist different possibilities. So-called pseudo-spin valves [37,38] comprise two ferromagnetic layers that switch their

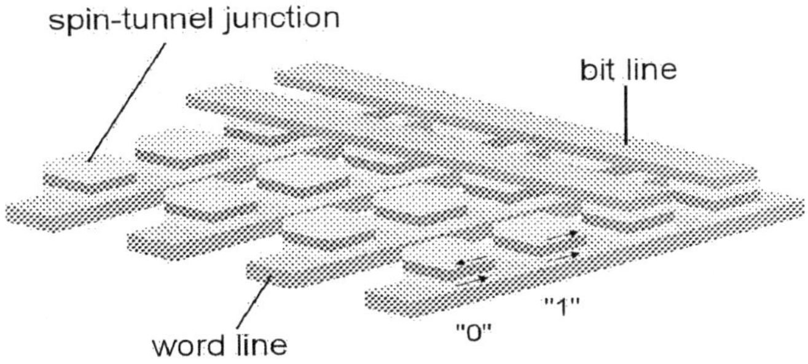

Figure 21. Schematic view of an MRAM array

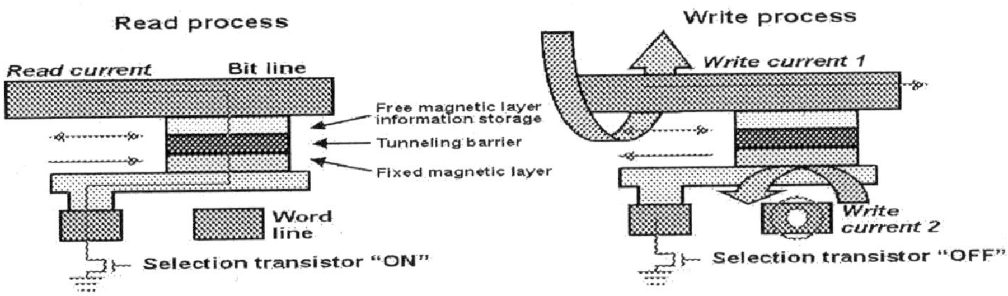

Figure 22. Operating principle of reading (left) and writing (right) an MRAM cell

magnetization direction at different magnetic fields (cp. Figure 5b); this can be accomplished by using layers of different magnetic materials, or layers of the same material but of different thickness.

In another category of materials, a.o. (exchange-biased) spin-tunnel junctions [39], the magnetization direction of one of the magnetic layers is so rigid that it can be considered fixed under normal operation conditions; as discussed earlier for sensor applications this can, for example, be achieved by using exchange biasing [17—19] (cp. Figure 5d) or an artificial antiferromagnet [14] (cp. Figure 5d).

The addressing of the MRAM is done by means of an array of crossing lines (see Figures 21 and 22). Writing a certain cell is equivalent to setting a magnetization in the desired direction (for example, magnetization to the left means '0' and magnetization to the right means '1'). By applying a current pulse to a bit line and a word line a magnetic field pulse is induced. Only the MRAM cell at the crossing point of both lines experiences the maximum magnetic field (i.e. the vectorial addition of the fields induced by both current pulses) and its magnetization is reversed; all other MRAM cells below the bit or word line are exposed to the lower field that is caused by a single current pulse and will therefore not change their magnetization directions.

The read-out method depends on the type of MRAM. In the case of pseudo-spin valves a number of cells (N) can be connected in series in the word line, because the resistance of these completely metallic cells is relatively low. This provides the interesting advantage that only $1/N$ switching element (usually a transistor) is needed per cell. The associated disadvantage is that the relative resistance change is divided by N. The read-out is done by measuring the resistance of a word line (with the series of cells), while subsequently a small positive plus negative current pulse is applied to the desired bit line. The accompanying magnetic field pulses are between the switching fields of the two ferromagnetic layers; thus the layer with the higher switching field (the data-storing layer) will remain unchanged, while the magnetization of the other layer will be set in a defined direction and then be reversed. From the sign of the resulting resistance change in the word line it can be seen whether a '0' or a '1' is stored in the cell at the crossing point the word and the bit line.

In the case of spin valves with a fixed magnetization direction the data are stored in the other, free magnetic layer, which of course should not be disturbed by the read-out (Figure 22). In this case the absolute resistance of the cell is measured (if desired, differentially with respect to a reference cell). This cell is selected by means of a switching element (usually a transistor), which implies that in this case one transistor is required per cell.

Like other memory technologies MRAM also offers in principle the possibility to store multiple bits per cell [40].

4.3 *PRESENT STATUS OF MRAM*

In the last year the R&D activities concerning MRAM have shown an acceleration. In December 2000 IBM and Infineon announced a cooperation on MRAM development [41,42]. It was stated that MRAM products would be commercially available already in 2004; this would relate to 128 Mb or 256 Mb memories in 0.13 μm technology. This timing corresponds to the planning of Motorola who announced samples in 2003 and volume production in 2004. In June 2001 [43] it was even stated that volume production would be only 1.5 year away as Motorola and IBM speed up their development work.

Also major progress has been made in the field of MRAM demonstrators: at the ISSCC2001 Motorola presented a complete 256kb MRAM chip in 0.6 μm technology [44].

In the meantime USTC very recently claimed the demonstration of the first commercial MRAM on their web site [45], although samples are not yet available. While most others concentrate on TMR, USTC uses the GMR effect in their MRAM.

Other industrial companies active in MRAM research and development include Hewlett-Packard, NVE, Toshiba, NEC, Samsung, Read-Rite and Philips.

In [34] we already compared MRAM with other types of embedded non-volatile memory. MRAM is, however, not only a potentially interesting alternative for non-volatile memories like Flash and EEPROM, but also for volatile memories like SRAM and DRAM. For all of these conventional memories foreseen problems in scaling have been reported. Further, the use of MRAM is not on beforehand limited to only embedded memory.

It is commonly known that Flash technology faces problems, like the rapid increase in the number of masks, the limited endurance and the power consumption. MRAM is an interesting alternative since it is much faster and does not show the endurance limitations.

For DRAM problems are expected for further scaling, since the aspect ratio cannot continue to increase [46]. MRAM might be able to replace DRAM and has the additional advantage of non-volatility. This would enable the realization of "instant-on" computers and could drastically increase the standby time of portable devices.

Even the radiation hardness of MRAM may become important for commercial applications, since radiation in the operation environment is reported to cause significant problems for small SRAM devices [47].

Of course, one of the most intriguing promises of MRAM is that it may replace more than one type of conventional memory.

5. Magnetoresistive biosensors

The device technology that is developed for MRAM offers an interesting opportunity for the realization of a magnetoresistive biosensor. The main potential advantages of this new type of biosensor are compactness, low cost, high sensitivity and the possibility to detect thousands of different analytes with a single device.

The concept for a GMR-based biosensor was presented for the first time in 1998 by Naval Research Lab. [48]. The basic operation principle is shown in Figure 23 for the example of a DNA sensor. On a substrate GMR sensor elements are fabricated and on top of them DNA fragments are immobilized; these probe DNA fragments may be different for each GMR element. In the measurement DNA fragments with a coupling group (for example biotine) are brought in contact with the sensor. Fragments will hybridize with probe fragments if the complementary sequence is present; other fragments will be washed away. As the next step small magnetic beads (which are well-known in biotechnology, e.g. for separation techniques) covered with a receptor (e.g. streptividin) are introduced. These receptors will bind tightly with the coupling groups (i.e. the biotine on the hybridized DNA fragments) and as a result the presence of magnetic beads on GMR sensor elements will correspond with the presence of certain DNA fragments in the substance that is analyzed. So now only the presence of magnetic beads has to be measured in a quantitative way. This is done by applying a magnetic field (preferably perpendicular to the substrate so that it does not influence the magnetization directions of the GMR sensor elements) in order to magnetize the magnetic beads. The resulting stray field directed in the plane of the substrate from the magnetic beads can easily be detected by the GMR sensor elements. This may be done in a differential way using a GMR sensor element without DNA fragments as reference for the resistance measurement. The MRAM technology provides a way to produce a multitude of individually addressable sensor elements in a single device. For a detailed discussion of magnetoresistive biosensors the reader is referred to Refs. [48—51].

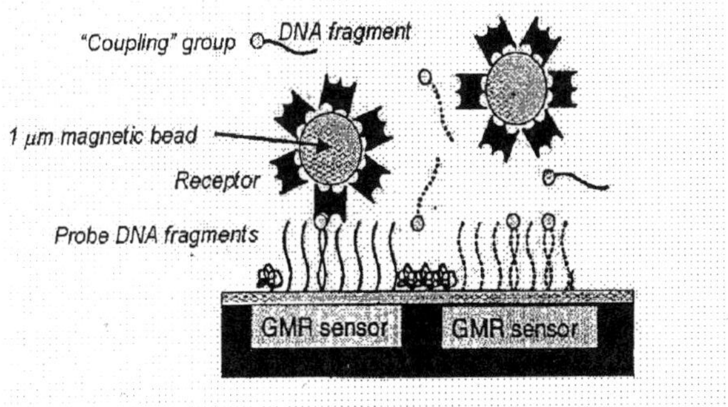

Figure 23. Operation principle of a GMR DNA sensor (after [49])

6. Conclusions

An introduction to magnetoresistance (AMR, GMR and TMR) effects was given, followed by a selection from the vast variety of sensor applications of these MR effects.

Further Magnetic Random Access Memory (MRAM) was discussed. Presently, there is a large interest in the development of MRAM, since this new type of solid-state memory may be able to replace several types of conventional memories. Because MRAM promises to be non-volatile, fast and dense, it could enable e.g. "instant-on" computers and portable devices with a much longer standby time. Although some weaknesses of MRAM still need to be overcome, it is clear that MRAM is expected to give revolutionary advantages over the currently used silicon memories. In particular, the low required number of extra masks and the fact that MRAM may be able to be substituted for "all" types of memories (also e.g. SRAM or DRAM) are very attractive, particularly for embedded memories and systems-on-a-chip.

The MRAM technology also offers possibilities for the realization of compact, high-sensitivity, low-cost, multi-analyte biosensors.

7. Acknowledgements

I gratefully acknowledge my colleagues D.J. Adelerhof, H.M.B. Boeve, B.K. Chong, H.J. Gassen, A.E.T. Kuiper, J.J.M. Ruigrok, G.H.J. Somers, W. Talen, F.W.M. Vanhelmont and J.B.A.D. van Zon for their valuable contributions.

References

1. Heremans, J. (1993), "Solid state magnetic field sensors and applications", *J. Phys. D: Appl. Phys.* **26**, 1149-1168.
2. Thomson, W. (1857), *Proc. R. Soc. London* **A8**, 546-550.
3. Binasch, G., Grünberg, P., Saurenbach, F. and Zinn, W. (1989), "Enhanced magnetoresistance in layered magnetic structures with antiferromagnetic interlayer exchange", *Phys. Rev.* **B39**, 4828-4830; Baibich, M.N., Broto, J.M., Fert, A., Nguyen Van Dau, F., Petroff, F., Etienne, P., Creuzet, G., Friederich, A. and Chazelas, J. (1988), "Giant magnetoresistance of (001)Fe/(001)Cr magnetic superlattices", *Phys. Rev. Lett.* **61**, 2472-2475.
4. See, for example, Berkowitz, A.E., Mitchell, J.R., Carey, M.J., Young, A.P., Zhang, S., Spada, F.E., Parker, F.T., Hutten, A. and Thomas, G. (1992), "Giant magnetoresistance in heterogeneous Cu-Co alloys", *Phys. Rev. Lett.* **68**, 3745-3748 and Xiao, J.Q., Jiang, J.S. and Chien, C.L. (1992), "Giant magnetoresistance in nonmultilayer magnetic systems", *Phys. Rev. Lett.* **68**, 3749-3752.
5. Hylton, T.L. (1993), "Limitations of magnetoresistive sensors based on the giant magnetoresistive effect in granular magnetic compounds", *Appl. Phys. Lett.* **62**, 2431-2433 and references herein.
6. Hylton, T.L., Coffey, K.R., Parker, M.A. and Howard, J.K. (1993), "Giant magnetoresistance at low fields in discontinuous NiFe-Ag multilayer thin films", *Science* **261** (1993) 1021-1024; Hylton, T.L., Coffey, K.R., Parker, M.A. and Howard, J.K. (1994), "Low field giant magnetoresistance in discontinuous magnetic multilayers", *J. Appl. Phys.* **75**, 7058-7060; Parker, M.A., Coffey, K.R., Howard, J.K., Tsang, C.H., Fontana, R.E. and Hylton, T.L. (1996), "Overview of progress in giant magnetoresistive sensors based on NiFe/Ag multilayers", *IEEE Trans. Magn.* **32**, 142-148.
7. Julliere, M. (1975), "Tunneling between ferromagnetic films", *Physics. Lett. A* **54A**, 225-226.
8. Moodera, J.S., Kinder, L.R., Wong, T.M. and Meservey, R. (1995), "Large magnetoresistance at room temperature in ferromagnetic thin film tunnel junctions", *Phys. Rev. Lett.* **74**, 3273-3276.
9. Hunt, R.P. (1971), "A magnetoresistive read out transducer", *IEEE Trans. Magn.* **7**, 150-154.
10. See, for example, Tsang, C.H., Fontana Jr., R.E., Lin, T., Heim, D.E., Gurney, B.A. and Williams, M.L. (1998), "Design, fabrication, and performance of spin-valve read heads for magnetic recording

applications", *IBM J. Res. Dev.* **42**, 103-116; Coehoorn, R., Kools, J.C.S., Rijks, Th.G.S.M. and Lenssen, K.-M.H. (1998), "Giant magnetoresistance materials for read heads", *Philips J. Res.* **51**, 93-124.

11. Ripka, P. (2001), *Magnetic sensors and Magnetometers*, Artech, Boston, U.S.A.
12. Philips Semiconductors (2001), "Semiconductor Sensors", Data Handbook SC17 or http://www.semiconductors.philips.com/products/standard/discretes/products/ga/index.html
13. Kuijk, K.E., Van Gestel, W.J. and Gorter, F.W. (1975), "The barber pole, a linear magnetoresistive head", *IEEE Trans. Magn.* **11**, 1215-1217.
14. Van den Berg, H.A.M., Clemens, W., Gieres, G., Rupp, G., Schelter, W. and Vieth, M. (1996), "GMR sensor scheme with artificial antiferromagnetic subsystem", *IEEE. Trans. Magn.* **32**, 4624-4626.
15. Differences with the general AF-coupled multilayer are, however, that the GMR ratio of the AAF subsystem is not important and that the AF-coupling in an AAF is preferably as strong as possible.
16. Dieny, B., Speriosu, V.S., Parkin, S.S.P., Gurney, B.A., Wilhoit, D.R. and Mauri, D. (1991), "Giant magnetoresistance in soft ferromagnetic multilayers", *Phys. Rev.* **B43,** 1297-1300.
17. Nogués, J. and Schuller, I.K. (1999), "Exchange bias", *J. Magn. Magn. Mater.* **192**, 203-232.
18. Stamps, R.L. (2000), "Mechanisms for exchange bias", *J. Phys. D: Appl. Phys.* **33**, R247-R268.
19. Kiwi M. (2001), "Exchange bias theory", *J. Magn. Magn. Mater.* **234**, 584-595.
20. Kools, J.C.S. (1996), "Exchange-biased spin-valves for magnetic storage", *IEEE Trans. Magn.* **32**, 3165-3184.
21. Rijks, Th.G.S.M., Folkerts, W., Kools, J.C.S., De Jonge, W.J.M. and Coehoorn, R. (1994), "Magnetoresistance in NiFe/Cu/NiFe/FeMn spin valves with low coercivity and ultrahigh sensitivity", *Appl. Phys. Lett.* **65**, 916-918.
22. Lenssen, K.-M.H., Kuiper, A.E.T. and Roozeboom, F. (1999), "Robust giant magnetoresistance material system for magnetic sensors", *J. Appl. Phys.* **85**, 5531-5533.
23. Lenssen, K.-M.H., Adelerhof, D.J., Gassen, H.J., Kuiper, A.E.T., Somers, G.H.J. and Van Zon, J.B.A.D. (2000), "Robust giant magnetoresistance sensors", *Sensors&Actuators* **A85**, 1-8.
24. Van Driel, J., Coehoorn, R., Lenssen, K.-M.H., Kuiper, A.E.T. and De Boer, F.R. (1999), "Thermal stability of Ir-Mn as exchange biasing material", *J. Appl. Phys.* **85**, 5522-5524; Van Driel, J., De Boer, F.R., Lenssen, K.-M.H. and Coehoorn, R. (2000), "Exchange biasing by $Ir_{19}Mn_{81}$: Dependence on temperature, microstructure and antiferromagnetic layer thickness", *J. Appl. Phys.* **88**, 975-982.
25. Lenssen, K.-M.H., Kuiper, A.E.T., Van den Broek, J.J., Van der Rijt, R.A.F. and Van Loon, A. (2000), "Sensor properties of a robust giant magnetoresistance material system at elevated temperatures", *J. Appl. Phys.* **87**, 6665-6667.
26. Van Zon, J.B.A.D. (2001), *personal communication*.
27. Adelerhof, D.J. and Geven, W. (2000), "New position detectors based on AMR sensors", *Sensors&Actuators* **A85**, 48-53.
28. Lenssen, K.-M.H. and Martens, P. (1998), "Input apparatus for a data processing system", *US Patent* **5831553**.
29. Lenssen, K.-M.H., Lambrechts, M.E.C. and Vos, G.J.J. (1999), "Input device", *WIPO publication* **WO 99/03060**.
30. Rapeli, J. and Lenssen, K.-M.H. (2001), "Communication system, a primary radio station, a secondary radio station, and a communication method", *US Patent* **6195559**.
31. Lenssen, K.-M.H., Somers, G.H.J. and Van Zon, J.B.A.D. (2001), "Magnetoresistive sensors for string instruments", *to be published*; patent pending.
32. Johnson, M. (2000), "Magnetoelectronic memories last and last ", *IEEE Spectrum* **37**, February 2000, 33-40.
33. Daughton, J. (2000), "Magnetoresistive Random Access Memory (MRAM)", http://www.nve.com.
34. Lenssen, K.-M.H., Dormans, G.J.M. and Cuppens, R. (2000), "Expectations of MRAM in comparison with other non-volatile memory technologies" in Proceedings Non-Volatile Memory Technology

Symposium 2000 (November 15-16, 2000, Arlington VA, U.S.A.), pp.26-31, Jet Propulsion Laboratory, California Institute of Technology.

35. Lenssen, K.-M.H. and Persoon, E.H.J. (2001), "Magnetic Random Access Memory (MRAM) and its prospects" in Proceedings Non-Volatile Memory Technology Symposium 2001 (November 7-8, 2001, San Diego CA, U.S.A.), Jet Propulsion Laboratory, California Institute of Technology, *to be published.*

36. Zhu, J.-G., Zheng, Y. and Prinz, G.A. (2000), "Ultrahigh density vertical magnetoresistive random access memory", *J. Appl. Phys.* **87**, 6668-6673.

37. Pohm, A.V., Everitt, B.A., Beech, R.S. and Daughton, J.M. (1997), "Bias field and end effects on the switching thresholds of 'pseudo spin valve' memory cells", *IEEE Trans. Magn.* **33**, 3280-3282; Everitt, B.A., Pohm, A.V., Beech, R.S., Fink, A. and Daughton, J.M. (1998), "Pseudo spin valve MRAM cells with sub-micrometer critical dimensions", *IEEE Trans. Magn.* **34**, 1060-1062.

38. Tehrani, S., Slaughter, J.M., Chen, E., Durlam, M., Shi, J. and DeHerrera, M. (1999), "Progress and outlook for MRAM technology", *IEEE Trans. Magn.* **35**, 2814-2819.

39. Parkin, S.S.P., Roche, K.P., Samant, M.G., Rice, P.M., Beyers, R.B., Scheuerlein, R.E., O'Sullivan, E.J., Brown, S.L., Bucchigano, J., Abraham, D.W., Lu, Y., Rooks, M., Trouilloud, P.L., Wanner, R.A. and Gallagher, W.J. (1999), "Exchange-biased magnetic tunnel junctions and application to nonvolatile magnetic random access memory", *J. Appl. Phys.* **85**, 5828-5833.

40. Zhang, R., Hassoun, M.M., Black Jr., W.C., Das, B. and Wong, K.A. (1999), "Demonstration of a four state sensing scheme for a single pseudo-spin valve GMR bit", *IEEE Trans. Magn.* **35**, 2829-2831; Jeong, W.-C., Lee, B.-I. and Joo, S.-K. (1999), "Three level, six state multilevel magnetoresistive RAM (MRAM)", *J. Appl. Phys.* **85**, 4782-4784.

41. Press releases from IBM and Infineon (2000), see, e.g. http://www.chips.ibm.com/news/2000/1206_infineon.html or http://www.infineon.com/news/press/012_020e.htm.

42. Reuters (2000),"IBM and Infineon unveil magnetic memory chips", http://www.cnn.com/2000/TECH/computing/12/07/ibm.reut/index.html.

43. Aslett, J. (2001), "Magnetic RAM to hit market a year early", *Electronics Times*, June 18, 2001, http://www.electronicstimes.com/story/OEG20010615S0027.

44. Naji, P.K., Durlam, M., Tehrani, S., Calder, J. and DeHerrera, M.F. (2001), "A 256kb 3.0V 1T1MTJ Nonvolatile Magnetoresistive RAM" in Digest of Technical Papers 2001 IEEE ISSCC (February 5-7, 2001, San Francisco CA, U.S.A.), paper 7.6, pp. 122-123, 438, IEEE, U.S.A. http://www.unionsemiconductor.com/VideoClips.htm.

45. Oishi, M. and Matsumoto, T. (2001), "US makers ready MRAM for post-DRAM era", *Nikkei Electronics Asia*, May 2001, 26-40.

46. Cataldo, A. (2001), "SRAM soft errors cause hard network problems", *EE Times*, August 17, 2001, http://www.eet.com/story/OEG20010817S0073.

47. Baselt, D.R., Lee, G.U., Natesan, M., Metzger, S.W., Sheehan, P.E. and Colton R.J. (1998), "A biosensor based on magnetoresistance technology", *Biosens. Bioelectron.* **13**, 731-739.

48. Edelstein, R.L., Tamanaha, C.R., Sheehan, P.E., Miller, M.M., Baselt, D.R., Whitman, L.J. and Colton, R.J. (2000), "The BARC biosensor applied to the detection of biological warfare agents", *Biosens. Bioelectron.* **14**, 805-813.

49. Miller, M.M., Sheehan, P.E., Edelstein, R.L., Tamanaha, C.R., Zhong, L., Bounnak, S., Whitman, L.J. and Colton, R.J. (2001), "A DNA array sensor utilizing magnetic microbeads and magnetoelectronic detection", *J. Magn. Magn. Mater.* **225**, 138-144.

50. Tondra, M., Porter, M. and Lipert, R.J. (2000), "Model for detection of immobilized superparamagnetic nanosphere assay labels using giant magnetoresistive sensors", *J. Vac. Sci. Technol.* **A18**, 1125-1129.

THE ELECTRON SPIN IN NANOELECTRONICS

Jo De Boeck, Vasyl Motsnyi, Liu Zhiyu, Jo Das, Liesbet Lagae, Roel Wirix-Speetjens, Hans Boeve, Wayne Hiebert, Willem Van Roy and Gustaaf Borghs

IMEC, Kapeldreef 75, B-3001 Leuven; Belgium

Abstract: While the silicon-based CMOS electronics is entering the nano-regime, it is preparing itself to accept new materials for increasing the functionality of the future IC's. One example is the integration of magnetic multilayer nanostructures for creating a new generation of Magnetic Random Access Memories. Proven functionality in magnetic devices with nanoscale dimensions is a key asset for their use in other types of integrated devices such as lab-on-chip biosensors. This chapter will describe the technological challenges to realise the full potential of the integrated magnetic structures in which the electron spin rules the characteristics. One recognised area of future nanoelectronics is that of the use of the electron spin in semiconductor components. This field has been coined spintronics. A second part of this chapter will be devoted to the description of some of the current challenges of spintronics. Most emphasis will be placed on the issue of injecting a spin-polarized current in a semiconductor heterostructure.

Key words: GMR / tunneling magnetoresistance / TMR / spin-tunnel junctions / multilayers / magnetic sensors / sensor applications / magnetic random access memory / MRAM / magnetoresistive biosensors / spintronics / spin-electronics / spin-injection

E. Buzaneva and P. Scharff (eds.), Frontiers of Multifunctional Nanosystems, 453–476.

1. Introduction: magnetoelectronic and spintronic nanotechnologies

1.1 CONVENTIONAL AND EMERGING NANO-TECHNOLOGIES

The silicon integration technology enters the nano-electronic regime in a few years from now, with the minimum feature sizes of integrated devices well below 100 nm. As explained in Table I below, there will be a continuous evolution in the IC functionality based on the Si IC technology backbone. In the coming years these developments will be based on the standard MOS (Metal Oxide Semiconductor) technology with introduction of some new device layouts and advanced materials (e.g. SiGe). Parallel with these developments there is a large research activity on the introduction of new concepts e.g. in the memory technology. Here we definitely enter the nano-electronic era, because the functionality of the device is dictated by the deep sub-micron size of the devices and by the 'new' physics that emerges on these length scales. Resonant tunneling and single electron devices are well known examples. In this chapter we will touch upon the potential of magnetoelectronics, where the electron spin rules the characteristics of the devices. We will describe the challenges of 'connecting' this novel magnetic nano-electronics to the silicon backbone. Needless to say that even though the shrinking of the devices will come to a stop, the Si integrated circuits will dominate with colours the IC-application consumer market for at least half a century to come. But for comfort-control electronics, applications in medicine, bio-electronics, ambient intelligence etc, there will be large opportunities to step off the paved road of Si electronics and explore new techologies, based on e.g. spintronics (with a spin-off in quantum computing) and electronics where the leading element is C rather than Si.

Table I Schematic diagram of the (r)evolution in nano-electronics in the coming decades.

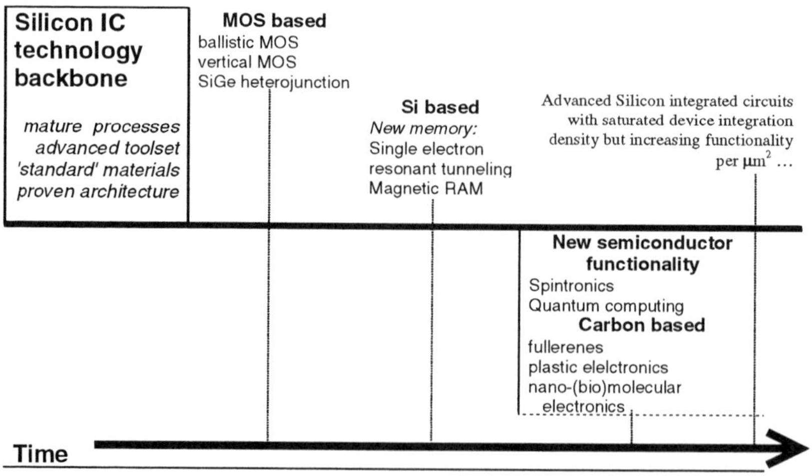

1.2 MRAM: AN ADD-ON NANOTECHNOLOGY

The field of magnetoelectronics[1] attempts to take advantage of the unique properties of (artificial) magnetic materials, in electronic components. The real interest came with the discovery of the Giant Magnetoresistance (GMR)[2,3] and Tunnel Magnetoresistance (TMR) [4] effects in multi-layered magnetic structures. We will first briefly explain the basic properties of these effects.

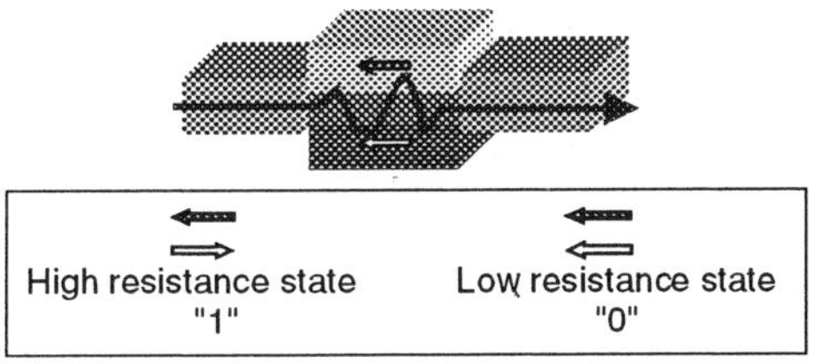

Fig 1. A Spin-valve structure. The resistance of this magnetic multilayer component depends on the orientation of the magnetizations of the ferromagnetic layers in the structure. The cause of this resistance difference lies in the spin-dependent electron transport properties in ferromagnets.

In its simplest form, a spin-valve is a tri-layer structure consisting of a soft magnetic layer and a harder magnetic layer, spaced by a noble metal. All layers are about five to ten nanometer in thickness. Such a tri-layer is deposited on an insulating carrier and is patterned to form a small strip. When the magnetic parts of the tri-layer have well defined macroscopic magnetic orientations, one can imagine two possible magnetic configurations: the magnetic moments in both layers are aligned or they are opposing.
In a ferromagnetic metal a majority of electrons has its spin aligned to the macroscopic magnetic orientation of the film. Importantly, the resistance to motion in a ferromagnetic is larger for a minority-spin electron. Hence, in a ferromagnetic metal, the density and transport properties of the electrons are spin-dependent.
Through the strip an electronic current is drawn, by applying a small voltage. While moving under the electric bias, the electrons cross the interfaces in the tri-layer many times in a random fashion, caused by scattering processes in the films. The electron can preserve its spin orientation even after experiencing some scattering events.
Now let's follow the motion of an electron through the tri-layer structure we called a spin-valve. We assume that the magnetic moments of both films are aligned. A majority-spin electron in one ferromagnetic layer will also be majority in the other. Such an electron experiences a low resistance to move in either ferromagnetic part of the tri-layer. The majority-spin electrons find their way rather easily through the strip leading to a low macroscopic resistive state for the tri-layer.

When the magnetic films have opposing magnetic moments, a majority electron in one film is minority in the other film. In this situation there is no preferred electron spin channel and the macroscopic electrical resistance of the strip is higher than when the moments are aligned.

A magnetic field can change the magnetic alignment in the spin-valve strip by rotating the softer magnetic layer with respect to the harder one. The electrical resistance of the strip will indicate the magnetic orientation of the layers. This effect is called magnetoresistance and the difference in resistance can be larger than 10 % at room temperature for spin-valves with a more complex layer stack.

The spin-valve can be designed such that the aligned and opposed magnetic configurations are stable in time, depending on magnetic hysteresis in the soft magnetic film. In this case, the resistance of the film will 'remember' the last applied field direction.

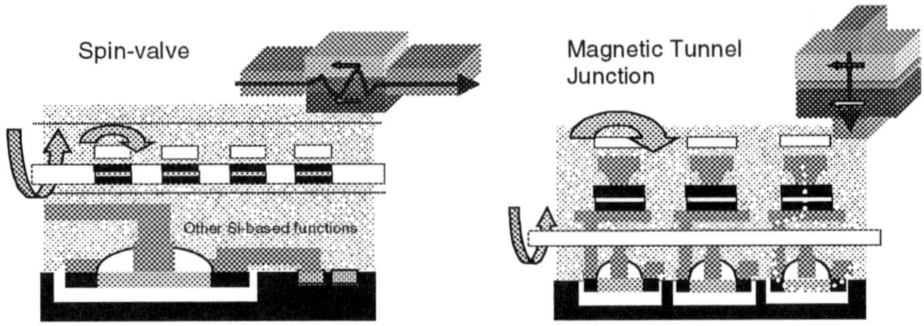

Fig. 2 Spin-Valve and Magnetic Tunnel Junctions have great potential for integration in magnetic Random Access Memories (MRAM). The operating principle of the MRAM will depend on the type of device used[5].

Other multilayer structures show comparable macroscopic magneto-resistance features. One example is a magnetic tunnel junction (Fig 2, right) in which two magnetic layers are separated by an insulating barrier, 1 to 2 nanometer thick. Under small electric bias, electronic current can tunnel through the barrier. The transport across this barrier will depend on the magnetic orientation of both magnetic layers. The probability of an electron passing is higher when both magnetic orientations are parallel. The difference in resistance for parallel versus opposed magnetic moments can be larger than 40 % at room temperature.

The unique features of the spin-valve and magnetic tunnel junction have lead to the development of magnetic memory elements [5-8] and industrial sensors[9].
The key part in each bit of a Magnetic Random Access Memory is a small magnetic element of which the magnetization can be pointed in one of two stable orientations. This will in practice be the magnetic layer in the stack which is most free to rotate, in other words: has the lowest corecivity. The "0" or "1" states will each correspond with

one of these orientations. The magnetic field required to change the magnetization orientation will be generated on chip in the close vicinity of the small magnetic element (Fig 2, field generated by current carrying conductors). When switching one bit, the others are not disturbed. No external fields are used. Once the magnetization of the bit is oriented, the information is written in the bit. The information will be stored as long as no other magnetic field is applied. This is the basis of the non-volatile storage of information in MRAM.

An electric current through a conductor will generate the magnetic field. The magnetic bits are arranged in a regular array. Along each column of the array runs a conductor for generating half of the magnetic field to switch a magnetic bit. The same is true for each row. The superposition of the two magnetic fields will result in writing of that bit located at the crossing of both conductors.

The reading or retrieving the stored information is done by sensing the resistance of the element (either in an non-destructive or destructive mode), since this resistance reflects the magnetic orientation of the written magnetic element.

The potential success of MRAM lies in the fact that the magnetic elements that represent the storage bits can be scaled down to very small dimensions (more so for the tunnel junction than for the spin-valve) without loosing their functionality. Also, the idea is to incorporate these magnetic nano-structures in a standard CMOS process without major changes in this process or the number of masks. Further the MRAM cells in principle have an infinite lifetime and the potential for very fast read-out. For a comparison of MRAM with various alternatives, we refer the reader to ref. 10. These features have triggered the interest of the industrial players in the (non-)volatile memory chip market. With the industrial interest in MRAM memory comes a very strict set of boundary conditions, inferred from the semiconductor technology roadmap[10]. For a better understanding of what it will take to fully exploit the potential of this magnetic nano-technology the reader is referred to section 2 of this chapter.

1.3 MAGNETS AND SEMICONDUCTORS IN ONE (HYBRID) DEVICE

Small ferromagnetic structures can generate static magnetic fields, called fringing fields, which can be used to modulate the electronic transport properties of low-dimensional electronic devices. Such structures are often referred to as *hybrid* magnetic/semiconductor devices, reflecting the combination of the two materials in a single device, each with its own functionality. The fringing fields reflect the magnetic orientation of the magnetic structure and can lead to a memory element[11]. The magnetic 'barriers' that are introduced by the fringing fields in the semiconductor, give rise to interesting magnetoresistance effects in a two-dimensional electron gas [12-16]. Further, there are magneto-optical effects that can be exploited in a magnetic/semiconductor hybrid device combination that could be quite interesting for opto-electronic applications [17-19]. These two categories of hybrid devices will not be further addressed in this chapter.

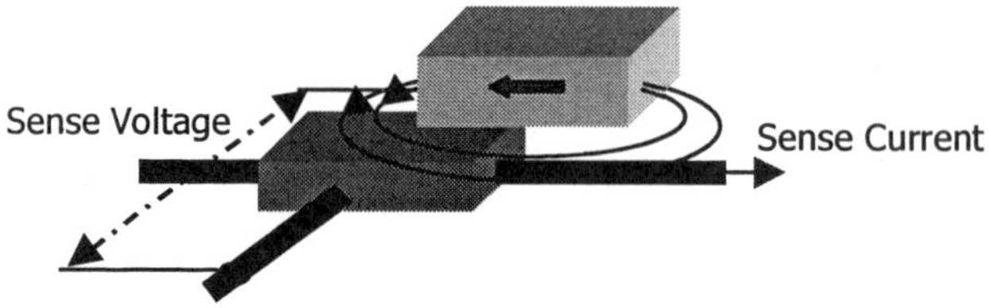

Fig. 3. Fringing field devices are an example of a hybrid ferromagnetic / semiconductor device in which the magnet and the semiconductor are combined in a non-contact fashion.

1.4 SPINTRONICS

An intriguing property of magnetic materials is the imbalance in the amount of up-spin vs. down-spin electrons taking part in the electrical transport in magnetic materials. In semiconductor based components, one is used to manipulate electronic charges with external inputs (voltage, light, magnetic field, ...). The manipulation of spins rather than charge in a semiconductor could lead to new spintronic components in which there are three different problem areas to distinguish - spin-injection, spin manipulation and spin detection – as discussed in section 3.1.

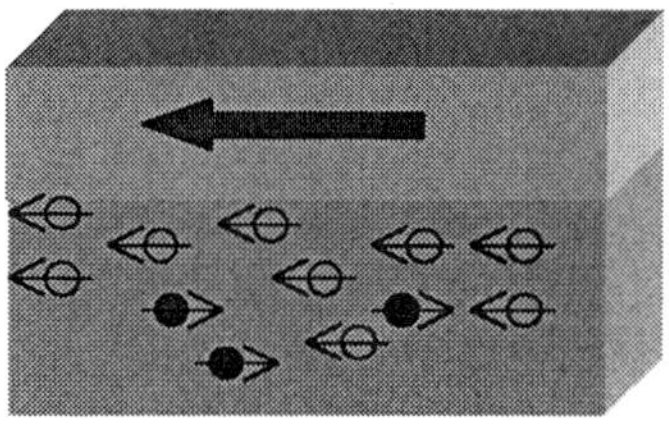

Fig 4. A magnetic layer making contact to a semiconductor, such that carriers can flow across the interface. The imbalance in majority versus minority spin numbers in the ferromagnet may lead to an injection of spin-polarized current in the semiconductor. The use of the spin in the semiconductor for novel devices is the area of spintronics.

The section 3 of this chapter will be devoted to issues dealing with this spintronics idea. We will explain some of the device concepts that have been described in the literature and emphasize the importance of the right choice of materials to realize efficient spin-injection. The most promising result being the injection by tunneling from a magnetic contact through a thin insulating barrier.

We will argue that the devices that are currently being fabricated are more test-vehicles to explore new physics than actual improved, performing new versions of the classic semiconductor devices from which they are deduced. These trials will lead to a better understanding of what is possible with the new degree of freedom offered by the electron spin manipulation. For this reason, there are new materials being developed such as magnetic semiconductors. Because of the importance of this class of materials for the future success of spintronics we devoted a small section (4) to it, concluding this chapter.

2. The challenges of MRAM nano-electronics

2.1 MAGNETIC TUNNEL JUNCTIONS AS BITS IN A CMOS CHIP

The use of magnetic multilayer devices in semiconductor IC memory components is probably the most promising near-term application area of semiconductor-based magnetoelectronics. The semiconductor base, in this case, is referring to the set of boundary conditions imposed on the integration of magnetic devices.

Focussing the attention to the integration of magnetic tunnel junctions (MTJ, a ferromagnet / insulator / ferromagnet stack, Fig 2 right), a set of boundary conditions for the successful realization of MRAM products is presented in Table II. The conditions are divided in two classes. The first class (I) is that of the intrinsic quality of the MTJ. Regardless of the integration process or product that is later going to host the MTJ, these characteristics can be independently optimized on single device or 'single-cell' scale. A second major class (II) is that of the parameters that are dictated by the technology and is arbitrarily subdivided in two areas of concern: (a) impact from scaling down to higher density products and (b) impact from materials choices in the integration process and more external factors.

Table II

Class I **Generic conditions**	**Class II** **Influenced strongly by integration technology generation** **(Roadmap and product vision)**	
	Size and spacing	Materials choice and other
Research on single MTJ (Cell)	***Research on small array***	***Development on production-viable platform***
Large MR	Resistance matching with transistors	Resistance uniformity
Resistance × area product		Etching process for MTJ
Electrical reliability	Cross-talk and flux keeping	Planarized surface smoothness
Tunnel barrier material, oxidation process, quality	Electrical model extraction	Temperature budget
Resistance to annealing above 450 °C	Memory architecture concepts	Materials compatibility issues and contamination control
New cell concepts	Switching field magnitude	Timing of coincident current pulses (low k , low R)
Switching process engineering (aspect ratio, shape, spin-transfer)	Switching field distribution	
Half-select and write robustness		Tool availability
Magnetic stability (anisotropy, volume)		Competitors' success Economic situation

In Class I problems are to be solved by fundamental understanding of the physics and optimization of the fabrication of the MTJ; for this type of research one does not need a state-of-the-art semiconductor fabrication facility, costs are moderate and progress can be fast. Class IIa is still very close to (and partly overlapping with) the generic MTJ issues. Much useful research can be done on small arrays with tools alien to an eventual production line but do require careful understanding of the technological approximation one makes when e.g. conducting tests on sub-micron MTJ's and their cross-talk. The materials and dimensions need to be carefully matched with the implementation of choice. For progress in Class II(b), major investment by industrial consortia is required, supported by a clear product vision and a convincing market analysis. Below we first review some of our results in the search for integrated magnetic elements, followed by a discussion of the Class II issues assuming one intends to fabricate an embedded memory, lined up with the semiconductor road map.

For an MRAM to operate fast and reliably, a component is required in each memory cell that can switch on and off the current through the magnetic memory element in that cell. In semiconductor technology, such component can be either a diode or a transistor. A tunnel junction, due to the vertical current direction in the device, is ideally suited for stacked integration. In case a diode is used as a non-linear element, the vertical stacking could lead to two terminal, vertical memory elements. From early implementations [20-22] critical success factors could be identified. Surface smoothness for instance, becomes a serious issue when the magnetic device is to be positioned on a processed chip. We found the quality of the MTJ to be acceptable with under-layer roughness up to 0.5 nm RMS. The situation is critical when planarization steps (such as Chemical Mechanical Polishing) are used after fabrication of a metal interconnect layer on chip. The roughness of the polished metal can easily exceed 1 nm RMS and the surface topography of the metal-to-oxide transitions are problematic from a viewpoint of MTJ integration.

When fabricating isolated MRAM cells for laboratory tests, the temperature budget during processing is limited to below 250 °C in order to avoid the degradation of the magnetic devices [23,24]. In the standard Si-integration technology the magnetic devices will have to sustain temperatures of 450 °C for about 30 minutes, since changes in standard processing recipes are not likely to be granted. Temperature stability of MTJ's up to 380 °C has been proven recently[25], maybe further improvement is possible.

The level of current send through the word-lines in the above implementations was on the order of 10 mA and more. These current levels are unacceptable for deep-submicron integration. This points to the need for efficient use of the generated magnetic flux and the implementation of flux-keepers in more advanced structures, also beneficial from a cross-talk point of view.

From an electronic standpoint, electrical matching of the semiconductor element [26] is required to obtain a maximum cell-output: voltage or current change. As a rule of thumb, the resistance should be of the same order as that of the implemented switch, typically 1 kΩ. This puts a maximum on the resistance-area product of the MTJ's of 1 - 0.05 kΩμm^2 and below for very high density. At the same time the resistance change should be as high as possible and values well above 40 % would be beneficial. Lower TMR values can be used at the expense of longer integration times in the read cycle. Fig 5 gives an overview of the resistance-area products vs the magnetoresistance of MTJ's fabricated with various oxidation methods.

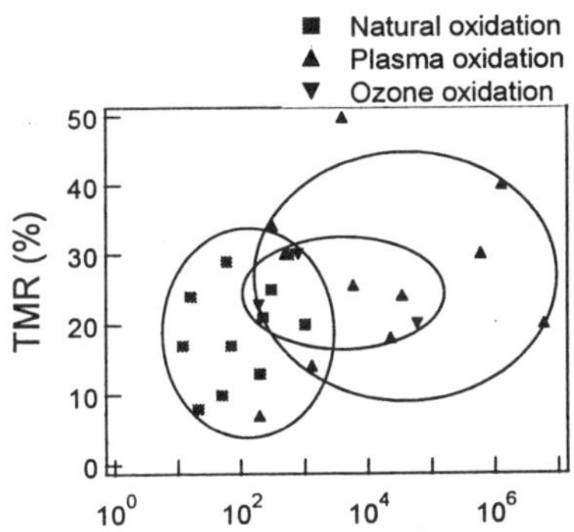

Fig 5 TMR vs resistance-area product for magnetic tunneljunctions fabricated with varius oxidation methods.

Fig 6 Scanning Kerr microscope snapshop during the switching process of a 20 x 4 μm^2 ferromagnetic element. The image is taken 20 nanoseconds after application of a magnetic field pulse. The pulse was too low in amplitude to fully reverse the element as can be seen from the presence of the magnetic domains in the film.

Currently much effort is focused on the switching of the magnetic elements. In well-designed test circuits, one can demonstrate through MR measurements that the intrinsic switching speed of the magnetic elements can be on the order of 1 ns at the expense of large swithing current[27]. Simultaneous study of the transients in MR [28] and microscopic domain dynamics [29,30] (see Fig 6 for a scanning Kerr effect image), supported by micromagnetic modeling, lead to a better understanding of the switching dynamics and the factors that determine the switching field distribution, a possible show-stopper in advanced, large scale MRAM implementations.

2.2 MAGNETIC NANO-TECHNOLOGY ON THE CMOS ROADMAP

Recently, MRAM implementations using the CMOS-technology have been reported [31-33]. These set the pace for development towards advanced MRAM products. In development and production, the boundary conditions are dictated by the product vision one has in developing the magnetic RAM. The conditions originate from the technology "generation" used for the integration process (see Table III).

To understand the impact this technology generation can have on the MRAM development, let's assume that it is the intention to build a so-called embedded memory. An embedded memory is a memory array that is implemented side-by-side with a processor and other circuitry on a single semiconductor chip. It is clear that all the circuits implemented on that same chip are fabricated with the same set of technological parameters. These are e.g. the choice of materials for the interconnect lines and for the inter-metal dielectrics, the minimum dimensions and aspect ratios for conductors and the minimum distances for placing various components on the chip. Also the level of current through the conductors, the resistivity of transistor structures and the operating voltages are quite rigidly defined by the so-called technology roadmap[10]. For a summary of some of the important parameters, see Table III.

Table III	2001	2004	2008	2014
"generation"	**130 nm**	**90 nm**	**60 nm**	**30 nm**
On-chip frequency	2 100 MHz	3 500 MHz	6 000 MHz	13 500 MHz
Local wire pitch	405 nm	295 nm	185 nm	95 nm
Local wire aspect ratio	1.5	1.6	1.9	2.3
Global wire pitch	850 nm	620 nm	390 nm	190 nm
Global wire aspect ratio	2.4	2.7	2.8	3.0
Effective k-value inter-metal dielectric	2.9 - 3.5	2.2 - 2.9	1.6	< 1.3
Conductor effective resistivity	2.2 μΩ-cm	2.2 μΩ-cm	< 2.2 μΩ-cm	< 2.2 μΩ-cm
Number of metal levels	8 - 9	9	9 - 10	10
Inter-metal dielectric material known ?	yes	?	no	no

Source: ITRS roadmaps 1999 and updated version 2000, values for integrated Logic

This roadmap helps to predict at what generation node MRAM may enter as an embedded memory product, taking into account the more mature technological competitors such as e.g. Ferro-*electric* RAM (FERAM or FRAM)[34]. The MRAM case is build on a better scalability, long lifetime, and lower complexity and the quest for a universal memory (fast and non-volatile). Assume that FERAM will be successful in products up to the generation with minimum feature size of about 100 nm, a technology generation projected to hit the market at around 2004 (see Table III). Of course, there will be a large effort to extend the use of FERAM one or a few more generations but the industry is willing to consider alternatives. MRAM? What we learn from the above is that the introduction generation of embedded MRAM should probably be at the "90 nm" node. From the table III one can read the boundary conditions for the MRAM development in terms of minimum feature size, interconnect line parameters, materials characteristics, etc. To return the investment, there should at least be two generations of MRAMs to follow the introduction technology. This leads to the requirement that 60 nm and 30 nm feature sizes for the magnetic devices have to be mastered, in respectively four and ten years after the introduction. To demonstrate the feasibility of such a set of generations, there is a need for deep-submicron magnetic tunnel junction fabrication tests and their related TMR and magnetic switching characteristics distribution assessment (Class I research items). Looking at the devices embedded in the 30 nm generation, with a target size of the memory cell being on the order of 10 λ^2 (9 × 10^{-5} μm^2), one could estimate that a maximum dimension for the MTJ of 50 nm on the side, without room for an aspect ratio different from 1. An acceptable value of magnetic stability $S = K_uV/kT$, to be determined by the allowed probability of an erroneous flip of a bit, has to be respected by having large enough magnetic volume and anisotropy in the bit.

Further, the integration process used for the embedded circuits is fixed in materials palette as mentioned above. The only problem is that for the generations beyond 90 nm there is a big uncertainty in what materials will be used as dielectrics in between the metal conductors. Hence, the magnetic devices have to be technologically matched to a currently unknown set of materials. One can safely assume that these inter-metal dielectric materials will require a curing step of 400 to 450 °C for about half an hour. It hence becomes a requirement that for successful MTJ integration in an embedded memory they need to be able to hold up during annealing up to 450 °C. One way out of this annealing problem could be to implement the magnetic devices as the final step in the integration process, on top of all-but-one interconnect levels. That will be rather difficult for embedded memory because at the global interconnect level the linewidth is much larger than at the local interconnect and maintaining high density schemes with one transistor/one magnetic device becomes impossible.

The current that can be driven is limited by the laws governing the electromigration and this limits magnitude of the write field. The field lines need to be restricted as much as possible to the elements on top or below the field generating conductors. At very small dimensions that is hard to maintain without tricks such as magnetic flux concentration. Embedding such flux concentration adds complexity (cost) to the integration process.

The delay in the lines limits the access time because the coincident current scheme for writing requires simultaneous arrival of two current pulses at the position of the bit to switch, at the crossing of the two pulse-carrying conductors. The timing control on the current pulse arrival will be difficult given the same line has to generate the half-field for the first and the last cell on its column or row. Increasing the pulse width will alleviate that problem, but at cost of larger power consumption during write.

From the above description it is clear that there are interesting opportunities for integrated magnetic elements, as well as formidable challenges. It is most likely that MRAM will appear as stand-alone memory products first, because this approach lifts some of the boundary conditions from the above mentioned discussion since the stand-alone memory products can be fabricated in a dedicated technology.

2.3 INTEGRATED MAGNETIC SENSORS FOR APPLICATIONS

It is important to realize that there will be opportunities for integrated magnetic elements in niche products. For instance small arrays of magnetic memory cells, implemented in non-standard technology could be quite useful in smart sensor applications. Also, the use of integrated magnetic sensors could be proven useful for e.g. detection of the presence of magnetically labeled bio-specimen [35]. The latter is illustrated in Fig 7, where an integrated magnetic sensor element is used to detect the presence of magnetic beads[36]. These beads can be used to label bio-molecules and they can serve two purposes: one is the transport of the biospecimen through a lab-on-chip using local magnetic field gradients, the other is the detection as described above. Fig 7 shows a layout of a proof-of-principle device where current conductors are tapered to realise a magnetic field gradient that guides magnetic labels that are dispersed over the chip towards the sensor area in the center. There, a spin-valve is integrated. When alternatingly the current is send through the conductors adjacent to the spin-valve, the magnetic particles that are present near the sensor will pass above the spin-valve. While passing, the stray field of the magnetic particles will change the magnetization in the spin-valve, leading to a read-out signal as shown in Fig. 7.

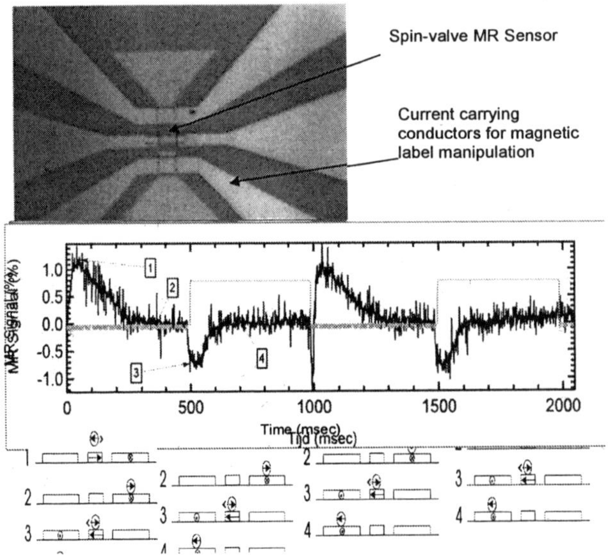

Fig. 7 Illustration of a proof-of-principle device for detecting the presence of magnetic labels using an integrated magnetic spin-valve sensor. Layout on top and a read-out signal example on the bottom[36].

3. Spintronic components

In the following sections we step away from the metallic devices used in the MRAM and sensor applications and focus our attention to the use of electron spins in semiconductor components, the area of spintronics.

3.1 INJECTION, MANIPULATION AND DETECTION OF SPINS

Whatever the type of spintronic component, the use of spins rather than charge always requires to face three different challenges.

One is the *creation* of a spin-ensemble in a semiconductor, or in other words, creation of a situation of majority and minority spins. This is most generally called spin-injection.

A second challenge is the *external control* over the spin-packet movement, coherence and lifetime on a (sub-)micron scale in the time frame of nano- to microseconds. Most of the experiments to date, that reveal properties of spin transport and lifetime, are performed using optical generation and in some cases resonant amplification of the spin-ensemble [37-41]. An example of spin-packet manipulation using an electric voltage was proposed in the spin-FET device [42] and is based on the Rashba effect. This will be briefly introduced further.

The third, is the *external observation* of a spintronic effect, be it a change in conduction of the device, a change in energy state, a difference in (polarization of) light-output or another discernable effect that could be used to interpret or 'read', the outcome of the device function. In most cases today this is done using an electroluminescent III-V device, based on the selection rules [43] as schematically depicted in Fig 8.

When the conduction band is occupied with carriers of one spin orientation only, for example states with $m_J = +1/2$, then according to the selection rules for the magnetic quantum number ($\Delta m_J = \pm 1$), only transitions to the heavy hole (+3/2) and light hole (-1/2) are allowed. The heavy hole transition is circularly polarised in one direction (σ^-), while the light hole transition in the other (σ^+). Since the probability of heavy hole transitions is 3 times larger than the light hole ones, thus radiative recombination of 100% spin polarised electron ensemble ($P = (n\uparrow - n\downarrow)/(n\uparrow + n\downarrow)$) results in circular polarisation ($P^* = (I^{\sigma+} - I^{\sigma-})/(I^{\sigma+} + I^{\sigma-})$) emission of 50%.

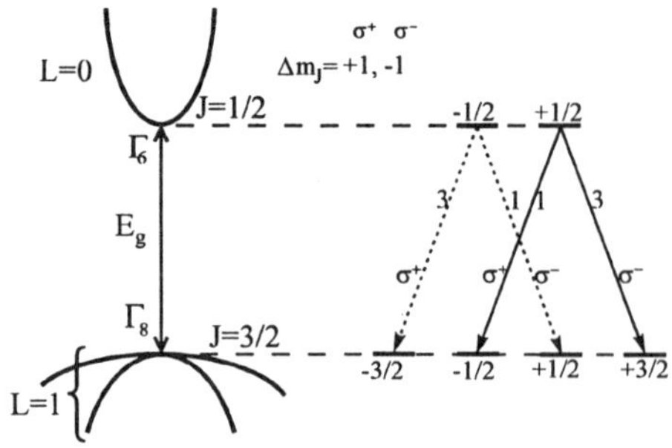

Fig 8 Band structure of GaAs, appropriate magnetic quantum numbers m_J and allowed transitions together with their probabilities.

It was shown by optical means that a current from an magnetic (Nickel) needle across a vacuum barrier into GaAs is spin-polarized [44]. The geometry in this experiment, which makes use of a Scanning Tunneling Microscope, is as follows. The injector needle's magnetization is along its length axis, pointing perpendicular to the semiconductor surface. Carriers were injected from the tip over a vacuum barrier into the surface depletion zone of a GaAs wafer. The spin orientation direction of the injected electrons will be also perpendicular to the semiconductor surface. The ensemble of electrons injected in the semiconductor will have a net polarization, provided that the spin-information is not lost while traversing the vacuum / semiconductor interface. Due to the spin-polarization of the injected electrons, their radiative recombination in the depletion region results in circulair polarized light emission from the semiconductor. The degree of polarization of the luminescence is a measure of the injected spin-imbalance, taking into account the spin rotation processes and spin-relaxation in the depletion region.

3.2 BUILDING ON EXISTING SEMICONDUCTOR DEVICE CONCEPTS

Fig 9 shows some devices structures that have been described in literature as magnet/semiconductor devices. Although they are copies of some well known semiconductor devices, the magnetoelectronic counterparts do not deliver an improved performance rather they deliver different functionality or can be used as a test-vehicle for the study of the physical properties and the materials combinations.

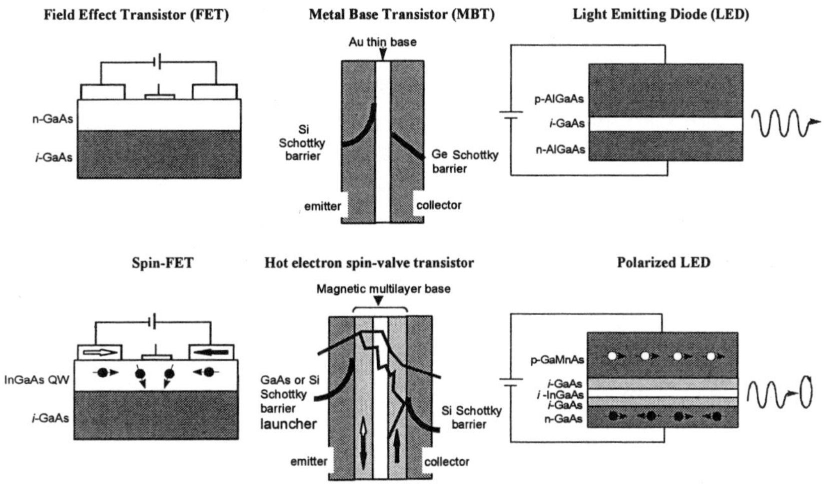

Fig 9 Selection of proposed magnetoelectronic devices with their standard semiconductor counterpart.

We will discuss some aspects of the spin-FET device fabrication because it is didactical in understanding the most important issues for future spintronic device fabrication as far as spin-injection is concerned. Also, the spin-LED offers the most natural way of assessing the injected spin-polarization and shall appear in the discussion below. The metal-base / spin-valve transistor will not further be discussed in this chapter. For a treatment of this device we refer to the literature[45-50].

3.2.1 The spin-FET concept

A FM/SC/FM device that resembles the structure of a Field Effect Transistor (FET) was proposed by Datta and Das [42]. Their Spin-FET is schematically represented in figure 9. It relies on spin injection from a ferromagnetic source contact and detection at a ferromagnetic drain contact. A gate is present to impose a surface electric field that will cause a precession of the injected spins. The analysis of this mechanism is as follows.

An effective magnetic field is felt by a moving electron in a two-dimensional electron gas (2-DEG, by convention in the x-y plane) when an perpendicular electric field Ez is present. The magnetic field is perpendicular to the electron motion and in the plane of the 2-DEG. Assuming that the electrons are confined in one subband, the spin degeneracy of the subband can be lifted by a spin-splitting energy caused by the interaction of the electron with the effective magnetic field.

In InAs or InGaAs heterostructures [51,52] the main origin of the spin-splitting energy is the surface electric field in the heterostructure, referred to as the Rashba mechanism [53,54]. This electric field can be modulated by an externally applied (gate-)voltage, which can change the band-bending and carrier density in the heterostructure.

The spin-orbit interaction parameter α is proportional to the expectation value of the electric field $\langle Ez \rangle$ perpendicular to the 2-DEG: $\alpha = b\langle Ez \rangle$. The energy associated with the Rashba mechanism is a linear function of this spin-orbit interaction parameter α and the Fermi wave number kF : $\Delta R = 2\ \alpha\ kF$. The precession angle $\Delta\theta$ of the electron spin in semiconductor heterostructures is dependent on α and is a function of the length of the channel between source (injector) and drain(detector), L [42], $\Delta\theta = 2m^* \alpha L / \hbar^2$.

Experimentally, the spin-orbit interaction parameter for a two dimensional electron gas (2-DEG) in a semiconductor heterostructure with sufficient mobility can be obtained from transverse magnetoresistance measurements. The experimental method relies on the occurrence of beat patterns in the Shubnikov – de Haas oscillations at moderate magnetic fields. In moderate fields the Zeeman energy contribution is smaller than the effect of the Rashba-mechanism.

3.2.2 The spin-switch concept

Making abstraction of the gate-voltage induced spin precession, a spin-dependent signature in the transport measurements is expected when the relative magnetization orientation of the source and drain is switched from parallel to anti-parallel. Such a geomctry is the equivalent of the all-metal spin-valve discussed above.

In order to make this spin-switch work, the channel mobility requirement is somewhat more relaxed, as long as the spin diffusion length in the channel is sufficiently large. This is the case for appropriately designed semiconductor (hetero)structures (spin-diffusion lengths of 1 to 10 μm). Nevertheless, devices with sub-micron channel length are desired to minimize the loss of spin-coherence during channel transit. This spin-disordering can be of particular importance for semiconductors with a large spin-orbit coupling, associated with the crystal asymmetry and in channels with a large numbers of magnetic scatterers where the number of spin-flip events is large.

3.2.3 Technological aspects of FM/SC spin-injection structures

A respectable set of ‘show-stoppers’ for the realization of a spin-FET or spin-switch can be pictured.

The magnitude and the sign of the polarization of the injected current depends strongly on the detailed properties of the band structure and relative contributions to the conduction of carriers of the s, p, d (hybridized) bands [55], the ferromagnet's crystal orientation versus the current direction, and the spin-transparency of the ferromagnetic / semiconductor interface. Therefor, the ideal ferromagnetic current injector is a single crystal with controllable crystal orientation, having a well defined (single) magnetic domain structure. Clearly, these observations push the ideal contact size to the nanoscale regime, putting the same requirements on the lithographic tools as the in other fields of micro-electronics. Another reason for ultra small contact sizes is the spin-coherence of the injected current. For smaller contacts, however, it will become harder to switch the magnetization direction. The geometry and size difference in source and drain contacts can actually be used to realize different coercive fields.

Magnetic contacts cause fringing fields, either from edge states or domain walls. Although it is unlikely that the ferromagnetic contacts will be in close enough proximity to suffer from magnetostatic coupling effects, the fringing fields close to the contacts may result in spurious magnetoresistance signals and spin dephasing.

Also one has to consider the FM/SC interface. Given the importance of the control of the magnetic properties of the magnetic layers serving as contacts, and the need for an extremely well controlled interface (structural, magnetic).

The most important point actually may be the electronic nature of the contact: Ohmic (diffusive) or tunnelcontact. We will return to this topic in the discussion of the alternatives for spin-injectors.

3.3 SPIN-INJECTORS

The schematic below can be used to arrive at the arguments why some strateges for spin injection should or should not work successfully. In the figure the resistor schematic is given for a structure consisting of a semiconductor with two magnetic contacts, left and right.

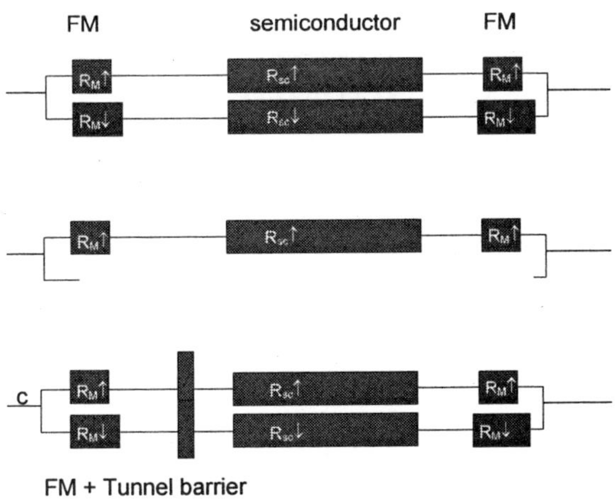

Fig. 10 Schematic representation of resistor models for different strategies for spin-injecting contacts to a semiconductor: a) metallic ferromanget makes ohmic, diffusive contact, b) half-metallic ferromagnet offers only one spin type for conduction and c) a tunnel barrier separates the ferromagnetic contact and the semiconductor.

In Fig 10a the contacts are diffusive or Ohmic. The spin-flip processes are neglected in the semiconductor and there is no difference in resistance for the spin-up and the spin-down channel of the electron spins in the semiconductor. There does exist a difference in resistivity for bot spin-channels in the ferromagnet, as we know, but this difference is so small in absolute value with respect to the resistance of the semiconductor that, in effect, there will be no observable difference in current for the spin up vs. the spin down channel. Hence, no spin-injection.

The situation of course changes for Fig 10b, when the ferromagnet has only one spin-band available for conduction: by definition there is spin-polarized current flowing, since only one polarization is available. Half metallic ferromagnets are such contact materials. One example is that of NiMnSb, which is a good candidate for epitaxial growth on GaAs, given the good correspondence of the crystal lattices [56,57]. The correct choice of interface with the semiconductor is important[58,59]. Another interesting alternative is CrO_2, that has recently received quite some attention [60] . A treatment of the half-metallic ferromagnetis is beyond the scope of this chapter.

Another option (c) is to use a tunnel barrier in between the ferromagnet. In that case, the current again becomes spin-polarized[61]. The actual physical mechanisms underlying these oversimplified arguments are complex and we will treat only that of case (a) in a little more detail. We will touch upon these different scenarios from the point of view of materials used to realise them.

3.3.1 Ohmic contact injectors

The electrical injection of spin polarized electrons in a semiconductor can be achieved in principle by driving a current from a ferromagnetic metal, where current is known to be significantly spin polarized, into the semiconductor via ohmic conduction. But for a diffusive Ohmic contact there are fundamental limitations to the spin-injection from a metallic ferromagnet with ($P < 100\%$) into the semiconductor [62]. Most of the magnetic materials used for the contact have limited polarization. Together with the conductivity mismatch, this leads to very low efficiency in the injection process using diffusive contacts [63,64].

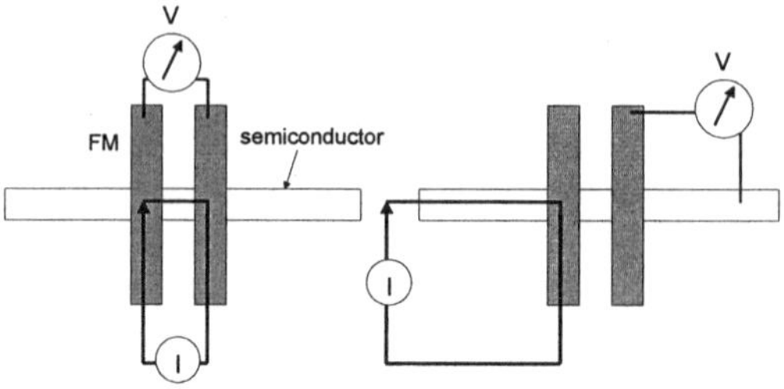

Fig 11. Local geometry (left) also 'classic spin-valve' geometry and non-local (right) geometry for spin-injection measurements. Shaded areas represent ferromagnetic materials on top of rectangular shaped semiconductor (open area).

In the experiments of Filip et al. [64], a multi-terminal lateral spin valve geometry is considered, as depicted in fig. 11. The valve exists of two NiFe, Co or Ni ferromagnetic metals on top of an InAs semiconductor. Two types of measurements are possible. In the first one, called the 'classic' spin valve geometry, the current is injected and taken out from the ferromagnetic electrodes. The voltage is measured between the same electrodes, giving a standard four terminal measurement of the junction. A second geometry, which we refer to as the non-local geometry, corresponds to injecting current from the semiconducting channel into the first ferromagnetic electrode and measure the voltage between the second ferromagnetic electrode and the semiconducting channel.

Due to current polarization in the injecting ferromagnet, at the interface a spin accummulation will form, which will extend over a characteristic spatial length scale given by the spin flip length. If a second ferromagnetic is present in the vicinity of this interface, it can be used as a spin sensitive voltage probe to detect this spin accumulation. This is similar to Johnson's potentiometric method, [65] used for detecting spin accumulation in Au. No electrical current is flowing between the injector and the detector electrodes. Therefore this geometry allows to suppress any 'spin independent' magnetoresistive contribution.

The resistance change between the parallel and the anti-parallel configuration of the magnetizations of the two electrodes in the 'classic' spin valve geometry is given by [66,67].

$$\Delta R_{SV} = 2R_{sq}\frac{\lambda_N}{\omega}\frac{\alpha_F^2}{(M^2+1)\sinh(L/\lambda_N)+2M\cosh(L/\lambda_N)}$$

with

$$M = 1+\frac{\sigma_F}{\sigma_N}\frac{\lambda_N}{\lambda_F}(1-\alpha_F^2)$$

$\alpha_F = \dfrac{\sigma_\uparrow - \sigma_\downarrow}{\sigma_\uparrow + \sigma_\downarrow}$ is the bulk polarization of the electrodes, $\sigma_{\uparrow\downarrow}$ are the spin up (down) conductivities, σ_N (σ_F) is the total conductivity in the semiconductor (ferromagnetic metal) and λ_N (λ_F) is the spin flip length in the semiconductor (ferromagnetic metal). R_{sq} is the square resistance of the semiconductor, L is the spacing between the two ferromagnets, and ω is the width of the channel. In the non-local configuration the signal is reduced by a factor of two

$$\Delta R_{NL} = \frac{1}{2}\Delta R_{SV}$$

The relevant range of parameters for ferromagnet/semiconductor/ ferromagnet junctions is $\sigma_F >> \sigma_F$ and $\lambda_N >> \lambda_F$, meaning that, for a spin polarization of the ferromagnet smaller than 100%, the conductivity mismatch correction factor M is large, $M >> 1$. Then the expected signal can be expressed as

$$\Delta R_{NL} = R_{sq}\frac{\lambda_N}{\omega}\frac{1}{\sinh(L/\lambda_N)}(\alpha_F/M)^2$$

i.e. the injection efficiency is reduced from α_F to α_F/M.

This shows spin valve signal is reduced due to the conductivity mismatch between the semiconductor and the ferromagnet. For NiFe on InAs Filip et al. calculated a reduction in spin efficiency M=90 corresponding to an absolute signal too low to be measured.

3.3.2 Tunnel junction injectors

A tunnel injector will be successful for spin-injection, as supported from theoretical [61] and experimental point of view. Ferromagnetic tips [44,68,69] and Al_2O_3 barriers[70,71] have been used in tunneling experiments. Spin-injection results with a Schottky barrier have been explained in the tunnel regime. [72,73].

The fabrication of a pin-hole free tunnel barrier with good interface properties is essential. The technology for the Al_2O_3 tunnel barrier fabrication is reasonably well established. Using a Schottky barrier can lead to unreliable devices due to intermixing and dead-layer formation during the device processing and operation.

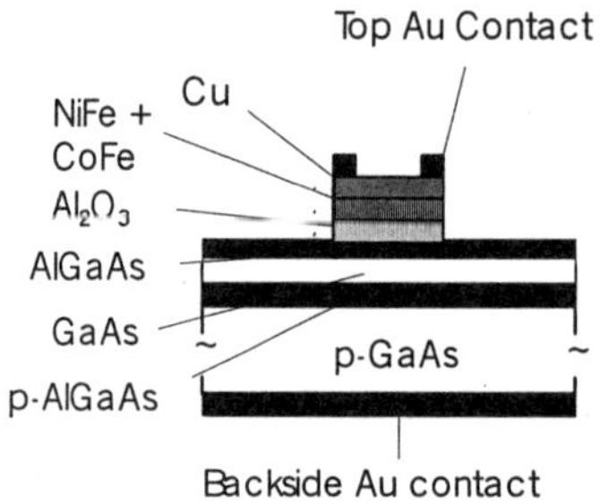

Fig 12 Schematic diagram of a polarized LED device with a ferromagnetic metal + Al_2O_3 tunnel barrier as a contact to the active semiconductor layer.

In recent experiments we fabricated the tunnel-injector structure on top of a semiconductor heterostructure that allows the observation of spin-injection. Since the spin-injection efficiency assessed through the optical response of the structure, we fabricated surface emitting LED components (see Fig 12). A GaAs / AlGaAs heterostructure was grown by MBE. The sample was unloaded and transferred to the sputtering tool to fabricate a 1.9 nm thick Al_2O_3 layer. As a magnetic top contact we sputtered NiFe or CoFe layers. We achieved spin injection in GaAs using these injector structures, the details of which are under preparation for publication at this moment [74].

4. Magnetic semiconductors

As an alternative to the use of metallic magnetic structures for spin-injectors there is a strong interest in the the use of paramagnetic II-VI BeMnZnSe semiconductor spin filters at low temperatures and high magnetic fields [75] and the development of magnetic

semiconductors such as GaMnAs [76-78]. Although the magnetic and electronic properties of the p-type GaMnAs thin films have been under study for some time[79], there still exists uncertainty about the exact nature of the magnetism and the routes to increase the temperature for paramagnetic to ferromagnetic transition, T_c [80].
Moderate spin-injection efficiencies have been obtained using GaMnAs as a spin-injector[81]. For devices it is important to attempt to incorporate magnetic semiconductor structures in heterostructures such as tunnel structures[82], 2-DEG structures[83] and Schottky barriers[84] and to control the magnetism with voltage[85]. From a materials point of view the challenges is the increase of the T_c to values above room temperature and the incorporation of magnetic structures in semiconductor heterostructures such that the spintronic properties of GaMnAs can be exploited.

For alternative magnetic semiconductor materials, first results on Mn and Fe doped GaN[79,86], ZnO[79,87] and TiO_2:Co [88] layers have been reported as well as more exotic materials such as CaB_6:La [89] - reported to be a semiconductor [90] - and $Cd_{1-x}Mn_xGeP_2$ [91]. It is expected that in the coming years progress in magnetic semiconductors with near-300K transition temperatures will be important.

5. Conclusions

This chapter has reviewed the critical issues in integrating magnetic thin films in different areas of magnetoelectronics. Magnetic materials have an important role to play in future nano-electronics. Via their success in "conventional" nano-electronic circuits such as deep sub-micron Magnetic Random Access memories, the magnetic thin films will enter in spintronics, a promising future direction of nanoelectronics.

Acknowledgments

The authors gratefully acknowledge the the collaboration with A. Filip and B. van Wees on the spin injection, with R. Sousa and P. Freitas on the MRAM issues. Parts of this work was funded under the EC contracts: TJ-MRAM, NM^2 and SPIDER. We further thank W. van de Graaf for MBE growth.

JD, L.L., KD acknowledge financial support from the I.W.T. (Belgium). WVR acknowledges financial support as a Postdoctoral Fellow of the Fund for Scientific Research Flanders-Belgium (F.W.O.).

References

1 G. Prinz, Science 282, 1660-1663 (1998).

2 M. N. Baibich, J. M. Broto, A. Fert, F. Nuygen Van Dau, F. Petroff, P. Etienne, G. Creuzet, A. Friederich, and J. Chazelas, Phys. Rev. Lett. 61, 2472-2475 (1988).

3 G. Binash, P. Grunberg, F. Saurenbach, and W. Zinn, Phys. Rev. B B39, 4828 (1989).

4 J. S. Moodera, L. R. Kinder, T. M. Wong, and R. Meservey, Phys. Rev. Lett. 74, 3273-3276 (1995).

5 J. De Boeck and G. Borghs, Physics World 12, 27-32 (1999).

6 J. M. Daughton, J. Appl. Phys 81, 3758-3763 (1997).

7 M. Johnson, IEEE Spectrum, 33 (2000).

8 J. Daughton, http://www.nve.com (2000).

9 K.-M. H. Lenssen, D. J. Adelerhof, H. J. Gassen, A. E. T. Kuiper, G. H. J. Somers, and J. B. A. D. van Zon, Sensors and Actuators 85, 1-8 (2000).

10 ITRS, http://public.itrs.net/ (2000).

11 M. Johnson, B. R. Bennet, M. J. Yang, M. M. Miller, and B. V. Shanabrook, Appl. Phys. Lett. 71, 974-976 (1997).

12 F. M. Peeters and P. Vasilopoulos, Phys. Rev. B 47, 1466-1473 (1993).

13 F. B. Mancoff, R. M. Clarke, C. M. Marcus, S. C. Zhang, K. Campman, and A. C. Gossard, Phys. Rev. B 51, 13369-13273 (1995).

14 P. D. Ye, D. Weiss, R. R. Gerhardts, M. Seeger, K. von Klitzing, K. Eberl, and H. Nickel, Phys. Rev. Lett. 74, 3013-3016 (1995).

15 H. A. Carmona, A. K. Geim, A. Nogaret, P. C. Main, T. J. Foster, M. Henini, S. P. Beaumont, and M. G. Blamire, Phys. Rev. Lett. 74, 3009-3012 (1995).

16 F. M. Peeters and J. De Boeck, in *Handbook of Nanostructured Materials and Nanotechnology; Vol. 3, Electrical Properties*, edited by H. S. Nalwa (Academic Press, 2000), p. 345-410.

17 M. Takenaka and Y. Nakano, Proc. of 11th Intern.Conf. on Indium Phosphide and related materials. may, 289-292 (1999).

18 W. Zaets and K. Ando, IEEE Photon. Technol. Lett. 11, 1012-1014 (1999).

19 M. Vanwolleghem, M. Leys, J. Das, L. Lagae, J. De Boeck, P. Van Daele, and R. Baets, Presented at LEOS 2001, San Diego, November (2001).

20 H. Boeve, J. Das, C. Bruynseraede, J. De Boeck, and G. Borghs, Electron. Lett. 34, 1754-1755 (1998).

21 J. Das, H. Boeve, J. De Boeck, and G. Borghs, in *Hybrid integration of spin-valves and MESFETS:technology test for future MRAM"*, Leuven, 1999, p. 272-275.

22 R. C. Sousa, P. P. Freitas, V. Chu, and J. P. Conde, Appl. Phys. Lett. 74, 3893-3895 (1999).

23 S. S. P. Parkin, K.-S. Moon, K. E. Pettit, D. J. Smith, R. E. Dunin-Borkowski, and M. R. McCartney, Appl. Phys. Lett. 75, 543-545 (1999).

24 S. Cordoso, P. P. Freitas, Z. G. Zhang, P. Wei, N. Barradas, and J. C. Soares, J. Appl. Phys. 89, 6650-6652 (2001).

25 Z. Zhang, S. Cardoso, P. P. Freitas, X. Batlle, P. Wei, N. Barradas, and J. C. Soares, J. Appl. Phys. 89, 6665-6667 (2001).

26 H. Boeve, C. Bruynseraede, J. Das, K. Dessein, G. Borghs, J. De Boeck, R. C. Sousa, L. V. Melo, and P. P. Freitas, IEEE Trans. Magnetics. 35, 2820-2825 (1999).

27 L. Lagae, S. Stoffels, J. Das, and J. De Boeck, unpublished (2001).

28 S. E. Russek, J. O. Oti, S. Kaka, and E. Y. Chen, in *High-speed characterization of submicrometer GMR devices*, Vancouver, 1998.

29 W. K. Hiebert, A. Stankiewicz, and M. R. Freeman, Phys. Rev. Lett. 79, 1134-1137 (1997).

30 M. R. Freeman, W. K. Hiebert, and a. Stankiewicz, J. Appl. Phys. 83, 6217-6222 (1998).

31 S. Tehrani, J. M. Slaughter, E. Chen, M. Durlam, J. Shi, and M. DeHerrera, IEEE Trans. Magn. 35 (1999).

[32] R. Scheuerlein, W. Gallagher, S. S. P. Parkin, A. Lee, S. Ray, R. Robertazzi, and W. Reohr, in *A 10 ns read and write non-volatile memory array using a magnetic tunnel junction and FET in each cell.*, San Francisco, USA, 2000, p. 128.

[33] M. Durlam, P. Naji, M. DeHerrere, S. Tehrani, G. Kerszykowski, and K. Kyler, in *Nonvolatile RAM based on magnetic tunnel junction elements.*, San Francisco, USA, 2000, p. 130.

[34] K.-M. H. Lenssen, G. J. M. Dormans, and R. Cuppens, in *Expectations of MRAM in comparison with other non-volatile memory technologies*, Arlington V.A., U.S.A., 2000.

[35] D. R. Baselt, G. U. Lee, M. Natesan, S. W. Metzger, P. E. Sheehan, and r. J. Colton, Biosensors and Bioelectronics 13, 731-739 (1998).

[36] L. Lagae, R. Wirix-Speetjens, J. Das, D. Graham, P. P. Freitas, G. Borghs, and J. De Boeck, J. Appl. Phys. to be published (2001).

[37] J. M. Kikkawa and D. D. Awschalom, Phys. Rev. Lett. 80, 4313-4316 (1998).

[38] I. Malajovich, J. M. Kikkawa, D. D. Awschalom, J. J. Berry, and N. Samarth, J. Appl. Phys. 87, 5073-5075 (2000).

[39] J. M. Kikkawa and D. D. Awschalom, Nature 397, 139-141 (1999).

[40] D. D. Awschalom and J. M. Kikkawa, Physics Today 52, 33-38 (June) (1999).

[41] D. Hagele, M. Oestreich, W. W. Ruhle, N. Nestle, and K. Eberl, Appl. Phys. Lett. 73, 1580-1582 (1998).

[42] S. Datta and B. Das, Appl. Phys. Lett. 56, 665-667 (1990).

[43] F. Meier and B. P. Zakharchenya, *Optical orientation*, Vol. 8 (North-Holland, Amsterdam, 1984).

[44] S. F. Alvarado and P. Renaud, Phys. Rev. Lett. 68, 1387-1390 (1992).

[45] D. J. Monsma, J. C. Lodder, T. J. A. Popma, and B. Dieny, Phys. Rev. Lett. 74, 5260-5263 (1995).

[46] D. J. Monsma, R. Vlutters, and J. C. Lodder, Science 281, 407-409 (1998).

[47] K. Dessein, H. Boeve, P. S. A. Kumar, J. De Boeck, J. C. Lodder, L. Delaey, and G. Borghs, J. Appl. Phys. 87, 5155-5157 (2000).

[48] R. Jansen, A. P. S. Kumar, O. M. J. van't Erve, R. Vlutters, P. de Haan, and L. J.C., Phys. Rev. Lett. 85, 3277-3280 (2000).

[49] K. Dessein, A. P. S. Kumar, L. Lagae, J. De Boeck, L. Delay, and G. Borghs, J. Magn. Magn. Mat. 226-230, 2081-2083 (2001).

[50] A. P. S. Kumar, R. Jansen, O. M. J. Van't-Erve, R. Vlutters, S. D. Kim, and J. C. Lodder, Physica C 350, 166-167 (2001).

[51] G. Lommer, F. Malcher, and U. Rössler, Phys. Rev. Lett. 60, 728-731 (1988).

[52] J. Luo, H. Munekata, F. F. Fang, and P. J. Stiles, Phys. Rev. B 38, 10142 (1988).

[53] E. I. Rashba, Sov. Phys. Solid State 2, 1109 (1960).

[54] Y. A. Bychkov and E. I. Rashba, J. Phys. C 17, 6039-6045 (1984).

[55] M. B. M.B Stearns, J. Magn. Magn. Mat. 5, 167 (1977).

[56] W. Van Roy, J. De Boeck, B. Brijs, and G. Borghs, Appl. Phys. Lett. 77, 4190-4192 (2000).

[57] W. Van Roy, M. Wojcik, E. Jedryka, S. Nadolski, O. Richard, B. Brijs, G. Borghs, and J. De Boeck, To be presented at JEMS 01, Grenoble, France (2001).

[58] C. T. Tanaka, J. Nowak, and J. S. Moodera, J. Appl. Phys. 86, 6239-6242 (1999).

[59] G. A. de Wijs and R. A. de Groot, Phys. Rev. B B64, 020402 (2001).

[60] Xiao and Chien, (2001).

[61] E. I. Rashba, Phys. Rev. B 62, R16267-R16270 (2000).

[62] G. Schmidt, D. Ferrand, L. W. Molenkamp, A. T. Filip, and B. J. van Wees, Phys. Rev. B 62, R4790-R4793 (2000).

[63] F. G. Monzon and M. L. Roukes, J. Magn. Magn. Mat. 198-199, 632-635 (1999).

[64] A. T. Filip, B. H. Hoving, F. J. Jedema, B. J. van Wees, B. Dutta, and G. Borghs, Phys. Rev. B 62, 9996-9999 (2000).

[65] M. Johnson, Phys. Rev. Lett. 70, 2142-2145 (1993).

[66] P. Van Son and e. al., Phys. Rev. Lett. 58, 2271 (1987).
[67] T. Valet and A. Fert, Phys. Rev. B 48, 7099 (1993).
[68] S. F. Alvarado, Phys. Rev. Lett. 75, 513-516 (1995).
[69] C. J. Hill, X. Cartoixa, R. A. Beach, S. D.L., and T. C. McGill, Cond-Mat/0010058 (2000).
[70] M. J. W. Prins, H. van Kempen, H. van Leuken, R. A. de Groot, W. Van Roy, and J. De Boeck, J. Phys.: Condens. Matter 7, 9447-9464 (1995).
[71] V. Motsnyi, W. Van Roy, J. Das, E. Goovaerts, G. Borghs, and J. De Boeck, in *Ferromagnetic metal / tunnel barrier / semiconductor devices for optical detection of spin-polarized current injection into a semiconductor*, Grenoble, France, 2001, p. (submitted).
[72] H. J. Zhu, M. Ramsteiner, H. Kostial, M. Wassermeier, H.-P. Schonherr, and P. K.H., Phys. Rev. Lett. 87, 016601 (2001).
[73] A. Hirohata, Y. B. Xu, C. M. Guertler, and J. A. C. Bland, J. Appl. Phys. 87, 1-3 (2000).
[74] V. F. Motsnyi, V. I. Safarov, J. De Boeck, J. Das, W. Van Roy, E. Goovaerts, and G. Borghs, Submitted (2001).
[75] R. Fiederling, M. Keim, G. Reuscher, W. Ossau, G. Schmidt, A. Waag, and L. W. Molenkamp, Nature 402, 787-790 (1999).
[76] H. Bender, A. Van Esch, W. Van Roy, R. Oesterholt, J. De Boeck, and G. Borghs, in *Structural characterization of InMnAs and GaMnAs epitaxial magnetic films grown by MBE on GaAs*, 1995 (IOP Publishing), p. 293-296.
[77] H. Ohno, A. Shen, F. Matsukura, A. Oiwa, A. Endo, S. Katsumoto, and Y. Iye, Appl. Phys. Lett. 69, 363-365 (1996).
[78] A. Van Esch, J. De Boeck, L. Van Bockstal, R. Bogaerts, F. Herlach, and G. Borghs, J. Phys. Cond. Matter 9, L361-L367 (1997).
[79] F. r. r. see:, Physica E 10 (2001).
[80] T. Dietl, H. Ohno, F. Matsukura, J. Cibert, and D. Ferrand, Science 287, 1019-1022 (2000).
[81] Y. Ohno, D. K. Young, B. Beschoten, F. Matsukura, H. Ohno, and D. D. Awschalom, Nature 402, 790-792 (1999).
[82] M. Tanaka and Y. Higo, J. Appl. Phys. 89, 6745-6747 (2001).
[83] Z. Liu, V. V. Moshchalkov, G. Borghs, and J. De Boeck, to be published (2001).
[84] Z. Liu, H. Boeve, W. Van Roy, S. Nemeth, V. V. Moshchalkov, G. Borghs, and J. De Boeck, J. Cryst. Growth 227-228, 867-873 (2001).
[85] H. Ohno, D. Chiba, F. Matsukura, T. Omiya, E. Abe, T. Dietl, Y. Ohno, and K. Ohtani, Nature 408, 944-946 (2000).
[86] H. Akinaga, S. Nemeth, J. De Boeck, L. Nistor, H. Bender, G. Borghs, H. Ofuchi, and M. Oshima, Appl. Phys. Lett. 77, 4377-4379 (2000).
[87] T. Fukumura, J. Zhengwu, M. Kawasaki, T. shono, T. Hasegawa, S. Koshihara, and H. Koinuma, Appl. Phys. Lett. 78, 958-960 (2001).
[88] Y. Matsumoto, M. Murakami, T. Shono, T. Hasegawa, T. Fukumura, M. Kawasaki, P. Ahmet, T. Chikyow, S.-y. Koshihara, and H. Koinuma, Science 291, 854-856 (2001).
[89] D. P. Young, D. Hall, M. E. Torelli, Z. Fisk, J. L. Sarrao, J. D. Thompson, H.-R. Ott, S. B. Oseroff, G. R.G., and R. Zysler, Nature 397 (1999).
[90] H. J. Tromp, P. van Gelderen, P. J. Kelly, G. Brocks, and P. A. Bobbert, Phys. Rev. Lett. 87,, 016401 (2001).
[91] G. A. Medvedkin, T. Ishibashi, T. Nishi, K. Hayata, Y. Hasegawa, and K. Sato, Jpn. J. Appl. Phys. 39, L949-L951 (2000).

SUBJECT INDEX

List of Participants

Aksenov, Victor
Joint Institute for Nuclear Research
6, J.Kuri St.
141980 Dubna
Moscov region
Russia
PHONE: 09621 65 796
FAX: 09621 65 484
E-mail: aksenov@nf.jinr.ru

Artyukh, Valery
A.V. Palladin Institute of
Biochemistry of the
Ukrainian National Academy of
Sciences,
Department of Biochemistry of
Sensory and
Regulatory Systems,
9 Leontovicha Str.,
01030 Kiev,
Ukraine
PHONE: +38 044 2294743
E-mail: nstarodub@hotmail.com

Belousov, Igor
Institute for Metal Physics,
Ukrainian National Academy of
Sciences
Vernadskii Ave.36,
01680, Kiev, Ukraine
PHONE: +38 044 444 10 31
E-mail: belousov@pwp.ru.kiev.ua

Brabec, Viktor
Institute of Biophysics,
Academy of Sciences of the Chech
Republic,
CZ-61265 Brno
Czech Republic
PHONE: +420 5 41517174
FAX: +420 5 41211293
E-mail: brabec@ibp.cz

Brust, Mathias
The University of Liverpool,
Centre for Nanoscale Science
Department of Chemistry
Liverpool L69 7ZD
UK
PHONE: +44 151 794 3554
FAX: +44 151 794 3588
E-mail : M.Brust@liv.ac.uk

Buzaneva, Eugenia
National Taras Shevchenko University
64, Vladimirskaya Str.,
01033 Kiev
Ukraine
PHONE: +38 044 266 05 82
FAX: +38 044 234 61 66,
E-mail: evb@mail.univ.kiev.ua

Dovbeshko, Galina
Institute of Physics
Department of Physics of Biological
Systems
Prospect Nauki 46,
Kyiv, 03039
Ukraine
PHONE: +38 044 265-73-64
FAX: +38 044 265-15-89
E-mail: galyna@ukma.kiev.ua

Gaburro, Zeno
INFM and Dipartimenti di Fisica
Universita'di Trento
Via Sommarive 14
I-38050 POVO Trento
Italy

PHONE: +39 0461 88 2034 2054
FAX: +39 0461 88 2034 1694
E-mail: gaburro@science.unitn.it

Galanov, Boris.
Institute for Problems in Materials Science, National Academy of Science
3 Krizhyzyanovsky Str.
03142, Kiev
Ukraine
PHONE: 380-44-444 01 02

Gogotsi, Yury
Drexel University
Department of Materials Engineering
LeBow Bldg., Room 431
3141 Chestnut St.
Philadelphia, PA 19104,
USA
PHONE: 1-215-895-6446
FAX: 1-215-895-6760
E-mail: gogotsi@drexel.edu

Gorchynskyy, Alexandr
National Taras Shevchenko University
64, Vladimirskaya Str.,
01033 Kiev
Ukraine
PHONE: +38 044 266 05 82
FAX: +38 044 234 61 66,
E-mail: evb@mail.univ.kiev.ua

Gurin, Valeriy
Physico-Chemical Research Institute
Belarusian State University
Leningradskaja str., 14
Minsk 220080
Belarus
PHONE: 375-172-095199
FAX: 375-172-264696
E-mail: gvs@fhp.bsu.unibel.by

Holiney, Roman
Institute of Semiconductor Physics,
National Academy of Science
45, Prospect Nauki
252028 Kiev,
Ukraine
PHONE: +38 044 265 83 17
E-mail: holiney@isp.kiev.ua

Horbach, Jürgen
Institut fuer Physik
Johannes Gutenberg-Universitaet Mainz
Staudinger Weg 7
D-55099 Mainz
Germany
PHONE: +49 (0) 6131/392-3641
FAX: +49 (0) 6131/392-5441
E-mail: horbach@komath.physik.uni-mainz.de

Huczko, Andrzej
Warsaw University
Department of Chemistry
Str. Pasteur, 1
02-093 Warsaw
Poland
PHONE: 48 22 822 02 11
FAX: 48 22 822 59 96
E-mail: ahuczko@chem.uw.edu.pl

Karachevtseva, Lyudmila
Institute of Semiconductor Physics NASU,
45 Nauki Prsp., Kyiv-03028,
Ukraine,
PHONE: +38 (044) 265 9815.
FAX: +38 (044) 265 8342,
E-mail: kartel@mail.kar.net

Kalchenko, Vladimir
Institute of Organic Chemisrtry,
National Academy of Sciences,
2, Murmanskaya str. ,
252028 Kiev, Ukraine
E-mail: Kalchenko@hotmail.com

Kepich, Tibery
Scientific Chief
The Scientific and Training Center
"Physical and Chemical Material
Science"of the Kiev Taras
Shevchenko University
and National Academy of Science
of Ukraine
6, Glushkova str.
Kiev, Ukraine
Tel: +38 044 266 92 30
E-mail: kty@gluk.org

Komaromi, Istvan
University of Debrecen
Medical and Health Science Center
Department of Clinical Biochemistry
and Molecular Pathology
Nagyerdei krt. 98
H-4032 Debrecen
Hungary
PHONE: (+36)-52-431-956
FAX: (+36)-52-417-631
E-mail: komaromi@jaguar.dote.hu

Koshets, Igor
Institute of Semiconductor Physics
NASU,
45 Nauki Prspect,
252028 Kyiv,
Ukraine,
PHONE: +38 (044) 265 18 27.
E-mail: Koshets@hotmai.com

Kovtyukhova, Nina
152 Davey Lab.
Chemistry Department
Pennsylvania State University
University Park, PA 16802
USA
PHONE: 1 (814) 8639797
FAX: 1 (814) 8638403
E-mail: nina@babylon.chem.psu.edu

Kukla, Andry
Institute of Semiconductor
Physics, National Academy of
Sciences
45, Prospect Nauki,
252028 Kiev Ukraine,
Phone:+(380)(44)2651827
E-mail: Kukla@hotmail.com

Kurzydlowski, Krzysztof Jan
Warsaw University of Technology
Department Materials Science and
Engineering
Str. Woloska, 141
02-507 Warsaw
Poland
PHONE: 48 22 660 85 29
FAX: 48 22 660 85 14
E-mail: kjk@inmat.pw.edu.pl

Lazarouk, Serguei
Belorusian State University of
Informatics and Radioelectronics
Microelectronics Dept.
P.Browky str. 6
220027 Minsk,
Belarus
PHONE : 375-172-39 88 69
FAX: 375 172 31 09 14
E-mail: serg@cit.org.by

Lenssen, Kars-Michiel H.
Philips Research Laboratories
(WA12)
Prof. Holstlaan 4
NL-5656 AA Eindhoven
The Netherlands
PHONE.: +31-(0)40-2742754
FAX: +31-(0)40-2744282
E-mail: Kars-Michiel.Lenssen@philips.com

Leonowicz,, Marcin
Warsaw University of Technology
Faculty of Materials Science and Engineering
Woloska 141; 02-507 Warsaw
FAX: (48-22) 6608514
TEL: (48-22) 6608450
E-mail: mkl@inmat.pw.edu.pl

Levkovetz, Inna.
A.V. Palladin Institute of
Biochemistry of UNAS
9 Leontovicha Str.
01030, Kiev
Ukraine
PHONE: 380-44-229 47 43
FAX : 380-44-2296365
E – mail: nstarodub@hotmail.com

Lytvyn, Petro
National Academy of Sciences
of the Ukraine
Institute of Semiconductor Physics
45, prospect Nauky,
03028 Kiev
Ukraine
PHONE: 380 044 265 59 40
FAX: 380 044 265 54 30
E-mail: plyt@isp.kiev.ua

Makarova, Tatiana
Ioffe Physical Technical Instutute
Polytekhnicheskaya 26
194021 St.Petersburg
Russia
PHONE: (812) 247 99 57
FAX: (812) 247 91 23
E-mail:
tatiana.makarova@physics.umu.se

Małgorzata, Przybyt,
Technical Univiversity of Lodz,
Department of Food Chemistry and Biotechnology
Institute of General Food Chemistry,
ul. Stefanowskiego 4/10,
90-924 Lodz,
Poland
PHONE: (00) 48 42 631 34 26
FAX: (00) 48 42 636 28 60
E-mail: mprzybyt@snack.p.lodz.pl

Mardanyan, Sona
H.Buniatian Institute of Biochemistry
of Armenian NAS
5/1, P.Sevak St.
Yerevan 375014
Republic of Armenia
PHONE: (374 2) 281 840
E-mail: biochem@ipia.sci.am

Mark, Geza
Research Institute for Technical
Physics & Material Science
H-1525 Budapest, P.O. Box 49
Str. Konkoly Thege u.29-33
H-1121 Budapest
Hungary
PHONE: (+36 – 1) 275 4993
FAX: (+38 – 1)392 2226
E-mail: mark@sunserv.kfki.hu

Matveeva, Ludmila
Institute of Semiconductor Physics,
National Academy of Sciences
45, Prospect Nauki,
252028 Kiev, Ukraine
PHONE: +38 044 265-83-17,
E-mail: matveeva@isp.kiev.ua

Motsnyi, Vasyl
IMEC, MAP (MCP/NMC) Division
Kapeldreef 75, B-3001 Leuven
Belgium
PHONE: 32 16 281298
FAX: 32 16 281501
E-mail: motsnyi@imec.be

Nakajima, Atsushi
Department Bof Chemistry,
Faculty of Science and Technology,
Keio University,
3 – 14 – 1 Hiyoshi, Kohoku – ku,
Yokohama, 223 –8522
Japan
PHONE: +81 45 566 1712
FAX: +81 45 566 1697
E – mail:
nakajima@sepia.chem.keio.ac.jp

Pirogova, Ljudmila
A.V. Palladin Institute of
Biochemistry of the
Ukrainian National Academy of
Sciences,
Department of Biochemistry of
Sensory and
Regulatory Systems,
9 Leontovicha Str.,
01030 Kiev,
Ukraine
PHONE: +38 044 2294743
E-mail: nstarodub@hotmail.com

Pogoryelov, Yevgen
Institute for Magnetism NAS of
Ukraine
36-b Vernadsky Blvd.,
03142, Kiev
Ukraine
PHONE: (+380) 44 444 9593
FAX: (+380) 44 444 1020
E – mail: pogor@imag.kiev.ua

Popova, Galina
National Taras Shevchenko
University
64, Vladimirskaya Str.,
01033 Kiev Ukraine
PHONE: +38 044 266 05 82
FAX: +38 044 234 61 66,
E-mail: evb@mail.univ.kiev.ua

Prylutskyy, Yuri
Biophysical Department,
Kiev National Taras Shevchenko
University
64 Volodymirska St.,
01033 Kyiv
Ukraine
PHONE: 38 044 266 22 76
E-mail: prylut@biocc.univ.kiev.ua

Ratner, Buddy
University of Washington Engineered
Biomaterials,
Seattle, WA 98195
USA
PHONE: 1 (206) 616-9781
FAX: 1 (206) 616-9763
e-mail:
ratner@uweb.engr.washington.edu

Remenyuk, Anna
Ioffe Physical Technical Instutute
Polytekhnicheskaya 26
194021 St.Petersburg,
Russia
PHONE. (812) 247 99 57
FAX (812) 247 91 23
e-mail: arem@pop.ioffe.rssi.ru

Repnytska , Olena
Institute of Physics Department of Physics of Biological Systems
Prospect Nauki 46,
03039 Kyiv, Ukraine
tel: 265-73-64
fax:265-15-89
e-mail: olena@iop.kiev.ua

Romanov, Sergey
Institute of Material Science and Department
of Electrical and Information Engineering
University of Wuppertal
Gauss-strasse 20
42097 WUPPERTAL
Germany
PHONE: (+49) (0) 202 439-2232
FAX: (+49) (0) 202 439-3037
E-mail: romanov@uni-wuppertal.de

Scharff, Peter
Technische Universitat Ilmenau
Fakultät für Mathematik und Naturwissenschaften
Institut für Physik / FG Chemie
Postfach 100565
98684 Ilmenau
Germany
PHONE: +49 (0) 3677 69 3602
FAX: +49 (0) 3677 69 3271
E-mail: scharff@physik.tu-ilmenau.de

Semchuk, Olexander
Institute of Surface Chemistry
17 Naumova Street,
03164 Kiev,
Ukraine
PHONE: 380(44) 444 96 97
FAX: 380(44) 444 3657
E – mail : inchem@mail.kar.net

Shirshov, Yuriy
Institute of Semiconductor Physics,
National Academy of Sciences
45, Prospect Nauki,
252028 Kiev
Ukraine,
PHONE: +(380) (44) 2651827
E-mail: y_shirshov@hotmail.com

Shmyryeva, Alexandra
National Technical University of Ukraine "KPI"
Pr. Pobeda 37, Faculty of Electronics,
Microelectronics Department, Corpus 12,
252056 Kiev,
Ukraine
PHONE: +38 044 2294743
E-mail: nstarodub@hotmail.com

Shtogun, Yaroslav
Kiev National Taras Shevchenko University
64, Vladimirskaya Str.,
01033 Kiev, Ukraine
PHONE: +38 044 266 05 82
FAX: +38 044 234 61 66,
E-mail: evb@mail.univ.kiev.ua

Starodub, Nikolaj
A.V. Palladin Institute of
Biochemistry of UNAS

9 Leontovicha Str.
01030, Kiev
Ukraine
PHONE: 380-44-229 47 43
FAX : 380-44-2296365
E – mail: nstarodub@hotmail.com

Starodub, Valentyna
N.V. Organon
P.O. Box 20,
NL-5340 BH Oss
The Netherlands
PHONE.: +31-(0) 412-66-3229
FAX +31-(0) 412-66-2572
E-mail: v_starodub@yahoo.com

Stukalov, Oleg
Laboratory of Semiconductors Physics
Institute of Solid State and
Semiconductors Physics
National Academy of Sciences of
Belarus
17 P.Brovki Str.,
220072 Minsk,
Belarus
FAX: 375 17 284 08 88
PHONE: 375 17 284 11 82
E-mail: nanom@ifttp.bas-net.by

Torbicz,, Wladyslaw
Institute of Biocybernetics and
Biomedical Engineering
Polish Academy of Sciences
O2-109 Warsaw, Ks. Trojdena 4
Poland
PHONE: +48 22 659 55 34
FAX: +48 22 659 55 34
E-mail: wtorbicz@ibib.waw.pl

Waring Michael
Department of Pharmacology
University of Cambridge
Tennis Court Road
Cambridge CB2 1QJ
England
PHONE (+) 44 1223 334034
FAX (+) 44 1223 334040
E-mail: mjw11@cus.cam.ac.uk

Yaminsky, Igor
Physical Department of
Moscow State University
Vorobyovy Gory
119899 Moscow
Russia
PHONE: +7 095 9301009
FAX :+7 095 9301009
E-mail: spm@genebee.msu.su

Zhavnerko, Gennady
Institute of Molecular and Atomic
Physic
National Academy of Sciences of
Belarus
F.Skaryna Ave. 70
220072 Minsk
Belarus
PHONE: +375 17 284 04 50
FAX: +375 17 283 00 30
E-mail: zhavn@ns.ichnm.ac.by

Zherebeskyy, Danila
National Taras Shevchenko
University
64, Vladimirskaya Str.,
01033 Kiev,Ukraine
PHONE: +38 044 266 05 82
FAX: +38 044 234 61 66,
E-mail: evb@mail.univ.kiev.ua